Walter Lawry Buller, Johannes Gerardus Keulemans

A History of the Birds of New Zealand

Second Edition, Vol. 1

Walter Lawry Buller, Johannes Gerardus Keulemans

A History of the Birds of New Zealand
Second Edition, Vol. 1

ISBN/EAN: 9783744762700

Printed in Europe, USA, Canada, Australia, Japan

Cover: Foto ©berggeist007 / pixelio.de

More available books at **www.hansebooks.com**

A HISTORY

OF THE

BIRDS OF NEW ZEALAND.

BY

SIR WALTER LAWRY BULLER, K.C.M.G.,
D.Sc., F.R.S.,

F.L.S., F.G.S., F.R.G.S., F.R.C.I., HON. F.S.Sc.;

'OFFICIER DE L'INSTRUCTION PUBLIQUE' DE LA FRANCE;
GALILEIAN MEDALLIST OF THE FACULTY OF NATURAL SCIENCES, ROYAL UNIVERSITY, FLORENCE;
CORRESPONDING MEMBER OF THE ZOOLOGICAL SOCIETY OF LONDON, OF THE AMERICAN ORNITHOLOGISTS' UNION,
AND OF THE ORNITHOLOGICAL SOCIETY OF VIENNA;
MEMBER OF THE BRITISH ORNITHOLOGISTS' UNION, OF THE PERMANENT INTERNATIONAL
COMMITTEE ON ORNITHOLOGY, OF THE ANTHROPOLOGICAL INSTITUTE OF GREAT BRITAIN,
AND OF THE NEW-ZEALAND INSTITUTE.

SECOND EDITION.

VOLUME I.

LONDON:
PUBLISHED (FOR THE SUBSCRIBERS) BY
THE AUTHOR,
8 VICTORIA CHAMBERS, VICTORIA STREET, WESTMINSTER, S.W.
1888.

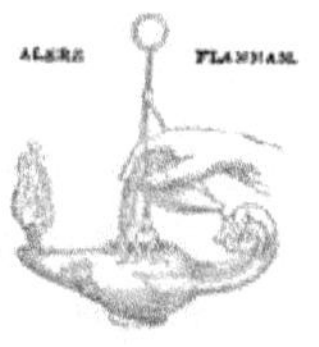

PRINTED BY TAYLOR AND FRANCIS,
RED LION COURT, FLEET STREET.

Dedication.

TO MY SONS,

WALTER LEOPOLD

AND

ARTHUR PERCIVAL,

NOW AT THE UNIVERSITY OF CAMBRIDGE,

BOTH OF WHOM WERE BORN IN NEW ZEALAND,

AND SPENT THEIR EARLIER YEARS

AMID THE SCENES AND SURROUNDED BY THE NATURAL OBJECTS

DESCRIBED IN THE FOLLOWING PAGES,

With much Affection,

I DEDICATE THIS BOOK.

ORIGINAL PROSPECTUS.

It has been remarked by a celebrated naturalist that "New Zealand is the most interesting ornithological province in the world;" and in a qualified sense this is no doubt true. The last remnant of a former continent, and, geologically considered, probably the oldest country on the face of our globe, it contains at the present day the only living representatives of an extinct race of wonderful Struthious birds.

Within recent historic times this circumscribed area, scarcely equal in extent to that of Great Britain, was tenanted, to the entire exclusion of Mammalia, by countless numbers of gigantic brevipennate or wingless birds, of various genera and species, the largest attaining to a stature nearly twice that of a full-grown Ostrich. These colossal ornithic types have disappeared; but their diminutive representatives (the different species of *Apteryx*) still exist, in diminished numbers, in various parts of the country: and these are objects of the highest interest to the natural-historian. But apart from this view of the subject the avifauna of New Zealand presents many special features of considerable interest. A large proportion of the genera are peculiar to the country; while some of the forms are perfectly anomalous, being entirely without a parallel in any other part of the world.

Under the changed physical conditions of the country, brought about by the operations of colonization, some of these remarkable forms have already become almost, if not quite, extinct, and others are fast expiring. It has been the author's desire to collect and place on record a complete life-history of these birds before their final extirpation shall have rendered such a task impossible; and it will be his aim to produce a book at once acceptable to scientific men in general and useful to his fellow-colonists.

It may be mentioned that the author's official position in New Zealand, during a period of more than twelve years, has enabled him to visit nearly every part of the country, while his frequent intercourse with the various native tribes has been highly favourable to such an object as the present undertaking.

The work will comprise an introductory treatise on the ornithology of New Zealand, a concise diagnosis of each bird in Latin and English, synoptical lists of the nomenclature, and a popular history and description of all the known species—and will contain coloured illustrations, by Keulemans, of all the more interesting or characteristic forms. It will be published in five Parts, each containing not less than seven coloured lithographs, comprising altogether about seventy figures of New-Zealand birds.

London, January 1872.

PREFACE TO FIRST EDITION.

The study of Ornithology has always been a source of intense enjoyment to me; and to write a history of the Birds of my native country was one of the day-dreams of my early boyhood. In maturer years my intervals of leisure, during an active official life in the colony, have been chiefly devoted to the collection of materials for such an undertaking; and the result is now presented to the public in a form which will, I trust, be acceptable to both the scientific and the general reader.

With what amount of success I have executed my self-imposed task it is not for me to decide. I am conscious, however, of having bestowed much honest labour upon it; and the highly favourable manner in which it has been reviewed, as well as the numerous letters of commendation and approval which I have received from persons in every way competent to form a judgment, give me reason to believe that my efforts have not been misdirected.

As a proof that I have spared myself no trouble to make the work complete I may mention that, without a single exception, the descriptions of the species have been taken from specimens actually before me, and that every measurement given throughout the book has been made or verified by myself. The life-histories are, for the most part, records of my own observations during a number of years; and I have endeavoured to make them as truthful as possible. It will be seen, however, that I have not failed to avail myself of the notes of other local naturalists, whose contributions are, in every instance, duly acknowledged.

I take this opportunity of expressing my gratitude to the Colonial Government for having granted me a prolonged leave of absence, on the most liberal terms, for the purpose of visiting England to superintend the publication of my work. To the authorities of the British Museum my thanks are due for the facilities which have been afforded me of studying the contents of perhaps the finest collection of Birds in the world, and to the gentlemen having charge of that department for their unvarying courtesy and attention—even my application to be allowed to remove the rare *Notornis* from its hermetically closed case, for the purpose of examination, having been readily complied with.

In working out the historical synonymy of the species I have found the Library of the

Zoological Society of great service; and in consulting authorities I have received valuable assistance from Mr. R. B. Sharpe, the late librarian, whose long connexion with the Society has made him familiar with the bibliography of the subject. The excellent lists already published by Mr. G. R. Gray and Dr. Otto Finsch had rendered this part of my task a comparatively light one; but all the references have been carefully verified, and the chronology given for the first time; while numerous synonyms have been added, and the whole of the nomenclature critically examined and revised.

To my brethren of the British Ornithologists' Union I hereby tender my acknowledgments for the readiness with which they have at all times given me the benefit of their opinions and judgment on doubtful points, or lent me specimens for comparison.

In conclusion I have only to state that, in consideration of the generous assistance accorded to me by the New-Zealand Government, I have presented the whole of my collection of Birds, on which the descriptive letterpress is chiefly founded, to the Colonial Museum at Wellington, where it will in future be accessible for purposes of reference.

W. L. B.

London, March 1873.

NOTICE OF THE NEW EDITION.

THE success which attended the Author's first edition of 'The Birds of New Zealand' (published in 1873, and containing comparatively few illustrations) has induced him to enter upon a more ambitious undertaking. Limited as that was to an impression of 500 copies, the whole edition was privately subscribed for; and the drawings on the stones, from which Mr. Keulemans had produced the inimitable Plates, were then erased. Published at *Five Guineas*, the price rapidly rose till, in a few years, a copy fetched £20 at public auction in New Zealand; then £21 in London (at the sale of Sir William Jardine's library); and, finally, in Melbourne, the extraordinary price of £37 10*s*. Even within the last few months, with the new edition well in progress, a second-hand copy reached £26 at Mr. Sotheby's sale-rooms.

The interval of thirteen years since elapsed has been spent by the Author in New Zealand, where he has enjoyed exceptional opportunities for obtaining fresh specimens and extending his knowledge of this remarkable avifauna.

This work will be issued in Thirteen Parts (to Subscribers only) at the price of *One Guinea* each, or *Twelve Guineas* for the whole if paid in advance.

Each part (except the last) will contain facsimiles of four beautiful coloured drawings by Mr. Keulemans, the birds being represented as they appear in life, with accessories drawn from the native flora of the country. These will be highly finished pictures in the best style of modern art, all the colour-stones being drawn either by or under the immediate direction of Mr. Keulemans himself. Specimens of these Plates, exhibited at the last *Soirée* of the Royal Society, were pronounced by 'The Times' reviewer "absolutely perfect."

A figure will be given of every form peculiar to New Zealand; and the enlarged size (Imperial

Quarto) will enable the artist to group the sexes or allied species together wherever it may be found desirable.

The final Part will contain a General Introduction, profusely illustrated with woodcuts, the List of Subscribers, and a complete Index to the whole work.

London, September 30, 1887.

Extracts from Professor Newton's Address to the Biological Section of the British Association (Manchester, 1887).

"When on a former occasion (at Glasgow in 1876) I had the honour of addressing a Department of this Section, I pointed out the enormous changes that were swiftly and inevitably coming upon the fauna of many of our colonies. The fears I then expressed have been fully realized. I am told by Sir Walter Buller that in New Zealand one may now live for weeks and months without seeing a single example of its indigenous birds, all of which, in the more settled districts, have been supplanted by the aliens that have been imported; while further inland these last are daily extending their range at the cost of the endemic forms. A letter I have lately received from Sir James Hector wholly confirms this statement, and I would ask you to bear in mind that these indigenous species are, with scarcely an exception, peculiar to that country, and, from every scientific point of view, of the most instructive character. They supply a link with the past that once lost can never be recovered. It is therefore incumbent upon us to know all we can about them before they vanish. . . . . The forms that we are allowing to be killed off, being almost without exception ancient forms, are just those that will teach us more of the way in which life has spread over the globe than any other recent forms; and for the sake of posterity, as well as to escape its reproach, we ought to learn all we can about them before they go hence and are no more seen. . . . . . One thing to guard against is the presumption that the fauna originated within its present area, and has been always contained therein. Thus, I take it, that the fauna which characterizes the New-Zealand Region—for I follow Professor Huxley in holding that a Region it is fully entitled to be called—is the comparatively-little changed relic and representative of an early fauna of much wider range."

PREFACE.

As stated in the Prospectus, this new and enlarged edition of 'The History of the Birds of New Zealand' is the outcome of a very general and rapidly increasing demand for a second issue. Its publication was commenced in June 1887, and the Author hopes to have the second volume completed by the end of February 1889.

Owing to the favourable reception accorded to this new 'History' in the Australasian Colonies, and the consequently increased number of Subscribers, the Author found himself in the gratifying position of being able to reduce the price of each Part from *One Guinea and a half* (as announced in the original prospectus) to *One Guinea*; but, as already notified, the edition will be strictly limited to 1000 copies, of which only about 250 will be available for Europe and America.

Although the Author has adhered to the general method and style of the former edition, he ventures to hope that the alterations and additions in the body of the work fairly represent, so far as New Zealand is concerned, the great advance which has been made in Ornithological Science during the present decade. The book itself is on a larger scale, being Imperial instead of Royal Quarto, and the Plates, instead of being hand-coloured lithographs, have been produced by the more costly but more exact and satisfactory process of printing in colours. It is generally admitted that nothing so perfect in colour-printing has hitherto been attempted; and the Author feels that special thanks are due to the talented artist, Mr. J. G. Keulemans, and to his able assistant, Mr. F. van Iterson, for the fidelity with which the drawings on stone have been executed; also to Mr. Otto Mayer (of the firm of Judd & Co.) for the faithful care and attention bestowed on the printing of the Plates, from first to last, so as to ensure the best possible finish.

To the Subscribers in England and the Colonies, and particularly in New Zealand, who have responded so liberally to the announcement of a New Edition, the Author tenders his grateful acknowledgments; for without such support he would never have undertaken so costly an enterprise. He would fain hope that the honest and patient labour which he has devoted to the work will be deemed a fitting return for their generous confidence.

W. L. B.

Inner Temple, London,
March 1888.

INTRODUCTION.

The first published list of the birds of New Zealand was drawn up by the late Mr. G. R. Gray of the British Museum, and appeared in 1843 in the Appendix to 'Dieffenbach's Travels.' This enumeration contained the names of eighty-four recorded species; but many of these were of doubtful authority, and have since been omitted. In the following year the same industrious ornithologist, in the 'Voyage of H.M.SS. Erebus and Terror,' produced a more complete list, embracing the birds of New Zealand and the neighbouring islands, accompanied by short specific characters, and illustrated by twenty-nine coloured figures, many of them of life-size. In July 1862 he published in 'The Ibis' a revision of this synopsis, with the newly-recorded species added, including, moreover, the birds inhabiting the Norfolk, Phillip, Middleton's, Lord Howe's, Macaulay's, and Nepean Islands. This enumeration contained altogether 173 species, of which 122 were said to occur in New Zealand and the Chatham Islands. In the 'Essay on the Ornithology of New Zealand,' written by myself at the request of the Exhibition Commissioners, in 1865, and afterwards published by the New-Zealand Institute *, eleven additional species were recorded; and in a paper which I communicated to the Wellington Philosophical Society in August 1868 † I gave the names of fourteen more. A few other species have since been added to the list; while, on the other hand, it has been found necessary to strike out several which had been admitted on insufficient evidence.

My first edition of the present work, published in 1873, contained descriptions of 147 species; and in my 'Manual of the Birds of New Zealand,' prepared at the request of the Colonial Government in 1882, twenty-nine more species were added to the list. The present edition does not profess to add many more to the number; but the classification and nomenclature have been revised, and a far more complete history has been given of each species than was possible before, seeing that I have, for a further period of fourteen years, enjoyed favourable opportunities for becoming better acquainted with the subject.

In the Introduction to the former edition it was stated that I had considered it necessary to omit the following species, there being no satisfactory evidence of their having occurred in New Zealand, viz.:—*Halcyon cinnamomina*, *Anthochæra carunculata*, *Gerygone igata*, *Rhipidura motacilloides*, *Aplonis zealandicus*, *A. caledonicus*, *Ortygometra fluminea*, *O. crex*, *Nesonetta aucklandica*, *Anous stolidus*, *Procellaria incerta*, *P. mollis*, *Dysporus piscator*, *Phalacrocorax sulcirostris*, and *Aptenodytes pennantii*. Further research has shown that *Gerygone igata* is only a synonym of *G. flaviventris*;

* Trans. N.-Z. Instit. 1868, vol. i.   † *Ibid.* pp. 105–112.

that *Nesonetta aucklandica* belongs legitimately to our list; and that *Anthochæra* (rectius *Acanthochæra*) *carunculata* and *Aptenodytes pennantii* (=*A. longirostris*), being of occasional occurrence, have now an undoubted claim to a place in our recognized Avifauna.

I ought perhaps here to refer to a species mentioned in the former Introduction as a newly-discovered addition to the New-Zealand Avifauna, but now omitted from our list. It was introduced by me in the following terms:—"In a country possessing such forms as *Notornis* and *Porphyrio* we might naturally look for the occurrence also of *Tribonyx*. Both of the latter are known to have a wide geographic range, while *Notornis*, which is a strictly local form, appears to combine in some measure the characters of each, being allied to *Porphyrio* in the form of its bill and in its general colouring, and to *Tribonyx* in the structure of its feet; while in the feebleness of its wings and the structure of its tail it differs from both. The recent discovery, therefore, in the South Island, of an example of *Tribonyx mortieri* which has been brought to England, and is now living in the Zoological Society's Gardens, is a very interesting fact in geographic natural history.

"The former acquisition by the Society of a similar bird, in July 1867, led to the discovery by Dr. Sclater that the species figured and described by Mr. Gould in his 'Birds of Australia' under that name was not the true *Tribonyx mortieri* of Du Bus (Bull. Acad. Sc. Brux. vii. p. 214), but a distinct bird, characterized by its smaller size and by the absence of white stripes on the wing-coverts. Dr. Sclater accordingly proposed the name of *Tribonyx gouldi* for the latter species (Ann. N. H. 1867, xx. p. 122), and gave the following distinguishing characters for *T. mortieri*:—'Major; alis albo striatis; plaga magna hypochondriali alba.'

"The bird now in the 'Gardens' was brought home (with other birds from New Zealand) by Mr. Richard Bills, and purchased by the Society on the 21st October, 1872. I am informed by the late owner that it was captured on the shores of Lake Waihora, in the Province of Otago, by a party of men who hunted it down with dogs. When first brought to him at Dunedin it was very wild and shy; but it soon became reconciled to confinement, and when he exhibited the bird to me in London it was perfectly tame and would feed from the hand" *.

Professor Hutton, having made the necessary inquiries on the spot, satisfied himself that the story was a pure invention, and that the dealer had purchased the bird in Dunedin, where it had doubtless been brought from Australia.

After the appearance of my first edition Dr. Otto Finsch, who had previously written several papers on the subject, contributed to the 'Journal für Ornithologie' (1874, p. 167) an admirable article entitled "Zusätze und Berichtigungen zur Revision der Vögel Neuseelands," which every student ought to consult.

* "*Descr.* ♀. Crown and sides of the head, nape, hind neck, back, and rump brownish olive, washed more or less with chestnut; wing-coverts greyish olive, shading into brown, each feather with a white streak down the centre; throat, fore neck, breast, and sides of the body dark ashy grey, passing into slaty black on the abdomen and under tail-coverts, where the plumage is slightly tipped or freckled with grey; the overlapping feathers on the flanks pure white in their apical portion, forming a conspicuous mark on each side of the body; under wing-coverts dull blackish brown, and all largely tipped with white; quills blackish brown, the secondaries brownish olive on their outer webs; tail-feathers black, the middle ones tinged with brown on their outer margins. Irides bright crimson, with a paler rim surrounding the pupil; bill greenish yellow, lighter towards the tip; legs and feet pale plumbeous tinged with yellow, the claws black. Total length 16·5 inches; extent of wing 25; wing, from flexure, 8; tail 4·5; bill, along the ridge 1·5, along the edge of lower mandible 1·4; tarsus 2·75; middle toe and claw 3·25; hind toe and claw 1·1."

In 1875 there appeared a new edition of the 'Voyage of the Erebus and Terror,' with an Appendix from the pen of Mr. R. Bowdler Sharpe, containing valuable notes on many of the species, and giving illustrations of some birds not figured in the earlier issue.

The most recent work containing notices of New-Zealand birds is Mr. Seebohm's on 'The Geographical Distribution of the Charadriidæ'*, where there is an excellent plate by Keulemans representing *Charadrius obscurus* in full summer plumage.

With regard to the changes I found it necessary, in my first edition, to make in the generally accepted nomenclature, my explanation was a simple one. While fully admitting the advantages of the rule "*quieta non movere*" in the case of names which had obtained universal currency, I considered it better, in undertaking a general revision of the whole subject, to apply the strict principle of modern nomenclature, and, in all cases where the subject was free from doubt, to adopt the oldest admissible title. I knew that we could not look for any finality in the generic appellations so long as the science was a progressive one; but I was desirous of giving something like fixity and permanence to the specific names; and with this view I endeavoured, so far as I could, to rectify all existing errors—altering the names entirely in cases where it appeared to me that wrong ones had hitherto been employed, and correcting obvious classical defects in others—substituting, for example, *Hymenolæmus* for *Hymenolaimus*, and *antipodum* for *antipodes*. In no instance did I introduce any change without very careful consideration and research; and the fact that the authorities in the British Museum adopted, with scarcely a single exception, my corrections and identifications in the classification of the New-Zealand birds in the national collection, may, I think, be accepted as a proof that I exercised proper judgment in this respect.

In the present edition some other corrections of a trivial kind have been made in the nomenclature, and in every instance I have given what I venture to think are sufficient reasons for the proposed changes. For example, no ornithological student will object to the rectification of *albicilla* into *albicapilla*, or the substitution of *Limosa novæ zealandiæ* for the museum name of *Limosa baueri*, originally published without any description.

In my classification I have departed considerably from the system followed in my first edition. This was inevitable in order to keep pace with the progress of Ornithological science. I may state that I have in general, and as far as practicable, adhered to the scheme of arrangement adopted in 1883 by a Committee of the British Ornithologists' Union for the classification of the birds of Great Britain. But wherever I have thought it necessary to make any alteration in the arrangement of the Ordinal groups, I have not hesitated to do so. For example, I have made the Orders GALLINÆ and COLUMBÆ follow ACCIPITRES instead of being placed after STEGANOPODES, as I consider this an equally natural arrangement and better suited to the proposed division of my work into two volumes, the first closing with the last-mentioned Order, and the second opening with the LIMICOLÆ. I shall treat the Order

* Doubtless it is easy enough to discover "blunders and omissions" in any book that professes to treat exhaustively of the birds of a particular country, or the members of any special group or division; but Mr. Seebohm seems to have been exceptionally unfortunate in his references to New-Zealand species. He says of *Charadrius obscurus* that "it breeds in the mountains, descending to the coast in winter;" he describes *Anarhynchus frontalis* as an "inland species;" and he confounds *Himantopus novæ zealandiæ* in winter plumage with the Australian Stilt under the novel title of "*Himantopus leucocephalus pictus*." It is not easy, however, to find fault with an author who fairly and openly says:—"If I have criticised the work of my fellow ornithologists too severely, I ask their pardon and hope that they will pay me back in my own coin by correcting my blunders with an unsparing hand."

GAVIÆ as naturally coming next, instead of being divorced by the interposition of PYGOPODES, as proposed by the B.O.U. Committee *. GRALLÆ, HERODIONES, STEGANOPODES, and TUBINARES will then follow in the order named; and I shall place PYGOPODES after ANSERES, closing the great Carinate division with the specialized group of IMPENNES or PENGUINS. After that, and concluding the work, will come a history of the Ratite forms in New Zealand (the various species of *Apteryx*), interesting not only on account of their low development but, as already explained, in respect of their relation to the extinct STRUTHIONES †.

In my arrangement of the genera composing the great Order of PASSERES I have for the most part followed the now well-beaten track of modern systematists; but in some instances I have ventured to depart from it, giving my reasons in every case. For example, I have followed Professors Parker and Newton in placing the Corvidæ at the head of the Order instead of the Turdidæ, and I have accordingly commenced my history of our Avifauna with an account of the New-Zealand Crow. It must be acknowledged, however, that *Glaucopis*, instead of being a typical Crow, betrays certain strongly aberrant characters, and it is possible that we may hereafter have to alter its exact location. In the present unsettled state of Ornithological nomenclature I am anxious to avoid, as far as possible, the multiplication of names; but *Glaucopis* may prove to be one of those abnormal Antipodean forms of a very ancient fauna—generalized types though existing in a specialized form—which have no analogues or representatives in the Northern Hemisphere. In this event it must ultimately become the type of a new Family, to which the name of *Glaucopidæ* might be appropriately applied. At page 30 I have given my reasons for removing our two species of Thrush from the typical Turdidæ and placing them in a new Family under the name of Turnagridæ. So far, however, as the New-Zealand Ornis is concerned, alterations of this kind will not affect the generic arrangement of the groups in their mutual relation to one another.

But, as remarked in my former Introduction, any system of classification, however excellent in itself, or ably conceived and elaborated, must of necessity be a provisional or tentative one, so long as our knowledge of the structural character and natural affinities of the vast majority of species continues so imperfect as it confessedly is at present. When the anatomy of every known bird on the face of the globe has been as fully investigated as that of the Rock-Dove (*Columba livia*) was by the late Professor Macgillivray, and its life-history becomes as thoroughly known, then, but not till then, will it be possible to devise a system of arrangement absolutely true to nature. The aim and purpose of all classification being to aid the memory in its effort to comprehend and master the complex and ever varied productions of nature, or, in other words, to assist the mind by a ready association of ideas

* "Prof. Parker long ago observed (Trans. Z. S. v. p. 150) that characters exhibited by Gulls when young, but lost by them when adult, are found in certain Plovers at all ages, and hence it would appear that the Gaviæ are but more advanced *Limicolæ*. The Limicoline genera *Dromas* and *Chionis* have many points of resemblance to the *Laridæ*; and on the whole the proper inference would seem to be that the *Limicolæ*, or something very like them, form the parent-stock whence have descended the *Gaviæ*, from which, or from their ancestral forms, the *Alcidæ* have proceeded as a degenerate branch."—*Enc. Brit.* vol. xviii. p. 45.

† Professor Newton, in his able article "Ornithology" in the 'Encyclopædia Britannica,' in treating of the recent and existing forms of toothless Ratitæ, says:—"Some systematists think there can be little question of the *Struthiones* being the most specialized and therefore probably the highest type of these Orders, and the present writer is rather inclined to agree with them. Nevertheless the formation of the bill in the *Apteryges* is quite unique in the whole Class, and indicates therefore an extraordinary amount of specialization. Their functionless wings, however, point to their being a degraded form, though in this matter they are not much worse than the *Megistanes*, and are far above the *Immanes*—some of which at least appear to have been absolutely wingless, and were thus the only members of the Class possessing but a single pair of limbs."

in the grand study of Creation, it follows that the method of arrangement which best subserves this practical end is the right one to adopt. But we must be content to see our carefully elaborated systems swept away one after another, till, perhaps, in the distant future some gifted mind shall arise, who, with the constructive energy of a second Cuvier, may be able to fashion, from the more complete materials at his command, a system perfect in all its parts and destined to endure till time shall be no more.

In portraying the manners and habits of the various species I have been careful to omit nothing that seemed calculated to elucidate their natural history. It has been said that a zoologist cannot be too exact in recording dates and other apparently trivial circumstances in the course of his observations, and that it is better to err on the side of minuteness than of vagueness, because an observer is scarcely competent to determine how far an attendant circumstance, trivial in itself, may afterwards be found to enhance the value of a recorded fact in science when viewed in relation to other facts or observations. It must be borne in mind, however, that we are as yet only imperfectly acquainted with many of the native species, and that probably, in the history of all that are here treated of, new facts or new features of character will hereafter come to light. It is extremely difficult to cultivate an intimate acquaintance with birds that are naturally shy and recluse, and especially so in a thinly peopled country, where they rarely cross the path of man and must be assiduously sought for in bush, swamp, and jungle. While relying generally on my own opportunities for observation, I have not failed to avail myself of the kind assistance of others; and in the body of the work numerous acknowledgments will be found of information furnished by correspondents in various parts of the country, who, amid the multifarious duties and engagements of a colonial life, have found time to take notice of the natural objects around them.

Before passing on to a consideration of the existing Avifauna it may be useful to take a rapid survey of the Families and Genera known to us by their fossil remains as having formerly inhabited New Zealand, or roamed over the continent of which these islands are the only remnants at the present day. These ornithic relics of a bygone time have been interpreted, restored, and classified with marvellous felicity by Professor Sir Richard Owen in his 'Memoirs of the Extinct Wingless Birds of New Zealand'*. These memoirs had appeared, from time to time, since the year 1838, in successive Parts of the 'Transactions of the Zoological Society of London,' and following the example of Baron Cuvier, who thus reprinted his numerous detached papers under the title of 'Recherches sur les Ossemens Fossiles de Quadrupèdes,' the venerable Professor has collected his many exhaustive Memoirs and combined them with additional matter and general remarks in two splendid volumes, replete with illustrations. In the prospectus announcing this work he did me the great honour to say that "his purpose, long entertained, was strengthened by the appearance and favourable reception of an excellent and comprehensive work on the existing Birds of New Zealand, to which the present Volumes may be deemed complementary." The volumes, thus modestly announced, commence with an Introductory Notice of the circumstances which led to the discovery and restoration of the extinct Avifauna of New Zealand; the descriptions which follow are accompanied by illustrations of the natural size of the fossils, together with reduced views of the restored skeletons on which the several genera and species have been founded; the whole is preceded by an illustrated Anatomy of the existing wingless bird (*Apteryx australis*), which, as the Professor states, is the nearest ally of the

* Quarto, 1878, 2 vols.

extinct *Dinornis*, and is followed by notices of the food, footprints, nests, and eggs of the Moas, the Maori traditions relating to these gigantic birds, the causes and probable period of their extirpation, and a speculation on the conditions influencing the atrophy of the wings in flightless birds—to all of which the learned author has appended Supplementary Memoirs on the Dodo, Solitaire, and Great Auk, with evidences of other extinct birds in Australia and Great Britain.

THE ANCIENT AVIFAUNA.

The first Moa-bone of which we have any record was a mere fragment of a femur six inches in length, with both extremities broken off, which was brought to England in 1839, and offered for sale at the Royal College of Surgeons by an individual who stated that he had obtained it in New Zealand from a native who told him that it was the bone of a great Eagle. Professor Owen, on its being first submitted to him, assured the owner that "it was a marrow-bone like those brought to table wrapped in a napkin"; but on subsequent and more critical examination he arrived at the conviction that it had in reality come from a bird, that it was the shaft of a thigh-bone, and that it must have formed part of the skeleton of a bird as large as, if not larger than, the full-sized male Ostrich, with this very striking difference, that whereas the femur of the Ostrich, like that of the Rhea and the Eagle, is "pneumatic," or contains air, the huge bone, of which a fragment was now submitted to him, had been filled with marrow like that of a beast. The price asked for this unique specimen was only ten guineas, and although Professor Owen strongly recommended its acquisition, the Museum Committee declined to purchase the "unpromising fragment." Much against the advice of his scientific contemporaries Owen insisted on publishing his conclusions, announcing boldly—"So far as my skill in interpreting an osseous fragment may be credited, I am willing to risk the reputation for it on the statement that there has existed, if there does not now exist, in New Zealand a Struthious bird, nearly, if not quite, equal in size to the Ostrich."

After the publication of Professor Owen's paper the bone was purchased by Mr. Bright, M.P. for Bristol, and many years subsequently came into the possession of the British Museum, where this historic relic is now carefully preserved.

More than three years elapsed before any confirmatory evidence was received from New Zealand; and then came a letter from the Rev. W. Cotton to Dr. Buckland, followed by another from the Rev. W. Williams, giving an account of the discovery of large numbers of these fossil remains and accompanied by a box of specimens, which triumphantly established the accuracy of Owen's prevision. The specimens transmitted by Mr. Williams were, as a matter of course, confided by Dr. Buckland to the learned Professor for determination; and these materials, scanty as they were, enabled him to define the generic characters of *Dinornis*, as afforded by the bones of the hind extremity. An examination of a second and richer collection sent home by Mr. Williams, together with three additional specimens lent by Dr. (afterwards Sir John) Richardson of Haslar Hospital, enabled him to discriminate six distinct species of the genus, ascending respectively from the size of the Great Bustard to that of the Dodo, of the Emu, of the Ostrich, and finally attaining a stature far surpassing that of the last-named biped.

The first of these was a Cursorial bird which, on account of the agreement of its tibia in its general characters with the same bone in the larger species, he referred at that time to the genus *Dinornis*, but which subsequent investigations proved to belong to another genus, characterized by the presence of a strong hind toe, for which the name of *Palapteryx* was proposed. As this bird had something of the appearance of the Great Bustard, he called it *Dinornis otidiformis*.

It may be here mentioned that in the Ostrich, Rhea, and Cassowary there is no vestige of a hind toe or hallux.

The next was a three-toed Struthious bird differing from the other species of *Dinornis* in its relatively shorter and broader metatarsus, in which characters it appeared closely to resemble the extinct Dodo (*Didus ineptus*) of the Isles of France and Rodriguez; and as it could not have been greatly superior in size to that bird, he named it *Dinornis didiformis*. Judging by its skeleton, this bird stood a little under four feet in height, or of intermediate size between the Cassowary and the Dodo. In the metatarsal of this bird, as with the larger species of *Dinornis* to be presently mentioned, there was not the slightest trace of the articulation of a fourth posterior toe, the generic distinction from *Didus* and *Apteryx* being thus distinctly indicated.

The next species described, which appears to have attained the average height of the Ostrich (about seven feet), with a more robust and stronger build, he named *Dinornis struthioides*, and pointed out characters which placed the fact of its being a good and true species beyond all cavil or doubt.

Another species, which attained the height of nine feet, he provisionally named *Dinornis ingens*; but, as will appear further on, this bird was also subsequently referred to the genus *Palapteryx*. Then came the discovery of a still larger form, standing ten feet in height if not more, which he distinguished as *Dinornis giganteus*. A fair idea of the size of this gigantic bird, in comparison with the stature of an ordinary-sized man, may be obtained from the accompanying woodcut, which is a reduction from the lithograph forming the frontispiece to my first edition *. The representation of the skeleton is from a photograph of the magnificent specimen in the Canterbury Museum, and the figure of the Maori, clothed in a dogskin mat and "wrapt in contemplation," is taken from the portrait of the old Ngapuhi chief, Tamati Waka Nene, as given in Angas's 'New Zealanders illustrated.'

* Copies of this interesting plate have appeared in Kennedy's 'New Zealand,' Sir Julius Vogel's 'Handbook of New Zealand,' and the Rev. James Buller's 'Forty Years in New Zealand.'

Yet another form, with a stature of about five feet, had to be discriminated, and this Owen named *Dinornis dromioides*, on account of its similarity in size to the Emu.

Not content with this large addition to the hitherto known Struthious birds of the world from one small area of land, the learned Professor made a happy forecast of further discoveries yet in store, for he then wrote:—

"Already the heretofore recorded number of the Struthionidæ is doubled by the six species of *Dinornis* determined or indicated in the foregoing pages: and both the Maori tradition of the destruction of the Moa by their ancestors and the history of the extirpation of the Dodo by the Dutch navigators in the Isles of Maurice and Rodriguez, teach the inevitable lot of bulky birds unable to fly or swim, when exposed, by the dispersion of the human race, to the attacks of man. We may therefore reasonably anticipate that other evidences await the researches of the naturalist, which will demonstrate a further extent of the Struthious order of Birds anterior to the commencement of the present active cause of their extinction."

Among the most important contributors to the history of *Dinornis* at this early period were the Rev. William Colenso, F.R.S., who not only collected specimens of the bones, but published a very interesting memoir on the subject in the 'Tasmanian Journal' (vol. vii., 1843), and the Rev. Richard Taylor, who, in 1844, wrote as follows:—

"Early in 1843 I removed from the Bay of Islands to Wanganui, and my first journey was along the coast of Waimate. As we were resting on the shore near the Waingongoro stream, I noticed the fragment of a bone which reminded me of the one I found at Waiapu. I took it up and asked my natives what it was. They replied 'a Moa's bone; what else? Look around and you will see plenty of them.' I jumped up, and to my amazement I found the sandy plain covered with a number of little mounds entirely composed of Moa-bones; it appeared to me to be a regular necropolis of the race. I was struck with wonder at the sight, but lost no time in selecting some of the most perfect of the bones. I had a box in which my supplies for the journey were carried; this I emptied, and filled with the bones instead, to the amazement of my followers, who exclaimed 'What is he doing? What can he possibly want with these old Moa-bones?' One suggested 'hei rongoa pea' (to make medicine perhaps); to this the others consented, saying 'koia pea' (most likely)."

Other stray collections continued to arrive from time to time, till at length Mr. Percy Earl, in 1846, unearthed from the turbary deposits of Waikouaiti and sent to England a more extensive series of bones than any other collector had succeeded in bringing together. These collections all found their way, more or less directly, into Professor Owen's hands, and he was thus enabled to rectify or confirm many of his former deductions. He was also enabled to add several new species. One of these was *Dinornis casuarinus*, nearly agreeing in size with *D. dromioides*, and combining the stature of the Cassowary with more robust proportions and especially more gallinaceous characters in the feet. A mutilated femur of this bird he had previously regarded as belonging to a young individual of the last-named species, and when he afterwards corrected the error he pointed out that it was a mistake on the safe side, "the caution which refrains from multiplying specific names on incomplete evidence being less likely to impede the true progress of zoological science than the opposite extreme." The most abundant of the remains collected by Mr. Earl belonged to this species (*D. casuarinus*); but there were also in the collection bones of another very remarkable species, which was named *Dinornis crassus*, in allusion to the strength of its osseous frame. It was intermediate in size between *Dinornis ingens* and *D. struthioides*, and, with a stature equal to that of the Ostrich, the femur and the tarso-metatarsus of this bird present double the thickness in proportion to their length. Of this species Prof. Owen writes:—"It must have been the strongest and most robust of Birds, and may be said to have represented the pachydermal type and proportions in the feathered class." A third new species, following next, in order of size, to *Dinornis didiformis*, and strictly confined in its range to the North Island, was named *Dinornis curtus*.

These more complete materials contained indubitable proof that *Dinornis dromioides* possessed, in common with *D. ingens*, the character of a distinct hind toe. Among the true forms of *Dinornis* this member was reduced (as in the *Apteryx*) to a high-placed hallux of diminutive size and functionless character, the attachment of this rudimentary toe being merely ligamentous. In most of the skeletons of *Dinornis* hitherto found there was no trace whatever of a hallux; but Professor Owen has, with every show of probability, ascribed this absence to the extremely small size of these bones and the ease with which they could be overlooked or lost rather than to their non-development, although at an earlier date he was inclined to make it a character of generic importance.

In 1850, Sir George Grey, who had been actively collecting Moa-bones in the district lying under Tongariro mountain, forwarded his collections to the British Museum; and two years later, Professor von Hochstetter, the naturalist attached to the Expedition of the Imperial Austrian frigate 'Novara,' who had undertaken a topographical examination of the North Island, obtained a rich collection from the same locality.

Most of these remains were found to belong to *Palapteryx ingens*, of which the annexed imaginary sketch is given in Prof. Hochstetter's 'Neu-Seeland' (1863, p. 438).

Up to this period of our narrative the remains discovered appear to have belonged exclusively to birds of the Struthious Order; and, as Professor Owen had on more than one occasion explained, the existing *Apteryx*, notwithstanding the inferiority of size, modified structure of the palate, and different proportions of the beak, was the nearest living representative of these extinct and comparatively ancient forms *.

But new discoveries of a most interesting kind were yet in store for the great comparative anatomist, by which he was afterwards able to demonstrate further links of connection between the extinct types and still existing forms.

In 1852–56 it fell to the lot of Mr. Walter Mantell (at that time a Government Land-Purchase Commissioner) to explore the Moa-bone deposits at Waingongoro, in the North, and at Waikouaiti, in the South Island, and the extensive collections which he then made and transmitted to England not only "excited the delight of the natural philosopher, and the astonishment of the multitude," but, having been deposited in the British Museum, these new materials, in hitherto unknown abundance, enabled the Professor not only to verify some of his former conclusions, but to establish the characters of several new genera †. No doubt the most important result was the discovery of *Dinornis elephantopus*, "a species which, for massive strength of the limbs, and the general proportion of breadth or bulk to height of body, must have been the most extraordinary of all the previously restored wingless birds of New Zealand, and unmatched, probably, by any known recent or extinct member of the class of Birds" ‡.

The excellent woodcut on the next page, showing this skeleton as articulated in the British Museum, is copied by permission from Dr. Thomson's 'Story of New Zealand,' p. 32.

But, in addition to this splendid Moa, the collection contained other very interesting novelties. Among these were *Aptornis*, the giant prototype of the existing Woodhen (*Ocydromus*), and notably the fossil remains of *Notornis mantelli*, a huge Coot, of which three recent or living examples were

* As to the affinities of the *Apteryx*, deducible from its anatomy, Prof. Owen says:—"Commencing with the skeleton, all the leading modifications of that basis of its structure connect it closely with the Struthious group. In the diminutive and keelless *sternum* it agrees with all the known Struthious species, and with these alone. The two posterior emarginations which we observe in the *sternum* of the Ostrich are present in a still greater degree in the *Apteryx*; but the feeble development of the anterior extremities, to the muscles of which the *sternum* is mainly subservient, as a basis of attachment, is the condition of a peculiarly incomplete state of the ossification of that bone of the *Apteryx*; and the two sub-circular perforations which intervene between the origins of the pectoral muscle on the one side, and those of a large inferior dermocervical muscle on the other, form one of several unique structures in the anatomy of this bird."

† "The *kainga* (at the stream now known as Awamoa) which we found in 1852 afforded further unmistakable proof of the coexistence of man with the Moa—the bones and egg-shells of *Dinornis* and its kindred, mixed with remains of every available variety of bird, beast, and fish used as food by the aborigines, being all in and around the umus (or native ovens) in which they had been cooked. Although my collection from this place reached England in 1853, it remained unopened until after my arrival there in 1856, when I caused it to be conveyed to the crypts of the British Museum, and there unpacked it in the presence of the great authority on our gigantic birds, Professor Owen. With the exception of two small collections which were selected for me by Professor Owen, and which I gave, one to the Museum of Yale College, U.S., and the other to that of the Jardin des Plantes, the whole of this collection is now in the British Museum. The fragments of egg-shells from these umus varied in size from less than a quarter of an inch of greatest diameter to three or four inches. These, after careful washing, I had sorted, and having, with some patience, found the fragments which had originally been broken from each other, and fitted them together, I succeeded in restoring at least a dozen eggs to an extent sufficient to show their size and outline. Six or seven of the best of these I gave to the British Museum after their purchase of the collection: one is in the Museum of the College of Surgeons; the rest, including one very beautiful egg with a polished ivory-like surface, are still in my ownership somewhere in England. Some idea of the labour entailed by this attempt to rehabilitate eggs may be gathered from the fact that several of those restored consisted of between 200 and 300 fragments. I may add that in the markings, size, and so forth (making allowance for the alteration of the former towards the ends of the eggs) I made out about twenty-four varieties, of which I have specimens."—*Mantell.*

‡ "By the side of the metatarsus of *Dinornis elephantopus*, that of *D. crassus* shrinks to moderate if not slender dimensions. But the peculiarities of the elephant-footed *Dinornis* stand out still more conspicuously when the bones of its lower limbs are contrasted with those of *D. giganteus*."—*Owen.*

afterwards obtained, and which, there is every reason to hope, still exists in the high tablelands and remote fastnesses of the South Island *.

The abundant remains of *Aptornis* enabled the Professor to discriminate two species, namely *Aptornis otidiformis* (originally, from the examination of a single bone, referred to *Dinornis*) and the still larger *Aptornis defossor*, of such size and strength that, to quote his own words, "the civil engineer might study perhaps with advantage the disposition of the several buttresses, beams, and arched plates which support the iliac roof of the pelvis, and strengthen the acetabular walls, receiving the pressure of the thigh-bones in this huge and powerful Woodhen."

I may here mention that an *Aptornis* skull, dug up by Mr. W. W. Smith at Albury, near Oamaru, affords slightly larger measurements than any hitherto recorded of *A. otidiformis*, and that Sir Richard Owen, to whom I presented the specimen, wrote to me saying:—"The facial part exceeds in size that of figs. 2 and 3 (pl. xliii. of the 4to vols.), but so little as not to support a distinct species, unless the rest of the skeleton affords corroborative characters †. The specimen you have kindly presented and which, for your sake, I shall value while the brief remainder of life lasts, is evidently, from its specific gravity, from a bird that has long passed away. I should, however, rejoice if confirmatory characters justified me in introducing to our Zoological Society an *Aptornis bulleri*."

Then followed the discovery, in succession, of *Dinornis geranoides*, *D. gravis*, *D. rheides*, and *D. robustus*. Of the last-named species there is now an almost perfect skeleton in the museum at York, in a remarkable state of preservation, with portions of the integuments and quill-part of the feathers still attaching to the sacrum, and the legs still preserving some of the ligaments and inter-articular cartilages. With this valuable series, which will be found fully described by Mr. Allis in the 'Proc. Linn. Society' (vol. viii. p. 50), were found the rudimentary wing-bones which had so long been an object of diligent search. The skeleton is probably that of a male bird, because lying immediately beneath it, buried in the heap of sand with which the remains were covered, were the bones of four young ones, presumably the whole of a clutch. This deduction as to the sex seems a fair one from analogy, inasmuch as the male *Apteryx*, and indeed that sex in the majority of Struthious birds, alone performs the duty of incubation.

* Dr. Meyer, from a careful comparison of the bones, concludes that the South-Island bird is a distinct species, for which (in his 'Abbildungen von Vögel-Skeletten') he has proposed the name of *Notornis hochstetteri*.

† In relation to measurements Prof. Owen says:—"These may perhaps be deemed by some ornithologists to be slight or trivial differences; yet, taken in connection with the greater breadth and thickness of the bone, in proportion to its length, they unquestionably support the conclusions of specific distinction deducible from those proportions."

Subsequently another species, coming from the extreme North, was determined by Prof. Owen, and named *Dinornis gracilis*, on account of the remarkable length and slenderness of its legs.

But there seemed to be practically no limit to the ornithic wonders revealed by the Post-pliocene deposits of New Zealand. Professor Owen had already well nigh exhausted the vocabulary of terms expressive of largeness by naming his successive discoveries *ingens*, *giganteus*, *crassus*, *robustus*, and *elephantopus*, when he had to employ the superlative in *Dinornis maximus* to distinguish a species far exceeding in stature even the stately *Dinornis giganteus*. In this colossal bird, as the Professor has well remarked, some of the cervical vertebræ almost equal in size the neck-bones of a horse! The skeleton in the British Museum, even in an easy standing posture, measures eleven feet in height, and there is evidence that some of these feathered giants attained to a still greater stature.

A fair idea may be gained of its proportionate size from the accompanying woodcut, which appeared some years ago in 'The Illustrated London News,' representing the entire left leg of a Moa (now in the Madras Government Museum) obtained by Major Michael, of the Madras Staff Corps, from the Glenmark swamp, about 40 miles from Christchurch, where it was found *in situ*, at a depth of four feet, by a party of workmen who were cutting a drain. The measurements are:—Femur 1 ft. 6 in.; tibia 3 ft. 3 in.; tarsus 1 ft. 8 in.; outer toe 9¾ in.

The corresponding right leg was exhumed a considerable time afterwards, when Mr. Fuller was conducting a search on behalf of the Canterbury Museum, and this specimen, with the phalanges complete, is now in my private collection.

In November 1878, Mr. H. L. Squires of Queenstown, South Island, obtained and forwarded to the British Museum the head of a Moa with a continuous part of the neck, with the trachea enclosed and covered by the dried integument, and exhibiting even the sclerotic bone-ring of the dried eyeballs; also the bones of both legs with the feet covered by the dried skin, with some feathers adhering to it, and with the claws intact.

It was this specimen that enabled Sir Richard Owen to characterize his *Dinornis didinus*, and we may imagine the delight with which the veteran scientist embraced this opportunity of examining, for the first time, a specimen in which the characters could be studied as in a living or recent bird, and the value of his deductions from the study of single bones thereby tested, as well as the satisfaction with which he found his general conclusions so amply verified. This bird was scarcely larger than *Dinornis didiformis*, but presented characters of sufficient importance to separate it specifically from that form. The result of a close comparison of this dried head with that of existing Struthious birds was that "the Moa is found to repeat most closely, in the form and proportions of the beak, and in the shape, relative positions, and dimensions of the narial, orbital, and auditory apertures, the Emus of the Australian continent." Two points of considerable interest were established by this specimen, namely, the existence, at any rate in this species of Moa, of a strong hind toe with almost grasping power, and secondly the remarkable fact that the tarsus was feathered right down to the toes.

The other newly-discovered species (*D. parvus*) was founded on a nearly complete skeleton procured during the construction of a new road, about forty miles to the north-west of Nelson. The opportunity thus afforded of examining the entire osteology of a single bird was of extreme importance in the final determination of the generic characters. In size *Dinornis parvus* was scarcely superior to the Bustard (*Otis tarda*): and, although the smallest known member of this race of Struthious birds, it had proportionately the largest skull of all the *Dinornithidæ*. On this curious fact Owen thoughtfully remarks that if the peculiarly nutritious roots of the common fern contributed, together with buds or foliage of trees, to the food of the various species of Moa, the concomitant gain of power in the locomotive and fossorial limbs does not appear to have called for a proportionate growth or development of brain or of bill.

But the turbary deposits of New Zealand had not yet yielded up the whole of their wonderful story of the past. In the year 1868, it was discovered that the Glenmark swamp was a veritable necropolis of extinct birds. It is said that portions of no less than eighteen skeletons were dug up from the spot whence Major Michael obtained his leg of *Dinornis maximus* and within an area of about ten square feet. Under the able direction of the late Sir Julius von Haast, and with indefatigable zeal, these fossil remains were exhumed literally by thousands, sent to the Canterbury Museum in waggon-loads, sorted and classified there, and then distributed among the museums of the world, producing in return, by a judicious system of interchange, some £20,000 worth of specimens of various kinds, and helping materially to place the Canterbury Museum in the proud position which it now occupies in the Colony.

Sir Julius von Haast worked out the collections which he had formed in a very painstaking manner, and published the results in an Address to the Philosophical Institute of Canterbury. His minute observations and measurements over a wide field of specimens had the effect of confirming in a very remarkable manner the conclusions arrived at previously by Sir Richard Owen as to generic and specific distinctions, although these were not unfrequently based on a single bone or fragmentary part of a skeleton.

But perhaps the most important discovery was that of the existence, contemporaneously with the Moa, of a gigantic bird of prey, far exceeding the Golden Eagle in size, to which Haast gave the name of *Harpagornis moorei*. The evidence of this furnished by the fossil remains was fully discussed by the discoverer in a paper published in the Transactions of the New-Zealand Institute *; and it has become a favourite theme of speculation whether the true function in life of this great Raptor was not to prey upon the smaller species of *Dinornis*, or the chicks or young broods of the more gigantic forms.

A second and smaller species was afterwards described, under the name of *Harpagornis assimilis*; but it is not unlikely that this was the male of *H. moorei*, the disparity in size, which is the only difference, being thus easily explained.

In addition to this wondrous store of bones, so intermixed and packed together that in some instances there were twenty-five or thirty specimens from different birds imbedded in one solid mass, Sir Julius Haast had afterwards the opportunity of examining another interesting collection of Moa-

* Vol. iv. pp. 192–196.

bones made by Mr. T. F. Cheeseman, from a cave at Pataua, near Whangarei, in the far North, and practically the same district from which the Rev. W. Cotton obtained the first bones of *Dinornis curtus* in 1844–45. In this collection there was an almost perfect skeleton of a species appreciably smaller than the last-named one; and, under the name of *Dinornis oweni*, Haast dedicated this to "the illustrious biologist to whom science in New Zealand is so much indebted" *.

In a paper published in 1874, Sir Julius Haast proposed a new classification of the extinct *Struthiones*, which, so far, does not appear to have met with general acceptance. He divided them into two Families, which he named respectively the *Dinornithidæ* and the *Palapterygidæ*, each with two genera, the former comprising *Dinornis* and *Meionornis*, and the latter *Palapteryx* and *Euryapteryx*. He made the total absence of hind toe or hallux the distinguishing character of the first-named Family, thus following the broad line by which Owen had already differentiated his genera *Dinornis* and *Palapteryx*. He ventured, moreover, to characterize his Family *Palapterygidæ* as one in which the anterior limbs are entirely absent; but his conclusions on this head are far from being decisive. It would appear more likely, from the analogy of the case, that in those species in which the wings are supposed to have been wholly absent they existed only in a very rudimentary form, and that the small bones have perished, leaving no trace behind for the modern student of palæontology. It seems to be placed beyond doubt that in all the so-called Wingless Birds, by long-continued disuse of the anterior limbs through many successive generations, these organs had become enfeebled and ultimately atrophied and dwarfed to the condition of mere rudiments, as is now conspicuously apparent in the existing species of *Apteryx*. Professor Owen has suggested that in the case of *Dinornis* "the degree of atrophy, which seems to have been carried to a total loss of the limb-appendages of the scapulo-coracoid arch, implies the operation of the influence of disuse through a period of pre-Maori æons greatly exceeding the time during which the Lamarckian law has operated on the Cassowary, the Rhea, and the Ostrich."

Following this came the discovery by Sir James Hector of the remains of an extinct Goose, of very large if not gigantic proportions, and undoubtedly flightless. This proved to be the bird a few detached bones of which Professor Owen had previously referred to a genus "hitherto unknown to science," and supposed to be of the Struthious Order, for which he proposed the name of *Cnemiornis calcitrans*. The first tolerably complete skeleton of this Anserine form, which was certainly contemporaneous with the colossal Moas, was obtained by the Hon. Captain Fraser in the Earnsclough caves, and was afterwards presented by him to the British Museum. Another coeval species determined by Professor Huxley, was the giant Penguin (*Palæeudyptes antarcticus*), of which the bones were discovered by Mantell in the Oamaru limestone in 1849. To the same species are doubtless referable the fossil remains more recently found by Mr. James Duigan at Hokitika. These were discovered imbedded in a reef exposed only at low water and forming part of the Seal Rock, a bold headland which protects the anchorage of Woodpecker Bay. The bones were thoroughly mineralized, resembling the condition in which fossil reptilian bones are usually found †.

Even now, although the Post-pliocene bone-deposits of New Zealand, both North and South, have been pretty thoroughly explored, new species of Wingless Birds are being from time to time added to the list. During the last seven years Professor Owen has characterized two new species from the

* Trans. Zool. Soc. vol. xii. p. 171.

† Trans. N.-Z. Instit. vol. iv. pp. 341–346.

South Island as *Dinornis didinus* and *D. parvus*; and he has suggested that another, which may ultimately turn out to be new, might be appropriately named *Dinornis huttoni*, in compliment to the discoverer.

As already mentioned Sir Julius Haast added *Dinornis oweni* to the list of species; and he likewise discovered an extinct form of *Apteryx*, far exceeding in size those existing at the present day, to which he gave the name of *Megalapteryx hectori*.

Doubtless other forms, perhaps as interesting and remarkable as any yet brought to light, remain entombed to reward the zeal and enterprise of the future explorer.

Bearing on the question of the geographical relations of the New-Zealand Avifauna, one very important fact presents itself to us. In the same way that, as at the present day, certain well-marked species in the North Island are represented in the South Island by closely allied but yet specifically distinct forms, so it was also with the extinct Avifauna. The strong-limbed Moas with bulky frames were *Dinornis gravis*, *D. crassus*, *D. elephantopus*, *D. robustus*, and *D. maximus*, and these appear to have been strictly confined to the South Island. Six species having proportionately thin leg-bones and a slighter frame, namely, *Dinornis didiformis*, *D. dromioides*, *D. gracilis*, *D. struthioides*, *D. ingens*, and *D. giganteus*, appear to have been restricted in their range to the North Island. *Dinornis rheides*, which appears to have inhabited both Islands, is intermediate in the strength and thickness of its limbs; and two species remarkable for their small size—*Dinornis geranoides* and *D. curtus*—have hitherto been found only in isolated localities.

As remarked by Sir Richard Owen, in one of his latest Memoirs, *Dinornis maximus* is specially remarkable for its great size and strength even in a race of giants. One specimen exhibits such extreme measurements that Owen has suggested the existence of a yet taller species, for which he selects the provisional name of *Dinornis altus*.

Having had opportunities of examining very large series of bones, exhibiting an almost continuous gradation of size from the largest to the smallest, my own belief is that some at least of the birds differentiated above are mere varieties or conditional states of one and the same species; but their discrimination is not the less interesting and important from a scientific point of view. Even Professor Hutton, whose paper "On the Dimensions of *Dinornis* Bones" (Trans. N.-Z. Inst. vol. vii. pp. 274–278) goes far to establish this view, especially as to the impossibility of defining any strict line of demarcation between *Dinornis elephantopus* and *D. crassus*, is constrained to add:—"Still, notwithstanding all that I have said, I am convinced that it will be necessary to retain the names both of *crassus* and *elephantopus* to mark both ends of the series as characterized by the proportions of the metatarsus, the length of which in *D. crassus* is more than four times the breadth of the middle of the shaft, while the length is less than four times the breadth in *D. elephantopus* and *D. gravis*."

It is clear that Owen has not founded his species of different stature on a mixture of old and young birds, as has been alleged by some naturalists, because in the Canterbury Museum are exhibited not only young bones of each species, from the chick to the full-grown bird where (to take only one bone as a guide) the tarsal epiphysis of the metatarsus is not yet quite anchylosed *, but also of such species a series of specimens, generally showing two distinct sizes, from which it may be reasonably

* "We possess, amongst others, the leg-bone of a specimen of *Dinornis maximus* which is in size only second to the largest bones we have, but in which this immature character in the metatarsus is not yet quite effaced."—*Von Haast.*

inferred, by the analogy of other Struthious birds, that these represent the male and female of each species.

Not the least interesting fact connected with these giant Wingless Birds is that they have passed away within the historic period. The remains of all the species mentioned above have been discovered intermingled with human bones; they have been found, calcined and chopped, amid the rejectamenta of old Maori feasts in the ancient kitchen-middens of both Islands—facts which, quite apart from Maori tradition, prove incontestably that they were coeval with the early native inhabitants, and that their final extirpation was accelerated, if indeed it was not occasioned, by human agency.

The only question remains—At what period of history did they cease to exist? The late Sir Julius von Haast, who had devoted years of study to the subject, came to the conclusion that the extinction of the various species of *Dinornis* dates back perhaps a thousand years, and that the association with man, as proved by the numerous kitchen-middens and cave-habitations which he himself explored, had relation to a prehistoric or autochthonous race which, in the remote past, inhabited New Zealand *. He wrote an elaborate paper, on "Moas and Moa-hunters," in support of this contention; but I do not think this view of the subject has obtained much support. To my mind the evidences of the comparatively recent existence of, at any rate, several species of *Dinornis* are overwhelming. The circumstance already mentioned of the discovery of a skeleton with a portion of the skin and feathers attached, in such a climate as that of New Zealand, is entirely opposed to the theory of remote antiquity.

Then, again, the comparatively recent date of the bones of even the larger species is attested by their chemical condition and the large amount of animal matter they contain. As compared with a recent tibia of the Ostrich, containing 26·51 of animal matter, the fossil femur of *Dinornis didiformis* has been found to contain 25·99. According to another comparative analysis, a recent femur of the Ostrich contained 34·86 of animal matter, and a fossil femur of *Dinornis struthioides* 37·86. As Professor Owen has already remarked, this superabundance of animal matter in the bone of the extinct bird is due chiefly to the fact of its being a marrow bone, whilst that of the Ostrich contains air.

With many of these buried skeletons are found little heaps of crop-stones, of a kind that are now met with only at a distance of forty or fifty miles from the place of interment. I have in my possession a very interesting collection of these "gizzard-stones," consisting of quartz-pebbles, carnelians, and

* Even this champion for the great antiquity of the Moa would appear to have latterly somewhat changed his views on this subject. I have a letter in my possession from him stating that having read the report of my speech in the Native Land Court, as Counsel for the Ngatiapa in the Rangatira case, with the Appendix containing an account of Apa-hapai-taketake and the pet Moa of the Ngatituwharetoa tribe—a story accepted by both the contending parties as true—he felt almost constrained to abandon the ground he had so persistently taken up.

The following is an extract from the evidence given on that occasion by the leading witness, Hue Te Huri:—"I have heard the name of Apa-hapai-taketake an ancestor of the Ngatiapa tribe. He was the original source of the quarrel. Apa-hapai-taketake stole a Moa, which was a pet bird of the Ngatituwharetoa. While doing so he fell over a cliff and broke his thigh and was thenceforward nicknamed 'Hapakoki' ('Hop and go one'). He got off with the Moa in spite of this. Then Ngatituwharetoa heard of it, and they went down upon his place and carried off his wife Hinemoatu in payment (utu) for the Moa which he had stolen. Then Hapakoki in great wrath went and seized the kumaras of Kaweru; and Ngatituwharetoa, in equal wrath, made an attack on the Ngatiapa. Then the Ngatiapa left and came to Maunganui, on the Upper Rangitikei; for all this happened at Putauaki, near the Awa-o-te-Atua, in the Bay of Plenty. The Ngatituwharetoa pursued them and attacked them at Maunganui. Ngatiapa moved on south and settled on the north-east side of the Taupo Lake; but they were followed up and again attacked, and they again moved on to Tawhara-Papauma and Moturoa, south of Taupo, and close to Rotoaira, on the edge of the lake of that name."

pieces of chert, all worn and polished by attrition, showing that they had been in use for a considerable time *.

The Hon. Walter Mantell, whose opinions on this point are entitled to every consideration, assigns a higher antiquity to the Moa-bones that are found under the stalagmite which forms the flooring to certain limestone caves, similar in character to those bone-caves in which traces of the early animals that inhabited Great Britain have been preserved to us; and he has declared his conviction that the more ancient species of Moa were extirpated by a race which inhabited New Zealand before the arrival of the aboriginal Maoris †.

Many of these bones have been found under a considerable depth of fluviatile deposits which may be of Quaternary or even of Pliocene age.

There can be doubt, however, from the evidences already mentioned, that some of the species, even of the largest stature, existed contemporaneously with the ancestors of the present race; and Mr. Mantell himself, during his early explorations in the South Island, discovered, drawn upon the walls of a cave in the Waitaki valley, a rude likeness of the Moa by some aboriginal artist of a bygone generation, painted with red ochre on the face of the rock, probably soon after the arrival of the first Maori immigrants.

Mr. Dallas, who, in 1865, described (Proc. Zool. Soc.) the feathers of *Dinornis robustus*, was the first to establish the fact that the feathers in some of the species of Moa possessed a large accessory plume or double shaft, as in the Emus and Cassowaries of Australia and the Indian Archipelago. In these feathers "the barbs consist of slender flattened fibres, bearing long silky and very delicate barbules, without any trace of barbicels."

In 1870 some feathers were found by Mr. S. Thomson, at the junction of Manuherekia with the Molyneux river, in association with Moa-bones under fifty feet of sand. Captain Hutton, in a letter to Professor Owen, described the feathers as being "quite fresh in appearance," and as having "lost none of their colouring." The largest of these is 7 inches in length, and gradually widens from ·25 of an inch at the base to rather more than ·75 at the tip, where it is broadly rounded off. "The lower half is downy, the barbs having unconnected barbules, and is of a brownish-grey colour. In the upper half the barbs are rather distant, unconnected, and without barbules. The brownish grey of the lower part passes gradually into black, which colour it keeps as far as the rounded tip, which is pure white, forming a narrow segment of a circle." It would seem from this description that these

* Mr. Frederick Chapman (to whom I am indebted for these 'Moa-stones') writes:—"When we came upon the ground disturbed by the wind (the soil being shifting sand) we soon found a number of distinct groups of gizzard-stones. It was impossible to mistake them. In several cases they lay with a few fragments of the heavier bones. In all cases they were in distinct groups; even where they had become scattered, each group only covered a few square yards of ground, and in that space lay thickly strewn. . . . . . The peculiar feature of the stones was that they were almost all opaque white quartz pebbles. In one place I found a small group of small pebbles of different colour, more like the few brown water-worn pebbles which may be picked up hereabouts. These lay with a set of bones much smaller than the very large bones I found with most of the clusters of pebbles. I did not gather these brown pebbles, as I thought it uncertain whether they were gizzard-stones or not, though it is possible that the species to which the smaller bones belonged was not so careful in selecting white stones. A glance at the pebbles lying about in the surrounding country showed that the quartz-pebbles were not collected here. . . . . . Mr. Murdoch and I collected three sets of pebbles, and these I can pronounce complete or nearly so. It is beyond question, too, that each set belongs to a distinct bird. No. 1 weighs 3 lb. 9 oz.; No. 2 weighs 4 lb.; while No. 3 weighs no less than 5 lb. 7 oz. This giant set contains individual stones weighing over 2 oz.; indeed, I have picked out 8 stones weighing almost exactly 1 lb." (Trans. N.-Z. Inst. vol. xvii. pp. 173, 174.)

† Trans. N.-Z. Inst. vol. i. pp. 18, 19.

feathers belonged to a different species of Moa to that in the York Museum; but as none of the bones were obtained it was impossible to say to which of those enumerated above. Hutton's drawing of the feather (as given at p. 442 of the 'Memoirs') does not accord very well with his description. The figure given on plate cxiv. (*l. c.* vol. ii.) from the same correspondent seems to be more accurate.

At a somewhat later date other Moa-feathers, in an equally fresh condition, were found in a locality between Alexandra and Roxburgh; and these, according to Hutton, are distinguished by the presence of barbules to the tips, from which it may be inferred that they belonged to a less typical Struthious form.

In 1871 Dr. (now Sir James) Hector described a remarkable specimen from the same district, being the neck of a Moa, apparently of the largest size, upon the posterior aspect of which the skin, partly covered with denuded feathers, was still attached by the shrivelled muscles and ligaments*. This unique specimen was found by a gold-miner in a cave, or under an overhanging mass of mica-schist, and is now in the Colonial Museum at Wellington.

In 1874 Professor Hutton described the right foot of *Dinornis ingens* "with the whole of the skin and muscles of the posterior side well preserved." It was found by Mr. Allen in a deep crevice among mica-schist rocks in the Knobby Ranges, in the provincial district of Otago. Of this specimen a figure, one-fourth the natural size, appeared in 'Nature' (Feb. 11, 1875). Through some inexplicable mistake the specimen is stated therein to be in the Natural History Museum at Paris; whereas,

* Sir James Hector, writing on this subject, says:—"The above interesting discoveries render it probable that the inland district of Otago, at a time when its grassy plains and rolling hills were covered with a dense scrubby vegetation or a light forest growth, was where the giant wingless birds of New Zealand lingered to latest times. It is impossible to convey an idea of the profusion of bones which, only a few years ago, were found in this district, scattered on the surface of the ground, or buried in the alluvial soil in the neighbourhood of streams and rivers. At the present time this area of country is particularly arid as compared with the prevalent character of New Zealand. It is perfectly treeless—nothing but the smallest sized shrubs being found within a distance of sixty or seventy miles. The surface-features comprise round-backed ranges of hills of schistose rock with swamps on the top, deeply cut by ravines that open out on basin-shaped plains formed of alluvial deposits that have been everywhere moulded into beautifully regular terraces, to an altitude of 1700 feet above the sea-level. That the mountain-slopes were at one time covered with forest, the stumps and prostrate trunks of large trees, and the mounds and pits on the surface of the ground which mark old forest land, abundantly testify, although it is probable that the intervening plains have never supported more than a dense thicket of shrubs, or were partly occupied by swamps. The greatest number of Moa-bones were found where rivers debouch on the plains, and that at a comparatively late period these plains were the hunting-grounds of the aborigines, can be proved almost incontestably. . . . . . Still clearer evidence that in very recent times the natives travelled through the interior, probably following the Moa as a means of subsistence, like natives in the countries where large game abounds, was obtained in 1865-6 by Messrs. J. and W. Murison. At the Manistoto plains bones of several species of *Dinornis*, *Aptornis*, *Apteryx*, large Rails, *Stringops*, and other birds are exceedingly abundant in the alluvium of a particular stream, so much so that they are turned up by the plough with facility. . . . . A desire to account for the great profusion of Moa-bones on a lower terrace shelf nearer the margin of the stream, led the Messrs. Murison to explore the ground carefully, and by excavating in likely spots they found a series of circular pits partly lined with stones, and containing, intermixed with charcoal, abundance of Moa-bones and egg-shells, together with bones of the dog, the egg-shells being in such quantities that they consider that hundreds of eggs must have been cooked in each hole. Along with these were stone implements of various kinds, and of several other varieties of rock, besides the chert which lies on the surface. The form and contents of these cooking-ovens correspond exactly with those described by Mantell in 1847, as occurring on the sea-coast; and among the stone implements which Mantell found in them, he remembers some to have been of the same chert which occurs *in situ* at this locality, fifty miles in the interior. The greater number of these chert specimens found on the coast are with the rest of the collection in the British Museum. . . . . The above facts and arguments in support of the view that the Moa survived to very recent times are similar to those advanced, at a very early period after the settlement of the colony, by Walter Mantell, who had the advantage of direct information on the subject from a generation of natives that has passed away. As the first explorer of the artificial Moa-beds, his opinion is entitled to great weight. Similar conclusions were also drawn by Buller, who is personally familiar with the facts described in the North Island, in an article that appeared in the 'Zoologist' for 1864."

as a matter of fact, it has never left New Zealand and is now in the Otago Museum. This woodcut had appeared originally in 'La Nature'; and by the courtesy of the editor of that journal I am enabled to reproduce it here. Some excellent anatomical notes on this foot, by Dr. Coughtrey, were published in the 'Transactions of the New-Zealand Institute' (vol. vii. p. 269).

Two years later Mr. Taylor White discovered in some caves in the Wakatipu district (South Island) Moa-feathers of two very distinct types. Some of these feathers are now in my possession; they are in a high state of preservation, the colours being perfectly fresh, and many of them have both shafts quite entire. The largest of them measures nearly 6½ inches in length, and is of a uniform pale yellowish white, being the only feather of this kind out of more than a hundred collected. It is single-shafted, there being no sign whatever of the former attachment of an accessory plume; the barbs are rather distant, unconnected, and filamentary or hair-like, and are placed at such an angle with the shaft as to give a maximum breadth of about an inch and a half in the middle portion of the feather, the width diminishing towards the tip and tapering downwards almost to the base of the tube, there being no downy part *. This unique feather is evidently a dorsal one, and probably helped to form the loose uropygial fringe or lower mantle in one of the smaller species of Moa. Another feather belonging to the same bird, and measuring nearly five inches in length, is of a similar filamentary character, but is furnished with an accessory plume only ·25 of an inch shorter than the main one; the former being dark brown with black margins, and the latter of a uniform brownish-yellow colour. There are smaller feathers, all of them single-shafted, with more distant, rigid, and shortened barbs, in which the shaft is of a transparent yellow colour, like polished amber. These, I should infer from their character, are from the neck of the bird. The rest of the feathers in this group, some of which are double-shafted, are deeply webbed with silvery-brown down for about two thirds of their basal

* A representation of this feather appeared in the Trans. N.-Z. Institute (vol. xviii. pl. ii.); but the figure is a misleading one, as it represents the barbs thickly furnished with barbules, whereas in the feather itself they do not exist at all.

extent, reddish brown in their apical portion, with whitish tips. It is probable that all these feathers belonged to *Dinornis casuarinus*, bones of which species were found in association with them in the Wakatipu cave, together with fragments of egg-shell of a pale green colour. The feathers from the Queenstown cave are of an entirely different type, and these may perhaps have belonged to *Dinornis didinus*. They measure from four to five inches in length; from the base for more than two thirds of their extent they have thick downy webs, of a uniform width of half an inch and of a greyish-brown colour, darker towards the shaft, the barbs having minute, thick-set barbules; then follow long, unconnected filaments, of a still darker brown, which run into a compact apical web of dark purple-brown, tipped with yellowish brown. Many of these feathers have an accessory plume, but this is always downy in its whole extent, which scarcely exceeds half the length of the main shaft. On placing a number of these feathers together they present a soft, glossy appearance and look as fresh as if plucked yesterday from the body of a living bird.

But a still more recent instance is afforded by the very interesting specimen of the Moa's foot in the University Museum at Cambridge, obtained in the Hector Ranges, Otago, in 1884. It was brought to England by Mr. W. J. Branford, who stated that he had himself found it in a cave where, as he believed, there was the entire skeleton of this bird and some more beside.

Professor Newton having kindly lent me this unique specimen for the purpose of having it photographed, I submitted it to Sir Richard Owen, who unhesitatingly pronounced it the metatarsal of *Dinornis elephantopus* *.

The bone is in a perfectly fresh condition, with about four square inches of dark brown integument, having a tuberous surface and with underlying dried tendons of a maximum thickness of 1·5 inch, adhering to the proximal extremity and representing the true heel; the *astragalus* (or a bone that performs the same function in the *Apteryx*, though not hitherto recorded in *Dinornis*) is in position, two of the phalanges are still articulated to the metatarsal by means of dried ligament, and a portion of the tough covering of the sole, nearly half an inch in thickness and of a yellowish-brown colour, is still attached to the lower surface.

Another piece of concurrent testimony was afforded by the discovery, about the year 1860, of a perfect Moa's egg, from which the contents had been extracted through an artificially bored hole on one side. It was found in an ancient Maori sepulchre at Kaikoura, and in such a position, in relation to the skeleton, as to suggest the idea that it had been placed in the hands of a corpse buried in a

* In the foregoing pages I have made free use of Sir Richard Owen's name in connection with the successive discoveries of the Dinornithidæ; I have stated his views as they were developed from time to time, and I have given publicity to one of his letters to myself. Under these circumstances, I thought it right to submit the proof-sheets to him before going to press. The learned Professor returned them to me without a single alteration, but accompanied by a letter which I am glad (with his permission) to place on permanent record, the more so as he assures me that the active work of his life is ended and his last contribution made to the Royal Society, of which body he has been so distinguished an ornament for upwards of half a century.

"Sheen Lodge, Richmond Park,
22nd March, 1888.

"My dear Buller,—

"Seldom have I enjoyed a morning more, in the quiet period of my existence, than during the perusal of the sheets of your 'Introduction,' so kindly submitted to me.

"Conclusions and inferences which had escaped my memory have been brought back, and I seem to be repeating or living again an active period of my zoological life.

"The Moa-bone from Hector Ranges, Otago, is a metatarsal of *Dinornis elephantopus*. . . . . .

"I do not recall anything that I could add which would heighten the pleasure your friendly visit has given me.

"Ever yours,
"Richard Owen."

"*Sir Walter Buller, K.C.M.G., F.R.S.*"

sitting posture. This unique egg was brought to England, and sold by auction at Stevens' sale-rooms, where it fetched only £105, and came into the possession of the late Mr. Dawson Rowley of Brighton, who figured and fully described it (under *D. ingens*) in his 'Ornithological Miscellany' *.

In 1866 two more eggs were discovered in the alluvial sandy loam of the Upper Clutha plains, Otago. One of these was two feet from the surface, the other only about a foot apart from it and three inches deeper. Of the first and more perfect one pieces were fitted together, making nearly one complete side of the egg, which was estimated to measure 8·9 inches in length by 6·1 in breadth. It contained the bones of an embryo chick, which are now preserved in the Colonial Museum. The shell had been eroded by the solvents of the soil, but on the granular surface so produced the characteristic linear air-pores were distinctly visible. The shell yielded 0·9 per cent. of organic matter, showing that it had not been long enough in the soil to part with all its soluble constituents †.

No one who takes the trouble to examine the skeleton of *Dinornis parvus* which now stands in the Palæontological gallery of the British Museum, exhibiting bleached but not fossilized bones, some of them still retaining their inherent "grease," will be able to resist the conclusion that the bird to which they belonged was living at a comparatively recent date.

The well-marked footprints of the Moa in the sandstones of Poverty Bay—models of which are now to be seen in most of our public museums—are interesting historically, but their presence in such a formation is quite consistent with the alleged antiquity of the bird. The case is different, however, with the round cakes of excrement collected by Mr. Taylor White, with other Moa remains, in the Wakatipu cave, the condition of which was such that undigested fragments of fern-stalk and other vegetable matter could still be detected ‡.

Further evidence of the comparatively recent existence of *Dinornis* is afforded by the fact that mixed with its remains are the bones of many species of birds still inhabiting New Zealand. Among these I may mention the following genera:—*Nestor, Stringops, Platycercus, Himantopus, Hæmatopus, Limosa, Larus, Diomedea, Rallus, Porphyrio, Anas, Phalacrocorax*, and *Eudyptula*.

With respect to the much reiterated assertion that the present race of Maoris have no traditions relating to the Moa, I would state that their ancient folk-lore, their historical *waiatas* (or songs), and their proverbial sayings are full of allusions, more or less direct, to this bird. The late Judge Maning, probably the best modern authority on the traditions of the Maoris, has left on record § a very full history of the Moa, as derived from these sources. According to that account the Moas still existed in considerable numbers when the first Maori immigrants arrived, from 500 to 600 years ago. There was little or no excitement in hunting these birds, because of their sluggish habits. They were destroyed wholesale by setting the grass and scrub on fire, the Maoris killing in this manner vast numbers more than they could use, or even find, when these fires spread to any great distance. Thus

* This egg was submitted to me for examination soon after its exhumation, and I made the following notes:—It is perfectly oval and measures 9·3 inches in length by 7 inches in breadth; of a pale cream-colour, stained on one side with yellowish brown as if it had been smeared with the yolk, an appearance which may have been due to contact with decomposed animal matter. The egg has a solid appearance, the surface looking more like half-polished Moa-bone than egg-shell, and its thickness is about that of a new shilling. The entire surface is covered with short linear air-pores, or minute puncta, as if made with the point of a pen-knife, and disposed longitudinally; being filled with darker matter than the shell they present the character of pencilled markings, varying in extent from mere points to lines one tenth of an inch in length; there are some irregular dark markings on one portion of the egg having the appearance of obscure marbling, but these do not seem to be inherent in the shell.

† Hector in Proc. Zool. Soc. 1867, p. 993.

‡ Trans. N.-Z. Instit. vol. viii p. 99.

§ *Loc. cit.* pp. 102–103; *cf.* John White, *loc. cit.* p. 79.

persecuted, the Moas rapidly diminished in numbers, and finally became extinct. These traditions all agree in representing the Moa as living on fern-root, and as being inactive in its habits but fighting fiercely in self-defence. "As inert as a Moa" is a saying in use to the present day; and the name "moa," still applied to a small patch of cultivation, has allusion to the manner in which these birds scratched up and harrowed the fern-root grounds.

From what we know of the range and habits of the *Struthiones* in other parts of the world, it cannot be supposed that the extinct race of Moas, comprising twenty, if not more, species or varieties, some of them attaining to colossal dimensions, were always confined within the geographic limits of modern New Zealand. The Ostrich inhabits the arid deserts of Africa, the Rhea (of which there are two, if not three, species, each occupying a separate district) is spread over a great portion of America, extending from Patagonia to Peru, two species of Emu and a Cassowary occupy the Australian continent, the range of each being well defined, and eight other species of Cassowary are limited to New Britain, New Guinea, and the islands of the Indian Archipelago, each inhabiting a separate area. It may be safely assumed that the Moas of the remote past roamed over a wide continent now submerged, and that when, by the gradual subsidence of their domain beneath the waters of the Great Pacific, they were driven as it were into a corner and overcrowded, the struggle for existence became a severe one and the extinction of the race commenced; that the more unwieldy giants, thus "cabined and confined," were the first to succumb; and that the smaller species, perhaps in course of time differentiated from their ancestors by the altered physical conditions of their environment, continued to live on till their final extirpation by man within recent historic times.

Professor Owen compares New Zealand to one end of a mighty wave of the unstable and ever-shifting crust of the earth, of which the opposite end, after having been long submerged, has again risen with its accumulated deposits in North America, showing us in the Connecticut sandstones of the Permian period the footprints of the gigantic birds which trod its surface before it sank; and he surmises that the intermediate body of the land-wave, along which the *Dinornis* may have travelled to New Zealand, has progressively subsided, and now lies beneath the Pacific Ocean. But Professor Hutton, in his treatise 'On the Geographical Relations of the New Zealand Fauna,' considers it necessary to account for the phenomenal number of Struthious species inhabiting New Zealand, as compared with the other much larger areas of the earth's surface. He supposes the existence of an ancient continent, with one or two species of *Dinornis*; then, by some convulsion of nature, this continent sinks beneath the ocean, leaving its mountain-ranges exposed, in the form of islands, and the only refuge for the surviving Moas; after a sufficiently long period to allow of specific changes, there is an elevation of the land and the differentiated birds are mingled together; then follows the final subsidence, when New Zealand as the central mountain-chain becomes a "harbour of refuge" for them all. In support of this bold hypothesis he refers to the remarkable fact of five or six distinct species of Cassowary inhabiting isolated localities extending from New Britain and New Guinea to the Molucca Islands. His general conclusion is thus expressed:—"The distribution, therefore, of the Struthious birds in the Southern Hemisphere points to a large Antarctic continent stretching from Australia through New Zealand to South America, and perhaps on to South Africa. This continent must have sunk, and Australia, New Zealand, South America, and South Africa must have remained isolated from one another long enough to allow of the great differences observable between the birds of each country being brought about. Subsequently New Zealand must have formed part of a smaller continent, not connected either with Australia or South America, over which the Moa roamed. This must have been followed by a long insular period, ending in another continent still disconnected from Australia and South America, which continent again sank, and New Zealand assumed somewhat of its present form."

Mr. A. R. Wallace, commenting on this, in his 'Island Life,' says:—" If New Zealand has really gone through such a series of changes as here suggested, some proofs of it might perhaps be obtained in the outlying islands which were once, presumably, joined with it. And this gives great importance to the statement of the aborigines of the Chatham Islands, that the *Apteryx* formerly lived there, but was exterminated about 1835. It is to be hoped that some search will be made here and also in Norfolk Island, in both of which it is not improbable remains either of *Apteryx* or *Dinornis* might be discovered. So far we find nothing to object to in the speculations of Captain Hutton, with which, on the contrary, we almost wholly agree; but we cannot follow him when he goes on to suggest an Antarctic continent uniting New Zealand and Australia with South America, and probably also with South Africa, in order to explain the existing distribution of Struthious birds. . . . The suggestion that all the Struthious birds of the world sprang from a common ancestor at no very remote period, and that their existing distribution is due to direct land communication between the countries they now inhabit, is one utterly opposed to all sound principles of reasoning in questions of geographical distribution. . . . . We have direct proof that the Struthious birds had a wider range in past times than now. Remains of extinct Rheas have been found in Central Brazil, and those of Ostriches in North India; while remains, believed to be of Struthious birds, are found in the Eocene deposits of England. As the intervening sea appears to be not more than about 1500 fathoms deep it is quite possible that such an amount of subsidence may have occurred. It is possible, too, that there may have been an extension northward to the Kermadec Islands, and even further to the Tonga and Fiji Islands, though this is hardly probable, or we should find more community between their productions and those of New Zealand. A southern extension towards the Antarctic continent at a somewhat later period seems more probable, as affording an easy passage for the numerous species of South-American and Antarctic plants, and also for the identical and closely allied freshwater fishes of these countries."

Professor von Haast, in his 'Anniversary Address to the Philosophical Institute of Canterbury' (1874), objected to the above theory on the ground that the geological record of these islands, so far as we are acquainted with it, does not warrant our assuming such repeated changes in the level of the land. He thus states the case:—" An unfortunate country, such as New Zealand, of which a good number of the species of its fauna and flora show great resemblance to other species from distant countries, has to be dipped down and brought up again a great many times in order to establish connexions in various directions, so that a bird or fish, a shell, insect, or centipede might cross from the one to the other, moreover, without allowing any other species from the same country to pass." But Professor Hutton, with a much broader grasp of the subject, returns to the discussion in an able article 'On the Origin of the Fauna and Flora of New Zealand' ('Journal of Science,' 1884, vol. ii. parts 1 and 6), in which, after qualifying his former theory by abandoning the idea of an extensive Antarctic continent, and substituting a South-Pacific continent connecting New Zealand with South America, he defends his views with considerable force of argument *.

* Contending all through that in Miocene times New Zealand was represented by a cluster of twenty or more islands, on which the various species of Moa were probably developed, Professor Hutton thus sums up his conclusions:—"the general results then are that in early Mesozoic times New Zealand, Eastern Australia, and India formed one biological region, land probably

THE EXISTING AVIFAUNA.

Having given the reader a rapid glance at the extinct genera and species, it may be useful now to take a general view of the existing Avifauna, for the purpose of indicating the points in which it differs from that of every other zoological region on the earth's surface, and of showing the close relation of some of our present forms to the types that have passed away.

The leading feature in the Ornithology of New Zealand is thus expressed by a very accomplished zoological writer:—"Recent birds being divided into two great and trenchantly marked groups, of very unequal extent, the smaller of these groups (the *Ratitæ*) is found to contain six most natural sections, comprising, to take the most exaggerated estimate, less than two score of species, while the larger group (the *Carinatæ*), though perhaps not containing more natural sections, comprehends some ten thousand species. Now, two out of the six sections of this small group are absolutely restricted to New Zealand; and these two sections contain considerably more than half of the species known to belong to it. Thus, setting aside the Carinate birds of our distant dependency (and some of them are sufficiently wonderful), its recent Ratite forms alone (some twenty species, let us say) may be regarded as the proportional equivalent of one tenth of the birds of the globe—or numerically, we may say, of an avifauna of about one thousand species" *.

A perusal of the following 'History' will show that the Avifauna of New Zealand possesses other distinguishing features of a very striking character, a full review and discussion of which would occupy many pages; but some of the more prominent of these may be here mentioned, more, however, in the way of general indication than with the intention of exhaustive treatment.

The feature that first strikes the general ornithologist is the comparatively large number of apterous birds, or species in which the anterior limbs are so feebly developed as to be absolutely useless for purposes of flight. Conspicuous among these are the four species of *Apteryx*, in which the wings are reduced to mere rudiments; next in order of development come the various species of *Ocydromus*, of which I shall have something to say further on, and the remarkable Ocydromine form,

---

extending continuously from New Zealand to New South Wales and Tasmania. At the close of the Jurassic period the New-Zealand Alps were upheaved and the geosynclinal trough between New Zealand and Australia was formed. During the Lower Cretaceous period a large Pacific continent extended from New Guinea to Chili, sending south from the neighbourhood of Fiji a peninsula that included New Zealand. Nearly all the southern part of America was submerged. Western Australia and Eastern Australia formed two large islands lying at some distance from the continent. This continent supported dicotyledonous and other plants, insects, land-shells, frogs, a few lizards, and perhaps snakes and a few birds, but no mammals. In the Upper Cretaceous period New Zealand became separated and reduced to two small islands: the South Pacific continent divided in the middle between Samoa and the Society Islands and—the eastern portion being elevated while the centre sank—it ultimately became what we know now as Chili, La Plata, and Patagonia. In the Eocene period elevation commenced in our district; Eastern Australia was joined to New Guinea, which stretched through New Britain to the Solomon Islands. New Zealand was also upheaved and extended towards New Caledonia, but the two lands were divided by an arm of the sea. The mainland of New Guinea had by this time been invaded from the north by a large number of plants, birds, lizards, snakes, &c., which migrated south into Eastern Australia and a few passed over the New-Caledonia channel and reached New Zealand. But still no mammals. In the Oligocene period New Zealand again gradually sank, carrying with it the sparse flora and fauna it had received, and in Miocene times was reduced to a cluster of islands; Eastern Australia all this time receiving constant additions to its fauna and flora through New Guinea. In the Pliocene period elevation once more took place; New Zealand extended towards the Kermadec Islands, and the continent of Australia was formed; after which subsidence again occurred in New Zealand."

* 'Nature,' July 18, 1872.

distinguished by Hutton under the generic name of *Cabalus*, from the Chatham Islands; then *Notornis*, the huge brevipennate Rail already mentioned; a small flightless Duck (*Nesonetta aucklandica*), strictly confined to the Auckland Islands; and, finally, the well-known Ground-Parrot (*Stringops habroptilus*), in which the sternum is almost devoid of a keel. The explanation in all these cases is sufficiently obvious. In a country like New Zealand where there have been no indigenous Mammalia, and consequently few birds of prey, species that habitually seek their food on the ground have no inducement to take wing, and from long disuse, continued perhaps through countless generations, lose the Carinate character of the sternum, and with it the faculty of flight, for without the keel on the breast-bone to give attachment to the great pectoral muscles the wings, however ample they may be in their outward development, are practically useless for purposes of flight.

Taking the Carinate division of our Avifauna, another very prominent characteristic is the number of endemic genera and species. The families, with a few exceptions to be hereafter mentioned, are the same that occur in other parts of the world; but when we come to examine the subordinate groups, the specialization is at once apparent. Out of a total of 88 genera, 47 belong to the *Limicolæ*, *Herodiones*, and the five web-footed Orders, and these, being in a sense cosmopolitan, may for the present be put out of sight. Of the remaining 41 genera, 21 are strictly peculiar to New Zealand. But even in the other more widely-spread genera there are many species that are not known elsewhere. Thus, out of a total of 181 species, composing the present list of our CARINATÆ, no less than 93 are strictly endemic. Even among the most diffuse Orders there are genera restricted in their range to the New-Zealand rivers or coasts, or to the outlying islands. Thus among the Limicolæ we have two strictly peculiar genera, *Thinornis* and *Anarhynchus*, and among the Anseres two more, namely, *Hymenolæmus* and *Nesonetta*.

Of the former, *Thinornis* belongs really to the Chatham Islands, for although *T. novæ zealandiæ* is comparatively common there, only straggling flocks are met with, at uncertain intervals, on the New-Zealand coast; and of the latter, *Nesonetta* is confined exclusively to the Auckland Islands, the only known species, *N. aucklandica*, never having been met with elsewhere. The other two genera I have instanced, *Anarhynchus* and *Hymenolæmus*, are restricted to New Zealand, never having been met with on the outlying islands.

Undoubtedly the most remarkable bird we have among the Waders is the Wry-billed Plover (*Anarhynchus frontalis*), in which, as the name implies, the bill is asymmetrical, being always turned to the right, a modification of structure admirably fitted to the bird's peculiar habits of feeding. The curvature in the bill is congenital, being equally present in the embryo chick, although not so fully developed; and this fact furnishes a beautiful illustration of the law of adaptation and design that prevails throughout the whole animal kingdom. A bird endowed with a straight bill, or with an upcurved or decurved one, would be less fitted for the peculiar mode of hunting by which the *Anarhynchus* obtains its living, as must be at once apparent to any one who has watched this bird running rapidly round the boulders that lie on the surface of the ground and inserting its scoop sidewise at every step, in order to collect the insects and their larvæ that find concealment there. But there is another feature in the natural history of this species that is deserving of special notice, which is this: the fully adult bird is adorned with a black pectoral band, which, in the male, measures ·75 of an inch in its widest part. Now it is a very curious circumstance that this band is, as a rule, far more conspicuous on the right-hand side, where, owing to the bird's peculiar habit of feeding, there is less necessity for concealment by means of protective colouring *. This character is constant in all the specimens of the male bird that I have examined, although in a variable degree, the black band being generally about one third narrower and of a less decided colour on the left side of the breast,—from which we may, I think, reasonably infer that the law of natural selection has operated to lessen the colouring on the side of the bird more exposed to Hawks and other enemies whilst the *Anarhynchus* is hunting for its daily food. There can be no doubt that a protective advantage of this sort, however slight in itself, would have an appreciable effect on the survival of the fittest, and that, allowing sufficient time for this modification of character to develop itself, the species would at length, under certain conditions of existence, lose the black band altogether on the left-hand side.

Commenting upon the above remarks, in my first edition, the accomplished Editor of 'The Ibis' (Mr. Salvin) indulged in the following reflections:—

"It would appear that the peculiarly shaped bill would only be an efficient weapon for obtaining food in this way so long as the bird walked one way round the stone, *i. e.* bearing to the off side or from west to east! The wider portion of the pectoral band would thus be always next the stone, and more hidden than the narrower or left portion. Has running round stones always the same way been the cause which enabled those birds which practised it to survive and transmit this habit to their offspring? and has their success been further promoted by the tendency to reduce the exposed side of their pectoral band, a secondary sexual character? Or has the process been reversed and the protection given to those birds which ran one way round stones, keeping the prominent portions of their pectoral bands from sight, tended to produce the curvature of the bill? The development of both characters seems to hang upon the birds acquiring the habit of running only one way round stones" †.

It seems to me that the more correct way of putting it is that the bird must, under any circumstances, keep the stone around which it is feeding to the right; for, in whatever way the habit may have been acquired, it is obvious that inasmuch as the curvature of the bill is always to the right it can only serve as an efficient scoop when the bird is in the left position in relation to the stone.

I do not propose to enter here into a discussion of the theory which a consideration of these facts seems necessarily to involve; but such cases as this can be rationally accounted for only on Darwinian principles, and I see myself no difficulty whatever in reconciling this view of the evolution of species by means of natural selection with a belief in the unity of design in Creation, and with the acceptance of the great truths of revelation. It is not a question of the Creation itself, as divinely revealed to man, but as to the plan and method of the Creation; and when, instead of the old literal interpretation of Sacred Scripture, we understand by the "six days" of the Mosaic record so many vastly extended geologic epochs, every difficulty in the way of orthodox belief disappears ‡.

* Mr. Seebohm states that in the two specimens which his collection contains this unsymmetrical character of the pectoral band is not observable, but he does not give the sex; and it is a curious fact, for which I do not pretend to offer any explanation, that in the female bird, in which the pectoral zone is quite inconspicuous, the peculiarity I have mentioned is hardly noticeable, if it is not entirely absent. As to the feature itself in relation to the male bird, I can only say that I have never met with a single example in which it was not more or less manifest; indeed the first to call my attention to it was Sir James Hector, with whom, years ago, I examined the fine series of specimens in the Colonial Museum, and with the result I have stated.

† Ibis, 1873, p. 99.

‡ "Allowing that Almighty Power has worked by constant laws, we have to consider the lapse of time during which our globe may have revolved in its orbit, in a condition approximating to the present, *i. e.* capable of sustaining vegetable and

To quote the language of one of the ablest and most liberal-minded of our theologians:—"Science discloses the method of the World, Religion its cause, and there is no conflict between them, except when either forgets its ignorance of what the other alone can know."

The next point to be noticed is the comparative abundance, in comparison with the rest of our Avifauna, of Rails, Ducks, and Cormorants. The first-named group contains in addition to *Notornis* and its allied form, *Porphyrio*, four, if not five, species of *Ocydromus*, three of *Rallus*, two of *Ortygometra*, and the diminutive Ocydromine representative in the Chatham Islands. Of Ducks, New Zealand possesses 11 species, belonging to ten genera, this number being far in excess of the proportion of Anseres to the general number of birds in other countries of similar extent. Of these Ducks, seven species are endemic, two of them (*Nesonetta* and *Mergus*) being confined to the small area of the Auckland Islands. Of Shags or Cormorants, including two at present doubtful forms, there are no less than fourteen species, of which eight, if not nine, are endemic, so that New Zealand in this respect takes the palm against all competitors. Some of the species, too, are of singular beauty, whereas in all other parts of the world the members of this family are noted for their plain faces and sombre plumage.

Seeing that New Zealand is so rich in Cormorants, it is indeed remarkable that there is no indigenous species of *Plotus*, a form so characteristic of Australia on the one hand and South America on the other. I have already recorded the occurrence of *Plotus novæ hollandiæ* as a straggler, which serves only to accentuate the inexplicable fact of its not being a native.

The entire absence of Woodpeckers might have been expected, as these birds do not extend beyond Celebes, never having been met with in the Moluccas or in Polynesia, New Guinea, or Australia. But it is difficult to account for the non-appearance of Swifts and Swallows, except as occasional visitants from Australia.

On the other hand, the Parrots are well represented. Besides the very typical *Stringops habro-*

---

animal life upon it. We have to allow time for those forgotten migrations of our race, for the previous rise of their religions and other cultured ideas in the East, and for the possible transmutation of animals from the saurians &c., revealed by geological investigations, to the present species. The several thousand years which have elapsed since some of the existing species were preserved as mummies in Egypt appear to have effected no change. But when we contemplate even 10,000 years as relatively a long period, are we not somewhat in the natural state of error in which the mind of an ephemeral summer-day's insect might fall if able to reflect and form estimates of time from the duration of its own existence? Living for one day after its rise from the chrysalis, it might conceive sixty days a long period for the life of the man who can crush it, just as we, able to live towards a century, have allowed about sixty centuries only for the duration of humanity upon the earth. The insect might fancy the statement wickedly preposterous if informed that our existences might extend to some 20,000 times the duration of its day. As a simile, it does not seem an irrational proportionate comparison by 'rule of three' to say that, as the insect's one day is to the 25,550 days of man, so may the human 70 years be to 1,788,500 years for the life of the world, past and future, after the completion of its primary formations. If we allow about a fourth of these for the past changes of species (viz. 400,000 years), and about the thirtieth part (viz. 56,000 years) for man's growth from infancy, from crude civilization to our present state of scientific culture, the computation seems reasonable in the light of scientific facts. It is at all events more consonant with them than our old dogmatic chronology."—*Cradle-land of Arts and Creeds.*

"From a consideration of the possible sources of the heat of the sun, as well as from calculations of the period during which the earth can have been cooling to bring about the present rate of increase of temperature as we descend beneath the surface, Sir William Thomson concludes that the crust of the earth cannot have been solidified much longer than 100 million years (the maximum possible being 400 millions), and this conclusion is held by Dr. Croll and other men of eminence to be almost indisputable. . . . . . So far as the time required for the formation of the known stratified rocks, the hundred million years allowed by physicists is not only ample, but will permit of even more than an equal period anterior to the lowest Cambrian rocks, as demanded by Mr. Darwin—a demand supported and enforced by the arguments, taken from independent standpoints, of Professor Huxley and Professor Ramsay."—*Island Life.*

ptilus, already mentioned, we have seven species belonging to the genera *Platycercus* and *Nestor*, all of which are peculiar to New Zealand and her satellites.

As compared with the Avifauna of Australia, the paucity of species is particularly noticeable in the following well-distributed families, namely, *Sylviidæ*, *Campephagidæ*, *Muscicapidæ*, *Alcedinidæ*, *Columbidæ*, and *Tetraonidæ*.

The families belonging exclusively to New Zealand are five in number—the *Turnagridæ*, *Xenicidæ*, *Nestoridæ*, *Stringopidæ*, and *Apterygidæ*—and, as already indicated, possibly a sixth represented by the remarkable genus *Glaucopis*. The great development of the *Procellariidæ*, or Family of Petrels, is a feature which New Zealand shares in common with Australia and Southern Polynesia. The South Pacific is the great nursery, so to speak, of this extensive Family, and no less than 33 species have, from time to time, been recorded on the New-Zealand coasts or from the surrounding seas. These include nearly all the known species of Albatros, and a number of oceanic birds of considerable interest, although as a rule not conspicuous for their beauty. Some of these have a range extending over both hemispheres; others are confined to apparently small tracts of ocean; while others again are migratory within certain degrees of latitude and longitude. Altogether they comprise a well-defined group of Birds (raised now to the dignity of an Order, under the name of *Tubinares*), whose economic and domestic history, owing to their pelagic and semi-nocturnal habits, has not yet been fully investigated or recorded.

The occasional capture in New Zealand of such tropical forms as *Phaethon rubricauda* and *Tachypetes aquila*, although interesting occurrences *per se*, cannot be regarded, in any strict sense, as a feature in the Avifauna.

Of Meliphagine birds New Zealand possesses a fair number in the genera *Prosthemadera*, *Anthornis*, *Pogonornis*, and, in a lesser degree, in *Zosterops* and the brush-tongued *Nestor*, all of which are endemic; but the honey-eating genera of Australia, such as *Ptilotis*, *Meliphaga*, and *Tropidorhynchus*, are entirely absent. *Acanthochæra carunculata* has occurred in a wild state, but only as an extremely rare straggler from the *Eucalyptus*-brushes of its native country.

Among the *Limicolæ* there are several species which touch at New Zealand in their seasonal migrations to and from the higher latitudes of the Eastern Hemisphere, or make this country their winter residence. Dr. Otto Finsch, as far back as 1867, in the Notes appended to his German translation of my 'Essay on the Ornithology of New Zealand,' expressed his surprise that such species as *Strepsilas interpres*, *Totanus incanus*, and *Tringa canutus* had not been recorded among these seasonal migrants. Since that time all of these, as well as *Phalaropus fulicarius*, *Numenius cyanopus*, and *Tringa acuminata*, have been added to the list. The two most remarkable instances, however, of this class are, on the one hand, the occasional occurrence of the Eastern Golden Plover (*Charadrius fulvus*), whose range extends over Australia, New Guinea, the islands of the Indian Archipelago, and Polynesia, and northwards to its breeding-grounds in Siberia and Kamtschatka, and, on the other, the regular autumnal migration of the Bar-tailed Godwit (*Limosa novæ zealandiæ*), which goes northwards to breed in the high latitudes of Eastern Asia. To my mind, in the whole romance of natural history there is nothing to be compared with this seasonal migration of the Godwit. This bird is the Eastern representative of the European *Limosa lapponica*, to which it bears a close resemblance; and, like that species, it has a very extensive geographical range. Both of them are migratory in their respective hemispheres; and while the other species breeds in the high Northern latitudes of Europe, and returns in winter to North-west and East Africa, our bird spends a portion of the year in Siberia, and visits, in the course of its migration, the islands of the Indian Archipelago, Polynesia, Australia, and New Zealand! Towards the end of March, or beginning of April, large flocks may be seen at the far North taking their departure from our country. Rising from the beach in a long line and with much clamour, they form into a broad semicircle, and mounting high in the air, take a course due north: sometimes they rise in a confused manner, and, after circling about at a considerable height in the air, return to the beach to reform, as it were, their ranks, and then make a fresh start on their distant pilgrimage. After foreign travels and adventures which the pen of Audubon alone could do justice to, the flocks begin to reappear at the north during the first week of November, and then rapidly disperse along the coast.

This subject of the seasonal migration of certain birds is a very wide one and full of interest. There is probably nothing in the whole field of ornithological research more remarkable than this traditional habit, acquired no doubt by experience accumulated through countless generations. The same unerring instinct which guides the Ground-Lark to her nest under some particular tussock in the midst of a wilderness, miles in extent, of exactly similar tussock, or which enables the sea-bird to single out her own eggs from among the thousands clustered together on the bare rock or sandy beach, likewise guides the movements of the migrant when the time comes round for its annual pilgrimage.

We have in New Zealand two species of Cuckoo belonging to different genera—both migratory and both parasitic. One of these (the Long-tailed Cuckoo), which is a native of the Society Islands, visits this country in the summer and breeds with us, entrusting the task of rearing its young to a little Warbler not larger than an English Wren. It arrives, year after year, during the second week of October, and leaves us again before the end of February—this migratory habit, persevered in through long generations, having become a necessary part of its natural existence. In the whole range of ornithological biography, there is perhaps nothing more marvellous than this punctual annual migration across some fifteen hundred miles of ocean. The other species, known as the Shining Cuckoo, visits us from Australia, performing its journey of a thousand miles with the same wonderful precision as to dates of arrival and departure, my register showing only a maximum variation of five days during a continuous period of ten years. Curiously enough, this mild little caterpillar-hunter entrusts the rearing of its young to the same bird that performs that friendly office for its predatory congener four times its size. But apart from these regular summer visitants, with which most colonists are familiar, we have numerous instances of eccentric and casual migration which are indeed very curious. The history of the little *Zosterops*, or Blight-bird, is a case in point. This migrant crossed Cook's Strait, for the first time within the memory of man, in the winter of 1856, coming over in numerous flocks, as if to explore the country; then retired for two years, and reappeared in greater numbers than before in the winter of 1858, since which time it has been a permanent resident in the North Island, breeding in every district, and becoming more plentiful every year. This migration was no doubt induced, in the first instance, by a scarcity of some particular food-supply in the South Island, which must have occurred again two years later. The exceptional feature, however, in this case is, that after the second migration the natural impulse to return home had lost its influence.

In Australia we have several records of non-migratory birds performing a kind of exodus from their own part of the country, swarming into some distant region, where they have remained for five or even ten years, and then disappearing as suddenly as they had come. Take, for example, the beautiful little Warbling Parrakeet (*Melopsittacus undulatus*), which, prior to 1838, was so rare in the southern parts of Australia that only a single example had been sent to Europe, but arrived in that year in countless multitudes. Or take the case of the Australian Moorhen (*Tribonyx ventralis*). This bird, although not endowed with any extraordinary powers of flight, acting under some mysterious influence, left its home in the remote interior and visited South Australia in 1840, coming in such countless myriads that whole fields of corn were trodden down and destroyed in a single night, and the streets and gardens of Adelaide were alive with them. But the casual occurrence with us of migratory species from Australia is even more singular, because it seems impossible to assign any definite cause. In March, 1851, a flight of the Australian Tree-Swallow appeared at Taupata, near Cape Farewell; ten years later they were observed again at Wakapuaka, near Nelson, and a specimen obtained; and after a further lapse of fully twenty years another flight—from which a specimen is now in my possession—appeared for several days in succession in the outskirts of Blenheim. More recently, the Press Association announced the appearance of "Swifts" at the White Cliffs, near Taranaki, and on receiving the only specimen that was shot, I found it to be the true Australian Swift (*Cypselus pacificus*), a bird common enough on the Hunter but migratory northward, and believed by most naturalists to be identical with the species inhabiting China and Amoorland. The two instances of the occurrence in New Zealand, after an interval of twenty years, of the Australian Wattle-bird (*Acanthochaera carunculata*), and more recently, in both North and South Islands, of the well-known Australian Roller (*Eurystomus pacificus*), are cases in point; and other instances might be given of the mysterious, overpowering impulse, under the influence of which certain birds, without any apparent motive, perform almost incredible aerial journeys without a break of any kind.

Another remarkable feature in the New-Zealand Avifauna is the inherent tendency to albinism *. The condition itself is no doubt due to the absence of the colouring-pigment in the feathers; but the difficulty is to find any sufficient cause for this in a temperate climate like that of New Zealand. In India, as is well known, the tendency is in the opposite direction, melanism being of very frequent occurrence.

Strange to say, there is the same tendency to albinism in the imported birds. Albino Sparrows

* In the body of the present work will be found carefully recorded instances of albinism, more or less pronounced, in the following species, viz.:—*Glaucopis wilsoni*, *G. cinerea*, *Heteralocha acutirostris*, *Creadion carunculatus*, *Myiomoira macrocephala*, *Anthus novae zealandiae*, *Anthornis melanura*, *Prosthemadera novae zealandiae*, *Platycercus novae zealandiae*, *P. auriceps*, *Nestor meridionalis*, *Spiloglaux novae zealandiae*, *Sceloglaux albifacies*, *Circus gouldi*, *Carpophaga novae zealandiae*, *Haematopus longirostris*, *H. unicolor*, *Himantopus novae zealandiae*, *Limosa novae zealandiae*, *Larus dominicanus*, *Ocydromus earli*, *O. australis*, *Porphyrio melanonotus*, *Ardea sacra*, *Phalacrocorax novae hollandiae*, *Ossifraga gigantea*, *Anas superciliosa*, *A. chlorotis*, *A. gibberifrons*, *Podiceps rufipectus*, *Apteryx australis*, *A. mantelli*, and *A. oweni*.

To the above list Mr. Kirk has recently added *Myiomoira toitoi*, having described ('Ibis,' 1888, p. 42) a specimen in the possession of Mr. J. H. Drew of Wanganui, in which the only indication of the normal colouring is a small patch of faint grey on one of the primaries, the whole of the remaining plumage being pure white.

In my account of *Anthus novae zealandiae* I have stated (at p. 64) that albinos, more or less pure, are of common occurrence. In the above-cited communication Mr. Kirk says of this species:—"While travelling through the bush on the east coast of the Wellington province, I came on a Maori plantation, and was shown by one of the natives a Ground-Lark exhibiting a tendency both to albinism and melanism. The following is a description, jotted down in my pocket-book:—Top of head, and down as far as a line through the eye, dull black; the whole of the body and wings, with the exception of the two outer primaries, were a delicate creamy white; the outer primaries retained the normal greyish-brown colour. The outside tail-feathers, which in an ordinary specimen would be white, were in this case jet-black. This bird, which was one of the most curious freaks of nature I ever saw, had been tamed, would come when called and allow itself to be picked up and examined, as though conscious of deserving attention on account of its extraordinary and fantastic dress. I endeavoured to effect a purchase, but without success, the Maoris appearing to set great store by their pet."

are far more common than they are in their native country, and even the Sky-Lark (*Alauda arvensis*) not unfrequently exchanges its sober dress for a yellowish-white one. In illustration of this I brought to England two specimens of the latter, one of which I presented to the British Museum, the other to the Natural-History Museum at Cambridge.

Among the Parrots I have recorded some beautiful crimson and yellow varieties, and in the case of *Platycercus novæ zealandiæ* a single instance of cyanism. But the only New-Zealand birds in which I have ever detected any tendency whatever towards melanism, and then only in a slight degree, were *Anthornis melanura* and *Miro albifrons*.

Many travellers in New Zealand have remarked on the notable absence of bird-life, especially in the woods; and at certain seasons of the year this is indeed very noticeable. But, as fully explained in my history of the Wood-Pigeon at page 232, the relative abundance or scarcity of birds is entirely regulated by the food-supply, which, in turn, is governed by the seasons. At all times, however, in winter and summer alike, the New-Zealand woods, whether alive with birds or not, possess an indescribable charm owing to their evergreen character. In my several accounts of their feathered inhabitants I have, as the reader will perceive, never lost an opportunity of paying my tribute to the luxuriant beauty of these woods; but I have always felt that it was quite impossible to do full justice to the subject *.

In the strictest sense of the term, New Zealand is without "song-birds"; but such species as the Tui, the Korimako, and the Piopio possess vocal powers of a very respectable kind, the compass and variety of their notes adding greatly to the charm of the New-Zealand woods. For example, the North-Island Thrush (*Turnagra hectori*) has many notes exactly resembling those of its English namesake. As fully explained at pp. 28, 29 this handsome species is rapidly dying out and will soon be but a memory of the past. But with the disappearance of this native Thrush, the English songster is fast becoming established in the country, frequenting the outskirts of the bush in the neighbourhood of European settlements and supplying to the loyal colonist yet another link of attachment to "dear old England."

Setting aside, however, their claim to the highest order of song, the birds of New Zealand do not fail, especially in the early morning, to make their native woods echo with delightful music, "each one giving out his own notes without any regard for the others, the score having evidently been written for the whole, since the innumerable strains make one divine harmony." In the midst of this melody of song, the harsh cry of the Kaka calling to its fellows will sometimes for a moment break the spell, but the performers, heedless of the discordant note and with bursting throats, continue their morning concert, till, as if by common consent, they cease altogether and disperse in quest of their daily food.

Another feature not to be lost sight of in considering the present condition of the New-Zealand Avifauna is the rapid way in which it is being affected, and in some instances effaced, by the intro-

* Mr. W. R. Ringland, of Belfast, who accompanied me through one of the northern forests, in the summer of 1884, thus graphically describes it in his 'Notes of Travel':—

"This bush scenery is indeed very wonderful. The enormous cabbage-trees, the gorgeous creepers clinging in a green network to the tall pines, the dense undergrowth of shrubs, the tree-ferns, the great kauris, and the exquisite tints of the whole mass of riotous vegetation are beautiful beyond description. Then the strange silence, unbroken even by the whir of a bird's wing, the unchanging sameness of the bush, that confuses you until you cannot tell how far you have travelled, the charred tree-trunks on either side of the road that have been burnt down to clear a passage, and the oppressive loneliness of it all, tell that you are far away from the beaten track of travel, and far into the heart of Maoriland."

duction of the so-called useful and ornamental birds from other parts of the world. The conditions of existence are very favourable to the establishment and increase of many of the imported forms, and, as a consequence, the indigenous species are being displaced and supplanted by them to a very alarming degree. The colony of New Zealand, like every other new country, has been, from time to time, possessed as it were by a rage for acclimatization; and the zeal for the introduction of novelties has not always been tempered by the judgment which comes of experience. The Author must himself plead guilty to having been accessory to the importation of the House-Sparrow in 1865, having in that year, on behalf of the Wanganui Acclimatization Society, advertised in the London newspapers offering a reward of £100 for 100 pairs of House-Sparrows delivered alive in the colony. The advertisement and the importation alike succeeded; and at the present day myriads of these birds in all parts of the country attest the fact, and in the grain-season especially they elicit even from their strongest partisans the admission that they are not an "unmixed blessing." While, however, admitting myself that the "Sparrow nuisance" does exist in rather an aggravated form, I do claim on behalf of this bird full credit for its strictly insectivorous habits at a certain season of the year, and I have never lost an opportunity, in spite of the odium, of putting in a plea for the poor persecuted *Passer domesticus* *.

* As with all questions of this kind, there is much to be said for and against the Sparrow, and numerous experiments have been made by friends and foes for the purpose of demonstrating the actual truth of the case. The following newspaper record contains the result of one of these experiments, and, so far as my observation goes, the weight of evidence is invariably in favour of the bird:—"A hundred and eighteen Sparrows have been offered upon the altars of science. As was the case with the Pagan sacrifices, their entrails have been carefully inspected, in order to furnish guidance to the inquirers. But it has not been in search of the cabalistic information to be derived from quaint contortion, or the credited, though impossible, absence of the heart, or some other vital organ, that the sacrificial knife has been bared. The contents of the stomachs of the victims have been examined, tabulated, recorded. Three culprits alone, out of this hecatomb of the favourites of Cytherea, were proved, by the unsparing search, guilty of having lived for the past four-and-twenty hours upon grain. In fact, there were *three* thieves out of the 118; all the other victims had worked, more or less, for their living. Beetles, and grubs, and flies, and *larvæ* of all obnoxious kinds had been their diet. In 75 of the birds, infants of all ages, from the callow fledgling to the little Pecksy and Flapsy that could just twitter along the ground, hardly any but insect remains were detected. What would the starved and industrious pioneers who have reared their wonderful temple and city by the Great Salt Lake have given for the aid of an army of English Sparrows against that greater and more formidable host of grasshoppers which thrice all but annihilated the settlement?"

To give the other side of the argument, and to show that the prejudice against the Sparrow and its consequent punishment is not confined to New Zealand, I may quote the following newspaper account of its status in Australia:—"Rome once owed its salvation to a Goose, but it has been reserved for the Sparrow in these degenerate modern days to threaten a flourishing young State with serious loss, if not, as the farmers assert, absolute ruin. Rabbits have for some years played an important part in directing legislation in some of the Australasian Colonies, and now in South Australia the Sparrow is becoming a power in the land, and calls for all the machinery of special Acts of Parliament to keep it within bounds. The bird, which only a few years ago such efforts were made to acclimatize in Australia, and whose first arrival was hailed with greater enthusiasm than would now be displayed on the landing of a Bend Or, a Duchess, or a prize merino, is now doomed to extermination—if that can possibly be achieved. So rapidly have the few pairs which were introduced a few years ago multiplied under the congenial skies and amid the luxuriant vegetation of the Australian Colonies, where there are few or none of the checks on their increase which exist in the Old Country, that the agriculturists complain of the serious injury done by them to their wheat and fruit crops, and have called upon the Government to devise some means of ensuring their destruction. . . . . Its work is done on a scale disheartening to the cultivator, and under conditions he cannot control, for the seed is taken out of the ground, the fruit-bud off the tree, the sprouting vegetable as fast as it grows, and the fruit ere it is ripe, and therefore before it can be bested and saved. Neither apricots, cherries, figs, apples, grapes, peaches, plums, pears, nectarines, loquats, olives, wheat, barley, peas, cabbages, cauliflowers, or seeds or fruits of any kind, are spared by its omnivorous bill; and all means of defence tried against its depredations, whether scarecrows, traps, netting, shooting, or poisoning, are declared to be insufficient to cope with the enemy. It remains to be seen whether the reward offered by the Government for the heads and eggs of these destructive little birds will result in any diminution in their numbers."

To my mind the popular outcry against the Sparrow is scarcely warranted by the actual state of the case. It is only at one particular period of the year, when the farmer's grain is "dead ripe" that this bird makes any inroad upon it. In large fields the loss is barely noticeable; but in the case of a small patch of grain, say an acre or two at the edge of the forest or in a bush-clearing, it naturally becomes a serious matter, because the Sparrows appear to concentrate their forces on such inviting spots, and to leave practically nothing but straw for the reaper. Hence, of course, the outcry and clamour on the part of the small farmer. But if people really knew how much the country is indebted to this much-abused bird, I venture to think that there would be a still louder outcry against the sinful practice, now so general, of poisoning Sparrows. It is a fact that on some farms they are poisoned in such numbers that the ground is literally strewed with their dead bodies, and labourers may be seen filling large baskets with them, and carrying them off in the confident belief that a great service is thus rendered to the farming industry of the country. But what are the facts? Is the Sparrow insectivorous, or not, in the strict sense of the term? Let us study it in the breeding-season, which extends in New Zealand from September to December or January. Each pair produces a brood of five young ones. These young birds are fed entirely and exclusively on animal food. Every five minutes or so during the long summer day one or other of the parent birds visits the nest carrying in its bill a caterpillar or a grub, a beetle, fly or worm, but never a grain of corn or fruit of any kind. Now let us consider what this means. Hundreds of thousands of Sparrows, all intent on the same business, having young ones at home that must have insect food of some sort! Every bush, every furrow, every inch of ground is hunted over and ransacked to supply that imperative demand. Millions of insects in all stages of development are daily passed into the insatiable throats of these young Sparrows. I would ask, what does this imply? How much direct benefit does not this bring to the husbandman? The answer is obvious. But look for a moment at the result. In former years the North Island, and especially the Auckland province, was periodically visited by a veritable plague of caterpillars. About once in three or four years the caterpillars came in legions and swept all before them. They would pass over a smiling field of young corn at night and leave scarcely a blade for the dews of morning. Whole districts were devastated in this manner, and the hopes of the farmer for the coming season hopelessly ruined. There was no means of openly meeting an insidious enemy of this kind. It was a moving army of atoms, and to attempt to meet and destroy it would have been a mere mockery *. Since the introduction of the Sparrow and other insectivorous birds the dreaded plague of caterpillars has disappeared. It has become, indeed, a mere matter of history. In their irritation at losing a handful of grain, the small farmers appear to be now bent on ruthlessly

* Under the sensational heading of "Trains stopped by Caterpillars," the following telegram once appeared in the colonial papers:—

"(UNITED PRESS ASSOCIATION.) Wanganui, February 13.

"The trains this morning and evening between Waverley and Nukumaru, on the way to Wanganui, were brought to a standstill through countless thousands of caterpillars on the rails. The officials had to sweep and sand the metals before the trains could proceed."

Another similar case is thus recorded in the 'Rangitikei Advocate':—"In the neighbourhood of Turakina an army of caterpillars, hundreds of thousands strong, was marching across the line, bound for a new field of oats, when the train came along. Thousands of the creeping vermin were crushed by the wheels of the engine, and suddenly the train came to a dead stop. On examination it was found that the wheels of the engine had become so greasy that they kept on revolving without advancing, as they could not grip the rails. The guard and the engine-driver procured sand and strewed it on the rails and the train made a fresh start, but it was found that during the stoppage caterpillars in thousands had crawled all over the engine and over all the carriages inside and out."

destroying the feathered friends to whose untiring efforts the very existence of their crops is in a large measure due *.

Then, again, to pursue the argument in another direction, if the Sparrow is fond of ripe grain it is still fonder of the ripe seeds of the variegated Scotch thistle. This formidable weed threatened at one time to overrun the whole colony. Where it had once fairly established itself it seemed well-nigh impossible to eradicate it, and it was spreading with alarming rapidity, forming a dense growth which nothing could face. In this state of affairs the Sparrows took to eating the ripe seed. In tens of thousands they lived on the thistle, always giving it the preference to wheat or barley. They have succeeded in conquering the weed. In all directions it is dying out, and simply because it has no chance of propagating itself in the only way possible, that is to say, by a dissemination of its seed. I would ask, is not this a benefit to the agriculturist of a kind to entitle the bird to the care and protection of the whole community?

It should be remembered, also, that the services of the Sparrow as a scavenger in our colonial streets are not to be despised. The droppings of the horses are turned over by these industrious little birds and scattered to the winds, and in a variety of other ways they contribute to the cleanliness and purity of our thoroughfares.

The resultant fact is that for all these inestimable benefits we must be prepared to pay something; and it seems to me that the small tithe of grain which the Sparrows levy at a time of the year when everything else fails them is a very moderate consideration indeed. But it is the old story over again of ignorant prejudice and popular clamour. In Hungary, as we are informed, the same indiscriminate crusade was carried on some years ago, and was persevered in till not a Sparrow remained; then, after sufficient time had elapsed to show what an error had been committed, the Government had to offer a bonus of so much per head for the birds in order to reestablish them in that country †.

* Even Mr. J. H. Gurney, Jun., whose pamphlet 'On the Misdeeds of the Sparrow' is the most recent contribution to the subject, and who urges the necessity of keeping down this bird, feels bound to say:—"It may be that in some exceptional seasons (when a great plague of insect-life shall again occur), as in 1574, when it is said cockchafers gathered in such numbers on the banks of the Severn as to prevent the working of the water-mills, and in 1688 when they formed a black cloud in Galway, which darkened the sky for a league, destroying vegetation so completely as to change summer into winter ('Wild Birds' Protection Report,' p. 170), Sparrows will do good. Bearing this in mind no one should advocate their extirpation." He candidly says "that they mix the corn with considerable quantities of wild seeds, including, be it freely admitted, the destructive knot-grass and corn-bindweed; but even then they take corn by preference." And he concludes: "Although it is desirable to keep them down at all times, it should be remarked that the mischief done by them at harvest-time is 20-fold greater than at seed-time."

† Thus writes the accomplished historian Michelet:—"The 'miserly agriculturist' is the accurate and forcible expression of Virgil. Miserly, and blind, in truth, for he proscribes the birds which destroy insects and protect his crops. Not a grain will he spare to the bird which, during the winter rains, hunted up the future insect, sought out the nests of the larvæ, examined them, turned over every leaf, and daily destroyed myriads of future caterpillars; but sacks of corn to the adult insects, and whole fields to the grasshoppers which the bird would have combated! With his eye fixed on the furrow, on the present moment, without sight or foresight; deaf to the grand harmony which no one ever interrupts with impunity, he has everywhere solicited or approved the laws which suppressed the much-needed assistant of his labour, the insect-destroying bird. And the insects have avenged the bird. It has become necessary to recall in all haste the banished. In the island of Bourbon, for example, a price was set on each Martin's head; they disappeared, and then the grasshoppers took possession of the island, devouring, extinguishing, burning up with harsh acidity all that they did not devour. The same thing has occurred in North America with the Starling, the protector of the maize. The Sparrow even, which attacks the grain, but also defends it—the thieving, pilfering Sparrow, loaded with so many insults, and stricken with so many maledictions—it has been seen that without him Hungary would perish; that he alone could wage the mighty war against the cockchafers and the myriad winged foes which

But the same popular prejudice was for a long time directed against the Common Pheasant. Gradually the country settlers were won over to a due appreciation of this valuable bird *.

In addition to those already mentioned, the following English birds may now be considered permanently established in the country:—the Common Thrush, Blackbird, Sky-Lark, Greenfinch, Linnet, Chaffinch, Redpoll, Goldfinch, and Starling. Some years ago a number of Rooks were imported by the Auckland Acclimatization Society, but they do not appear to have spread far beyond the district in which they were first liberated.

In addition to two species of Quail, we have imported very successfully from Australia the Indian Minah and the Native Magpie, both of which are useful and ornamental birds.

Many other species have been introduced, and have appeared to thrive in their new home, although they cannot yet be looked upon as fairly established.

I am not aware that any serious effort has been made to introduce Owls of any kind, but this is a matter well worth the attention of the local Acclimatization Societies. In 1873 I sent out from England a pair of Wood-Owls (*Syrnium aluco*). They arrived safely at Napier, and after recruiting their strength were turned loose in a distant part of the Province. The Hon. Mr. Ormond, as superintendent of the Province, gave orders for their protection under the Act; but notwithstanding all these precautions, the unfortunate immigrants fell victims to popular prejudice.

In some of the principal lakes in both islands the Australian Black Swan (*Chenopis atrata*), the first of which were introduced into the North Island by myself, about the year 1864, is now to be seen in considerable flocks, often numbering many hundreds. They appear to associate freely with the Grey Duck (*Anas superciliosa*), but it is an undeniable fact that on waters where this Swan

---

reign in the low-lying lands; his banishment has been revoked, and the courageous militia hastily recalled, which, if not strictly disciplined, are not the less the salvation of the country."

The Sparrow in New Zealand has an able and ever-ready champion in Mr. W. T. L. Travers, the well-known barrister, who thus attacked a proposal in the Colonial Legislature to exterminate it:—"War is to be waged against the Sparrows, under the authority of Parliament. The following short extracts show the wisdom brought to bear in discussing the question. The Hon. Mr. Chamberlain says that the Hawk is the natural enemy of the Sparrow, a deduction, no doubt, from the name 'Sparrow-hawk' applied to one species of Hawk, but no New-Zealand Hawk that I know of ever touches a Sparrow. Mr. Oliver tells us that it was a mistake to introduce the Sparrow, and so does Mr. Gray. Mr. Miller says that none but the agriculturist was fit to discuss the question, and drew a comparison between the Sparrow and the Starling, which was about as appropriate as if he had attempted to compare the Sparrow with the elephant. Mr. Acland said that the Sparrow did not destroy insects. Mr. Holmes read some extracts in support of his opinions against the Sparrow, and I can supply him with any quantity more of the same kind, emanating from equal ignorance of the subject. It would be well if hon. gentlemen, in dealing with this question, would take the trouble to read the evidence given before a committee of the House of Lords on the subject of Sparrow-clubs in England, and if they should still entertain any respect for the intelligence of that august body, they would probably be disposed to change the opinions above expressed. Not many years ago the agriculturists of Hungary succeeded in getting the Sparrow proscribed by law, and he disappeared from the land. Within five years from that time the Government were compelled to spend 250,000 rix dollars in reintroducing him from other countries. In the North Island, and in the northern parts of the South Island, the cultivation of valuable deciduous trees was practically impossible until the large snails had been greatly reduced in numbers, and if Mr. Acland had seen, as I and many others have, the Sparrow actively engaged in destroying these creatures and devouring them, he might probably change his opinion. The nestling Sparrow cannot eat hard food, and careful observation has shown that a pair of parent Sparrows will bring upwards of 3000 insects to the nest in the course of a single day to feed its brood."

* A practical farmer thus writes to one of the newspapers:—"As much has been written and said for and against this beautiful bird, I will add my experience on the subject. On the one hand the Pheasants completely cleared a patch of maize for me; but on the other hand, when, some time after, I shot one of the depredators, its crop was found to contain about half-a-pint of fragments of black crickets. I have therefore resolved for the future to endeavour to scare them away from my crops, but on no account to exterminate them."

establishes itself the Ducks rapidly disappear. It is said also to be very inimical to the presence of the White Swan *.

Another bird that bids fair to be well acclimatized is the Cape or Egyptian Goose (*Chenalopex ægyptiaca*). Just before I left the colony one of these Geese was shot on Te Aute Lake, and submitted to me as a supposed addition to the New-Zealand Avifauna. Recognizing the species, and being satisfied that the individual bird was a wild one, I wrote to Sir George Grey for the purpose of ascertaining whether he had brought any of these Geese from the Cape. The information in reply was exactly what I had expected. Sir George Grey brought eight or ten of these birds with him to the Colony in 1860. They bred freely at the Kawau, and many of them crossed over to the mainland. Judge Rogan informed Sir George that he had seen as many as four shot at the Kaipara during his residence there. The fact that it has already found its way to the Hawke's Bay district shows how this species is establishing itself in a country where certainly all the conditions are favourable to its existence.

ORIGIN OF THE NEW-ZEALAND AVIFAUNA.

I have already said enough about the ancient and existing forms of bird-life in New Zealand to convince the most casual reader that we have within this comparatively small area a very remarkable ornithological province. In some respects it is quite unique, and, taken altogether, it is perhaps, to the student of biological history, the most interesting insular district on the face of our globe. In his admirable work on 'The Geographical Distribution of Animals,' Mr. Wallace has given, in a large woodcut, an ideal scene in New Zealand, representing some of its more singular forms. Referring to this, he says, "no country on the globe can offer such an extraordinary set of birds as are here depicted"; and in his elaboration of the subject, he has thrown more light than any previous writer on the origin and development of these peculiar ornithic types.

Looking to the fragmentary character of the New-Zealand fauna generally—the almost total absence of Mammalia and Amphibia, the phenomenal development of wingless birds that existed till quite recent times and are now represented by the various species of *Apteryx*, the highly specialized forms of non-volant Rails, besides the many other endemic genera of land-birds, and the great paucity of reptiles and insects—we must conclude that it is but the remnant of an ancient fauna, perhaps the most ancient in the world, which formerly occupied a very much wider area of the earth's surface.

Professor Newton, in his Address to the British Association last year, called the attention of naturalists generally to the extreme interest which attaches to every portion of this unique fauna. Remarking on its origin and development he says:—"One thing to guard against is the presumption

* It is popularly supposed that the Black Swan and the White Swan will not live together on the same waters; but the fact is that no systematic attempts, so far as I am aware, have yet been made to acclimatize the White Swan, either in Australia or New Zealand. Years ago, Baron von Mueller showed me a small flock of White Swans commingling with their dark cousins on a fine sheet of water in the Melbourne Acclimatization Gardens. A few tame pairs have been placed on ponds and ornamental waters in the South Island, and these have bred freely enough notwithstanding the constant presence of the Black Swan. In the North Island the experiment has not yet been tried. Sir George Grey was unfortunate enough to lose one of the beautiful pair presented to him by Her Majesty, or the North Island might have been ultimately stocked from Kawau. I am now arranging to send out some of these noble birds as a present to the Ngatiraukawa tribe, in order that they may be placed on the Horowhenua Lake, where the other species is already established, and it will be interesting to note their future history.

that the fauna originated within its present area, and has been always contained therein. Thus I take it that the fauna which characterizes the New-Zealand Region—for I follow Professor Huxley in holding that a Region it is fully entitled to be called—is the comparatively little changed relic and representative of an early fauna of much wider range; that the characteristic fauna of the Australian Region exhibits in the same way that of a later period; and that of the Neotropical Region of one later still." He points out that the indigenous species are with scarcely an exception peculiar to the country, and from every scientific point of view of the most instructive character; and he urges the importance of their closest study, because the Avifauna is now being fast obliterated by colonization and other agencies, and with it will pass into oblivion, unless faithfully recorded by the present generation, a page of the world's early history full of scientific interest.

The biological problems which the peculiar fauna and quasi-tropical flora of New Zealand suggest can only be met and reasonably explained on the hypothesis of a former land-connection between these islands and the northern or tropical portion of Australia; the severance, by submersion of the intervening land, having taken place at a period anterior to the spread of Mammalia over this portion of the earth's surface. Mr. Wallace has, I think, made it perfectly clear that this ancient land-connection was with North Australia, New Guinea, and the Western Pacific Islands, rather than with the temperate regions of Australia. At p. 443 of his 'Island Life' he gives a reduced map showing the depth of the sea around Australia and New Zealand, as established by the most recent soundings. From this it is manifest, as he points out, that there is a comparatively shallow sea, or, in other words, a submarine bank, at a depth of less than 1000 fathoms, indicating the additional land-area that would be produced if the sea-bottom were elevated 6000 feet. This submerged plateau, if we may so term it, presents a remarkable conformation, extending in a broad mass westward and then sending out two great arms, one reaching to beyond Lord Howe's Island, while the other stretches over Norfolk Island to the great barrier reef, thus forming the required connection with tropical Australia and New Guinea. It is argued that the ancient land-connection thus indicated, with perhaps, at a still more remote epoch, a connection with the great Southern continent by means of intervening lands and islands, will explain many of the difficult zoological problems that New Zealand presents.

This theory, while it accounts for the introduction into New Zealand by a north-western route, in very ancient times, of the Struthious type of birds, from which all the known species of *Dinornis* and *Apteryx* may have descended, explains too the tropical character of much of the New-Zealand flora, which is somewhat anomalous considering the temperate climate of New Zealand as we know it. Mr. Wallace states, as the result of careful research, that there are in New Zealand thirty-eight thoroughly tropical genera of plants, thirty-three of which are found in Australia, and, with a very few exceptions, in the northern or tropical portions only. To these may be added thirty-two more genera of plants which, though chiefly developed in temperate Australia, extend also into the tropical or subtropical portion of it, and which, it may reasonably be inferred, reached New Zealand by the same route. But to make this line of reasoning perfectly intelligible, it ought to be mentioned that the geological history of Australia shows it to have been for an immense period of time divided into an Eastern and a Western island, in the latter of which only the largely peculiar flora of temperate Australia—distinguished by its Eucalypti, Proteas, and Acacias—was developed, and where alone the marsupial Mammalia had their home. At this period, according to the above theory, New Zealand was in connection with the tropical portion of the Eastern island alone. This important geological fact will therefore account for the non-introduction into New Zealand, along with the

g

ancestors of the Moas and Kiwis and the tropical plants referred to, of the marsupial fauna and the peculiar temperate flora so characteristic of Australia as we now know it *.

Sir Joseph Hooker, undoubtedly the ablest and most accomplished of living botanists, referring to an apparently insoluble enigma in the relations of the flora of New Zealand with that of Australia, thus expresses himself in the Introduction to his well-known 'Flora of Australia':—

"Under whatever aspect I regard the flora of Australia and New Zealand, I find all attempts to theorize on the possible causes of their community of feature frustrated by anomalies in distribution, such as I believe no two other similarly situated countries on the globe present. Everywhere else I recognize a parallelism or harmony in the main common features of contiguous floras, which conveys the impression of their generic affinity, at least, being affected by migration from centres of dispersion in one of them, or in some adjacent country. In this case it is widely different. Regarding the question from the Australian point of view, it is impossible in the present state of science to reconcile the fact of *Acacia*, *Eucalyptus*, *Casuarina*, *Callitris*, &c. being absent in New Zealand, with any theory of trans-oceanic migration that may be adopted to explain the presence of other Australian plants in New Zealand; and it is very difficult to conceive of a time or of conditions that could explain these anomalies, except by going back to epochs when the prevalent botanical as well as geographical features of each were widely different from what they are now. On the other hand, if I regard the question from the New-Zealand point of view, I find such broad features of resemblance and so many connecting links that afford irrefragable evidence of a close botanical connection, that I cannot abandon the conviction that these great differences will present the least difficulties to whatever theory may explain the whole case."

It will be seen that the theory of which an outline has been given, while accounting in a rational manner for the marked peculiarities of the New-Zealand fauna, offers at the same time a probable solution of some of the strange anomalies of its flora in relation to that of Australia.

Mr. Wallace has explained that, in zoology, discontinuity in the areas of distribution must be accepted as an indication of antiquity, and that the more widely the fragments are scattered the more ancient we may, as a rule, take the parent group to be. "Thus the marsupials of South America and Australia are connected by forms which lived in North America and Europe; the camels of Asia and the llamas of the Andes had many extinct common ancestors in North America; the lemurs of Africa and Asia had their ancestors in Europe, as did the Trogons of South America, Africa, and

* "If we examine the geological map of Australia, we shall see good reason to conclude that the eastern and the western divisions of the country first existed as separate islands, and only became united at a comparatively recent epoch. This is indicated by an enormous stretch of Cretaceous and Tertiary formations extending from the Gulf of Carpentaria completely across the continent to the mouth of the Murray River. During the Cretaceous period, therefore, and probably throughout a considerable portion of the Tertiary epoch, there must have been a wide arm of the sea occupying this area, dividing the great mass of land on the west—the true seat and origin of the typical Australian flora—from a long but narrow belt of land on the east, indicated by the continuous mass of Secondary and Palæozoic formations already referred to, which extend uninterruptedly from Tasmania to Cape York. Whether this formed one continuous land, or was broken up into islands, cannot be positively determined; but the fact that no marine Tertiary beds occur in the whole of this area renders it probable that it was almost, if not quite, continuous, and that it not improbably extended across to what is now New Guinea. At this epoch, then, Australia would consist of a very large and fertile western island, almost or quite extra-tropical, and extending from the Silurian rocks of the Flinders range in South Australia to about 150 miles west of the present west coast, and southward to about 350 miles south of the Great Australian Bight. To the east of this, at a distance of about 250 or 400 miles, extended, in a north and south direction, a long but comparatively narrow island, stretching from far south of Tasmania to New Guinea; while the crystalline and Secondary formations of Central North Australia probably indicate the existence of one or more large islands in that direction."—*Island Life*.

tropical Asia. But besides this general evidence, we have direct proof that the Struthious birds had a wider range in past times than now. Remains of extinct Rheas have been found in Central Brazil *, and those of Ostriches in North India, while remains believed to be of Struthious birds are found in the Eocene deposits of England; and the Cretaceous rocks of North America have yielded the extraordinary toothed bird, *Hesperornis*, which Professor O. Marsh declares to have been a 'carnivorous swimming Ostrich.' As to the second point, we have the remarkable fact that all known birds of this group have not only the rudiments of wing-bones, but also the rudiments of wings, that is, an external limb bearing rigid quills or largely developed plumes. In the Cassowary these wing-feathers are reduced to long spines like porcupine-quills, while even in the *Apteryx* the minute external wing bears a series of nearly twenty stiff quill-like feathers †. These facts render it probable that the Struthious birds do not owe their imperfect wings to a direct evolution from a reptilian type, but to a retrograde development from some low form of winged birds, analogous to that which has produced the Dodo and the Solitaire from the more highly developed Pigeon type. Professor Marsh has proved that, so far back as the Cretaceous period, the two great forms of birds—those with a keeled sternum and fairly developed wings and those with a convex keel-less sternum and rudimentary wings—already existed side by side: while in the still earlier *Archæopteryx* of the Jurassic period we have a bird with well-developed wings, and therefore probably with a keeled sternum. We are evidently, therefore, very far from a knowledge of the earlier stages of bird-life, and our acquaintance with the various forms that have existed is scanty in the extreme; but we may be sure that birds acquired wings, and feathers, and some power of flight before they developed a keeled sternum, since we see that bats (with no such keel) fly very well. Since, therefore, the Struthious birds all have perfect feathers, and all have rudimentary wings, which are anatomically those of true birds, not the rudimentary fore legs of reptiles, and since we know that in many higher groups of birds—as the Pigeons and the Rails—the wings have become more or less aborted, and the keel and the sternum greatly reduced in size by disuse, it seems probable that the very remote ancestors of the Rhea, the Cassowary, and the *Apteryx* were true flying birds, although not perhaps provided with a keeled sternum or possessing very great powers of flight ‡. But in addition to the possible ancestral power of flight, we have the undoubted fact that the Rhea and the Emu both swim freely, the former having been seen swimming from island to island off the coast of Patagonia. This, taken in connection with the wonderful aquatic Ostrich of the Cretaceous period discovered by Professor Marsh, opens up fresh possibilities of migration; while

* Reinhardt is of opinion that "the ancient and the modern *Rhea* are of one and the same species."—*Ibis*, 1882, p. 332.

† "See Buller's illustration in Trans. N.-Z. Instit. vol. iii. plate 12 *b*. fig. 2."

‡ "Professor Marsh has shewn that there is good reason for believing that the power of flight was gradually acquired by Birds, and with that power would be associated the development of a keel to the sternum, on which the volant faculty so much depends, and with which it is so intimately correlated that, in certain forms which have to a greater or less extent given up the use of their fore-limbs, the keel, though present, has become proportionally aborted. Thus the Carinate type would, from all we can see at present, appear to have been evolved from the Ratite. This view receives further support from a consideration of the results of such embryological research as has already been made—the unquestionable ossification of the Ratite sternum from a smaller number of paired centres than the Carinate sternum, in which (with the doubtful exception of the Anatidæ) an additional, unpaired centre makes its appearance. Again, the geographical distribution of existing, or comparatively recent, Ratite forms points to the same conclusion. That these forms—Moa, Kiwi, Emeu and Cassowary, Rhea, and finally Ostrich—must have had a common ancestor nearer to them than is the ancestor of any Carinate form seems to need no proof. If we add to these the *Æpyornis* of Madagascar, the fossil Ratitæ of the Siwalik rocks, and the as yet but partially recognized *Struthiolithus* of Southern Russia, to say nothing of *Gastornis*, the evidence is stronger still. Scattered as these Birds have been or are throughout the world, it seems justifiable to consider them the survivals of a very ancient type, which has hardly undergone any essential modification since the appearance of Bird-life upon the earth—even though one at least of them has become very highly specialized."—*Prof. Newton in Enc. Brit.* vol. xviii. pp. 43, 44.

the immense antiquity thus given to the group and their universal distribution in past time, renders all suggestions of special modes of communication between the parts of the globe in which their scattered remnants *now* happen to exist altogether superfluous and misleading" *.

In his last-named work, Mr. Wallace divides all known islands into two classes, "Continental" and "Oceanic." The former are always more varied in their geological formation—containing both ancient and recent stratified rocks—are rarely remote from a continent, and always contain some land Mammalia, also Amphibia and representatives of the other classes of animals in considerable variety. The "Oceanic" islands are usually far removed from continents and are always separated from them by very deep seas, are entirely without land Mammalia or Amphibia, but are generally well stocked with birds and insects and with some reptiles. Now New Zealand, which is undoubtedly "Continental" in its geological formation, also in the existence of the submerged bank already described connecting it in ancient times with North Australia and New Guinea, is as decidedly "Oceanic" in its zoological character, except as regards its wingless birds and the remarkable tuatara lizard (*Sphenodon punctatum*), which is said to constitute *per se* a distinct order of Reptilia of extreme antiquity. Mr. Wallace therefore terms New Zealand and the Celebes, where the conditions are somewhat similar, "Anomalous islands;" but *Ancient continental* may be perhaps a more convenient term.

As already explained, at the time of the supposed land-connection to the North-west, the Marsupial fauna could not have reached the eastern land now forming part of Australia; but it seems very probable that, at this early period, tropical Australia was tenanted by some Struthious kind of bird, perhaps volant in its character, which had reached this land, by way of New Guinea, through some ancient continental extension. If this theory, so well propounded by Mr. Wallace, is the true one, then the Cassowaries of New Guinea, the Emus of Australia, the extinct *Dromornis* of Queensland, and the Moas and Kiwis of New Zealand are doubtless the modified descendants of this ancestral type. "The total absence (or extreme scarcity) of mammals in New Zealand obliges us to place its union with North Australia and New Guinea at a very remote epoch. We must either go back to a time when Australia itself had not yet received the ancestral forms of its present marsupials and monotremes, or we must suppose that the portion of Australia with which New Zealand was connected was then itself isolated from the mainland, and was thus without a mammalian population. . . But we must on any supposition place the union very far back, to account for the total want of identity between the winged birds of New Zealand and those peculiar to Australia, and a similar want of accordance in the lizards, the freshwater fishes, and the more important insect groups of the two countries. From what we know of the long geological duration of the generic types of these groups we must certainly go back to the earlier portion of the Tertiary period at least in order that there should be such a complete disseverance as exists between the characteristic animals of the two countries, and we must further suppose that, since their separation, there has been no subsequent union or sufficiently near approach to allow of any important inter-migration, even of winged birds, between them. It seems probable, therefore, that the Bampton shoal west of New Caledonia, and Lord Howe's Island further south, formed the western limits of that extensive land in which the great wingless birds and other isolated members of the New-Zealand fauna were developed. Whether this early land extended eastward to the Chatham Islands and southward to the Macquaries we have no means of ascertaining; but as the intervening sea appears to be not more than 1500 fathoms deep,

* 'Island Life,' by Alfred Russel Wallace, pp. 451, 452.

it is quite possible that such an amount of subsidence may have occurred. It is possible, too, that there may have been an extension northward to the Kermadec Islands, and even further to the Tonga and Fiji Islands, though this is hardly probable, or we should find more community between their productions and those of New Zealand. A southern extension towards the Antarctic continent at a somewhat later period seems more probable, as affording an easy passage for the numerous species of South American and Antarctic plants and also for the identical and closely allied freshwater fishes of these countries. The subsequent breaking up of this extensive land into a number of separate islands—in which the distinct species of Moa and Kiwi were developed—their union at a later period, and the final submergence of all but the existing islands, is a pure hypothesis, which seems necessary to explain the occurrence of so many species of these birds in a small area, but of which we have no independent proof" *.

In a preceding section I have already mentioned that, as a rule, the species of *Dinornis* which, in former times, inhabited the North Island were different in character from their contemporaries in the South Island, although the two areas of land are only separated by a strait scarcely eighteen miles across in its narrowest part. The same feature is maintained to the present day in the existing Avifauna, clearly showing that each island has a biological history of its own. Thus the Saddle-back (*Creadion carunculatus*) of the North is represented in the South by *C. cinereus*, a closely-allied species, but differing in the colour of its plumage; *Turnagra hectori* (now almost extinct) is represented by *T. crassirostris*, a species that will soon follow suit, although still plentiful in certain localities; the Weka (*Ocydromus earli*) is represented by several other closely-related species (*O. australis*, *O. fuscus*, and *O. brachypterus*) so closely resembling the northern bird both in appearance and habits that they are called "Wood-hens" by the settlers of both islands and by them, as well as by the natives, are generally regarded as identical; the Popokatea (*Clitonyx albicapilla*) is represented by another species (*C. ochrocephala*) differing in colour, but so closely allied to it that the Maoris apply the same name to both; the Toutouwai (*Miro australis*), to which precisely the same remark applies, is represented by *M. albifrons*, and *Glaucopis wilsoni* by *G. cinerea*, distinguishable only by the colour of its ornamental wattles. Another case in point is furnished by the two representative species of *Apteryx*, the North Island bird being characterized by a different structure of plumage to that of the well-known *Apteryx australis* inhabiting the South Island. Till of late years it was believed that *Apteryx oweni*, which differs entirely from both of these species in the grey colour and mottled appearance of its plumage, was confined to the colder districts of the South Island; but in 1876 I communicated to the Wellington Philosophical Society the discovery of this bird near the summit of the Tararua mountains on the north side of Cook's Strait, where it was found frequenting the stunted vegetation immediately below the snow-line †. The existence of this species was entirely unknown to the Maoris of the North Island, and its occurrence under the conditions I have mentioned is a very interesting fact in geographical distribution.

Analogous cases of representative species in more or less widely separated areas are of frequent occurrence in other parts of the world. "The cause of this" (writes Mr. Wallace) "is very easy to understand. We have already shown that there is a large amount of local variation in a considerable number of species, and we may be sure that were it not for the constant intermingling and intercrossing of the individuals inhabiting adjacent localities this tendency to local variation would soon form distinct races. But as soon as the area is divided into two portions, the intercrossing is stopped,

* 'Island Life,' p. 454. † Trans. N.-Z. Instit. vol. viii. pp. 193–194.

and the usual result is that two closely allied races, classed as representative species, become formed. Such pairs of allied species on the two sides of a continent, or in two detached areas, are very numerous; and their existence is only explicable on the supposition that they are descendants of a parent form which once occupied an area comprising that of both of them,—that this area then became discontinuous,—and, lastly, that, as a consequence of the discontinuity, the two sections of the parent species became segregated into distinct races or even species."

In his 'Geographical Distribution of Animals' Mr. Wallace treats New Zealand and her satellites as forming a subregion of Australia. The Australian, or "great insular region of the earth," is divided by him into four subregions, distinguished as the Austro-Malayan, Australian, Polynesian, and New Zealand. The last-named subregion is made to include Norfolk Island, Phillip and the Nepean Isles, Lord Howe's Island and the Kermadec Isles on the north, the Chatham Islands on the east, the Auckland, Macquarie, Emerald, Campbell, Antipodes, and Bounty Islands on the south and south-east.

Other prominent writers on the subject have claimed for New Zealand full recognition as a separate biological province, quite distinct from Australia and every other region of the earth. My own study of the subject having brought me to the same conclusion, I propose to examine here, very briefly, the grounds upon which Mr. Wallace links New Zealand to Australia as contiguous sections of one biological region. He admits, of course, that there is a "wonderful amount of speciality," but he contends that "the affinities of the fauna, whenever they can be traced, are with Australia or Polynesia."

If we take Mr. Wallace's own table of the geographical distribution we find, on a careful analysis, that out of twenty-eight families stated to be common to Australia and New Zealand, three are included in error, namely Sittidæ, Dicæidæ, and Pandionidæ, thus reducing the number to twenty-five. Of these, fifteen are admitted by him to be cosmopolitan, and may therefore be discharged from the present inquiry. Of the remaining ten, four belong to the Old World, four to the Oriental, Ethiopian, Austro-Malayan, and Polynesian regions respectively, and one highly specialized family, the Spheniscidæ, to the south temperate regions, leaving thus only one family, the Paridæ, as restricted in its range to the two countries. This family is represented in New Zealand by a single genus, *Certhiparus*, about the true position of which there is considerable doubt, and this genus again is represented by a single species, so that, as regards the mere distribution of *families*, the argument altogether fails. Let us now examine the far more important question of identical or representative genera and species in the two countries, for this after all is the true test of a common origin. Of the twelve *genera* of Australian birds which he treats as belonging equally to New Zealand, it may be remarked that two (namely *Graucalus* and *Acanthochæra*), each of them represented by a single species, have only occurred in New Zealand as accidental stragglers, at very long intervals; that *Tribonyx*, as already explained at p. xiv, has never actually occurred in a wild state; and that *Orthonyx* and *Hieracidea* have, on further investigation of their characters, been replaced by two endemic genera, *Clitonyx* and *Harpa*. Of the remaining seven, two alone (*Gerygone* and *Sphenæacus*) are characteristic of Australia, the others ranging over a great part of the southern hemisphere; thus, *Platycercus* is spread over New Guinea and Polynesia, as well as Australia, *Rhipidura* extends to India, and *Zosterops* through Polynesia and the Malay Archipelago to India and Africa. Of the five *species* mentioned by Mr. Wallace as being identical in Australia and New Zealand, it may be mentioned that three (*Acanthochæra carunculata, Graucalus melanops*, and *Hirundo nigricans*) are among our rarest stragglers from abroad, and that the Shining Cuckoo (*Chrysococcyx lucidus*) is an annual migrant to and from Australia: thus leaving only *Zosterops cærulescens* to be accounted for, and this species has been sufficiently treated of already.

Mr. Wallace's strongest point is the Family Meliphagidæ, which is a very typical and well-distributed Australian group. But accepting, as I think we must do, his theory of the introduction of the ancestral types into New Zealand by way of tropical Australia and New Guinea, it is easy to account for the presence of this peculiar form in both countries, inasmuch as the Meliphagidæ have representatives as far north as the Sandwich Islands, whilst other members of the group are spread through the Austro-Malayan subregion, finding their extreme western limit in the Celebes. Supposing that the ancient type reached New Zealand by the north-western route, it then resolves itself into a mere question of time and "descent with modification."

Dr. Otto Finsch, who has written several interesting papers on New-Zealand Ornithology, appears to me to exaggerate very much the importance of this feature, for he accepts it as a proof of "far more intimate connection with Australia than one would suppose from the geographical position of the two countries." He is unable, however, to account for the absence of true Trichoglossi in New Zealand, seeing that this group is so strongly developed in the temperate parts of Australia.

It is a point of some significance that the Meliphagine genera in New Zealand are not very closely related to those of Australia, except in the case of *Pogonornis*, which approximates somewhat to *Ptilotis*, a decidedly subtropical genus. Apart from the true Honey-eaters, the only genera that Mr. Wallace specially refers to as related to peculiar Australian ones are *Miro* and *Myiomoira* (allied to *Petræca*), *Ocydromus* (allied to *Eulabeornis*), and *Hymenolæmus* (allied to *Malacorhynchus*). It seems to me, therefore, that he has not succeeded in establishing a co-ordinate relation between the avi-faunæ of these so-called subregions of Australia.

It is worthy of remark also that, with the exception of the highly developed Meliphagidæ, comprising four very distinct genera (and numbering altogether only five species), none of the New-Zealand families contain more than two genera, presenting a marked difference in this respect to the numerous subordinate groups among the birds of Australia.

Seeing that the Shining Cuckoo (*Chrysococcyx lucidus*) is met with in New Guinea, and probably further west, that it is likewise found in tropical Australia, and that it comes to us from the north, or north-west—for it always makes its appearance first at the extreme north—it is easy to understand that the migratory impulse has been inherited from time immemorial, and the more so as the closely-allied species (*C. plagosus*) is also a summer visitant to the temperate and southern portions of Australia. But it is very difficult to imagine why the Long-tailed Cuckoo (*Eudynamis taitensis*), which hibernates in the warm islands of the Pacific—the Friendly, Society, Marquesas, Fiji, and Samoa groups—ranging over more than 40° of longitude, should make its annual pilgrimage across 1500 miles of ocean to New Zealand *.

The Waders are, for the most part, cosmopolitan, and are therefore of little account when estimating the geographical relations of the avifauna.

* Mr. Wallace says, in his account of the Chatham Islands ('Island Life,' p. 454):—"It is stated that the *Zosterops* differs from that of New Zealand, and is also a migrant; and it is therefore believed to come every year from Australia, passing over New Zealand, a distance of nearly 1700 miles!" But this is evidently a *lapsus calami*, the bird intended being the *Chrysococcyx*. Prof. Hutton stated (Trans. N.-Z. Instit. vol. v. p. 225) that this happened in the case of *C. plagosus*; but I have shown elsewhere that he was wrong in his identification of the species, the Shining Cuckoo (*C. lucidus*) which annually visits the Chatham Islands being identical with the New-Zealand bird. *Chrysococcyx plagosus*, distinguished by its narrower bill, has never been met with in New Zealand, and it would be strange indeed if this Australian species had occurred in the Chatham Islands to the eastward.

One of the most widely distributed species is the Eastern Golden Plover (*Charadrius fulvus*), which, at all times rare in New Zealand, is plentiful in Australia, and spreads itself over the Polynesian Islands and the Indian Archipelago, westward to Ceylon, and northward to Siberia and Kamtschatka, where it rears its young.

Several of our Ducks are common to Australia, but it is well known that this Order is a very diffuse one in all parts of the world. Our common Grey Duck (*Anas superciliosa*), for example, extends its range into Tasmania and Australia, over a large portion of Polynesia, and as far north as the Sandwich Islands; whilst the White-winged Duck (*Anas gibberifrons*) is met with, not only in Australia, but in New Caledonia and the Indian Archipelago. The genus *Hymenolæmus*, represented by our peculiar Mountain Duck, is closely related to an Australian one, and our Shoveller (*Rhynchaspis variegata*) is a representative species to that inhabiting Australia and Tasmania, the two forms being very closely allied. Two other Ducks, however (*Dendrocygna eytoni* and *Nyroca australis*) are so rare with us that they may fairly be regarded as Australian stragglers. Even where the species is peculiar to New Zealand, the genus to which it belongs may be a widely spread one: for example, *Fuligula novæ zealandiæ* belongs to a genus which has representatives in the northern parts of America, in Europe and in Asia, and our splendid *Casarca variegata* represents a genus which is almost cosmopolitan.

One of the most puzzling of these occurrences is the Little Bittern (*Ardetta pusilla*), which, although decidedly rare, has been met with on the west coast and in the southernmost part of the South Island. Both this and our common species (*Botaurus pœciloptilus*) are birds of feeble wing; yet they are identical with the species inhabiting temperate Australia, showing that they must have preserved their individuality as species for a very long period of time. The same remark applies to our *Porphyrio melanonotus*, and, in a lesser degree, to *Rallus philippensis* and *Ortygometra affinis*, which are very closely related to *R. pectoralis* and *O. palustris* respectively.

When we come to compare our avifauna with that of the Polynesian "subregion" there is still less resemblance, for the only genera common to both are the two referred to above, whilst the only species mentioned by Mr. Wallace as identical is our other migratory Cuckoo (*Eudynamis taitensis*). It is true that he questions the fact of these Cuckoos being migratory at all, and endeavours to account for their disappearance in winter by suggesting that "in a country which has still such wide tracts of unsettled land, they may only move from one part of the islands to another." But quite apart from the lengthened form of the wing in both of these Cuckoos, which at once proclaims them "birds of passage," the fact of their seasonal arrival in and departure from our country, as fully recorded in my account of each species, is well attested, and forms an essential part of their natural history.

Besides the genera of occasional or accidental occurrence (*Acanthochæra* and *Hirundo*) and the migratory Cuckoo already mentioned, the only groups of land-birds common to New Zealand and Polynesia are *Platycercus*, *Carpophaga*, and *Zosterops*, and the widely spread genera *Rhipidura*, *Halcyon*, and *Circus* *.

* It may be worth noting that I have remarked the following similarity between the names employed in the Fijian and Maori languages for the same or corresponding birds;—

Fijian.		*Maori.*
Kawakawaa.	=	Kawekaweo (*Eudynamis taitensis*).
Lulu.	=	Ruru (an Owl).
Kaka (a kind of Parrot).	=	Kaka (*Nestor meridionalis*).
Toa (any fowl-like bird).	=	Moa (*Dinornis*).
Toro.	=	Toroa (an Albatros).
Kula (a red Parrot).	=	Kaka-kura.

The Cormorants are evidently adapted by nature to a cold or temperate climate, for as we advance towards the Tropics they disappear, and it is said that not a single species is to be found in the whole of Polynesia.

In New Caledonia and the New Hebrides, which form a sort of transition ground into Australia proper and the Papuan group, we have the same genera, and in addition thereto, inhabiting New Caledonia, a flightless Rail, allied to the New-Zealand Woodhen.

The Chatham Islands to the east of New Zealand, the Auckland Islands, and the other scattered islets to the south and south-east are so obviously related to New Zealand geographically, besides coming within the political limits of the Colony, that I have included their birds in the present work. It is interesting to notice, however, that these islands nearly all contain one or more peculiar species, showing that the isolation has been of sufficiently long duration to allow of this development. Thus in the Chatham Islands and the adjacent islets there are seven peculiar species, namely, *Anthornis melanocephala*, *Gerygone albofrontata*, *Miro traversi*, *Sphenœacus rufescens*, *Rallus dieffenbachii*, *Cabalus modestus*, and *Phalacrocorax featherstoni*.

In the Auckland Islands, lying about 300 miles to the south of New Zealand, the three species mentioned by Mr. Wallace as peculiar (*Anthus aucklandicus*, *Platycercus aucklandicus*, and *P. malherbii*) have been proved to have no existence as valid species; but, as already mentioned, this small area contains two species of Duck (*Nesonetta aucklandica* and *Mergus australis*) hitherto not met with elsewhere; also a Snipe (*Gallinago aucklandica*) and a species of Rail (*Rallus brachypus*), both of which are supposed to be peculiar to these small islands.

From Macquarie Island, still further south, we have the handsome *Phalacrocorax nycthemerus* and possibly a new species of Rail; from Campbell Island, so far as our present knowledge extends, another fine Cormorant (*P. magellanicus*) and a peculiar Penguin; from the Snares the unique *Eudyptes atrata*, described by Prof. Hutton; and from Antipodes Island the interesting Ground-Parrakeet (*Platycercus unicolor*) lately discovered by Captain Fairchild. A small Hawk received by me from Macquarie Island is undoubtedly the same as our *Harpa ferox*, and the Rail which Prof. Hutton has distinguished as *Rallus macquariensis* seems to me to be merely a local race of *R. philippensis*, if at all separable from that species. There can be no doubt, therefore, of the propriety of including even this remote island in the New-Zealand region.

The case is different, however, with the islands to the north of New Zealand. The instances mentioned by me at p. 24 of the present volume make it abundantly clear that at some period there was a land-connection with Lord Howe's Island, Norfolk Island, and the Nepean group, and possibly with the Kermadec Islands * to the eastward; but, owing to the introduction from time to time of a colonist population, so to speak, from the nearer continent, the avifauna of these islands is decidedly more Australian than New Zealand. With the exception of *Nestor productus* and *Notornis alba*, all the species of land-birds inhabiting Norfolk Island and the Nepean group belong to Australian genera; and of the sixteen recorded species, all but three occur also in various parts of Australia.

The same remarks apply to Lord Howe's Island lying midway between Norfolk Island and Australia. With the exception of *Ocydromus sylvestris*, all the birds belong to well-known Australian

* Mr. J. F. Cheeseman, who accompanied the Annexation expedition to the Kermadec Islands last year, has lately communicated to the Linnean Society (through Sir Joseph Hooker) a report on the flora of these islands. He mentions incidentally that the land-birds found there, which were few in number, appeared to belong to New-Zealand species, but he does not state what these birds were.

types, and the species themselves are identical, with the exception of *Zosterops strenuus* and *Z. tephropleurus*, both of which, strange to say, are peculiar to this small island.

To summarize the results, it may be mentioned that out of sixty-nine species of "land-birds" (excluding the Herons and Bitterns) only eleven have a wider range than New Zealand. Of these exceptions five are only accidental stragglers from Australia; two are annual migrants; and the remaining four are *Zosterops cœrulescens*, *Rallus philippensis*, *Ortygometra tabuensis*, and *Porphyrio melanonotus*. But, what is even more remarkable still, out of thirty-four genera, after making a similar elimination to the above, not less than twenty-two are strictly endemic, showing at a glance how restricted is the character of the New-Zealand Avifauna.

That the Ornis of New Zealand may have been, from time to time, affected by casual immigration from Australia is probable enough, for, as we have seen, even during recent years, many individual cases of the kind have been recorded at irregular intervals; and it is rather matter for surprise, on this ground, that there is not a stronger family likeness, so to speak, between the indigenous birds of the two countries at the present day.

CONTENTS OF VOL. I.

Before concluding this Introduction it may be well to offer one or two general observations on the Families and Genera treated of in the present volume, which closes with the New-Zealand Wood-Pigeon (*Carpophaga novæ zealandiæ*).

The number of species described is fifty-five, and these have been referred to twenty-three Families and thirty-five Genera. Of the former four, and of the latter seventeen, are strictly endemic or peculiar to the New-Zealand Avifauna.

Of the fifty-five species all but eight are endemic, being found only in New Zealand and the adjacent islands. Of the exceptions one is *Zosterops cœrulescens*, whose erratic history has already been noticed, two are migratory birds (*Eudynamis taitensis* and *Chrysococcyx lucidus*), which only spend the summer with us, and five are occasional stragglers from the continent of Australia, not one of which has ever been known to breed with us. Indeed, in estimating the character of the Avifauna it is hardly fair to take count of these accidental visitants—such birds, for example, as the Australian Swift, which has been recorded only once in the history of the Colony and may never reappear, or the Australian Honey-eater, which has been recorded twice; so that, adopting this view, the number is reduced to one. It will thus be seen at a glance that the so-called "land-birds" are, almost without exception, characteristic of the country. Even in the case of *Zosterops*, which I have treated as identical with the Australian bird, there is some ground for regarding the New-Zealand form as a distinct local race. The late Mr. Gould and myself had no difficulty in picking out two of our birds from a whole case of Australian specimens, so manifest was the difference in the tone of coloration. While accepting therefore the identity of the species, I would point out that the difference I have mentioned can only be accounted for on the supposition that the birds have been separated for a considerable length of time. This tends to support my location of the species in the south-west region of the South Island, before it came northwards, and is therefore opposed to Professor Hutton's theory * that it arrived quite recently from Australia.

* 'New-Zealand Magazine,' January 1876, p. 96.

Coloured Illustrations are given of the following species:—

The Blue-wattled Crow (*Glaucopis wilsoni*).
The Orange-wattled Crow (*Glaucopis cinerea*).
The Huia (*Heteralocha acutirostris*), male and female.
The Saddle-back (*Creadion carunculatus*).
The Jack-bird (*Creadion cinereus*).
The North-Island Thrush (*Turnagra hectori*).
The South-Island Thrush (*Turnagra crassirostris*).
The North-Island Tomtit (*Myiomoira toitoi*).
The South-Island Tomtit (*Myiomoira macrocephala*).
The North-Island Robin (*Miro australis*).
The South-Island Robin (*Miro albifrons*).
The Grey Warbler (*Gerygone flaviventris*). Figured on same Plate as the Long-tailed Cuckoo.
The White-head (*Clitonyx albicapilla*).
The Yellow-head (*Clitonyx ochrocephala*).
The New-Zealand Creeper (*Certhiparus novæ zealandiæ*).
The Fern-bird (*Sphenœacus punctatus*).
The New-Zealand Pipit (*Anthus novæ zealandiæ*).
The Pied Fantail (*Rhipidura flabellifera*).
The Black Fantail (*Rhipidura fuliginosa*).
The Silver-eye (*Zosterops cærulescens*).
The Bell-bird (*Anthornis melanura*), male and female.
The Tui or Parson bird (*Prosthemadera novæ zealandiæ*), adult and young.
The Stitch-bird (*Pogonornis cincta*), male and female.
The Bush-Wren (*Xenicus longipes*).
The Rock-Wren (*Xenicus gilviventris*).
The Rifleman (*Acanthidositta chloris*), male and female.
The New-Zealand Kingfisher (*Halcyon vagans*), adult and young.
The Long-tailed Cuckoo (*Eudynamis taitensis*), adult and young.
The Shining Cuckoo (*Chrysococcyx lucidus*), with young in Warbler's nest.
The Yellow-fronted Parrakeet (*Platycercus auriceps*).
The Red-fronted Parrakeet (*Platycercus novæ zealandiæ*).
The Orange-fronted Parrakeet (*Platycercus alpinus*).
The Kaka Parrot (*Nestor meridionalis*), with variety "Kaka-Kura."
The Kea Parrot (*Nestor notabilis*).
The Kakapo or Owl Parrot (*Stringops habroptilus*), with Alpine variety.
The Morepork (*Spiloglaux novæ zealandiæ*).
The Laughing-Owl (*Sceloglaux albifacies*).
The New-Zealand Harrier (*Circus gouldi*), adult and young.
The Quail-Hawk (*Harpa novæ zealandiæ*), adult and young.
The New-Zealand Quail (*Coturnix novæ zealandiæ*).
The New-Zealand Pigeon (*Carpophaga novæ zealandiæ*).

I have endeavoured to make the technical part of the work as exhaustive and exact as possible. After the diagnostic characters of each species (rendered, according to the usual custom, in Latin), I have given full descriptions of both sexes, with their seasonal changes of plumage (if any), followed by an account of the young, commencing with the nestling, or fledgling, and noting the various adolescent states of plumage in the progress of the bird towards maturity. Under the head of 'Varieties,' I have been careful to record every appreciable departure from the normal character that has come under my notice during an acquaintance with this peculiar Ornis extending over the best part of my life.

The measurements of each bird described are given in inches and decimals. In taking the extreme length my rule has always been to measure from the tip of the bill, following its curvature (if any) to the end of the tail. The advantage of this plan is that by deducting the measurements of the culmen and the tail, which are given separately, the exact length of the body may be ascertained. The same rule has been followed in regard to the claws wherever measurements are given.

In order to make the descriptions intelligible to the ordinary reader, some knowledge is essential of the names usually applied to the various parts of a bird and to the feathers which cover them. To supply an index to the descriptive terms commonly employed throughout the present work, it may be useful to reproduce here, on a slightly reduced scale, the diagram given in my 'Manual of the Birds of New Zealand,' the outline selected for the purpose being that of one of our commonest species.

REFERENCES.—1, forehead; 2, crown or vertex; 3, hind head; 4, nape; 5, lore or loral space; 6, eye (shaded margin iris); 7, ear-coverts; 8, hind neck; 9, side of neck; 10, back or dorsal region; 11, rump or uropygium; 12, upper tail-coverts; 13, tail-feathers or rectrices; 14, primaries or quills; 15, secondaries; 16, larger wing-coverts; 17, lesser wing-coverts (including "median"); 18, carpal flexure, or bend of wing; 19, scapulars; 20, chin; 21, throat; 22, fore neck; 23, breast; 24, abdomen; 25, vent; 26, under tail-coverts; 27, tibial plumes; 28, cere; 29, ridge of upper mandible or culmen; 30, lower mandible; 31, tarsus; 32, middle toe and claw; 33, hind toe and claw, or hallux.

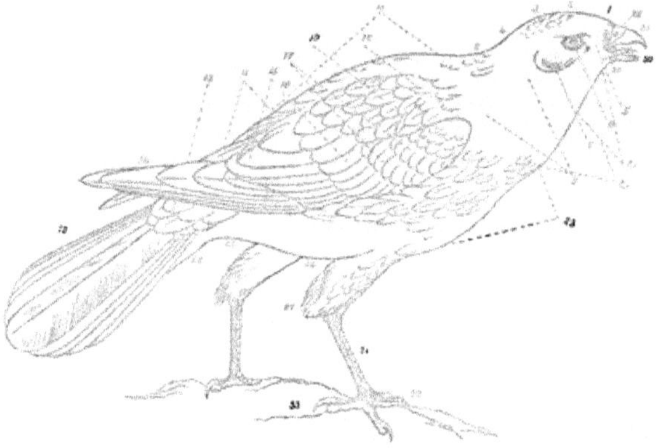

Outline of New-Zealand Harrier (*Circus gouldi*).

CLASSIFICATION.

In order to show at a glance the scheme I have adopted for the systematic arrangement of the existing Avifauna of New Zealand, I shall give here a Synopsis of the classification, with the superficial characters of each genus as at present defined.

As the characters of the genera are given in their entirety, I have thought it unnecessary to overload this section by adding the characters of the Orders and Families, which may be obtained from any text-book *.

I do not underrate the importance of the internal organs for determining generic distinctions. "But" (as Dr. Günther says, in his Preface to vol. vii. of the British Museum Catalogue of Birds) "it seems to me that investigations in the latter direction must lead to more numerous subdivisions than Ornithologists are inclined to admit at present."

Of every endemic group, except the Owls (which do not differ widely from the genus [illegible]), I have given a woodcut in illustration of one or more of the characters.

* For the generic characters I have, for the most part, relied on Mr. G. R. Gray's 'Genera of Birds,' in which work, although somewhat out of date, the definitions, taken as a whole, are marvellously correct. Many of the genera have since been split up by other ornithologists, but the broad lines remain; and while following these I have not hesitated to introduce such alterations and modifications as I deemed necessary.

Subclass CARINATÆ.

Order PASSERES.

Family **CORVIDÆ.** Crows.

Genus GLAUCOPIS, *Gmelin.* Endemic.

Generic characters.—*Bill* short, strong, with the culmen elevated at the base, and suddenly curved from the base to the tip, which is entire; the sides compressed, and the gonys lengthened and slightly arched; the nostrils basal, lateral, pierced in a membranous channel, and the opening partly concealed by the frontal plumes. *Wings* short and rounded, with the sixth and seventh quills equal and longest. *Tail* moderately long and rounded, with the shaft of each feather ending in a bristly point. *Tarsi* long, longer than the middle toe, and strongly scutellated in front, with one lengthened scale. *Toes* moderate, the lateral ones unequal and free at their base, the outer toe the longest; the hind toe very long and strong, and all armed with strong curved claws.

Family **STURNIDÆ.** Starlings.

Genus HETERALOCHA, *Cabanis.* Endemic.

Gen. char.—*Bill* long, arched, and acutely pointed, with the culmen, lateral margins, and gonys curved to the tips; much produced in the female, forming a sexual character; the sides compressed; the nostrils basal, lateral, and placed in a short, broad, membranous groove, which is mostly covered by the projecting plumes, leaving the opening small and exposed. *Wings* long and rounded, with the fifth, sixth, and seventh quills nearly equal and longest. *Tail* rather long, broad, and somewhat rounded. *Tarsi* much longer than the middle toe robust, and curved in front, with slightly divided broad scales. *Toes* long and robust, with the inner toe shorter than the outer and free at the base; the outer united at its base; the hind toe two thirds the length of the tarsus, and armed with a very long, strong, curved, acute claw; those of the fore toes long, curved, and acute.

Genus CREADION, *Vieillot.* Endemic.

Gen. char.—*Bill* longer than the head and rather straight, with the culmen flattened and sloping and the sides compressed to the tip, which is depressed and obtuse; the lateral margins straight, and angulated near the base; the gonys long and ascending; the nostrils lateral, and placed in a membranous groove, which is mostly clothed with short feathers, with the opening suboval. *Wings* short and rounded, with the first quill short, and the fourth, fifth, and sixth equal and longest. *Tail* long and rounded. *Tarsi* nearly as long as the middle toe, and covered in front with almost entire scales. *Toes* long; the lateral toe unequal, with the outer united at the base; the hind toe long and strong; the claws long, curved, and very acute.

Family **TURNAGRIDÆ.** Thick-billed Thrushes.

Genus TURNAGRA, *Lesson.* Endemic.

Gen. char.—*Bill* short, broad, and elevated at the base, with the culmen curved and the sides compressed to the tip, which is emarginated; the lateral margins much curved, and the gonys long and ascending; the nostrils basal, with the opening anterior, rather rounded, and slightly covered with a few bristles and plumes. *Wings* moderate and rounded, with the fifth and sixth quills equal and longest. *Tail* long, broad and rounded. *Tarsi* longer than the middle toe, strong, and covered in front with broad scales. *Toes* long and strong, with the outer toe longer than the inner, and united at its base; the hind toe long, strong, and armed with a strong curved claw.

Family **SYLVIIDÆ.** Warblers.

Genus MIRO, *Lesson.* Endemic.

Gen. char.—*Bill* two thirds the length of the head, slender, straight, higher than broad, sides compressed, the culmen slightly curved, the gonys long and ascending; nostrils basal, the opening rather large and suboval. *Wings* moderate, extending to half the length of the tail, rounded and concave, with the first quill very short, the third nearly as long as the fourth, which is the longest, the fifth and sixth scarcely shorter. *Tail* moderate, rather broad and even, the feathers cut sharply off at their tips. *Tarsi* very long and slender.

Genus MYIOMOIRA, *Reichenbach*. Endemic.

Gen. char.—The same as in *Miro*, except that the *bill* is shorter, being only one third the length of the head, narrow and sharp-pointed, the *wings* longer, extending for two thirds the length of the tail; and the claw of hind toe weaker.

Genus GERYGONE, *Gould*. New Zealand, Australia, New Guinea, and Indo-Malayan Islands.

Gen. char.—*Bill* moderate, slender and straight, with the culmen slightly curved, and the sides compressed to the tip, which projects beyond the lower mandible; the gonys long and ascending; the nostrils basal and in a membranous groove, with the opening linear. *Wings* rather short and rounded, with the first quill very short, and the third nearly as long as the fourth, which is the longest; fifth and sixth scarcely shorter. *Tail* long and rather rounded. *Tarsi* twice the length of the middle toe, slender, and covered in front with an entire scale. *Toes* moderate, with the inner toe shorter than the outer, which is united at its base; the hind toe long, and armed with a moderately strong, curved claw.

Family **PARIDÆ**. Tits.

Genus CERTHIPARUS, *Lafresnaye*. Endemic.

Gen. char.—*Bill* moderate, with the culmen curved and the sides compressed to the tip, which is entire, and the gonys long and slightly ascending; the nostrils lateral, placed in a groove, with the opening lunate, and partly concealed by the projecting frontal plumes. *Wings* moderate and rounded, with the fifth quill the longest. *Tail* long and rounded. *Tarsi* much longer than the middle toe, and broadly scutellated in front. *Toes* long, with the lateral ones equal; the hind toe long and strong, the claws moderate, slightly curved and acute.

Family **TIMELIIDÆ**. Grass Warblers.

Genus CLITONYX, *Reichenbach*. Endemic.

Gen. char.—*Bill* half as long as the head, robust, with the culmen curved and the gonys ascending; the tip of the upper mandible projecting over the lower; the nostrils basal, with a large suboval opening. *Wings* rather long, reaching to the middle of the tail, much rounded, with the fifth and sixth quills equal and longest. *Tail* rather long, broad, and rounded, the feathers slightly incurved, and the shafts more or less denuded at their tips. *Tarsi* much longer than the middle toe, and protected anteriorly by broad scales. *Toes* strong, and armed with well-curved, acute claws, that of the hind toe specially so.

Genus SPHENŒACUS, *Strickland*. New Zealand and Australia.

Gen. char.—*Bill* short, and more or less strong, with the culmen more or less curved, and the sides compressed to the tip, which is entire or slightly emarginated; the gonys long and ascending; the gape furnished with very short weak bristles; the nostrils basal, placed in a membranous groove, with the opening lunate, exposed and partly closed by a scale. *Wings* short and rounded, with the fourth and fifth quills equal and longest. *Tail* long, graduated on the sides, with more or less filamentous webs. *Tarsi* rather longer than the middle toe, strong, and covered in front with broad scutellations. *Toes* lengthened and slender, with the lateral toes nearly equal, the outer united at its base; the hind toe long, and armed with a long claw.

Family **MOTACILLIDÆ**. Pipits.

Genus ANTHUS, *Bechstein*. Cosmopolite.

Gen. char.—*Bill* more or less straight and slender, with the culmen almost straight or slightly curved, and the sides compressed to the tip, which is emarginated; the lateral margins straight and inflected; the gonys long and ascending; the nostrils lateral, placed in a short broad groove, with the opening rounded and partly closed by a membrane. *Wings* moderate, with the first three quills equal and longest. *Tail* moderate and emarginated. *Tarsi* longer than the middle toe, rather slender, and covered in front with broad transverse scales. *Toes* long and rather slender; with the lateral toes equal, and the outer one slightly united at its base; the hind toe long; the claws of the anterior toes rather short and curved, and that of the hind toe very long and acute.

Family **CAMPEPHAGIDÆ**. Caterpillar-eaters.

Genus GRAUCALUS, *Cuvier*. Africa, Oriental Region, and Australia.

Gen. char.—*Bill* short, and broad at the base, with the culmen rather depressed, slightly curved, and the sides gradually compressed to the tip, which is emarginated; the gonys long and slightly ascending; the gape furnished with a few short bristles; the nostrils basal, lateral, rounded, and concealed by the frontal plumes. *Wings* moderate, with the first quill short, the second shorter than the third, and the third more or less shorter than the fourth, which is the longest. *Tail* long, broad, and rounded on the sides. *Tarsi* short, the length of the middle toe, and covered in front with broad scales. *Toes* moderate, the inner toe shorter than the outer, which is united at its base; the hind toe moderate and broad, padded beneath; the claws moderate, compressed, and curved.

Family **MUSCICAPIDÆ**. Flycatchers.

Genus RHIPIDURA, *Vig. & Horsf.* New Zealand, Australia, New Guinea, India, and Indo-Malayan Islands.

Gen. char.—*Bill* moderate, broad at the base, and narrowing towards the end, with the culmen rather depressed and curved to the tip, which is emarginated; the lateral margin straight; the gonys long and slightly ascending, and the gape furnished with numerous lengthened bristles; the nostrils basal, lateral, and partly covered by the plumes and bristles. *Wings* long and rather pointed, with the first quill short and the fourth and fifth the longest. *Tail* lengthened, broad, and graduated. *Tarsi* longer than the middle toe, and covered in front with broad scales. *Toes* short, with the outer one longer than the inner, the hind toe long, and the claws moderate, curved, compressed, and acute.

Family HIRUNDINIDÆ. Swallows.

Genus PETROCHELIDON, *Cabanis*. Peculiar to Old World.

Gen. char.—*Bill* short, strong, broad at the gape, gradually compressed on the sides; the culmen elevated at the base, and slightly curved to the tip; the nostrils basal, rounded, and exposed, without a superior membrane, the aperture longitudinal or oval. *Wings* long, with the first quill the longest. *Tail* square or only slightly emarginate. *Tarsi* longer than the middle toe and clothed with plumes. *Toes* long, not feathered, the lateral ones unequal; the claws moderate and curved.

Family MELIPHAGIDÆ. Honey-eaters.

Genus ZOSTEROPS, *Vig. & Horsf.* New Zealand, Australia, South Africa, India, and the Malay Archipelago.

Gen. char.—*Bill* moderate, and slightly curved, with the culmen curved, and the sides compressed to the tip, which is acute and emarginated; the gonys long and slightly ascending; the gape furnished with very short weak bristles; the nostrils basal, and placed in a broad groove, with the opening closed by a lunate scale. *Wings* moderate, with the first quill very small, and the fourth and fifth equal and longest. *Tail* moderate, broad and slightly emarginated in the middle. *Tarsi* rather longer than the middle toe, and covered in front with broad scales. *Toes* rather long; with the outer toe rather longer than the inner and united at its base; the hind toe long, strong, and armed with a long curved claw.

Genus PROSTHEMADERA, *Gray*. Endemic.

Gen. char.—*Bill* long, rather slender, broad and elevated at the base, with the culmen and lateral margins curved and the sides compressed to the tip, which is slightly emarginated and acute; the gonys long and curved; the nostrils basal, large, in a broad membranous groove, and the opening covered by a prominent membranous scale. *Wings* moderate, with the fifth and sixth quills equal and longest, and the third, fourth, fifth, and sixth more or less emarginated in the middle of the inner webs. *Tail* long, broad, and rounded on the sides. *Tarsi* as long as or longer than the middle toe, and covered in front with transverse scales. *Toes* moderate, with the inner toe shorter than the outer, which is united at its base; the claws long, slender, curved, and very acute.

Genus ANTHORNIS, *Gray*. Endemic.

Gen. char.—This genus differs from the preceding one in the form of the *wings*, which are moderate, with the first quill short and pointed; the second shorter than the third, acutely pointed in the male, and emarginated and narrowing into a long point in the female; the third rather shorter than the fourth, fifth, and sixth, which are equal, longest, and rounded at the ends.

Genus POGONORNIS, *Gray*. Endemic.

GEN. CHAR.—*Bill* moderate, very slender, and much compressed on the sides, with the culmen and lateral margins gradually curved to the tip, which is strongly emarginated; the gonys long and curved; and the gape furnished with lengthened slender bristles; the nostrils basal, large, and placed in a large groove, with the opening linear, oblique, and covered by a membranous scale. *Wings* moderate, with the fourth quill the longest. *Tail* moderate and emarginated. *Tarsi* long and robust, and covered in front with transverse scales. *Toes* long, with the outer longer than the inner, and united at its base; the hind toe long and strong; the claws long, compressed, and acute.

Genus ACANTHOCHÆRA, *Vig. & Horsf.* Peculiar to Australia.

GEN. CHAR.—*Bill* long, rather slender, broad, and elevated at the base, with the culmen and lateral margins curved, and the sides compressed to the tip, which is slightly emarginated and acute; the gonys long and curved; the nostrils basal, large, in a broad membranous groove, and the opening covered by a prominent membranous scale. *Wings* moderate and rounded, with the first four quills graduated, and the fifth and sixth equal and longest. *Tail* long, broad, and graduated on the sides. *Tarsi* as long as, or longer than, the middle toe, and covered in front with transverse scales. *Toes* moderate, with the inner toe shorter than the outer, which is united at its base; claws long, slender, curved, and very acute.

Family **XENICIDÆ**. Dwarf Pittas.

Genus XENICUS, *Gray*. Endemic.

GEN. CHAR.—*Bill* moderate, more or less straight, moderately narrow at the base, and compressed to tip; culmen slightly curved at the apex; margin straight; gonys angulated one third of its length, and advancing towards the tip, and straight to the base; nostrils sunk in a short broad groove, with the opening large, oval, and partly closed by a membrane. *Wings* short, rounded, with the third, fourth, and fifth quills nearly equal and longest. *Tarsi* lengthened, slender, longer than the middle toe, covered by an entire scale. *Toes* rather long, slender; inner toe free at the base, the outer one connected nearly to the first joint of the middle toe; claws long, curved, and very acute.

Genus ACANTHIDOSITTA, *Lafresnaye*. Endemic.

Gen. char.—*Bill* long, straight, and very slender, with the culmen straight and slightly curved at the tip, the sides compressed, and the gonys long and gradually advancing upwards; the nostrils basal, lateral, and placed in a deep, broad groove, with the opening linear and near the culmen. *Wings* moderate, with the third and fourth quills the longest, the first shorter than the second, which is shorter than the third and fourth. *Tail* short and rounded. *Tarsi* shorter than the middle toe, and covered in front with an almost entire scale. *Toes* long and very slender, the lateral toes unequal, the outer longest and united at its base, the hind toe nearly as long as the middle one; the claws long, compressed, and curved.

Order PICARIÆ.

Family CYPSELIDÆ. Swifts.

Genus CYPSELUS, *Illiger*. Warmer parts of the World.

Gen. char.—*Bill* short and depressed, with the gape very wide, and the sides gradually compressed to the tip, which is curved; the nostrils basal, lateral, and large, with the opening longitudinal, on each side of the culmen, and the margins beset with small feathers. *Wings* lengthened, with the second quill longest. *Tail* moderate, forked or uneven. *Tarsi* very short, and feathered to the base of the toes. *Toes* all directed forwards, short, thick, and armed with short, curved, and compressed claws.

Family CORACIIDÆ. Rollers.

Genus EURYSTOMUS, *Vieillot*. The warmer parts of the Old World, Australia, New Guinea, and the Malay Archipelago.

Gen. char.—*Bill* strong, depressed and broad at the base, sides much compressed towards the tip, which is hooked; nostrils basal, oblique, partly covered by a plumed membrane. *Wings* long and pointed, reaching to end of tail; second quill the longest. *Tail* moderate and even. *Tarsi* shorter than middle toe, and covered with transverse scales. *Toes* long, united at the base; hind toe long; claws moderate, curved, and acute.

Family ALCEDINIDÆ. Kingfishers.

Genus HALCYON, *Swainson*. Africa, India and its Archipelago, Australia, New Zealand, and the islands of the South Pacific.

Gen. char.—*Bill* long, broad at the base, sometimes depressed, with the sides gradually compressed, and the culmen more or less straight to the tip, which is acute; the lateral margins usually straight, and the gonys more or less straight and ascending; the nostrils basal and lateral, placed in a small membranous space, with the opening small, longitudinal, and partly concealed by the projecting plumes. *Wings* moderate, with the first quill long, and the third the longest. *Tail* moderate, and rounded on the sides. *Tarsi* very short, rather slender, and covered in front with transverse scales. *Toes* moderate and unequal, with the outer toe long and united to the third joint, and the inner to the second joint, of the middle toe; the claws moderate, compressed and acute.

Family **CUCULIDÆ.** Cuckoos.

Genus EUDYNAMIS, *Vig. & Horsf.* Oriental Region, Australia, New Zealand, and Polynesia.

Gen. char.—*Bill* long, broad, with the culmen curved, and the sides compressed to the tip, which is slightly emarginated; the gonys short and angulated; the nostrils basal, lateral, and placed in a short membranous groove, with the opening large and exposed. *Wings* moderate, with the fourth and fifth quills equal and longest. *Tail* lengthened and rounded. *Tarsi* rather short, robust, and covered in front with broad scales. *Toes* unequal, the outer anterior toe the longest.

Genus CHRYSOCOCCYX, *Boie.* Warmer portions of the Old World.

Gen. char.—*Bill* broad, and rather depressed at the base, with the culmen curved, and the sides gradually compressed towards the tip, which is entire and acute; the gonys long and arched; the nostrils basal, lateral, and placed in a short, broad, membranous groove, with the opening round and exposed. *Wings* lengthened and pointed, with the third quill the longest. *Tail* long and graduated, or even, and the outer feathers on each side shorter than the others. *Tarsi* very short, feathered below the knee, and the exposed part covered with broad scales. *Toes* unequal; the outer anterior toe the longest, and united to the inner one at the base.

Order PSITTACI.

Family **PLATYCERCIDÆ.** Parrakeets.

Genus PLATYCERCUS, *Vigors.* New Zealand, Polynesia, Australia, and New Guinea.

Gen. char.—*Bill* moderate, with the sides swollen, and the culmen rounded, and arched to the tip, which is sometimes obtuse; the lateral margins curved and slightly dentated, or entire; the gonys broad, rather biangular on the sides, and curved upwards; the nostrils basal, lateral, exposed, and rounded, and placed in a small rounded cere near the culmen. *Wings* moderate and concave, with the first quill shorter than the second and third, which are nearly equal and longest, and the webs of the first four quills suddenly dilated near the base. *Tail* lengthened, broad, and nearly even, or much graduated, with the feathers towards the tip more or less narrowed and rounded or pointed. *Tarsi* shorter than the middle toe, and covered with minute scales. *Toes* moderate, much padded beneath, the outer anterior one the longest; and the claws long, compressed, curved, and acute.

Family **NESTORIDÆ.** Nestors.

Genus NESTOR, *Wagler.* Endemic.

Gen. char.—*Bill* much lengthened, the sides compressed, especially near the culmen, which is rounded and much arched to the tip, which is long and acute; the base of the lower mandible partly hidden by the projecting feathers and the sides rather compressed, with the gonys nearly flat and ascending towards the tip; the nostrils moderate, rounded, and placed in the cere. *Wings* long and pointed, with the third and fourth quills the longest. *Tail* moderate, and nearly even at the end, with the feathers firm and broad, and the shafts prolonged beyond the web. *Tarsi* as long as the inner anterior toe and covered with small scales. *Toes* moderate, the two outer ones the longest, and all covered with small irregular scales.

Family **STRINGOPIDÆ.** Owl Parrots.

Genus STRINGOPS, *Gray*. Endemic.

Gen. char.—*Bill* higher than broad, slightly compressed, and grooved on the sides; the culmen much curved to the tip, which is acute; the lateral margins dentated in the middle; the lower mandible with the gonys broad, rounded, and much grooved longitudinally, and the base of both mandibles covered by the basal feathers, with the shaft of each prolonged into hairs; the nostrils basal, lateral, large, and rounded. *Wings* rather short and rounded, with the fifth and sixth quills equal and longest. *Tail* moderate, weak, and much rounded, with the end of each feather rather pointed, and the shafts projecting beyond the web. *Tarsi* short, robust, and covered with rounded scales. *Toes* unequal, and covered with quadrate scales, except at the end of each toe, where the scales are transverse; the claws long, strong, and slightly curved.

Order STRIGES.

Family **STRIGIDÆ.** Owls.

Genus SPILOGLAUX, *Kaup*. The Indian Peninsula, Ceylon, China, Japan, the Malay Archipelago, Australia, New Zealand, and Madagascar.

Gen. char.—*Bill* short, partly concealed by the projecting plumes, the sides compressed, the culmen much arched to the tip, which is hooked and acute; the nostrils basal, lateral, and hidden by the frontal plumes. *Wings* rather long and pointed, with the first quill much shortened, the third and fourth quills equal and longest. *Tail* rather long and nearly even. *Tarsi* longer than the middle toe, and covered with plumes. *Toes* short, and covered with scattered hairs; the claws long, arched, and acute.

Genus SCELOGLAUX, *Kaup*. Endemic.

Gen. char.—Similar to *Spiloglaux*, but distinguished by its more developed tarsi, which are twice the length of the middle toe, and thickly feathered in their whole extent.

Order ACCIPITRES.

Family **FALCONIDÆ.** Hawks.

Genus CIRCUS, *Lacépède*. Most parts of the World.

Gen. char.—*Bill* moderate, elevated at the base of the culmen and arched to the tip, which is hooked, the sides compressed, and the lateral margins festooned; the nostrils large, oval, and partly concealed by the curved hairs of the lores. *Wings* long, with the third and fourth quills nearly equal and longest. *Tail* long and rounded on the sides. *Tarsi* long, slender, and compressed, the outer side covered with transverse scales, and the inner with small scales. *Toes* moderate, with the outer one longer than the inner; the claws long, slender, and acute.

Genus HARPA, *Bonaparte*. Endemic.

Gen. char.—*Bill* short, strong, with the culmen much arched from the base to the tip, which is acute; the sides compressed, the lateral margins strongly toothed near the tip; the nostrils placed in a short cere, naked, and rounded, with a central tubercle. *Wings* moderate, with the second and third quills nearly equal and longest. *Tarsi* lengthened, rather slender, and covered in front with rounded scales. *Toes* long, especially the middle toe, which is more than twice the length of the culmen, the lateral ones equal, the hind toe rather long; the claws moderately robust.

Order GALLINÆ.

Family **TETRAONIDÆ.** Quails.

Genus COTURNIX, *Mœhring*. Warmer and temperate parts of Old World, Australia, and New Zealand.

Gen. char.—*Bill* short, more or less elevated at the base and arched to the tip, which is obtuse; the sides compressed; the nostrils basal, lateral, and covered by a hard scale. *Wings* moderate, with the second, third, and fourth quills the longest. *Tail* very short, mostly hidden by the coverts, and pendent. *Tarsi* short, covered in front with divided scales, and unarmed. *Toes* moderate, united at their base, with the inner toe shorter than the outer; the hind toe short; the claws short, and slightly curved.

Order COLUMBÆ.

Family **COLUMBIDÆ.** Pigeons.

Genus CARPOPHAGA, *Selby*. India, the Malay Archipelago, Australia, New Zealand, and Polynesia.

Gen. char.—*Bill* moderate, slender, with the base depressed, the tip compressed and moderately arched, and the margin slightly sinuated; the nostrils placed in the soft basal portion of the bill, and forming a longitudinal slit. *Wings* moderate and pointed; with the second, third, and fourth quills nearly equal and longest. *Tail* lengthened, and generally rounded. *Tarsi* very short, and clothed with down below the knee. *Toes* strong, and broadly padded below; with the outer toe longer than the inner, and the hind toe much developed.

Order LIMICOLÆ.

Family **CHARADRIIDÆ.** Plovers.

Genus CHARADRIUS, *Linn.* Cosmopolite.

Gen. char.—*Bill* more or less short, robust, and straight; the culmen, for two thirds its length, usually depressed, and the tip vaulted and curved; the sides compressed, and furnished on both mandibles with a groove, which extends on the upper mandible for two thirds of its length; the nostrils basal, linear, and placed in a groove. *Wings* long and pointed, with the first quill the longest. *Tail* moderate, broad, and rounded. *Tarsi* longer than the middle toe, more or less slender, and covered in front with small reticulated scales. *Toes* three, moderate; the outer toe longer than the inner, and more or less united at the base by a membrane, the inner toe usually free; the hind toe wanting; the claws small, compressed, and slightly curved.

Genus THINORNIS, *Gray*. Endemic.

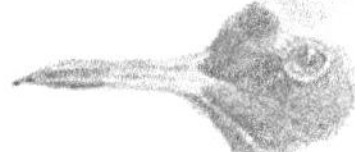

Gen. char.—*Bill* long, straight, and slender, with the apex scarcely vaulted and acute, the sides compressed, and both mandibles grooved; the nostrils lateral, placed in a groove that extends for two thirds the length of the bill, and the opening linear. *Wings* long and pointed, with the first and second quills nearly equal and longest. *Tail* long and rounded. *Tarsi* as long as, or shorter than, the middle toe, strong and covered with small scales. *Toes* three, more or less long and robust; with the outer toe rather longer than the inner, and united at the base by a membrane, and all margined on the sides; the hind toe wanting.

Genus ANARHYNCHUS, *Quoy & Gaim*. Endemic.

Gen. char.—Same as *Thinornis*, but with the bill asymmetrical, being always turned to the right.

Genus LOBIVANELLUS, *Strickland*. Australia.

Gen. char.—*Bill* moderate, and more or less strong, with the culmen depressed at the base and vaulted at the tip, the sides compressed and grooved; the nostrils lateral, basal, and placed in the groove of the upper mandible, which extends for two thirds its length, with the opening linear; the front and sides of the head lobed. *Wings* long and pointed; with the first, second, and third quills nearly equal and longest, armed at the flexure with a sharp spur. *Tail* moderate, broad, and even. *Tarsi* much longer than the middle toe, slender, and covered in front with divided broad scales. *Toes* four; the three anterior toes long and rather slender; the outer toe longer than the inner, and united at the base; the hind toe short and elevated.

Genus STREPSILAS, *Illiger*. Cosmopolite.

Gen. char.—*Bill* rather shorter than the head, straight and slightly depressed at the base, with the culmen straight, and the sides much compressed to the tip, which is truncated; the lateral margins of both mandibles curved upwards at the tip; the gonys moderate and ascending; the nostrils lateral, and placed in a membranous groove that extends half the length of the upper mandible, with the opening linear and longitudinal. *Wings* very long and pointed, with the first quill the longest. *Tail* moderate, and slightly rounded. *Tarsi* as long as the middle toe, robust, and covered in front with broad scales. *Toes* long, the outer toe rather longer than the inner, and both free at the base, and the sides of all margined by a narrow membrane; the hind toe elevated, with the tip resting on the ground.

Genus HÆMATOPUS, *Linnæus*. Most parts of the world.

GEN. CHAR.—*Bill* longer than the head, strong, straight, with the culmen slightly depressed at the base, and the apical portion much compressed to the tip, which is obtuse; the nostrils placed in a lateral membranous groove, which reaches nearly to the middle of the bill, with the opening linear. *Wings* long, with the first quill the longest. *Tail* moderate and even, or slightly rounded. *Tarsi* strong, longer than the middle toe, and covered with small reticulated scales. *Toes* moderate, strong; the lateral toes united to the middle toe by a basal membrane, especially the outer one; the claws strong, broad, and slightly curved.

Family SCOLOPACIDÆ. SNIPES.

Genus RECURVIROSTRA, *Linn.* Most parts of the world.

GEN. CHAR.—*Bill* very long and slender, with the culmen slightly depressed at the base, the sides grooved to the middle and compressed to the tip, which is gradually pointed; the nostrils lateral, and placed in the groove, with the opening linear and membranous. *Wings* long and pointed, with the first quill the longest. *Tail* short and rounded. *Tarsi* much longer than the middle toe, rather compressed, and covered in front with reticulated scales. *Toes* united together by an indented web; the outer toe rather longer than the inner; the hind toe extremely short; the claws short, compressed, and acute.

Genus HIMANTOPUS, *Brisson*. Most parts of the world.

GEN. CHAR.—*Bill* much longer than the head, very slender and straight, with the sides grooved to the middle and compressed towards the tip, which is acute; the nostrils basal, and placed in the groove, with the opening long, linear, and closed by a membrane. *Wings* long and pointed, with the first quill the longest. *Tail* short and nearly even. *Tarsi* very long, slender, and covered in front with reticulated scales. *Toes* moderate, and united at the base by a small membrane, especially the outer toe; the hind toe wanting; the claws small, compressed, and acute.

Genus PHALAROPUS, *Brisson*. Inhabits the northern regions of the globe, migrating to the more temperate climes during severe winters.

GEN. CHAR.—*Bill* as long as, or longer than, the head, more or less slender, but sometimes enlarged and depressed towards the tip, which is curved and acute; the sides grooved for nearly its whole length, in which groove the nostrils are placed, with the opening basal, linear, and partly closed by a membrane. *Wings* long and pointed, with the first and second quills equal and longest. *Tail* more or less short or rounded. *Tarsi* as long as, or longer than, the middle toe, rather robust and compressed. *Toes* long; the lateral toes united to the middle by a membrane that runs along the margin of each toe, which is more or less lobed; the hind toe moderate, elevated, and slightly margined by a membrane; the claws short and acute.

Genus GALLINAGO, *Leach*. Cosmopolite.

GEN. CHAR.—*Bill* long, straight, grooved, and compressed on the sides, and the culmen rather depressed near the tip, which is obtuse, and curved over that of the lower mandible; the nostrils basal, placed in the groove, with the opening oblong and exposed. *Wings* moderate and pointed, with the first and second quills equal and longest. *Tail* short and rounded. *Tarsi* moderate, shorter than the middle toe, strong, and covered in front with narrow transverse scales; the tibia bare for a short space above the tarsal joint. *Toes* long, the inner toe shorter than the outer, and free at their base; the hind toe moderate and elevated, with the claw long and curved.

Genus TRINGA, *Linn.* All the more genial parts of the world.

GEN. CHAR.—*Bill* as long as, or longer than, the head, straight, slender, with the sides compressed at the base, and rather dilated and depressed at the tip; the nostrils placed in a nasal groove, which extends to near the tip, basal, lateral, and longitudinal. *Wings* moderate and pointed, with the first quill the longest. *Tail* rather short and nearly even. *Tarsi* strong, rather long, and covered in front with transverse scales. *Toes* moderate, slightly united at the base of the outer toe, and all margined on the sides by a membrane; the hind toe very small and elevated.

Genus TOTANUS, *Bechstein*. Both Hemispheres, and especially in the temperate and northern portions.

GEN. CHAR.—*Bill* more or less long and strong, with the culmen straight or slightly curved, and the sides compressed to the tip, which is slightly curved and acute; the gonys long and slightly curved upwards; the nostrils linear, and placed in a membranous groove, which does not extend beyond half the length of the bill. *Wings* reaching beyond the end of the tail and pointed, with the first quill the longest. *Tail* moderate and nearly even. *Tarsi* as long as, or longer than, the middle toe, more or less slender, and covered in front with numerous very narrow scales. *Toes* long, slender, the anterior toes united by a membrane, especially the outer; the hind toe slender, elevated, and hardly touching the ground.

Genus LIMOSA, *Brisson*. Cosmopolite.

GEN. CHAR.—*Bill* long, rather slender, and more or less inclined upwards towards the tip, with the sides compressed and grooved on both mandibles for nearly their entire length; the nostrils lateral, basal, and placed in the groove, with the opening longitudinal and closed by a membrane. *Wings* long and pointed, with the first quill the longest. *Tail* short and even. *Tarsi* longer than the middle toe, rather slender, and covered in front with narrow transverse scales. *Toes* long; the outer toe united to the middle toe by a membrane as far as the first joint; the inner toe slightly united; the hind toe long, slender, and partly resting on the ground; the claws short and obtuse.

Genus NUMENIUS, *Latham*. Cosmopolite.

GEN. CHAR.—*Bill* more or less long, slender, and curved from the base, with the sides compressed and grooved for nearly its whole length; the tip of the upper mandible projecting over that of the lower, and rather obtuse; the nostrils basal, lateral, and placed in the lateral groove, with the opening longitudinal, and covered by a membrane. *Wings* long and pointed, with the first quill the longest. *Tail* short and even. *Tarsi* longer than the middle toe, slender, and covered in front with narrow transverse scales. *Toes* moderate, the lateral ones unequal and united at their base; the hind toe long, slender, and partly resting on the ground; the claws short and obtuse.

Order GAVIÆ.

Family **LARIDÆ**. GULLS.

Genus LARUS, *Linn*. All parts of the world except Polynesia.

GEN. CHAR.—*Bill* more or less strong, as long as or shorter than the head, straight, and laterally compressed, with the culmen straight at the base and arched to the tip, the gonys slightly angulated and advancing upwards; the nostrils lateral, with the opening near the middle of the bill, and longitudinal. *Wings* lengthened and pointed, with the first quill the longest. *Tail* moderate and even. *Tarsi* nearly as long as the middle toe, strong, and covered in front with transverse scales. *Toes* moderate, the anterior ones united by a full web; the hind toe short and elevated.

Genus STERCORARIUS, *Brisson*. The colder regions of both Hemispheres.

GEN. CHAR.—*Bill* moderate, straight, and strong, with the culmen straight, rounded, and covered with a membranous or horny cere; the apex curved, vaulted, and strong; the gonys much angulated and ascending; the nostrils placed in the fore part of the cere, narrow, and enlarging anteriorly. *Wings* lengthened and pointed, with the first quill the longest. *Tail* moderate and rounded, with the two centre feathers sometimes lengthened. *Tarsi* longer than the middle toe, strong, and covered in front with strong scales. *Toes* moderate and strong, the anterior one united by a full web; the hind toe very small and hardly elevated.

k

Family **STERNIDÆ.** Terns.

Genus STERNA, *Linn.* Cosmopolite.

Gen. char.—*Bill* more or less long, strong, with the culmen slightly curved to the tip, which is acute; the gonys straight, and half the length of the bill; the nostrils lateral, placed towards the middle of the bill, and longitudinal, with the frontal plumes advancing close to, or near, the opening. *Wings* very long and pointed, with the first quill the longest. *Tail* more or less long and generally forked. *Tarsi* more or less long and slender. *Toes* moderate, the two outer ones nearly equal, and the three anterior ones united by an indented web; the hind toe very short; the claws moderate, slightly curved, and acute.

Genus HYDROCHELIDON, *Boie.* Southern Hemisphere.

Gen. char.—*Bill* strong, short, with the culmen rather arched to the tip, which is acute; the sides compressed, and the gonys long, straight, and advancing upwards to the tip; the nostrils basal, lateral, and longitudinal, with the frontal plumes projecting to the opening. *Wings* long, with the first quill the longest. *Tail* moderate and slightly emarginated. *Tarsi* rather shorter than the middle toe and slender. *Toes* long, slender, the two outer toes equal and longest, the three anterior toes united only at the base, the web continuing along the inner margin of each toe; the hind toe moderate and slender; the claws also long and slender.

Order GRALLÆ.

Family **RALLIDÆ.** Rails.

Genus RALLUS, *Linn.* Cosmopolite.

Gen. char.—*Bill* longer than the head, slender, and straight, with the culmen slightly curved from the front of the nostrils, and the sides compressed to the tip, which is obtuse and slightly emarginated; the gonys long and slightly curved upwards; the nostrils placed in a membranous groove, which extends for two thirds the length of the bill, with the opening exposed and linear. *Wings* short, with the second and third quills equal and longest. *Tail* short and rounded. *Tarsi* moderate, shorter than the middle toe, and covered with transverse scales. *Toes* long and rather slender, the inner toe shorter than the outer, both free at their base; the hind toe short and slender; the claws short, compressed, and very acute.

Genus ORTYGOMETRA, *Linnæus.* Australia, New Zealand, and Polynesia.

Gen. char.—*Bill* shorter than the head, and more or less strong, with the culmen keeled, slightly curved, and the sides compressed to the tip, which is slightly emarginated; the gonys short and ascending; the nostrils lateral and placed in a membranous groove, with the opening exposed, linear, and near the middle. *Wings* moderate, with the second and third quills equal and longest. *Tail* short and graduated. *Tarsi* rather robust. *Toes* more or less long and slender, with the inner toe rather shorter than the outer, the hind toe very slender, and rather short; the claws moderate, compressed, and acute.

Genus OCYDROMUS, *Wagler.* Endemic.

Gen. char.—*Bill* rather long, and very strong, with the culmen slightly curved and the sides much compressed to the tip, which is slightly emarginated; the gonys short and ascending; the nostrils lateral, and placed in the fore part of a membranous groove, with the opening oval and exposed. *Wings* very short and rounded, with the fifth and sixth quills equal and longest; the secondaries and the coverts lengthened and very soft. *Tail* more or less lengthened, round and soft. *Tarsi* robust, shorter than the middle toe and covered with transverse scales. *Toes* long and strong, with the inner toe rather shorter than the outer, the hind toe short and rather slender; the claws moderate and rather acute.

Genus PORPHYRIO, *Brisson*. Most parts of the World.

Gen. char.—*Bill* short, very much elevated at the base, which is flat and broadly dilated on the forehead; the culmen much arched to the tip; the sides much compressed; the nostrils placed in a small nasal groove and rounded. *Wings* moderate, with the second, third, and fourth quills nearly equal and longest. *Tail* short and rounded. *Tarsi* long, shorter than the middle toe, and scutellated with broad transverse scales. *Toes* very long, slender, and free at their base, with the lateral ones unequal, the outer longest; the claws long, slender, and somewhat curved.

Genus NOTORNIS, *Owen*. Endemic.

Gen. char.—*Bill* somewhat shorter than the head; greatly compressed on the sides, both mandibles being much deeper than broad; tomia sharp, curving downwards, inclining inwards and slightly serrated; culmen elevated, much arched and rising on the forehead to a line with the posterior angle of the eye; nostrils round and placed in a depression near the base of the bill. *Wings* very short, rounded, and slightly concave; primaries soft and yielding, the first short, the third, fourth, fifth, sixth, and seventh equal and the longest. *Tail*-feathers soft, yielding, and loose in texture. *Tarsi* powerful, longer than the toes, almost cylindrical; very broad anteriorly, defended in front and on either side posteriorly by broad and distinct scutellæ; the spaces between the scutellæ reticulated. Anterior toes large and strong, armed with powerful hooked nails, and strongly scutellated on their upper surface; hind toe short, strong, placed somewhat high on the tarsus, and armed with a blunt hooked nail.

Genus CABALUS, *Hutton*. Confined to the Chatham Islands.

Rallus philippensis. *Cabalus dieffenbachii.*

Gen. char.—*Bill* longer than the head, moderately slender and slightly curved, compressed in the middle and slightly expanding towards the tip; nostrils placed in a membranous groove, which extends beyond the middle of the bill; openings exposed, oval, near the middle of the groove. *Wings* very short, rounded; quills soft, the outer webs as soft as the inner, fourth and fifth the longest, first nearly as long as the second; a short compressed claw at the end of the thumb. *Tail* very short and soft, hidden by the coverts. *Tarsi* moderate, shorter than the middle toe, flattened in front, and covered with transverse scales. *Toes* long and slender, inner nearly as long as the outer; hind toe short, very slender, and placed on the inner side of the tarsus; claws short, compressed, blunt.

NOTE.—This genus was established by Professor Hutton for the reception of a small form of flightless Rail, which he had previously described under the name of *Rallus modestus*. In my former edition I treated the bird as the young of *Rallus dieffenbachii*, an extremely rare form of Rail from the Chatham Islands, which Mr. G. R. Gray had originally placed in the genus *Ocydromus*. It has been clearly shown that *Cabalus modestus* has Ocydromine characters in its skeleton, and, whether an adult bird or not, it is undoubtedly right to separate it generically from *Rallus*.

Mr. Sharpe, in treating of *Cabalus dieffenbachii* (App. Voy. Ereb. & Terr. p. 29), says:—"In his latest article on the 'Birds of New Zealand,' Dr. Finsch believes in *Rallus modestus* of Hutton being a distinct species from *R. dieffenbachii*. I examined the type of Captain Hutton's species, and thoroughly believe it to be the young of the latter Rail. Perhaps Captain Hutton is right in referring this Rail to a genus or subgenus intermediate between *Rallus* and *Ocydromus*, and I have therefore, for the present, adopted his genus *Cabalus*."

Order HERODIONES.

Family **ARDEIDÆ.** HERONS.

Genus ARDEA, *Linn.* Most parts of the World.

GEN. CHAR.—*Bill* lengthened and more or less slender, with the culmen nearly straight to the tip, which is acute and emarginated, the sides compressed, and the lateral margins straight and sometimes serrated; the gonys moderate and ascending; the nostrils lateral, basal, and placed in a groove, which extends for more than half the length of the bill, with the opening linear, and closed by a membranous scale. *Wings* long, with the first quill nearly as long as the second and third, which are equal and longest. *Tail* rather short and even. *Tarsi* longer than, or as long as, the middle toe, rather slender, and covered in front with transverse scales, those near the toes large and of a hexagonal form. *Toes* long and rather slender; the outer toe longer than the inner, and united at the base; the hind toe long; the claws moderate, slight, curved, and acute.

Genus NYCTICORAX, *Stephens.* Most parts of the World.

GEN. CHAR.—*Bill* rather longer than the head, strong, with the culmen gradually curved, and the sides compressed to the tip, which is emarginated; the gonys long and ascending; the nostrils lateral and placed in a groove, with the opening linear and closed by a membranous scale. *Wings* long, with the first quill shorter than the second and third, which are equal and longest. *Tail* short and even. *Tarsi* as long as the middle toe, rather strong, and covered with large irregular scales. *Toes* long, rather slender; the outer toe longer than the inner, both united at their base, especially the former; the hind toe long, rather slender, and on the same plane with the others; the claws moderate, curved, and acute.

Genus BOTAURUS, *Stephens.* All parts of the World.

GEN. CHAR.—*Bill* long and straight, with the culmen straight, flattened at the base, and rounded and curved to the tip, which is strongly emarginated, and the sides compressed; the gonys short and ascending; the nostrils basal, and placed in a deep groove that extends for two thirds of the length of the bill, with the opening linear. *Wings* long, with the three first quills equal and longest. *Tail* short and even. *Tarsi* as long as the middle toe, rather strong, and covered in front with broad transverse scales. *Toes* very long and rather slender; the claws very long, slightly curved, and very acute.

Family **PLATALEIDÆ**. Spoonbills.

Genus PLATALEA, *Linn.* Most parts of the World.

Gen. char.—*Bill* lengthened, straight, thin, much depressed, and broadly dilated at the tip, which is spatula-formed, with a lateral groove commencing on the forehead, extending, in a parallel line with the edge, to the tip, which is slightly bent downwards; the nostrils basal and placed in a groove, with the opening oval and partly closed by a membrane. *Wings* long, and the second quill the longest. *Tail* short. *Tarsi* longer than the middle toe, rather slender, and covered with reticulated scales. *Toes* long, with the anterior toes much united at their base by a membrane, which extends along the sides of the toes to the tip; the hind toe long, rather elevated, and only partly resting on the ground; the claws short, scarcely curved, and obtuse.

Order STEGANOPODES.

Family **PELECANIDÆ**. Pouched Birds.

Genus PHALACROCORAX, *Brisson.* Cosmopolite.

Gen. char.—*Bill* moderate, straight, somewhat slender, with the culmen concave and suddenly hooked at the tip; the sides compressed and grooved; the nostrils basal, linear, placed in the lateral groove, and scarcely visible. *Wings* moderate and pointed, with the second and third quills the longest. *Tail* moderate, and rounded at its end. *Tarsi* short, one third shorter than the middle toe, much compressed, and covered with reticulated scales. *Toes* long, with the outer toe rather longer than the middle one, and all four united by a full web. The base of the lower mandible is furnished with a coriaceous pouch, which is capable of extension.

Genus PLOTUS, *Linn.* Various parts of America, Asia, and Africa, Australia and New Guinea.

Gen. char.—*Bill* longer than the head, straight, and very slender, with the sides much compressed to the tip, which is very acute, the lateral margins finely serrated, and the gonys long and slightly ascending; the nostrils basal, linear, and scarcely visible. *Wings* long, with the second and third quills equal and longest. *Tail* long and broad towards the end, which is rounded. *Tarsi* half the length of the middle toe, strong, and covered with small scales. *Toes* rather long, all united by a broad web; the outer toe as long as the middle one; the claws short, curved, and acute.

Genus DYSPORUS, *Illiger.* Cosmopolite.

Gen. char.—*Bill* longer than the head, robust, straight, broad at the base; with the sides compressed, and grooved towards the tip, which is slightly curved, and the lateral margins obliquely and unequally serrated; the nostrils basal, lateral, linear, placed in a lateral groove and almost invisible. *Wings* long, pointed, and tuberculated, with the first two quills the longest. *Tail* moderate and graduated. *Tarsi* short, one third shorter than the outer toe, rounded anteriorly and keeled posteriorly. *Toes* lengthened, the outer and middle ones nearly equal, and all four connected by a full membrane; the claws moderate and rather flat, that of the middle toe serrated, and the hind claw rudimental. Beneath the base of the lower mandible is a naked space, reaching towards the breast, which is capable of expansion.

Genus TACHYPETES, *Vieillot.* Confined to the Tropics.

Gen. char.—*Bill* longer than the head, broad at the base, with the culmen depressed, concave, and suddenly hooked and acute; the sides compressed and grooved; the lateral margins dilated on the sides near the base; the nostrils basal, lateral, linear, placed in the lateral groove, and scarcely visible. *Wings* extremely long and narrow, with the first two quills the longest. *Tail* very long and strongly forked. *Tarsi* very short, one third shorter than the outer toe, much compressed, and half covered with feathers. *Toes* long, all united by a strongly indented web, the lateral ones unequal, the outer one the longest, and the hind toe half the length of the middle one; the claws moderate and curved. The throat naked, and capable of being dilated into an extending pouch from near the tip of the lower mandible downwards to the breast.

Family **PHAËTHONIDÆ**. Tropic-birds.

Genus PHAËTHON, *Linn.* Tropical seas.

Gen. char.—*Bill* as long as the head, broad, and dilated at the base; with the culmen elevated, curved, and the sides much compressed to the tip, which is entire and acute; the lateral margins more or less serrated; the nostrils basal and lateral, with the opening linear, partly closed by a membrane, and exposed. *Wings* long and pointed, with the first quill the longest. *Tail* moderate and graduated, with the two middle feathers lengthened and linear. *Tarsi* shorter than the middle toe, strong, and covered with small scales. *Toes* long; the outer toe longer than the inner; the three anterior ones and the hind toe all united together by a broad membrane; the claws small, compressed, and acute.

Order TUBINARES.

Family **PROCELLARIIDÆ***. Petrels.

Genus DIOMEDEA, *Linnæus.* Colder parts of both Hemispheres, but more especially in the Southern Ocean.

Gen. char.—*Bill* longer than the head, very robust, straight; the sides compressed and longitudinally grooved, with the tip greatly curved and acute; the lateral margins dilated and curved; the culmen broad, convex, and rounded; the lower mandible weak, compressed, with the tip truncated; the nostrils placed near the base, in the lateral groove, covered by a tube which is short, widening and spreading anteriorly from the side of the bill, with the aperture somewhat rounded and open in front. *Wings* very long, very narrow, with the second quill the longest. *Tail* short and rounded. *Legs* short, strong, with the tarsi one fourth shorter than the middle toe, and the inner toe the shortest. The two lateral toes margined exteriorly by a narrow membrane; the web between the toes full and entire; the hind toe and claw entirely wanting; the claws short and obtuse.

Genus PELECANOIDES, *Lacép.* Southern Hemisphere.

Gen. char.—*Bill* shorter than the head, broad at the base, and much depressed; the sides swollen, grooved, and gradually compressed towards the tip, which is lengthened, compressed, arched, and acute; the lower mandible broad at the base and suddenly compressed at the tip, which is, with the gonys, arched and acute; the sides longitudinally grooved and deep; beneath is placed a membranous pouch, capable of extension; the nostrils basal, one fourth the length of the bill, flattened above, and forming two lengthened, sublinear, exposed apertures, placed side by side on the surface. *Wings* very short, with the first two quills nearly equal and longest. *Tail* short and rounded. *Tarsi* rather shorter than the middle toe, laterally compressed, and covered with small scales. *Toes* long; the outer nearly as long as the middle toe; the hind toe and claw wanting.

Genus PRION, *Lacépède.* Southern Hemisphere: generally observed between 30° and 70° south latitude.

Gen. char.—*Bill* the length of the head, broad or very broad at the base, depressed above; culmen nearly straight, laterally swollen, but gradually compressed towards the tip, which is arched, elevated, compressed and acute; the lateral margins dilated near the base, with a series of very fine laminæ running along the whole length internally rather above the margin; the lower mandible broad at the base, gradually compressed towards the tip, which is

* Upon a closer study of the subject, I have decided on recognizing a larger number of groups in this Family than were admitted into my first edition, or into my 'Manual,' published in 1882. This will add to the total number of genera indicated at p. xxxvii; but as none of these are endemic it will not affect my general argument.

The late Mr. Forbes, in his excellent account of the Petrels, in the 'Voyage of the Challenger' (Zool. vol. iv. pp. 1-61), recognized a separate family under the name Oceanitidæ, embracing the four closely allied genera *Garrodia*, *Oceanites*, *Pelagodroma*, and *Fregetta*, which form together a very compact section. I prefer, however, to retain the whole of these natural groups under the general denomination of Procellariidæ, leaving the proposed divisions to take rank as subfamilies.

much compressed, with the margin and gonys arched; the nostrils basal, tubular, elevated above the culmen, short, opening with two apertures in front. *Wings* moderate, pointed, with the first quill nearly equalling the second, which is the longest. *Tail* moderate, broad, and rounded at the end. *Tarsi* shorter than the middle toe, laterally compressed, and covered with small scales. *Toes* long, the outer nearly as long as the middle, and the hind toe nearly in the form of a broad, short, pointed claw.

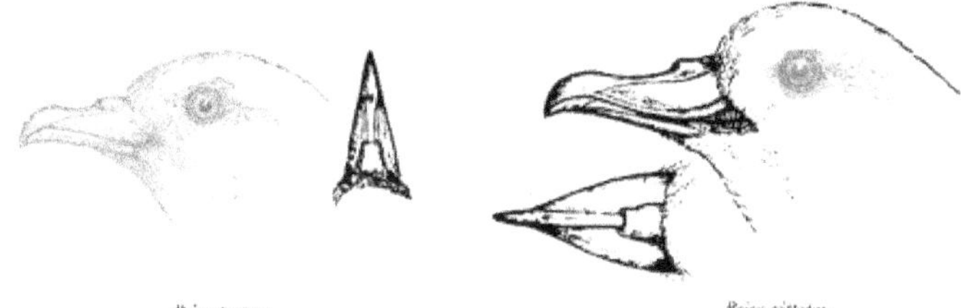

Prion turtur. *Prion vittatus.*

Genus HALOBÆNA, *Is. Geoff.* Southern Hemisphere.

Gen. char.—*Bill* nearly as long as the head, more or less broad at the base; the sides gradually compressed towards the tip, which is much elevated and arched, lengthened and acute; the upper mandible furnished near its edge with laminated serrations, but few and inconspicuous as compared with *Prion*; the lower mandible shorter than the upper, with the tip and gonys arched and acute; the nostrils basal, tubular, horny, elevated above the culmen, with the aperture double, frontal, and crescent-shaped. *Wings* long, pointed, with the first quill the longest, and the second scarcely shorter. *Tail* moderately long and truncated. *Legs* with the apical part of the thigh hardly naked. *Tarsi* shorter than the middle toe, laterally compressed, and covered with small scales. *Toes* long, with the outer as long as the middle toe, the inner shortest, and all united by a full web; the lateral toes margined exteriorly, the hind toe in the form of a large subtriangular claw.

Genus DAPTION, *Stephens.* Southern Hemisphere.

Gen. char.—*Bill* much dilated, unguis small and weak; inter-ramal space wide and partially naked; oblique sulci on inner face of cutting-edge of mandible; nasal tubes long. *Wings* long and pointed, with the second primary nearly as long as the first. *Tail* rather short, moderately rounded. *Tarsi* and toes as in *Œstrelata.*

Genus ŒSTRELATA, *Bonaparte.* Chiefly confined to Southern Hemisphere.

Gen. char.—*Bill* about as long as the tarsus, stout, compressed, higher than broad throughout, lateral outlines nearly straight, and converging to the unguis, which is much compressed; unguis very large and strong; outline of upper mandible very convex, rising almost immediately from the end of the nasal tubes, leaving but a very short and quite concave culmen proper; outline of lower mandible nearly straight, the gonys a little concave; sulci on both mandibles distinct. *Wings* rather long, extending beyond the tail when folded, and pointed; the second primary nearly as long as the first. *Tail*, which is composed of twelve feathers, long and much produced, sometimes almost cuneate, usually much rounded. *Tarsi* moderately compressed, and about as long as, or a little less than, middle toe; hallux short, sessile, conical, acute, and elevated.

Genus OSSIFRAGA, *Hombr. et Jacq.*

Gen. char.—*Bill* as long as, or rather exceeding, the tarsus, very robust; the nasal case very long, depressed, carinated, the aperture small. *Wings* of moderate length, reaching to end of tail. *Tail* moderately long and rounded. *Tarsi* short, being much less than the middle toe without its claw, compressed, stout, reticulated.

Genus THALASSŒCA, *Reich.* Southern Hemisphere.

GEN. CHAR.—*Bill* slightly shorter than the tarsus, higher than broad at the base, the commissure a little curved. *Wings* of moderate length, reaching to the end of tail. *Tail* short, more or less rounded, composed of fourteen feathers. *Tarsi* slender, compressed, reticulated, shorter than the middle toe; outer toe as long as the middle one; inner toe considerably shorter; hallux very short, being only observable as a stout, obtuse, subconical claw.

Genus PUFFINUS, *Brisson.* Both Hemispheres.

GEN. CHAR.—*Bill* as long as, or shorter than, the head, much compressed, and grooved obliquely on the sides; the tip lengthened, arched, suddenly hooked and acute; the lower mandible somewhat shorter than the upper, with the apical margin and gonys equally curved with the upper, the latter angulated beneath, and the sides longitudinally grooved; the nostrils basal, elevated above the culmen, opening obliquely in two tubes, placed side by side. *Wings* long, slender, somewhat acute, with the first quill the longest. *Tail* moderate and rounded, composed of twelve feathers. *Legs* moderate, with the apical part of the tibia naked. *Tarsi* compressed and equal in length to the middle toe. *Toes* long, the outer equal to the middle one, the inner shortest, and the lateral toes margined exteriorly by a narrow membrane.

Genus ADAMASTOR, *Bonaparte.* Southern Hemisphere.

GEN. CHAR.—*Bill* about three fourths the length of the tarsus, broad and stout at the base, narrowing regularly to the strong, very convex, compressed unguis; nasal tubes rather long, very broad, depressed, but vertically truncated at their extremity, and with an unusually thin septum. *Wings* rather short, the primaries broad and stout, the second as long as the first. *Tail* rather short and slightly cuneate. *Tarsi* shorter than the middle toe without its claw, outer toe larger than the middle.

Genus MAJAQUEUS, *Reich.* Southern Hemisphere.

GEN. CHAR.—*Bill* a little shorter than the head, about equal to the tarsus, stout, compressed, higher than broad at the base, the culmen rising immediately from the nostrils; unguis large, very convex, much hooked, commissure unusually curved; outline of lower mandible straight as far as the unguis; nasal tubes long, elevated laterally, obliquely flattened, carinated along the median line, apically truncated, with a considerable emargination; the nostrils circular. *Wings* comparatively long. *Tail* very short and subtruncated, the graduation of the lateral feathers being slight. *Tarsi* greatly abbreviated, being much shorter than the middle toe without its claw; outer toe, without claw, longer than the middle; tip of inner claw reaching to base of middle one.

Genus OCEANITES, *Keys. et Blas.* Almost cosmopolite on the high seas.

GEN. CHAR.—*Bill* shorter than the head, slender, weak, the sides much compressed, and slightly grooved, with the tip suddenly hooked and acute; the lower mandible shorter than the upper, the tip arched, with the gonys hardly angular beneath; the nostrils elevated above the culmen at its base, tubular, with a single aperture in front. *Wings* long and pointed, with the first quill much longer than the third, and the second the longest. *Tail* of moderate length and even. *Legs* long, slender, with the naked space of the tibia extensive. *Tarsi* longer than the middle toe, and ocreate in front. *Toes* rather short, the outer toe nearly equal to middle one, and the inner the shortest, with the claws rather narrow and pointed.

Genus PELAGODROMA, *Reich.* Southern Hemisphere.

GEN. CHAR.—Differs from *Oceanites* in having the second quill shorter than the third, the tail furcate, and the tarsi scutellated in front, with the nails broad and flattened, and the hallux in the form of a triangular claw.

Genus GARRODIA, *Forbes.* Southern Hemisphere.

GEN. CHAR.—Similar to *Pelagodroma*, but with somewhat shorter legs, and having the sternum posteriorly entire, instead of being excavated on its margin.

Genus FREGETTA, *Bonaparte.* Southern Hemisphere.

GEN. CHAR.—Differs from *Pelagodroma* in having the tarsi ocreate, the feet very short, with the nails peculiarly broad and blunt.

Order ANSERES.

Family ANATIDÆ. DUCKS.

Genus ANAS, *Linn.* Cosmopolite.

GEN. CHAR.—*Bill* longer than the head, higher than broad at the base, nearly of equal breadth throughout; the culmen nearly straight, and depressed to the tip, which is armed with a strong broad nail; the lamellæ of the upper mandible hardly visible beyond the lateral margin, strong, and widely set, especially near the middle; the nostrils placed near the base of the culmen, lateral and oval. *Wings* moderate and pointed, with the tertials lengthened and acute, and with the first quill the longest. *Tail* short and wedge-shaped. *Tarsi* shorter than the middle toe and compressed. *Toes* united by a full web; and the hind toe small and somewhat lobed.

Genus NESONETTA, *Gray.* Endemic.

GEN. CHAR.—*Bill* shorter than the head, the width and elevation at the base equal; the culmen gradually sloping to the tip, which is armed with a moderate-sized nail, the sides compressed and of equal breadth throughout; the lamellæ of the interior margins of the upper mandible small and widely set, strongest near the base; the nostrils near the base lateral and oval. *Wings* very short and pointed, with the second quill the longest. *Tail* short and wedge-shaped, with the end of the stem of each feather bare and rigid. *Tarsi* robust, about two thirds the length of the middle toe. *Toes* strong, with the outer toe shorter than the middle, and all the fore toes united by a full web; the hind toe short, elevated, and somewhat lobed.

Genus MERGUS, *Linn.* North Temperate regions; also Auckland Islands.

GEN. CHAR.—*Bill* as long as, or longer than, the head, straight, slender; the culmen elevated, and convex towards the tip, which is suddenly hooked and armed with a large broad nail; the lateral margins of both mandibles serrated with short and widely-set teeth, all pointing backwards; the nostrils lateral, placed near the base of the bill, oblong, pierced longitudinally in a membrane and pervious. *Wings* moderate and pointed, with the first and second quills of nearly equal length and longest. *Tail* moderate and graduated. *Tarsi* shorter than the middle toe. *Toes* moderate; the outer and middle ones of nearly equal length, and the three anterior ones united by a full web; the hind toe moderate and much lobed.

Genus CASARCA, *Bonaparte.* Europe, as well as Australia and New Zealand.

GEN. CHAR.—*Bill* as long as the head, nearly straight, the width equalling the height at the base, the anterior half depressed, and scarcely curved upwards at the tip, which is armed with a strong, broad nail; the basal part of the lateral margin straight, and the apical part slightly curved upwards; the lamellæ of the upper mandible prominent below the lateral margins, slender, and set rather widely apart; the nostrils suboval, near the base of culmen. *Wings* moderate, with the second quill the longest. *Tail* short and rounded. *Tarsi* robust, shorter than the middle toe. *Toes* long, and united by a full web; and the hind toe long, elevated, and lobed.

l

Genus DENDROCYGNA, *Swainson*. Most parts of the World, but migratory in their habits.

GEN. CHAR.—*Bill* long, higher at the base than broad, with the culmen sloping to the tip, which is armed with a strong, broad nail, and the lateral margins straight; the lamellæ of the upper mandible advancing below the lateral margins, slender, and set widely apart; the nostrils large, oval, and placed near the base of culmen. *Wings* short and rounded, with the second, third, and fourth quills the longest; the first quill with a deep notch in the middle, and the secondaries nearly as long as the quills. *Tail* moderate, and rounded at the end. *Tarsi* slightly shorter than the middle toe, robust. *Toes* long, the lateral ones united to the middle one by an indentated membrane, and the hind toe very long, elevated, and simple.

Genus RHYNCHASPIS, *Stephens*. Most parts of the World.

GEN. CHAR.—*Bill* longer than the head, narrowed at the base: the culmen straight, depressed, and the side much dilated for nearly half its length from the tip, which is furnished with a small hooked nail; the lamellæ of the upper mandible very prominent near the middle, slender and widely set; the nostrils placed near the base and culmen, lateral, and oval. *Wings* lengthened and pointed, with the first quill nearly as long as the second, which is the longest. *Tail* rather short and wedge-shaped. *Tarsi* much shorter than the middle toe. *Toes* united by a full web, and the hind toe very small and slightly lobed.

Genus FULIGULA, *Stephens*. Besides New Zealand, members of this genus inhabit the northern regions of Europe, Asia, and America, migrating to the temperate parts in winter.

GEN. CHAR.—*Bill* nearly as long as the head, broader at the base than high, the culmen gradually sloping to the tip, which is armed with a broad and strong nail; the sides dilated, especially anteriorly, where it is rounded, the lateral margins straight and curved upwards to the nail; the lamellæ of the upper mandible not prominent, and widely set; and the nostrils small, oblong, and near the middle of the bill. *Wings* moderate and pointed, with the first quill the longest. *Tail* short and rounded. *Tarsi* half the length of the middle toe and compressed. *Toes* lengthened and united by a full web.

Genus NYROCA, *Fleming*. Most parts of the World.

GEN. CHAR.—*Bill* as long as the head, higher at the base than broad; the culmen gradually sloping towards the tip, which is depressed, slightly dilated, and armed with a strong nail; the lamellæ of the upper mandible not prominent; and the nostrils oval and placed near the base. *Wings* lengthened and pointed, with the first two quills the longest. *Tail* short and rounded. *Tarsi* half the length of the middle toe and compressed. *Toes* lengthened and united by a full web.

Genus HYMENOLÆMUS, *Gray*. Endemic.

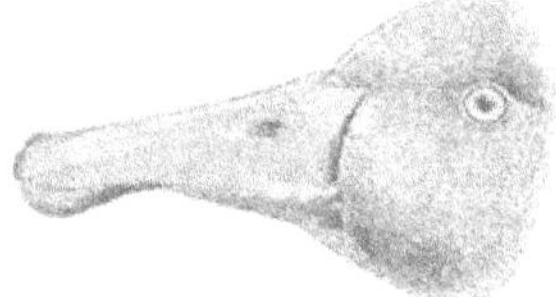

GEN. CHAR.—*Bill* as long as the head, equally compressed, elevated at the base, with the culmen for three fourths of its length straight and then slightly sloping to the tip; the sides shelving from the culmen to the lateral margins, of which the basal half is firm, and furnished with lengthened slender laminæ; the apical half of the margin composed of a soft flexible skin that hangs over the lower mandible, widening towards the tip, where it is truncate, and the nail not very prominent; the nostrils situated near the middle, and oval. *Wings* short, slender, with the first, second, and third quills nearly equal, but the second the longest; the shoulder armed with a short, blunt spur. *Tail* lengthened and composed of broad feathers, with the end rather rounded. *Tarsi* nearly as long as the middle toe, exclusive of the claw; the fore toes strong and fully webbed, and the hind toe moderate and strongly lobed.

Order PYGOPODES.

Family **PODICIPEDIDÆ.** GREBES.

Genus PODICEPS, *Latham.* Cosmopolite.

GEN. CHAR.—*Bill* more or less long, strong, straight, the culmen slightly curved at the tip, which is acute and entire; the sides much compressed, and the gonys short and advancing upwards to an acute point; the nostrils placed in a short groove, with the opening longitudinal and exposed. *Wings* short and pointed, with the first or sometimes the second quill the longest, and slightly emarginated near the tips. *Tail* short, not apparent. *Tarsi* shorter than the middle toe, much compressed, the anterior and posterior edges covered with small scales, which are serrated posteriorly, and the sides with transverse scales. *Toes* long, the outer the longest, depressed, margined on the sides, especially on the inner side, and united at the base to the middle toe; the hind toe short and strongly lobed; the claws short, very broad, flat, and obtuse.

Order IMPENNES.

Family **SPHENISCIDÆ.** PENGUINS.

Genus EUDYPTES, *Vieillot.* Southern Hemisphere.

GEN. CHAR.—*Bill* more or less long, straight, much compressed, and grooved on the sides, and the culmen rounded and curved at the tip, which is acute; the end of the lower mandible truncated, and the gonys moderate and advancing upwards; the nostrils linear, placed in the lateral groove, which extends for three fourths of the length of the bill; and the frontal plumes advancing to the opening. *Wings* imperfect. *Tail* long, and composed of narrow rigid feathers. *Tarsi* very short, much flattened, and covered with small scales. *Toes* long and strong, with the anterior ones united to the middle one by a web, the lateral toes unequal, the outer the longest; the hind toe very small, and united to the tarsus at the base of the inner toe; the claws strong, compressed, and slightly curved.

Genus EUDYPTULA, *Bonaparte.* Australia and New Zealand.

GEN. CHAR.—*Bill* moderate, much compressed, and strong, with the culmen rounded and curved at the tip, which is acute; the tip of the lower mandible suddenly truncated, and the gonys moderate and curved upwards; the nostrils rather rounded, and placed in the lateral groove near the middle of the bill. *Wings* imperfect, and covered with scale-like plumes. *Tail* very short. *Tarsi* very short, thick, flattened, and covered with small scales. *Toes* long, the lateral ones unequal and united to the middle toe by a web; the hind toe very small, and united to the tarsus at the base of the inner toe; the claws long, compressed, and slightly curved.

Genus APTENODYTES, *Forster.* High southern latitudes only.

GEN. CHAR.—*Bill* longer than the head, rather slender, compressed on the sides, slightly bent at the end, with the base of the upper mandible covered with short close-set plumes, and the side grooved to near the tip, which is acute; the lower mandible covered with a smooth naked skin; the nostrils linear, and placed in the lateral groove. *Wings* imperfect, and covered with scale-like plumes. *Tail* very short, and composed of narrow rigid feathers. *Tarsi* very short, flattened, and covered with short plumes. *Toes* rather short and depressed, the anterior ones united by a web; the hind toe very small, and almost entirely connected to the inner side of the tarsus; the claws large, depressed, and very slightly curved.

Subclass RATITÆ.

Order APTERYGES.

Family APTERYGIDÆ. Kiwis.

Genus APTERYX, *Shaw*. Endemic.

Gen. char.—*Bill* more or less lengthened, very slender, with the base covered by a bony cere, broad, and rather depressed; the culmen rounded, straight to near the tip, which projects over that of the lower mandible, and rather obtuse; the sides gradually compressed, and grooved towards the end; the gonys very long and slightly curved; the nostrils placed on each side at the tip, very small, and sublinear; the base of the bill furnished with lengthened hairs. *Wings* abbreviated and covered with feathers. *Tail* not apparent. *Tarsi* the length of the middle toe, very robust, and covered with variously sized scales, those of the inner and outer sides the smallest. *Toes* three before, with the lateral ones equal, and all covered above with broad scales; the hind toe very short, united to the tarsus, and armed with a long, strong, and rather acute claw.

GLAUCOPIS WILSONI.

(BLUE-WATTLED CROW.)

Glaucopis wilsoni, Bonap. Consp. Gen. Av. i. p. 368 (1850).
Callæas wilsoni, Gray, Ibis, 1862, p. 227.
Callæas olivascens, Pelz. Verh. zool.-bot. Gesellsch. Wien, 1867, p. 317, note.
Glaucopis olivascens, Finsch, J. f. O. 1870, p. 324.

Native name.—Kokako.

Ad. suprà schistaceo-cinereus, subtùs paullò cyanescens: loris cum vittâ frontali angustâ, regione oculari mentoque nigerrimis: facie laterali et gutture paullò canescentibus: fronte posticâ et supercilio indistincto albidis: carunculâ rictali ovali utrinque cyaneâ: remigibus et rectricibus nigricantibus dorsi colore lavatis: rostro et pedibus nigris: iride saturatè brunneâ.

Juv. dorso toto olivaceo-fusco: abdomine toto cum hypochondriis et subcaudalibus pallidè cinereo-brunneis: carunculis minoribus, pallidè cyaneis.

Adult male. General plumage dark cinereous or bluish grey, tinged more or less on the upper surface of the wings and tail and on the rump and abdomen with dull brown; a band of velvety black, half an inch broad, surrounds the base of the bill, fills the lores, and encircles the anterior portion of the eyes; immediately above this band and continued over the eyes light ashy grey, shading into the darker plumage; quills and tail-feathers slaty black. Irides blackish brown; bill and legs black. The wattles, which form a distinguishing feature in this bird, are, during life, of a bright ultramarine-blue; but they fade soon after death, and in the dried state become almost black. Total length 17·25 inches; extent of wings 20·5; wing, from flexure, 7·25; tail 7·75; bill, along the ridge 1·25, along the edge of lower mandible 1; tarsus 2·5; middle toe and claw 2·15; hind toe and claw 1·5.

Female. Similar to the male, but somewhat smaller and more deeply tinged with brown on the lower part of the back, rump, and abdomen. Total length 17 inches; extent of wings 19·75; wing, from flexure, 6·6; tail 7·25.

Young. The young of both sexes have the whole of the back and the upper surface of the wings and tail, as well as the sides of the body, dull olivaceous brown; the abdomen and under tail-coverts yellowish brown; the wattles smaller than in the adult and of a pale blue colour.

Nestling (only partially fledged). Frontal band very inconspicuous except in front of the eyes; wattles extremely small and of a pinky colour. The plumage as in the adult but duller, and the wing-feathers washed on their outer vanes with brown.

Note. Professor Hutton is of opinion that the female is "rather larger than the male;" but my observations lead me to an opposite conclusion. I must admit, however, that I have found the size somewhat variable in both sexes. The wattle is always appreciably smaller in the female. In a pair from Wainuiomata, that of the male measured ·75 of an inch in diameter, and that of the female only ·5, besides being less rounded in form.

Varieties. There is a fine albino specimen in the Colonial Museum, obtained in the Rimutaka ranges and presented by a settler, who had it alive for several months. The whole of the plumage is white, with a creamy tinge on the fore neck and underparts; the shafts of the quills and tail-feathers conspicuously whiter; the caruncles very small and colourless; bill horn-coloured; feet yellowish brown; the tail-feathers somewhat abraded at the tips.

Another abnormal specimen in the same collection (received from the Wairarapa district) has the entire plumage of a washed-out ash-grey colour, paler and tinged with brown on the quills and tail-feathers. There is an approach to the normal bluish-grey colour on the throat and towards the edges of the frontal patch, which is dull brown instead of velvety black; bill and feet brown; caruncles faded to the same colour.

Obs. As will be seen from the above synopsis, I am unable to admit the so-called *Glaucopis olivascens* to the rank of a distinct species. It was founded on a specimen obtained at Auckland by M. Zelebor, and the diagnostic characters by which it is distinguished from *G. cinerea* are the brownish-olive colour of the back, wings, and tail, the greyish olive of the underparts, its greater size, and the "dusky colour of the mouth-caruncles." As I have already shown, this description applies to the young of *G. wilsoni*. The dusky colour of the wattles is of no value as a specific character, because, as already mentioned, these appendages entirely change colour in dried specimens, leaving no trace of the original blue. Even in the living bird the colour of the wattles varies considerably in its tone, according to age and other physical conditions; and Dr. Hector has observed that when in confinement its wattles undergo remarkable variations, the exterior margin sometimes assuming a decided yellowish tinge, and again changing back into blue. Dr. Hector writes to me that of three specimens caught together, of which the sex was ascertained, two with olive-brown backs and very small wattles proved to be males, while the third, which had large wattles, of a deep blue colour, and only a slight tinge of brown on the upper parts, was unmistakably a female; and he expresses his belief that *Glaucopis olivascens* is the male of *G. wilsoni*. Accepting the result of Dr. Hector's dissection as conclusive evidence of the sex in each case, I should be inclined to pronounce his two brown-backed males birds of the first year, and the female an adult in full breeding-plumage. I may add that the bird from which my description of the adult male is taken was shot in company with two others (an adult female and a young male), all of which were carefully sexed by myself.

This singular representative of the Crow family is sparingly dispersed over the North Island, being very local in its distribution. It is met with more frequently in the wooded hills than in the low timbered bottoms, but its range is too eccentric to be defined with any precision. During many years' residence at Kaipara, north of Auckland, I never obtained more than five specimens, all of which were shot in the low wooded spurs of the Tangihua ranges. In particular localities, however, even further north, it is comparatively plentiful: for example, between the headwaters of the Wairoa and Whangarei rivers there are several strips of forest in which I never failed to meet with the Kokako; and in the Kaitara ranges in the Whangarei district it was, till within the last few years, rather abundant. I have heard of its occurrence in various parts of the Waikato district *, and in certain localities in the Hawke's Bay and Wellington provincial districts it is far from being an uncommon species. During the autumn months it is comparatively plentiful in the Mangorewa forest between Tauranga and Rotorua. The traveller, at this season, frequently meets with it hopping about along the road or among the bushy branches of *Solanum* on either side.

The Kokako is adorned with fleshy wattles of a brilliant blue colour, which spring from the angles of the mouth, and when the bird is in motion they are compressed under the chin. The first specimen obtained from the Tangihua ranges was a fine bird in full plumage; but the Maori who brought it had torn off the beautiful wattles and pasted them, by way of ornament, on his dusky cheeks.

The notes of the male are loud and varied; but the most noticeable one is a long-drawn organ-note of surpassing depth and richness. I have not been able to discover whether the female is

* The Maoris state that it is common at Taupo and at Maungatautari, one of those whom I questioned on the subject observing, "Where the range of the Huia ceases, that of the Kokako begins." Reischek met with several on the Great Barrier, but never saw it on the Little Barrier, nor on the Hen and Chickens. Lying off Cape Brett, the southernmost head of the Bay of Islands, there is a wooded islet called by the Maoris "Motukokako," in allusion to its having been at one time inhabited by this bird.

similarly endowed, but I have often heard two or more Kokakos, each in a different key, sounding forth these rich organ-notes with rapturous effect; and it is well worth a night's discomfort in the bush to be awakened at dawn by this rare forest music. I never hear it without being reminded of Waterton's saying of the pretty snow-white Campanero, that "Actæon would stop in mid-chase, and Orpheus himself would drop his lute" to listen to its toll. Another of its notes may be described as a loud cackle, while others, again, are scarcely distinguishable from those of the Tui, resembling the soft tolling of a distant bell; but it is only in the early morning that they can be heard to perfection. It has another note, which is very much like the mewing of a cat; but this is only occasionally heard, and then immediately before rain, indicating, it would seem, a highly sensitive nature.

In the pairing-season the male bird loves to display himself before the other sex, arching his neck, spreading his wings, and dancing round the mate of his choice in a very ludicrous manner. They manifest much mutual attachment, and often continue to associate in pairs long after the cares of reproduction have been got rid of and the brood of young ones have grown up and dispersed.

This species subsists chiefly on small fruits and berries, but, like all the members of the family to which it belongs, it will readily partake of insect food of every kind. I have sometimes found its crop distended with the ripe pulpy seed of the tatarama (*Rubus australis*), or with the berries of the kaiwiria (*Parsonia albiflora*) and kareao (*Rhipogonum scandens*); and it is said to feed also on the leaves of the thistle and wild cabbage. The branch depicted in the Plate is that of the native fuchsia, or kohutuhutu, the fruit of which forms a part of its favourite diet. When feeding, it often uses its feet, after the manner of a parrot.

Its wings are small and rounded, and its flight is consequently feeble and generally limited to very short distances. Its progression through the forest is usually performed by a succession of hops, the wings and tail being partially spread—a movement precisely similar to that of the Huia (*Heteralocha acutirostris*).

The stomach of this species consists of a very muscular sac, with a tough epithelial lining or integument, which peels off readily on being pulled, as with the fruit-eating Pigeons and some other birds. The plumage is beautifully soft and silky, owing to the peculiar texture of the feathers. The wattles are smooth and somewhat glossy, but their rich cerulean colour gradually fades out after death.

In disposition the Kokako inherits the true characteristics of the Crow family, being inquisitive, shy, and crafty. I purchased a live one from the Otaki natives in the winter of 1862, and as it shared my apartments for nearly a week (much to the discomfiture of my excellent landlady), I had a good opportunity of studying its habits and character. I was often much amused with the tricky manœuvres of this sprightly bird, and I regretted the accident which deprived me of so intelligent a companion. It generally remained concealed under a side table in a dark corner of the room: but in cold weather was accustomed to steal quietly to the inside of the fender, in order to get warmth from the fire. My presence had become familiar to it, but on the entrance of a stranger it would immediately spring out and hop away to its dark retreat under the corner table.

The bird represented in the Plate is one of a pair shot on the Poroporo ranges during the Huia-hunting expedition of which an account is given further on. They were found perched in the midst of a superb bunch of puawhananga (*Clematis indivisa*), and feeding with avidity on the white petals, stopping at intervals to coy with each other and converse in a low musical twitter. The mated pair, with their unique floral surrounding, formed a lovely picture of real nature.

On dissecting the male, I found the whole of the viscera and even the membrane and skin covering it stained to a vivid blue; and on opening the stomach, I found it crammed with comminuted vegetable matter of a perfectly black colour. On examining some of this matter after washing it in cold water, I found that it was in reality composed of *Clematis*-flowers, the change in colour being apparently due to the action of some acid in the bird's stomach.

Mr. Reischek found a nest of this species in a bunch of *Astelia*, the birds having simply made a round depression in the centre of the clump and placed a few dry twigs there. There were three young birds. Two of these sprang out of the nest on his approach, but were afterwards shot; the youngest he managed to catch before it could escape, and from this I have taken my description of the nestling. On another occasion he met with the nest near the wooded summit of the Waitakere ranges. It was a large irregular-shaped structure, composed of twigs and moss coarsely put together, and placed high up on a miro tree. The young birds (three in number) had just left the nest, but had not yet quitted the tree. They were shy and wary, and, on an alarm being sounded by one of the parent birds, they immediately secreted themselves in the thick foliage, from which it was found impossible to dislodge them. This was on January 3rd, which fixes approximately the breeding-season; although my son discovered a nest at Whangarei, containing three well-fledged nestlings, at a somewhat earlier date.

I agree in the opinion expressed by Mr. Kirk * that the egg brought to the Colonial Museum by Mikaera on October 20, 1885, and disposed of as the egg of the Huia †, is in reality that of the present species. Subsequent events have shown that Mikaera's testimony cannot be depended on; and no credence can be given now to his statement that it was "taken from a cavity in a dead tree." The egg contained a young bird, apparently just ready for extrusion, and both embryo and shell are now in the Museum collection. The egg is ovoido-conical in form, measuring 1·45 by 1·05 inches, and is of a pale stone-grey, irregularly stained, freckled, and speckled with purplish grey, the markings in some places running into dark wavy lines. The chick has the bill very stout, with the caruncles at the angles of the mouth well developed and of a flesh-white colour. The whole of the body is bare, with the exception of what appears (in spirit) to be strips of coarse, black, hair-like filaments, from one half to three quarters of an inch in length, but which are in reality tufts of extremely fine downy feathers. A strip of these filaments encircles the crown, a line passes down the course of the spine, and there is another along the outer edge of each wing and behind each thigh.

Accepting, as I do, the view so well formulated by Professor Parker, that "in all respects, physiological, morphological, and ornithological, the Crow may be placed at the head, not only of its own great series (birds of the *Crow-form*), but also as the unchallenged chief of the whole of the Carinatæ" ‡, I have, in my systematic arrangement of the New-Zealand ornis, accorded the foremost rank to the family Corvidæ, instead of placing the Turdidæ at the head of the list as is now the fashion with writers on Systematic Ornithology. Some doubts, however, having hitherto existed as to the true position of the genus *Glaucopis*, I was glad of the opportunity to place a skeleton of this species in the hands of Dr. Gadow, of Cambridge, in order that he might investigate its natural affinities. That gentleman made a critical examination of the bones, and compared them with those of *Strepera*, *Gymnorhina*, *Paradisea*, *Struthidea*, *Graucalus*, *Ptilonorhynchus*, *Heteralocha*, and *Sturnus*, with the following general result. He finds that *Glaucopis* is a Corvine form, being closely allied to the Austrocoraces, a group of birds which form a connecting-link between the true Corvidæ and the Laniidæ. It agrees with *Strepera*, and shows considerable similarity in structure with *Ptilonorhynchus*, although *Glaucopis* presents in its skull, sternum, and sacrum several characters which are peculiar to the genus. *Struthidea* agrees with *Glaucopis* by far less than might have been supposed, whilst *Graucalus* is still further removed, being apparently on the line through which *Glaucopis* reaches the Muscicapine forms. Dr. Gadow sums up the results of his investigation by saying that "if a Satin-bird could be induced to marry a Piping-Crow, their offspring might, in New Zealand, become a *Glaucopis*."

* Journal of Science, 1882-83, vol. i. p. 262.

† Trans. New-Zealand Instit. 1875, vol. viii. p. 192.

‡ "There are, of course, innumerable points in regard to the Classification of Birds which are, and for a long time will continue to be, hypothetical as matters of opinion, but this one seems to stand a fact on the firm ground of proof" (art. "Ornithology," Encycl. Brit., by Prof. Newton, F.R.S.).

GLAUCOPIS CINEREA.

(ORANGE-WATTLED CROW.)

Cinereous Wattle-bird, Lath. Gen. Syn. i. p. 364, pl. xiv. (1781).
Glaucopis cinerea, Gm. Syst. Nat. i. p. 363 (1788).
Cryptorhina callæas, Wagl. Syst. Av. *Cryptorhina*, sp. 5 (1827, ex Forster, MSS.).
Callæas cinerea, Forster, Descr. Anim. p. 74 (1844).

Native name.—Kokako.

Ad. similis *G. wilsoni*, vix saturatior, paullò minor: caranculis aurantiacis ad basin tantùm cyaneis distinguendus.

Adult. Similar in plumage to *G. wilsoni*, but with less of the brown tinge on the lower parts, and the tail-feathers blackish towards the tips. It is readily distinguished, however, by the colour of the wattles, which are of a rich orange, changing sometimes to vermilion, and blue at the base. Irides blackish brown; bill and feet black. Total length 16 inches; wing, from flexure, 6·25; tail 7; bill, along the ridge 1·25, along the edge of lower mandible 1; tarsus 2·5; middle toe and claw 2·15; hind toe and claw 1·5.

Partial albino. There is an interesting specimen in the Colonial Museum, which was obtained by Mr. Henry Travers at the foot of Mount Franklin, in the Spencer ranges, in January 1869. The general plumage as in ordinary specimens; hind head, sides and fore part of neck, and the whole of the back largely marked with pure white: one or two of the quills in each wing are either wholly or partially white, and there are a few scattered white feathers on the sides, abdomen, and thighs.

THIS species is the South-Island representative of *Glaucopis wilsoni*, to which it bears a general resemblance, except in the colour of its wattles and its rather smaller size. Like the North-Island species also, its distribution is very irregular: thus, in Otago, Dr. Hector found it very plentiful on Mount Cargill and in a strip of bush near Catlin river, but never in the intervening woods; while in the Nelson provincial district, as I am informed by Mr. Travers, its range is exclusively restricted to certain well-defined localities, although the berries on which it is accustomed to feed abound everywhere. It is said to be very abundant on some of the wooded ranges of Westland, and Sir J. von Haast has obtained numerous specimens from the Oxford ranges in the provincial district of Canterbury.

I ought to add that, in the summer of 1867, one of these birds was seen by Major Mair at Te Mu, near Lake Tarawera, in the North Island. He followed it for some distance, in the low scrub, and got near enough to obtain a good view and to observe its bright orange wattles.

The habits of this bird differ in no essential respect from those of the preceding species. Mr. Buchanan, of the Geological Survey, has mentioned to me a very curious circumstance frequently observed by himself at Otago: he has seen these birds travelling through the bush on foot, Indian fashion, sometimes as many as twenty of them in single file, passing rapidly over the ground by a succession of hops, and following their leader like a flock of sheep; for, if the first bird should have occasion to leap over a stone or fallen tree in the line of march, every bird in the procession follows suit accordingly!

I saw a pair of caged ones at Hokitika, in the possession of Mr. M'Nee, who told me that he had snared them in the woods with perfect ease. They were apparently quite reconciled to confinement, hopping from perch to perch in a very lively manner, and occasionally meeting to utter a low chuckling note, as if in confidential intercourse. I observed that they usually carried the wattles firmly compressed under the rami of the lower jaw.

One of the many interesting discoveries, since the publication of my first edition, has been the finding of the nest and eggs of the Orange-wattled Crow. The Canterbury Museum contains two nests of this bird, both of which were obtained at Milford Sound.

One is a massive nest, with a depth of eight inches, composed of rough materials, but with a carefully finished cup. The foundation consists of broken twigs, some of them a quarter of an inch in diameter, and placed together at all angles, so as to form a compact support; over this a layer of coarse moss and fern-hair, to the thickness of two inches or more; then a capacious well-rounded cup, lined with dry bents, intermixed with fern-hair. The general form of the nest is rounded, but at one end of it the twig foundation is raised and produced backwards, for what purpose can only be conjectured *. The other is of similar construction, composed of numerous broken twigs, intermixed with dry moss, and the projection is as conspicuous as in the former, extending some eight inches beyond the nest proper, which is about a foot in diameter. The cup-shaped depression is shallower than in the other, but has the same thick lining of dry grass. This nest was, I am informed, found among the branches of a totara overhanging a stream of water, in the month of January, and contained at that time young birds. The other nest also was discovered in the vicinity of water †.

Two eggs of this species, collected by Docherty on the west coast, were presented by Mr. Potts to the Canterbury Museum, where I had an opportunity of examining them. They are of a regular ovoido-conical form, one of them being slightly narrower than the other, measuring, respectively, 1·60 by 1·15, and 1·65 by 1·10 inches. They are of a dark purplish grey, irregularly spotted and blotched with dull sepia-brown. These spots and markings are thicker and more prominent at the larger end, and of various shades, the lighter ones fading almost to purple and presenting a washed-out appearance.

Mr. W. D. Campbell has published ‡ an account of two nests which he found, in the month of February, in the low bush which covers the river-flats of Westland. One of these nests contained an egg, and the other two nearly-fledged birds. The nests, which were built in the branches of the *Coprosma* scrub, about 9 feet above the ground, measured 15 inches externally, were somewhat loosely constructed of twigs and roots, and had a well-formed cup-shaped interior, lined with pine-roots and twigs. He kept the two young birds for some weeks in a cage for the purpose of studying their habits. During life their wattles were of a light rose tint, changing into a violet colour towards the base; but after death, when their skins were dried, the wattles assumed a dull orange tint.

* In connection with the above I may mention that in the Canterbury Museum there is a much larger nest, from Australia, exhibiting the same form of construction in a more pronounced degree. It was presented by the Baron A. von Hügel, who obtained it at Dandenong, Mount Victoria, and who assigns the structure to the Lyre-bird (*Menura superba*). It is composed chiefly of twigs and small sticks, some of them half an inch in diameter, laid together in a compact mass. The cavity is deep, rounded, and lined with soft fern-fronds, some of which are also interlaced with the framework of the nest. Its width on the outside is only 15 inches; but, owing to its extension backwards, its length is 2 feet 6 inches. The cup is situated at the proximal end, where the nest is more compact and somewhat raised, but without any appearance of a dome.

† The author of 'Out in the Open' describes, at p. 105, the finding of two nests, at heights varying from ten to seventeen feet from the ground, in the bush that fringes Milford Sound. This was in the month of January, and one of the nests contained two young birds, apparently just hatched. "They were partially clothed with slate-coloured down, which on the cranium stood up like a broad crest, or rather crown; the neck and underparts were quite bare; beak flesh-colour, with a greenish tinge about the point of the upper mandible; rictal membrane pale greenish, changing to blue; wattles rosy pink, like an infant's hand; legs and feet slatish anteriorly, dull flesh-colour behind; claws dull white. The old bird suffered a close inspection of its home and its inmates without uttering any alarm-cry or showing any signs of defending its young."

‡ Trans. New-Zealand Instit. 1879, vol. xii. pp. 249, 250.

HETERALOCHA ACUTIROSTRIS.

(HUIA.)

Neomorpha acutirostris, Gould, P. Z. S. 1836, p. 144 (♀).
Neomorpha crassirostris, Gould, P. Z. S. 1836, p. 145 (♂).
Neomorpha gouldi, Gray, List of Gen. of B. p. 15 (1841).
Heteralocha gouldi, Cab. Mus. Hein. Th. i. p. 218 (1850).

Native name.—Huia.

♂ undique sericeo-niger, sub certâ luce obscurè viridi nitens: caudâ conspicuè albo terminatâ: pileo carunculis magnis rotundatis lætè aurantiacis utrinque ad basin mandibulæ positis ornato: rostro valido, eburneo, versus basin cinereo: pedibus cinereis, unguibus corneis.

♀ mari similis, sed rostro longo valdè decurvato semper distinguenda.

Adult. The whole of the plumage is black, with a green metallic gloss; the tail with a broad terminal band of white. Bill ivory-white, darkening to blackish grey at the base. Wattles large, rounded, and of a rich orange-colour in the living bird. Tarsi and toes bluish grey; claws light horn-colour.

Male. Length 18·75 inches; extent of wings 22·5; wing, from flexure, 8; tail 7·5; bill, along the ridge 2·75, along the edge of lower mandible 2·75; tarsus 3; middle toe and claw 2·5; hind toe and claw 2.

Female. Length 19·5 inches; extent of wings 21; wing, from flexure, 7·5; tail 7·25; bill, along the ridge 4, along the edge of lower mandible 4·12; tarsus 3; middle toe and claw 2·25; hind toe and claw 1·75.

Young female. Differs from adult bird in having the entire plumage of a duller black, or slightly suffused with a brownish tinge, and with very little gloss on the surface. Under tail-coverts tipped with white, and the terminal white bar on the tail washed with rufous-yellow, especially in the basal portion. Wattles small and pale-coloured. Bill only slightly curved.

In another specimen in my possession, apparently a year older, the tail-coverts are without the margin, the white on the tail-feathers is purer, and the bill is perceptibly longer, with a darkened tip. In another, the tips of both mandibles are perfectly black for about half an inch in extent; the tail-feathers are only slightly stained with rufous, but instead of having an even white border the shafts are black to their tips, and the terminal bar has an emarginate edge.

Young male. On comparing a specimen in my collection with the above, the same general remarks apply, except that the under tail-coverts are not tipped with white at all, while the soft feathers on the lower part of the abdomen are largely tipped with pale rufous and white. The pale rufous wash on the tail-bar is likewise more conspicuous.

Varieties. The Maoris speak of a "red-tailed Huia," but I have no doubt that this is merely the condition of tail noticed above. A single tail-feather in my possession has the terminal band stained with rust-colour, and this would be described by a Maori as "red." They also say that the birds from the Ruahine range have a somewhat broader band on the tail than those from Tararua, the skins from the former locality being in much greater demand on that account. A specimen which came into my hands had a single white feather in the tail—not a feather of the full quality but aborted in its character, being short, narrow, and shaped like one of the outer primaries, although filling the place of an ordinary tail-feather. In another specimen there was a narrow white streak down the shaft-line of the middle feather. The most remarkable variety, however, is that known to the Maoris as a Huia-ariki. I have never seen but one of these birds, of which I have already published the following notice *:—

"I have received from Captain Mair some feathers which, in colour, have much the appearance of the soft grey plumage of *Apteryx oweni*, but which are in reality from the body of a Huia, being of extremely soft texture. I hope to receive the skin for examination, but in the meantime I will give a quotation from the letter forwarding the feathers:—Old Hapuku, on his death-bed, sent for Mr. F. E. Hamlin, and presented him with a great *taonga*. This has just been shown to me. It is the skin of a very peculiar Huia, an albino I suppose, called by the Hawke's Bay natives 'Te Ariki.' I send you a few feathers. The whole skin is of the same soft dappled colour, but the feathers are longer and softer. The bill is nearly straight, strong, and of full length. The wattles are of a pale canary-colour. The centre tail-feather is the usual black and white, while the others on each side are of a beautiful grey colour. These birds are well known to the Huia-hunting natives of Hawke's Bay; and to possess an 'Ariki' skin one must be a great chief. The specimen I have described was obtained in the Ruahine mountains."

The skin was afterwards sent to me, for examination, and was exhibited at a Meeting of the Wellington Philosophical Society. It is that of a male bird of the first year. The whole of the body-plumage is brownish black, obscurely banded or transversely rayed with grey; on the head and neck the plumage is darker, shading into the normal glossy black on the forehead, face, and throat. The tail-feathers are very prettily marked: with the exception of the middle one, which is of the normal character in its apical portion, they are blackish brown, irregularly barred and fasciated with different shades of grey, and with a terminal band of white; the under tail-coverts, also, are largely tipped with white, indicating adolescence.

Obs. In some adult examples of both sexes the white at the end of the tail is tinged more or less with rufous. It should be noted also that the brightness of the fleshy wattles depends, in some measure, on the health or condition of the bird; for during sickness they change to lemon-yellow. A recently killed specimen weighed 14½ oz. The palate and soft parts of the throat are bright yellow. The tongue is horny and slightly bifid at the tip. In fully matured examples the wattle measures nearly an inch across.

This is one of those anomalous forms that give to the New-Zealand avifauna so much special interest. Considerable difference of opinion has existed among naturalists as to its proper position in our artificial system. For many years it was placed, by common consent, among the Upupidæ, and that it possesses strong affinities to the Hoopoes is, I think, undeniable. Dr. Finsch proposed to group it in a separate family with *Glaucopis* and *Creadion*, under the name of Glaucopidæ; and Mr. Sharpe, in the British Museum Catalogue, has placed it with both of those forms in the family Corvidæ. According to my view, however, the investigation of its anatomy by the late Prof. Garrod leaves no doubt whatever that its natural place is among the Starlings.

The late Mr. Gould, who was the first to characterize the form, was deceived by the great difference in the shape of the bill, and treated the sexes as distinct species, naming them respectively *Neomorpha acutirostris* and *N. crassirostris*—a very natural mistake, "many genera even," as Mr. Gould observes, "having been founded upon more trivial differences of character." Mr. G. R. Gray, having determined their identity, proposed to substitute the specific name of *gouldi*, in compliment to the original describer; and his example has been followed by others; but I have deemed it more in accordance with the accepted rules of zoological nomenclature to adopt the first of the two names applied to the species by Mr. Gould; and the name *Neomorpha* having been previously used in ornithology, it becomes necessary to adopt that of *Heteralocha*, proposed by Dr. Cabanis for this form.

In November 1870, I communicated to the Wellington Philosophical Society a paper † containing

* Trans. New-Zealand Instit. 1878, vol. xi. p. 370.

† *Op. cit.* 1870, vol. iii. pp. 24–29.

all the information I could collect respecting the Huia, which was then a somewhat rare species. As will presently appear, the bird is now far more plentiful than formerly. But, in order to preserve its full history, I will reproduce here a portion of that paper:—

A well-known writer in 'Nature' (Dr. Sclater), in describing the peculiarity in the form of the bill that distinguishes it from the female, observes: "Such a divergence in the structure of the beak of the two sexes is very uncommon, and scarcely to be paralleled in the class of birds. It is difficult to guess at the reason of it, or to explain it on Darwinian or any other principles." In the absence of any published account of its habits, beyond mere fragmentary notices, I have thought the subject of sufficient interest to justify my placing before the Society the following complete account of all that I have been able to ascertain respecting it. The peculiar habits of feeding, which I have described from actual observation, furnish to my own mind a sufficient "reason" for the different development of the mandibles in the two sexes, and may, I think, be accepted as a satisfactory solution of the problem.

Before proceeding to speak of the bird itself, I would remark on the very restricted character of its habitat. It is confined within narrow geographical boundaries, being met with only in the Ruahine, Tararua, and Rimutaka mountain-ranges, with their divergent spurs, and in the intervening wooded valleys. It is occasionally found in the *Fagus* forests of the Wairarapa valley, and in the rugged country stretching to the westward of the Ruahine range, but it seldom wanders far from its mountain haunts. I have been assured of its occurrence in the wooded country near Massacre Bay *, but I have not been able to obtain any satisfactory evidence on this point. It is worthy of remark that the natives, who prize the bird very highly for its tail-feathers (which are used as a badge of mourning), state that, unlike other species which have of late years diminished and become more confined in their range, the Huia was from time immemorial limited in its distribution to the district I have indicated.

My first specimen of this singular bird (an adult female) was obtained in 1855, from the Wainuiomata hills, a continuation of the Rimutaka range, bounding the Wellington harbour on the northern side—the same locality from which Dr. Dieffenbach, nearly twenty years before, received the examples figured by Mr. Gould in his magnificent work 'The Birds of Australia.' I have since obtained many fine specimens, and in the summer of 1864 I succeeded in getting a pair of live ones. They were caught by a native in the ranges, and brought down to Manawatu, a distance of more than fifty miles, on horseback. The owner refused to take money for them; but I negotiated an exchange for a valuable greenstone. I kept these birds for more than a year, waiting a favourable opportunity of forwarding them to the Zoological Society of London. Through the carelessness, however, of a servant, the male bird was accidentally killed; and the other, manifesting the utmost distress, pined for her mate, and died ten days afterwards.

The readiness with which these birds adapted themselves to a condition of captivity was very remarkable. Within a few days after their capture they had become perfectly tame, and did not appear to feel in any degree the restraint of confinement: for, although the window of the apartment in which they were kept was thrown open and replaced by thin wire netting, I never saw them make any attempt to regain their liberty. It is well known, however, that birds of different species differ widely in natural disposition and temper. The captive Eagle frets in his sulky pride; the Bittern

* Mr. Kane informs me that when travelling, two years ago, in the South Island he saw several Huias in a forest lying between Nelson and Picton. He states that he was quite close to them, and could not possibly be mistaken in the bird, with which he is familiar. Mr. W. T. Owen, who is a very careful observer, assures me that he met with it on the other side of Nelson. If the range of the Huia does in reality extend across the Straits, it is a very remarkable fact in the geographical distribution of this much-restricted species. That it does occasionally wander far beyond the limits assigned to it in the North Island is certain, because in 1881 Mr. Ambrose Potts met with one near Te Hauopounga, in the Patea country. This was not an escaped bird, because the natives of the district knew nothing about it, and would scarcely credit the statement.

refuses food and dies untamable; the fluttering little Humming-bird beats itself to death against the tiny bars of its prison in its futile efforts to escape; and many species that appear to submit readily to their changed condition of life, ultimately pine, sicken, and die. There are other species, again, which cheerfully adapt themselves to their new life, although caged at maturity, and seem to thrive fully as well under confinement as in a state of nature. Parrots, for example, are easily tamed; and I have met with numerous instances of their voluntary return after having regained their liberty. This character of tamability was exemplified to perfection in the Huias.

They were fully adult birds, and were caught in the following simple manner. Attracting the birds by an imitation of their cry to the place where he lay concealed, the native, with the aid of a long rod, slipped a running knot over the head of the female and secured her. The male, emboldened by the loss of his mate, suffered himself to be easily caught in the same manner. On receiving these birds I set them free in a well-lined and properly ventilated room, measuring about six feet by eight feet. They appeared to be stiff after their severe jolt on horseback, and after feeding freely on the huhu grub, a pot of which the native had brought with them, they retired to one of the perches I had set up for them, and cuddled together for the night.

In the morning I found them somewhat recruited, feeding with avidity, sipping water from a dish, and flitting about in a very active manner. It was amusing to note their treatment of the huhu. This grub, the larva of a large nocturnal beetle (*Prionoplus reticularis*), which constitutes their principal food, infests all decayed timber, attaining at maturity the size of a man's little finger. Like all grubs of its kind, it is furnished with a hard head and horny mandibles. On offering one of these to the Huia, he would seize it in the middle, and, at once transferring it to his perch and placing one foot firmly upon it, he would tear off the hard parts, and then, throwing the grub upwards to secure it lengthwise in his bill, would swallow it whole. For the first few days these birds were comparatively quiet, remaining stationary on their perch as soon as their hunger was appeased. But they afterwards became more lively and active, indulging in play with each other and seldom remaining more than a few moments in one position. I sent to the woods for a small branched tree, and placed it in the centre of the room, the floor of which was spread with sand and gravel. It was most interesting to watch these graceful birds hopping from branch to branch, occasionally spreading the tail into a broad fan, displaying themselves in a variety of natural attitudes and then meeting to caress each other with their ivory bills, uttering at the same time a low affectionate twitter. They generally moved along the branches by a succession of light hops, after the manner of the Kokako (*Glaucopis wilsoni*); and they often descended to the floor, where their mode of progression was the same. They seemed never to tire of probing and chiselling with their beaks. Having discovered that the canvas lining of the room was pervious, they were incessantly piercing it, and tearing off large strips of paper, till, in the course of a few days, the walls were completely defaced.

But what interested me most of all was the manner in which the birds assisted each other in their search for food, because it appeared to explain the use, in the economy of nature, of the differently formed bills in the two sexes. To divert the birds, I introduced a log of decayed wood infested with the huhu grub. They at once attacked it, carefully probing the softer parts with their bills, and then vigorously assailing them, scooping out the decayed wood till the larva or pupa was visible, when it was carefully drawn from its cell, treated in the way described above, and then swallowed. The very different development of the mandibles in the two sexes enabled them to perform separate offices. The male always attacked the more decayed portions of the wood, chiselling out his prey after the manner of some Woodpeckers, while the female probed with her long pliant bill the other cells, where the hardness of the surrounding parts resisted the chisel of her mate. Sometimes I observed the male remove the decayed portion without being able to reach the grub, when the female would at once come to his aid, and accomplish with her long slender bill what he had failed to do. I noticed, however, that the female always appropriated to her own use the morsels thus obtained.

For some days they refused to eat anything but huhu; but by degrees they yielded to a change of food, and at length would eat cooked potato, boiled rice, and raw meat minced up in small pieces. They were kept supplied with a dish of fresh water, but seldom washed themselves, although often repairing to the vessel to drink. Their ordinary call was a soft and clear whistle, at first prolonged, then short and quickly repeated, both birds joining in it. When excited or hungry, they raised their whistling note to a high pitch; at other times it was softly modulated, with variations, or changed into a low chuckling note. Sometimes their cry resembled the whining of young puppies so exactly as almost to defy detection.

I had afterwards another captive Huia, which came from the *Fagus*-covered hills at Wainuiomata. This bird became very tame, knew me well, and always welcomed my approach by making a melodious chirping note. He was fond of fresh meat, chopping it up into very small pieces with his bill, making a sound like the tapping of a Woodpecker as he cut up his dinner on the floor of his cage. He ultimately made his escape, and although he remained about the gardens and shrubberies of Wellington for more than two months, consorting freely with the Indian-Minahs, and occasionally indulging in a flight over his old habitation, he seemed to prefer freedom to captivity, and remained at large; but disappeared at last, having probably fallen a victim to the catapult of some city larrikin.

Dr. Dieffenbach, in forwarding his specimens of the Huia to Mr. Gould, in 1836, wrote:—"These fine birds can only be obtained with the help of a native, who calls them with a shrill and long-continued whistle resembling the sound of the native name of the species. After an extensive journey in the hilly forest in search of them, I had at last the pleasure of seeing four alight on the lower branches of the trees near which the native accompanying me stood. They came quick as lightning, descending from branch to branch, spreading out the tail and throwing up the wings."

On the first occasion of my meeting with this species in its native haunts, I was struck by the same peculiarities in its manners and general demeanour. In the summer of 1867, accompanied by a friend and two natives, I made an expedition into the Ruahine ranges in search of novelties. After a tramp on foot of nearly twenty miles through a densely wooded country, we were rewarded by finding the Huia. We were climbing the side of a steep acclivity, and had halted to dig specimens of the curious vegetating caterpillar (*Sphæria robertsii*), which was abundant there. While thus engaged, we heard the soft flute-note of the Huia in the wooded gully far beneath us. One of our native companions at once imitated the call, and in a few seconds a pair of beautiful Huias, male and female, appeared in the branches near us. They remained gazing at us only a few instants, and then started off up the side of the hill, moving by a succession of hops, often along the ground, the male generally leading. Waiting till he could get both birds in a line, my friend at length pulled trigger; but the cap snapped, and the Huias instantly disappeared down the wooded gully. Then followed a chevy of some three miles, down the mountain-side and up its rugged ravines. Once more, owing to the dampness of the weather, the cap snapped, and the birds were finally lost sight of. I observed that while in motion they kept near each other, and uttered constantly a soft twitter. The tail was partially spread, while the bright orange lappets were usually compressed under the rami of the lower jaw.

We camped that night near the bed of a mountain rivulet, in a deep wooded ravine, and soon after dawn we again heard the rich notes of a Huia. Failing to allure him by an imitation of the call, although he frequently answered it, we crossed to the other side of the gully, and climbed the hill to a clump of tall rimu trees (*Dacrydium cupressinum*), where we found him. He was perched on the high limb of a rimu, chiselling it with his powerful beak, and tearing off large pieces of bark, doubtless in search of insects; and it was the falling of these fragments that guided us to the spot and enabled us to find him. This solitary bird, which proved, when shot, to be an old male, had frequented this neighbourhood (as we were informed by the natives) for several years, his notes being familiar to the people who passed to and fro along the Otairi track leading to Taupo. On asking a native how the Huia contrived to extract the huhu from the decayed timber, he replied "by digging with his pickaxe"—an expression which I found to be truthfully descriptive of the operation; and on dissecting this specimen I found an extraordinary development of the requisite muscles. The skin was very tough, indicating, probably, extreme age. The stomach contained numerous remains of coleopterous insects, of the kind usually found under the bark of trees, also one or two caterpillars.

On skinning the two sexes, it is at once apparent that the head of the male is formed on a different model to that of the female. In the latter the skin peels off very readily, but in the male the head seems too large for the neck. This difference is occasioned by the greatly developed muscles, forming a rounded mass or cushion on each side of the occiput, which enables this sex to wield his chisel in the effective manner described.

In October 1883, I made a special expedition into the mountain-forest in quest of the Huia; and as it will serve to complete my history of the species, I have transcribed the following narrative from my note-book:—

Taking the early train from Wellington to Masterton on the 9th, I met Captain Mair by appointment, and we forthwith made our arrangements for a start on horseback at daybreak. Instead of a fine day, as we had hoped, the morning opened with a heavy shower, which somewhat delayed our departure, and the day turned out drizzly. Our road lay through a bush and along a highway which had been formed but not metalled. The mire was knee-deep for the horses, and, for most part of the way, it was very toilsome work. The distance to be traversed was only twenty miles, the first four of which were over a hard road; but the shades of evening were closing in around us by the time we reached our camping-ground at the foot of the Patitapu range, and our Maori attendant (Rahui) had barely time to fix up our tent and collect "whariki" for bedding before thick darkness had set in. Our approach to this camping-place lay along the edge of a wooded ravine. On the opposite side from us there was a grove of tall manuka trees, several hundred acres in extent. Rahui informed us that this was a favourite resort of the Huia when feeding on the weta or tree-cricket (*Deinacrida thoracica*). The dull russet-green of the manuka bush was relieved on the sides of the ravine by those ever changing, ever beautiful, light-green tints so characteristic of our New-Zealand woods. Here and there a shapely rewarewa reared its tapering top, spangled all over with bunches of crimson flower, while along our path were fringes of the scented pukapuka with its dark green leaves, showing their silver lining as they yielded to the breeze, and covered with a profusion of cream-coloured inflorescence. At intervals might be seen a leafless kowhai laden with a wealth of beautiful golden blossom, and in the more open parts of the widening valley clumps of *Cordyline* with their waving crowns of green; whilst, adding immeasurably to the charm of the whole scene, the star-like clematis, in huge white clusters, hung everywhere in graceful festoons from the tangled vegetation. Down in the bed of the ravine, and hiding the babbling brook, the stunted overhanging trees were for the most part clothed in a luxuriant mantle of kohia, kareao, and other epiphytic plants.

Such was the spot in which we first heard the soft, whistling call of the Huia! Rahui imitated the cry, and in a few moments a fine male bird came across the ravine, flying low, taking up his station for a few seconds on a dead tree, and then disappearing, as if by magic, in the undergrowth below. Our guide continued to call, but the Huia was shy and would only respond with a low chirping note. But this was enough, and led us to where he was engaged, apparently grubbing among the moss on the ground. We shot the bird, which proved to be in beautiful plumage, and Rahui accepted this as an earnest of our success on the morrow.

Our camp was selected as only a native can select in the bush. The spot fixed upon was a gentle slope under the shadow of a three-stemmed tawhero (*Weinmannia racemosa*), sheltered all round by close-growing porokaiwiria, torotoro, and other shrubby trees, and the whole fenced in, as it were, by a thick undergrowth of bright green pukapuka, mixed with the still brighter mahoe, and protected in front by a perfect network of kareao vines, attached to and suspended from the higher trees. We soon had a roaring camp fire and some ribs of mutton roasting for supper. As the night closed in upon us we heard all round the solemn notes of the New-Zealand Owl: first, a distinct *kou-kou, kou-kou*; then in a weaker key (perhaps the responsive call of the female) *keo-keo-keo*; and then, in alternation, the alarm-note and the ever familiar cry of "more-pork."

Even after a pall of darkness had settled on the woods, some Tuis in the tall tree-tops kept up a delicious liquid song, like the measured tolling of a silver bell, and far into the night could be heard, at intervals, the low whistling note of the Kaka communing with his mate. Then all was quiet, the night being very dark, and nothing broke the stillness of the forest till the Huia-call of our native guide brought us to our senses in the early dawn. But the day turned out unpropitious. The drizzling rain continued and a strong breeze set in; so we determined to shift our camp to the other side of the range. Our road lay along the side of another ravine. We had not proceeded more than a mile when Rahui's call was answered from the other side. The bird's loud cry was presently succeeded by a whistling whimper, and then he came towards us, bouncing through the brushwood as if in a desperate hurry. Descending to the ground a few yards in front of us, he hopped along the surface, and then up the trunk of a prostrate tree, with surprising agility. My companion took a shot at him; but owing to the dampness his gun missed fire, and the bird, taking alarm, disappeared in an instant, all our efforts to recall him proving of no avail. On reaching the head of the valley, we tethered our horses and commenced the ascent of the range, which we found very steep. About half-way up, we rested on the ground. Rahui continued his call—a loud clear whistle—not much like the ordinary call of the bird, being louder and more shrill. In a few seconds, without sound or warning of any kind, a Huia came bounding along, almost tumbling, through the close foliage of the pukapuka, and presented himself to view at such close range that it was impossible to fire. This gave me an opportunity of watching this beautiful bird and marking his noble bearing, if I may so express it, before I shot him. While waiting to get the bird within proper range, I heard far below me the rich note of the Kokako, repeated several times. It is scarcely distinguishable from the call of the Tui, but is preceded by a prolonged organ-note of rare sweetness. My next shot was at an adult male Huia who came dashing up, with reckless impetuosity, from the wooded gully. Being anxious to obtain a perfect specimen, I risked a long shot and only wounded my bird. Down he went to the ground like an arrow, with a sharp flute-note of surprise or pain, and then darted off, kangaroo-fashion, covering the ground with wonderful rapidity, and disappeared in the tangle.

We found the descent of the range much easier than our toilsome climb. Remounting our horses we continued up the valley. At a turn in the road, at a spot hemmed in by a wooded amphitheatre of beautiful shapely trees (chiefly rata), we halted for a moment to gaze on the scene. On a tree, immediately in front of us, a pair of Wood-Pigeons were sitting side by side, showing off their ample white breasts under the rays of sunlight glancing through the rain-drops. Whilst we were looking at and admiring this little picture of bird-life, a pair of Huias, without uttering a sound, appeared in a tree overhead, and as they were caressing each other with their beautiful bills, a charge of No. 6 brought both to the ground together. The incident was rather touching, and I felt almost glad that the shot was not mine, although by no means loth to appropriate the two fine specimens. Before we reached our next camping-ground, at the foot of Poroporo, we had bagged another bird (a female of last year) who was unattended, and came up quite fearlessly to her doom.

After we had secured our horses and "refreshed the inner man," Rahui and I started again for Huias, whilst our companion remained to fish for eels in the creek near our camp. After we had walked about a mile, a bird answered our call, and immediately afterwards a pair of Huias alighted in a pukatea tree above us. I brought them down, with right and left, and then another bird (a young male) appeared on the scene. He exhibited great excitement and was evidently at a loss to know what it all meant. Uttering a low, sibilant cry, with a tender pathos, he hopped down lower and lower, till within a yard or two of my head. I could easily have knocked the pretty creature over with a stick, but had not the heart to do so. I was less scrupulous, however, about having him caught, and in far less time than I take to write it, Rahui had selected a long stick, fixed a noose at the end of it, and slipped it over the bird's head. The Huia nimbly jumped through the loop but was caught by the feet. On finding himself a captive, he uttered no sound, but, in the most practical way, at once attacked my hands with his bill, striking fiercely and repeatedly at a white-faced signet-ring. On the following day Rahui managed to snare another, which was fortunately a female, thus making a pair of young birds. They became at once reconciled to confinement, eating freely of the huhu grub, and resting very contentedly on a perch to which they had been attached by a thong of flax. The young of the first year has a low and rather plaintive cry, easily distinguished from all other sounds in the forest, and pleasant enough to the ear. Our third and last day turned out wet and stormy; but we nevertheless got some more Huias, our bag consisting altogether of sixteen birds, exclusive of the live ones.

The Huia never leaves the shade of the forest. It moves along the ground, or from tree to tree, with surprising celerity by a series of bounds or jumps. In its flight it never rises, like other birds, above the tree-tops, except in the depth of the woods, when it happens to fly from one high tree to another. The old birds, as a rule, respond to the call-note in a low tremulous whistle or whimper, and almost immediately afterwards answer the summons in person, coming down noiselessly and almost with the rapidity of an arrow. Occasionally a shy old bird refuses every allurement, and takes himself quietly off. These knowing ones are distinguished by the bird-hunters as Huia-paoke. Young birds answer the call, although somewhat feebly, but do not, as a rule, present themselves. With these, it is necessary to mark down the direction, and follow them up with gun or snare.

They are generally met with in pairs, but sometimes a party of four or more are found consorting together.

Its food consists largely, as already stated, of the huhu grub; but it also subsists on the weta and other insects of various kinds, and the berries of such trees and shrubs as hinau, porokaiwiria, poukaka, and karamu. In the stomachs of those which I opened I found hinau berries (*Elæocarpus dentatus*), orthopterous insects, caterpillars, and the remains of a large spider; and Mr. Drew informs me that birds skinned by him had been feeding on the green and brown *Mantis*.

Within its restricted habitat the Huia appears to maintain its position notwithstanding the wholesale slaughter of late years. To say nothing of the zeal of collectors, who obtain large numbers for the European markets, the natives annually kill a great many for the sake of their feathers. For example: a party of eleven natives went out for a month and scoured the wooded country lying between the Manawatu gorge and Akitio, and brought in 646 skins; and a party of three men obtained a considerable number near Turakirai on the south-western side of the Wairarapa Lake. Other instances of the kind might be given, all tending to show that the struggle for existence with the Huia is becoming a severe one. Already the fate of several species which, a few years ago, were plentiful enough in these woods has been decided. In the course of our expedition, which extended altogether 27 miles beyond Masterton, we travelled over a broad extent of broken, wooded country, and, to say nothing of Korimako and Pitoitoi (which have long since disappeared), we never saw or heard the notes of either the Piopio, the Tieke, or the Hihi, all of which birds were at one time more numerous even than the Huia. The *Zosterops* was everywhere abundant, also the Grey Warbler and Rifleman, and along the edges of the bush we found the Tomtit comparatively plentiful; the Parrakeet chased its mate through the tree-tops with sharp cries of "twenty-eight"; the Tui, in its playful flight, mounted high in the sunlight overhead; and among the tangle of the underwood the ever-present Flycatcher displayed its pretty fan-like tail. But, of course, the charm of these dark *Fagus*-woods was the beautiful bird for which we had expressly come, and of which we had secured so many fine specimens.

One of the birds shot on our last day was a sitting female. The whole of the abdomen was denuded of feathers, and the skin had a smooth or polished appearance, as if the bird had been incubating for some time. This was on October 12, and was perhaps a case of early nesting, as none of the other birds presented any such appearance. In the ovary was a cluster of eggs, the largest of which was scarcely equal to a No. 6 shot. The ovarial duct was much enlarged, from which it may be inferred that the egg had only lately been laid. Another point deserving of notice is that the bird was very fat, even the intestines being overlaid with thin layers; whereas most of the birds we shot were in rather poor condition. May we not fairly infer from this that the male bird attends upon and feeds the female during incubation?

In the generality of dried specimens, and in the published drawings that have hitherto appeared, the bill is of a yellowish horn-colour; but this, instead of being natural, is caused by the decomposition of the animal matter inside. I have succeeded in retaining the ivory whiteness of the bill, in preserved specimens, by treating them after the manner recommended by Waterton for preserving the bill of the American Toucan (see 'Wanderings,' p. 103)—that is to say, by removing with a sharp scalpel the whole of the inner substance, leaving nothing but the outer shell, which then retains its original appearance. The process is a tedious one; but the result amply repays the trouble. The wattles of the Huia are of a bright orange colour, and during life are usually carried half-curled inwards.

I have given elsewhere * a figure of the dried head of a Huia handed to me, many years ago, by a native who had been wearing it as an ear-ornament. This specimen, which is now in the University Museum at Cambridge, represents a more highly curved form of bill than is usually met with.

I have also described and figured † a curious deformity in the bill of this bird. The lower mandible, in this instance, having been at some time accidentally broken off, the upper mandible had considerably overgrown it, becoming somewhat thickened beyond the point of friction ‡.

In my former edition I mentioned that a live female Huia had been added to the collection of the Zoological Society. The cage containing it was kept in the "Parrot House," being placed between

* Trans. New-Zealand Institute, 1870, vol. iii. pl. iv. fig. 3.

† *Op. cit.* 1877, vol. x. p. 211.

‡ More curious still is the case of deformity recently described by the Rev. W. Colenso, F.R.S., in a paper read before the Hawke's Bay Institute on the 9th August, 1886 (not yet published), of which the author has kindly sent me a copy, from which I extract the following:—"The head exhibited is that of a female Huia, the upper mandible of its bill being greatly and strangely deformed. From about one inch or one-fourth of the normal length of the upper mandible from its base it suddenly rises and remains at an angle of 45°, forming a regular, ascending, sub-erect spiral of two large and equal curves of ·75 of an inch, open, interior diameter, not unlike a gigantic cork-screw, and reminding one of the spiral horn of the *Strepsiceros*. The total length of this deformed mandible, following the curves, is just six inches. It is flat above and devoid of nostrils, and the end or tip is sharply pointed. . . . . The lower mandible is 2·75 inches long, being very much shorter and not so much curved as this portion of the bird's bill is in the normal state. . . . . There is not the slightest indication of the upper mandible ever having been broken or bruised. . . . . From its strange configuration it appears to have been far more than merely useless, for it must have been always an obstacle in the way and the means of keeping the bird's mouth constantly open. How it could have managed to exist seems truly wonderful!" *Vide* woodcut on page 17.

In connection with this tendency to abnormal growth, I may mention a suggestive circumstance that has lately come under my notice. A male bird which I presented to the Zoological Society was fully adult when I brought it to England. For about a year, in its new home, it has been fed on soft food, the bill being thus deprived of the ordinary wear and tear incident to the natural habits of the species. As a consequence, the bill has far outgrown in length its normal proportions, and has assumed a somewhat curved form, resembling that of an immature female. The wattles have retained their rich orange colour, and the bird seems to be in perfect health.

Mr. T. W. Kirk mentions (Trans. N.-Z. I. vol. xii. p. 249) another curious instance of deformity in the bill of a female Huia, in the Museum Collection at Wellington, and gives a woodcut to illustrate it. In this case, it appears to have resulted from an accident, a shot having probably passed through and split the upper mandible immediately below the nostril.

those of a Toucan on one side and a Hornbill on the other; and I was assured by Mr. Bartlett, the Superintendent of the Gardens, that this bird (although without a mate of its own species) seemed perfectly happy and contented in the midst of these new surroundings. It was supplied with a mixed food, in which boiled eggs, fresh meat, and earthworms formed the principal ingredients; but its diet required careful regulation, to prevent scouring, to which the bird was very liable. It did not, however, long survive this condition of things, and ultimately succumbed, as I venture to think, to the tropical heat of its environment—the prosector's official report being that it had died "in a much emaciated condition, but without organic disease" *.

There is now living in the "Western Aviary" in the same Gardens a fine male bird which I brought to England in April 1886, and which had been in possession of the Wairarapa Maoris for nearly a year previous to my leaving the colony.

A study of this living bird has enabled my artist to depict the species in the highly characteristic attitudes shown in the Plate. The berries represented are those of the titoki (*Alectryon excelsum*), on which the Huia doubtless feeds, for although habitually insectivorous, I have often found in the stomach the kernels of the hinau and other berries; and Mr. Tone informs me that he once saw

* As stated in the Introduction to my former edition (page xvii) the loss to the collection was a gain to science, for the late Prof. Garrod had thus an opportunity of studying the osteology and anatomy of this singular form; and I quoted the following passages from his valuable paper on the subject read before the Zoological Society on the 21st of May, 1872:—

"The arrangement of the feathers is completely Passerine. The rhomboid saddle of the spinal tract does not enclose any ephippial space, therein differing from the Crow's and resembling the typical Starling's. There are nineteen remiges, of which ten are on the hand; they increase in size up to the fifth. The rectrices are twelve in number. The oil-gland is nude. . . . The gizzard is well developed. The intestines are 16 inches long, with the bile-ducts 1½ inches from the gizzard. The cæca are 1 inch from the cloaca and ⅛ inch long, being cylindrical. There is one carotid artery, the left. . . . The palate is strictly ægithognathous; that is, the vomer is truncate in front abruptly, and cleft behind; the postero-external angles of the palatines are produced; the maxillo-palatines are slender, and approach towards, but do not unite with, one another, nor with the vomer, which they partly embrace. There is no ossification in the nasal septum anterior to the vomer. The whole cranial configuration closely resembles that of *Sturnus*; but the mandible, instead of being bent upwards, is straight. Like it, the palatines are narrow and approximate; the antero-internal angles of the posterior portions of those bones are reduced and rounded off, as is sometimes the case with *Sturnus*. The vomer is completely truncated in front, and is not prolonged forwards at its external angles, as in *Corvus* and its allies. The zygoma is not so slender as in *Sturnus*; but the curves are similar. The articular surfaces on the quadrate bone for the mandible are proportionally very large. The anterior extremities of the pterygoid bones articulate with the sphenoidal rostrum much as in *Corvus*, meeting in the middle line behind the posterior extremities of the palatines for a short distance. The maxillo-palatines, in their approximate portions, are shorter from before backwards than in *Sturnus*, and much resemble those of *Corvus*. The antero-inferior processes of the orbit are large and spongy; they almost touch the zygoma. But the most characteristic portion of the skull of *Heteralocha* is the occipital region; and in this it presents a great exaggeration of the peculiarities of *Sturnus* and its allies. In *Corvus* and most Passerines the digastric muscles occupy a narrow space intervening between the auditory meatus and the mass of occipital muscles, not extending so high up the skull as the latter. The occipital ridge encloses a space elongated from side to side, and of but little depth. In *Sturnus* the digastrics are much broader, and they narrow the occipital space; they also extend up the skull to so great an extent that they nearly meet in the middle line above the origin of the biventres cervicis muscles; but in *Heteralocha* they are of still greater size, and, meeting above the middle line, they form a strong ridge, which extends for some distance into the parietal region vertically. This peculiar development of these muscles produces a corresponding change in the shape of the space enclosed by the occipital ridge. In *Heteralocha* it is almost circular, and it extends some way above the foramen magnum. In *Sturnus* there is an approximation to this condition. A vertical parieto-occipital ridge in many other birds closely resembles that of *Heteralocha*; but it is the median limit of the temporal fossa in most. Correlated with this extensive digastric origin is a large surface for its insertion. The angle of the mandible is prolonged directly backwards for this purpose, in a manner unique among Passerine birds, but well seen in the *Anatidæ*. In *Sturnus* the angle of the mandible is slightly prolonged backwards for a similar purpose. . . . In the sternum *Heteralocha* differs in no important point from *Sturnus*, except that the posterior notches tend to be converted into foramina, as observed by Mr. Eyton in his 'Osteologia Avium.' . . . . In conclusion, it may be stated that the anatomy of *Heteralocha* shows clearly that it is truly Passerine, and not related to *Upupa*, as was previously supposed by most authors. When examined more in detail its relation to the *Sturnidæ* is found to be very intimate, and its structure is clearly not closely allied to that of the *Corvidæ*. In its relation to *Sturnus* it seems to present an exaggeration of the peculiarities of that bird, which would place it at the head of the family."—*Proc. Zool. Soc.* 1872, pp. 643–647.

three of them, at Akitio, feeding with avidity on the ripe fruit of the kahikatea (*Podocarpus dacrydioides*).

The Maoris prepare the skin in a very primitive way: cutting off the wings and legs, they strip the body and then flatten the skin to dry between two sheets of totara bark, tied tightly round with native flax, taking special care to keep the tail-feathers unsoiled. The latter are much prized as head-plumes on festive occasions, and for the ornamentation of the dead. In former days very artistic boxes (papa-huia) were carved in relief as caskets for these precious feathers.

This species builds its nest in hollow trees, forming it of dry grass, leaves, and the withered stems of herbaceous plants, carefully twined together in a circular form, and lined with softer materials of a similar kind *. An egg was brought to me on the 11th October, 1877, by Mikaera of Wainuiomata, who stated that it was found by him *in utero* when engaged in skinning a Huia. As already mentioned on page 4, the testimony of this man is not very reliable; but there can be little doubt that this is in reality the egg of the Huia, for it agrees in general character with one subsequently received at the Colonial Museum and described by Mr. Kirk †. My specimen was perfectly fresh when brought to me, and the shell was of such extreme delicacy that it was fractured under the gentlest handling in blowing. It is ovoido-elliptical in form, measuring 1·8 inch in length by 1·1 inch in breadth, of a very delicate stone-grey, inclining to greyish white, without any markings except at the larger end, where there are, chiefly on one side, some scattered rounded spots of dark purple-grey and brown; towards the smaller end there are some obsolete specks, but over the greater portion of its surface the shell is quite plain.

The specimen described by Mr. Kirk is somewhat smaller, being 1·45 inch in length by 1·1 in its widest diameter, the shell "having a beautifully fine and delicate structure, and pure white without any trace of markings whatever." This egg was obtained by Mr. G. M. Hewson from the Maoris of Murimotu, who assured him that it was that of the Huia.

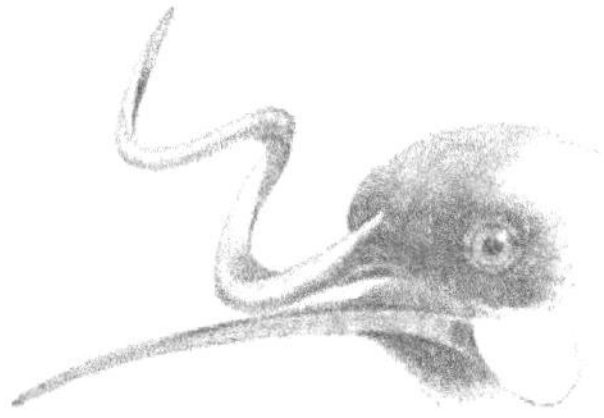

Abnormal growth of a Huia's bill (from a photograph). See footnote, p. 15.

* See an interesting account by Mr. Potts ('Zoologist,' 1884, p. 387) of a nest found in the cavity of an ancient hinau tree at Manawatu. On November 18th it contained one young bird. Another nest in the same neighbourhood contained three.

† Journal of Science, 1882–83, vol. i. p. 263.

CREADION CARUNCULATUS.

(THE SADDLE-BACK.)

Wattled Stare, Lath. Gen. Syn. iii. p. 9, pl. 36 (1783).
Sturnus carunculatus, Gm. Syst. Nat. i. p. 805 (1788, ex Lath.).
Creadion pharoides, Bonn. et Vieill. Enc. Méth. p. 874 (1823).
Icterus rufusater, Less. Voy. Coq. i. p. 649, pl. xxiii. fig. 1 (1826).
Xanthornus carunculatus, Quoy et Gaim. Voy. de l'Astr. i. p. 212, pl. 12. fig. 4 (1830).
Oxystomus carunculatus, Swains. Classif. of B. ii. p. 270 (1837).
Creadio carunculatus, Cab. Mus. Hein. Th. i. p. 213 (1850).

Native names.

Tieke, Tiraweke, Tirauweke, and Purourou.

♂ *ad.* nitidè niger: dorso cum tectricibus alarum, supracaudalibus et subcaudalibus latè ferrugineis: carunculis rictalibus miniatis: rostro et pedibus nigris; iride nigricanti-brunneâ.

♀ mari similis, sed minor et carunculis minoribus distinguenda.

Adult male. General plumage glossy black; back, wing-coverts, upper and lower tail-coverts bright ferruginous. Irides blackish brown; bill and legs black; wattles varying in tint from a clear yellow to a bright vermilion, being apparently affected by physical conditions, such as the health of the bird or the temperature of the weather. Total length 10 inches; extent of wings 12·5; wing, from flexure, 4; tail 3·5; bill, along the ridge 1·25, along the edge of lower mandible 1·4; tarsus 1·5; middle toe and claw 1·25; hind toe and claw 1·1.

Female. Of inferior size to the male, and having the wattles of a somewhat lighter colour.

Young. Has the colours of the adult, but with the tints duller and no sheen or gloss on the plumage; the wattles extremely small and of a pale yellow colour.

Obs. In the Natural-History Museum of the Jardin des Plantes, in Paris, I observed an adult specimen in partial albino plumage; and in the Canterbury Museum there is an example with a single white feather on the breast.

This bird derives its popular name from a peculiarity in the distribution of its two strongly contrasted colours, black and ferruginous, the latter of which covers the back, forms a sharply defined margin across the shoulders, and sweeps over the wings in a manner suggestive of saddle-flaps. The colours, in the male bird especially, are of so decided a kind as to attract special attention, to say nothing of the loud notes and eccentric habits of this remarkable bird. The bill is strong, sharply cut, and wedge-shaped, being well adapted for digging into decaying vegetable matter in search of larvæ, grubs, and insects, on which this species largely subsists. Berries, tender buds, and other vegetable substances likewise contribute to its support. From the angle of the mouth on each side there hangs a fleshy wattle, or caruncle, shaped like a cucumber-seed, and of a changeable bright yellow colour. The wings are short and feeble, and the flight of the bird, though rapid, is very laboured, and always confined to a short distance.

[The range of this species extends as far north as the Lower Waikato, beyond which district it is only rarely met with. It is numerous in the wooded ranges between Waikato Heads and Raglan, and is occasionally found in the neighbourhood of the Hunua coal-fields; but I have never heard of its occurrence in the Tauranga district, on the east coast, although I have an excellent ornithological correspondent there. In the summer of 1852 I obtained a pair at the Kaipara; but the bird was decidedly a *rara avis*, few of the natives in that part of the country being familiar with it. Captain G. Mair met with it once at Kaitaia, near the North Cape, and he afterwards saw a pair in the Maungatapere bush, near Whangarei. These are the only instances I can give of its occurrence on the mainland north of Auckland; but, strange to say, it is very plentiful on the Barrier Islands, in the Gulf of Hauraki. Mr. Layard was the first to notice its existence there, having shot a specimen on the Little Barrier, which he visited, in company with Sir George Grey, in 1863. He speaks of it (Ibis, 1863, p. 244) as "an apparently very rare bird;" but Captain Hutton, who visited these islands in December 1867, found it on both the Great and Little Barrier, and "very common" on the latter*. It is comparatively abundant in the wooded hills in the vicinity of Wellington and in those skirting the Tararua and Ruahine ranges; and it occurs also, and more plentifully, in many parts of the South Island.]

This species, formerly comparatively plentiful but now extremely scarce in the North Island, is very irregular in its distribution. In my first edition I endeavoured as above (within the brackets) to describe its range; but I omitted to mention that in one locality north of Auckland—a small wood at Kaitaia called Mauteringi, some three or four miles in extent—this bird was plentiful, although rarely ever met with in other parts of that district. Although never seen in the Bay of Plenty woods, it was, till within the last few years, numerous enough in the Ngatiporou country, where the natives were accustomed to regard it also as a bird of omen. A war-party hearing the cry of the Tieke to the right of their path would count it an omen of victory, but to the left a signal of evil. It is also the mythical bird that is supposed to guard the ancient treasures of the Maoris. The relics of the Whanauapanui tribe—*mere pounamus* and other heir-looms of great antiquity and value—are hidden away in the hollow of a tree at Cape Runaway, and it is popularly believed that the Tieke keeps guard over these lost treasures. According to Maori tradition, among these hidden things is a stone *atua*, which possessed at one time the power of moving from place to place of its own accord, but has since become inactive.

At the present time it is more plentiful on the Hen (a little wooded islet in the Hauraki Gulf) than anywhere else, a fact which may be attributable to the absence of wild cats; for on the Barrier Islands, where the cat has obtained a footing, this bird is nearly exterminated. On the Hen, according to Mr. Reischek, it is actually increasing in numbers. During his earlier visits they were only to be met with on the west and north-west sides of the island; on his last visit, after a lapse of only four years, they were to be heard and seen everywhere, being indeed the commonest bird on the island. They appeared to be of all ages; but neither here nor on the mainland did he ever meet with *Creadion cinereus*, which appears to be strictly confined to the South Island, where both species commingle.

The natives state that this species usually places its nest in the hollow of a tree, and they point to holes in well-known trees where the Tieke has reared its young for many years in succession. A pair is said to be still breeding in the hollow of the famous tree at Omaruteangi, known all over the country as "Putatieke" †. The bird is accordingly regarded with some degree of superstitious

* Trans. New-Zealand Inst. 1868, vol. i. p. 160.

† *Putatieke*: a renowned hinau tree in the Urewera country. It is supposed to possess miraculous attributes. Sterile women visit it for the purpose of inducing conception. They clasp the tree with their arms, and repeat certain incantations by way of invoking the *atua*.

reverence by the Arawa, who will not allow it to be wilfully destroyed. Those who have read Maori history will be familiar with the story of Ngatoroirangi and his sacred Tiekes of Cuvier Island. Hence the proverb, "Manu mohio kei Repouga," commonly applied to a man wise in council, and used in the sense of our own proverbial saying "Old birds are not to be caught with chaff."

Dr. Hector has informed me of a peculiarity in the habits of this species as observed by him in Otago. It is accustomed to follow the flocks of *Clitonyx ochrocephala* through the bush; but for what purpose it is difficult to imagine. Wherever he saw a flock of Yellow-heads there was invariably one of these Saddle-backs in attendance, mingling freely with them and, as it were, exercising a general supervision over the flock. He assures me that, during many months' residence in the woods, he had almost daily opportunities of verifying his observations regarding this very curious fact.

Active in all its movements, it seldom remains more than a few seconds in one position, but darts through the branches or climbs the boles of the trees, performing the ascent by a succession of nimble hops, and often spirally. It is naturally a noisy bird, and when excited or alarmed becomes very clamorous, hurrying through the woods with cries of "tiaki-rere," or a note like *cheep-te-te*, quickly repeated several times. At other times it has a scale of short flute-notes, clear and musical; but the most remarkable exhibition of its vocal powers takes place during the breeding-season, when the male performs to his mate in a soft strain of exquisite sweetness. This love-song is heard only on a near approach, and it is at first difficult to believe that so clamorous a bird could be capable of such tender strains.

When feeding its young the female has a different cry—a low, musical whistle, repeated once or twice. When the nest is invaded, or the safety of the young threatened, the male bird becomes very excited and utters his shrill cry with renewed energy and with quicker repetition.

The Plate represents the bird on a flowering branch of the pukapuka (*Brachyglottis repanda*); and I may here mention that in this and some other instances Mr. Keulemans has availed himself of my son's drawings of the New-Zealand flora.

Professor Hutton discovered the nest of this species on the Little Barrier Island. It was situated about two feet down the hollow stem of a dead tree-fern that had been broken off at the top, and from which he saw a Saddle-back emerge. The nest was roughly composed of stems of *Hymenophyllum* and dead fibres of nikau (*Areca sapida*), and lined with the fine papery bark of the *Leptospermum*; and it contained three eggs, which, at the time they were found (December 27th), had been slightly sat upon. One of these specimens was kindly forwarded to me and is now in the Colonial Museum; it measures 1·4 inch in length by 1 in breadth, and is white, marked and spotted, especially at the thicker end, with purplish brown of different shades.

An egg more recently received by the Canterbury Museum, from the West Coast, is of a rather elliptical form, measuring 1·2 inch in length by ·85 of an inch in its greatest width. It is of a delicate purplish grey, becoming lighter at the smaller end, and marked all over the surface, but more thickly at the larger end, with points, spots, and blotches of dark purple and brown.

I was informed by an intelligent Maori at Wellington that this bird is accustomed to repair, for many successive seasons, to the cavity in which it has once reared its brood, and that, although the number of eggs is generally three, he has occasionally found a nest containing four.

CREADION CINEREUS.

(JACK-BIRD.)

Creadion carunculatus (var.), Dieff. Report to N.-Z. Comp. (1844).
Creadion carunculatus (juv.), Hombr. & Jacq. Voy. Pôle Sud, Zool. iii. p. 12, fig. 4 (1853).
Creadion cinereus, Buller, Essay N.-Z. Orn. p. 10 (1865).
Creadion carunculatus (juv.), Finsch, Journ. f. Orn. 1867, p. 343; Hutton, Cat. of B. of N. Z. 1871, p. 17; Buller, Birds of N. Z. p. 149 (1873).

Native name.—Tieke.

Ad. cinerascenti-brunneus, subtùs pallidior: scapularibus alisque umbrino lavatis: supracaudalibus et subcaudalibus laetè rufescentibus: tectricibus alarum minimis rufo maculatis.

Adult. The entire plumage dark cinereous brown, paler on the underparts, and tinged with umber-brown on the wings and scapulars; the tips of the small wing-coverts and the entire upper and lower tail-coverts bright rufous.

Young. May be distinguished by the extreme smallness of the caruncles.

Obs. Individuals vary in the general tone of the plumage, some being greyish and others more strongly suffused with brown; the extent of the rufous markings on the wing-coverts is likewise variable, and in some examples they are entirely absent.

Mr. Potts has published* some interesting notes on six specimens in the Canterbury Museum, all in the plumage of *Creadion cinereus*, for the purpose of showing "how much variation may be met with in the young state of *C. carunculatus*." He admits, however, that these supposed young birds were "procured at different seasons of the year," which he accounts for on the supposition of an "extended breeding-season," or "that the adult state is not arrived at till the second year." It will be seen from what follows that this view is untenable.

In my 'Essay on the Ornithology of New Zealand,' published by command in 1865, I characterized and named what appeared to me then a new species of *Creadion* in the following terms:—"This species is of the size and general form of *C. carunculatus*, to which it bears a close affinity; but the colouring of the plumage is altogether different. The common species (the 'Saddle-back') is of a deep uniform black, relieved by a band of rufous brown, which occupies the whole of the back, and, forming a sharp outline across the shoulders, sweeps over the wing-coverts in a broad curve. In the present bird, however, the plumage is of a dark cinereous brown, paler on the underparts, and tinted with umber on the wings and scapularies; the upper and lower tail-coverts, and a few spots on the smaller wing-coverts, bright rufous. The wattles are of the same colour and shape as in *Creadion carunculatus*, but somewhat smaller."

My new species was at once attacked by Dr. Otto Finsch, who declared his belief that it was the young of *Creadion carunculatus*. Subsequently, in a paper which appeared in the 'Transactions of the New-Zealand Institute' (vol. v. p. 208), Dr. Finsch expressed his satisfaction that "Captain

* 'Out in the Open,' pp. 202, 203.

Hutton's examination of the types" had "shown *C. cinereus* to be undoubtedly the young of the above-named species." In my reply (*l. c.* vol. vi. p. 116) I explained that an examination by myself of a fine series of specimens in the Canterbury Museum *, showing what appeared to be transitional changes of plumage, had forced me to this conclusion, and that I had communicated the result to Captain Hutton long before the appearance of his 'Catalogue.' I was careful, nevertheless, to add the following qualifying passage:—"I confess, however, that the subject is still beset with some difficulty in my own mind. Supposing the plumage of *C. cinereus* to be the first year's dress of *C. carunculatus*, it seems to me quite inexplicable that the bird has never been met with in that state in the North Island. Captain Hutton suggests that this is due to the comparative scarcity of the species at the North. But during several years' residence in the province of Wellington I obtained probably upwards of fifty specimens, at various times, without ever detecting any sign of this immature condition of plumage. Admitting the comparative scarcity of the species, one would naturally suppose that the younger birds would be more likely to fall into the collector's hands than the fully adult ones. It may be suggested whether the condition of the Canterbury-Museum specimens has not possibly resulted from intercrossing; for we have not heard of any further examples of the kind being obtained. At any rate, till a specimen in the supposed immature dress has actually been taken in the North Island, the point cannot, I think, be considered finally set at rest."

The descriptive notes which I had made will be found at page 149 of my former edition, with a statement of the conclusion arrived at. But I then added:—"Mr. Buchanan has observed the so-called *C. cinereus* in Otago in the summer, and Captain Hutton saw four birds in this plumage near Collingwood in the month of August; while, in the North Island, I have obtained fully-coloured specimens of *C. carunculatus* all the year round. It is sufficiently obvious, therefore, that the former cannot be a seasonal state of plumage."

Strange to say, after a lapse of nearly fifteen years, the required evidence is forthcoming, and my *Creadion cinereus* recovers the specific rank so long denied to it.

In 1881 Mr. A. Reischek, a very careful observer, wrote to me as follows:—"About *Creadion cinereus* I have this to state: In December 1877, when I was on the west coast of the South Island, I shot about twenty of both kinds and of both sexes. What were supposed to be the young of *C. carunculatus* (your *Creadion cinereus*) I found on dissection to be fully adult birds, both male and female. My observations on this point were perfectly reliable. In December 1880 I stayed on the Hen (an island in the Hauraki Gulf) three weeks, and shot about thirty specimens of *Creadion carunculatus*, all of them being in the common saddle-back plumage. I could only determine the sex in each case by dissection, and what appeared to be the young birds differed only from the adult in having the wattles smaller and lighter in colour. I roamed over the whole island during my stay there, and never saw a bird in the plumage of your *Creadion cinereus*."

In 1882, and again in the early part of 1884, this naturalist re-visited the Hen, and on both occasions remained there a considerable time exploring every part of the island and collecting its productions. During his last visit he saw probably forty or more birds, all in the plumage of *C. carunculatus*, and collected many specimens of both sexes and all ages. On the Little Barrier he found the species scarce, and obtained only two specimens; while on the Chickens and Island of Kawau he did not meet with the bird at all. In some which he dissected the testes were almost microscopic, the only external differences between these and the old birds being that the plumage

* This series consists of four birds, all obtained in one locality:—No. 1 is in the plumage of *Creadion cinereus*, as described above: No. 2 presents a few black touches on the head and neck: No. 3 has some new black feathers between the crura of the lower mandible, also on the sides of the head and along the edges of the wings; the upper wing-coverts bright ferruginous; the half-grown new secondaries and tail-feathers perfectly black, the back and rump presenting indications of change: No. 4 is in the plumage of *C. carunculatus*, as described at page 18.

was not so glossy, and the wattles not so large or bright. In the adult male these ornamental appendages are of a beautiful orange colour, and in the adult female a little lighter. In the young birds they are still lighter and extremely minute.

To place the matter, however, beyond all doubt, he found, on the occasion of his last visit (on the 14th February), two adult birds feeding a young one, and was successful enough to secure all three birds, which he carefully preserved and marked. He was loth to part with these specimens; but, to enable me to demonstrate the specific value of *Creadion cinereus*, he handed all three birds over to me (marked respectively male, female, and young), and they are now in my collection.

In 1859 I found this species very abundant in the woods on Banks' Peninsula; but it has long since disappeared before the advancing tide of European settlement. It is still, however, comparatively plentiful on the western and south-western portions of the South Island.

Its habits are precisely similar to those of *Creadion carunculatus*, already described; and its mode of reproduction is the same *.

It has become the habit to speak of this bird as the Brown Saddle-back; but this is a misnomer, inasmuch as the absence of the "saddle" is its distinguishing feature. I have accordingly adopted the name of Jack-bird, by which it is known among the settlers in the South Island. Why it should be so called I cannot say, unless this is an adaptation of the native name "Tieke," the same word being the equivalent, in the Maori vernacular, of our "Jack."

That the two species occasionally interbreed is, I think, sufficiently evident from the specimens in so-called transitional plumage, in the Canterbury Museum, already specially mentioned. This is known to occur pretty often with the two allied species of Fan-tailed Flycatcher (*Rhipidura flabellifera* and *R. fuliginosa*) in the South Island, and, as there is every reason to believe, likewise in the case of our two species of Oyster-catcher, in both islands.

Under the head of Sturnidæ, Mr. G. R. Gray, in his 'List of the Birds of New Zealand,' published in 1862, included the genus *Aplonis*, with two species, *A. zealandicus* and *A. obscurus*. In my former edition, I omitted these birds altogether, as I had been unable to obtain any satisfactory evidence of their occurrence in New Zealand. In my 'Manual of the Birds of New Zealand' (published in 1882) I admitted *Aplonis zealandicus* on the authority of Dr. Finsch, who wrote:— "This is an excellent and typical species, which I had the pleasure of seeing in the Leiden Museum, being one of the typical specimens brought home by the 'Astrolabe' Expedition. Dr. Hartlaub informs me that there are three specimens in the Museum in Paris, all marked 'Tasman's Bay, New Zealand,' and collected by the French travellers." Further investigation, however, has satisfied me that it has no claim whatever to a place in the New-Zealand avifauna.

Last year I visited the Museum of the Jardin des Plantes in Paris for the express purpose of examining the type specimens referred to by Dr. Finsch; and, through the courtesy of Dr. Oustalet, the officer in charge of the Ornithological department, I had an opportunity of thoroughly investigating the subject.

* "For its nesting-place a hollow or decayed tree is usually selected; sometimes the top of a tree-fern is preferred. We found a nest in a dead tree-fern not far from Lake Mapourika, Westland. This was of slight construction, built principally of fern-roots, deeply woven into rather a deep-shaped nest with thin walls; for as the structure just filled the hollow top of the tree-fern, thick walls were unnecessary. Another nest, in a small-sized decayed tree in the Okarita bush, was in a hole not more than three feet from the ground. It was roughly constructed, principally of fibres and midribs of decayed leaves of the kiekie, with a few tufts of moss, leaves of rimu, lined with moss and down of tree-ferns; and it measured across, from outside to outside of wall, 12 inches 6 lines, cavity 3 inches diameter, depth of cavity 2 inches. The egg, measuring nearly 1 inch 4 lines through the axis with a breadth of $11\frac{1}{2}$ lines, sprinkled over with faint purplish marks, towards the broad end brownish purple, almost forming one large blotch."—*Out in the Open*, p. 202.

There are two specimens in the mounted collection, from the voyage of the 'Astrolabe,' labelled *Aplonis zealandicus*, Quoy & Gaim., but without any habitat being assigned to them, the words "New Zealand" on the label having been crossed out. On referring to the original entry in MM. Quoy and Gaimard's catalogue of the 'Astrolabe' collection, I found the following note under the No. relating to this species—"Vani koro (New Hebrides) et New Zealand." There seems to be no other authority than this for considering it a New-Zealand bird; and I have no doubt, in my own mind, that the true home of the species is in the New Hebrides, the addition of "New Zealand" being merely a mistake in the entry, especially as there is no locality named. It is not the kind of bird that would rapidly become extinct; and if the French travellers had met with it during their casual visit to New Zealand, it is fair to assume that the species would have been known to the inhabitants of the country. The specimen in the Leiden Museum being simply a duplicate from this collection, the same remarks apply to that also. For these reasons I again reject *Aplonis zealandicus* as a New-Zealand form; but as one species occurs on Norfolk Island and possibly another on Lord Howe's Island—within what is in reality the New-Zealand zoo-geographical region, although not within the scope of the present work—and as the claims of *Aplonis zealandicus* may again come up for discussion, I think it may be useful to place on record a full description of the species; and as there is much confusion in the nomenclature of this and the closely allied forms from Polynesia and Australia, I will add the result of my recent examination and identification of specimens both at Paris and in the British Museum.

As to the species itself being a good and valid one, I agree with Dr. Finsch, for although closely related to the other members of this confused group, the bright rufous colouring on its upper parts makes it readily distinguishable.

According to the views propounded by Mr. A. R. Wallace in his 'Geographical Distribution of Animals,' and now generally accepted, Norfolk Island, Phillip Island (or the Nepean group), Lord Howe's Island, and the Kermadec Isles represent the minimum extension to the northward of a continental area perhaps exceeding that of Australia in extent, of which New Zealand in ancient times formed a part. The existence at the present day, or till within a very recent date, of a species of Kaka Parrot (*Nestor productus*) on Phillip Island, of a form of Weka Rail (*Ocydromus sylvestris*) on Lord Howe's Island, and of the great brevipennate Rail (*Notornis alba*) on Norfolk Island, if not on Lord Howe's Island as well, indicates beyond doubt a former land connection, because it would be manifestly impossible for birds of this kind to traverse a wide extent of ocean. That the separation from each other of these distant habitats, by the submersion of the intervening land, took place at a very remote period, is sufficiently evident from the extreme specialization of the forms I have mentioned, although undoubtedly referable to the generalized New-Zealand types. From this point of view, it might be deemed advisable to include the birds inhabiting these various islands in the New-Zealand avifauna, which Mr. Wallace has already practically done by defining the boundaries of the New-Zealand "sub-region." It will be found, however, on a closer examination, that, owing probably to accidental transportation and occasional immigration of individuals, over a long period of time, the avifaunæ of these islands have acquired features more in common with Australia than New Zealand. This very instance, indeed, of the existence in Norfolk Island of *Aplonis fuscus* (although not mentioned by Wallace) betrays this fortuitous relation, if I may so term it, of its ornis to that of Australia and of Central Polynesia. I have therefore decided to confine myself, in the present work, to the islands which come within the political limits or jurisdiction of New Zealand, namely, the Chatham Islands on the east, the Auckland Islands, Campbell Island, Macquarie Island, and Antipodes Island on the south and south-east; and I shall only refer incidentally to the occurrence of allied forms in the remote islands to the north in my treatment of our local species. As the number of Plates is necessarily limited, I shall figure only birds that are actually found in New Zealand, but

I shall be careful to give an illustration of every endemic species. Birds that are common to other countries may or may not be figured, according to the circumstances of each particular case.

APLONIS ZEALANDICUS.—Two examples (in Paris): no sex stated; but one is slightly larger than the other, with the colours of the plumage a little brighter, and is presumably the male.

♂ *ad.* General plumage rufous-grey, darker on the upper parts and deepening to rufous-brown on the lower part of back, rump, and upper tail-coverts; from the anterior edge of the eye a dull black streak extending to the nostrils; the primaries bright rufous on their outer webs only, being blackish brown on their inner webs; large wing-coverts and bastard quills bright rufous; tail-feathers dark rufous-brown, with a rich vinous tinge on their outer edges; underparts lighter, the feathers of the breast and abdomen having obscure, narrow, greyish margins; flanks, vent, and under tail-coverts rufous-brown, mixed with tawny yellow, the feathers becoming lighter at the tips. Bill blackish brown, with a reddish tinge on the under mandible; legs and feet pale brown; claws yellowish brown. Total length 7·5 inches; wing, from flexure, 4; tail 2·5; bill, along the ridge ·75, along the edge of lower mandible ·75; tarsus ·8; middle toe and claw ·9.

♀ *ad.* Similar to the male, but with duller plumage, and of somewhat smaller size.

Obs. I am satisfied that *A. rufipennis*, Layard, from Vaté Island, New Hebrides, described in 'The Ibis,' 1881, p. 512, is this bird, and not *Calornis castoroides* as suggested by Canon Tristram.

ALLIED SPECIES. *Aplonis tabuensis*, Gmel. (= *A. vitiensis*, Layard, = *marginatus*, Gould, = *marginalis*, Hartl., = *marginata*, Cass., = *cassinii*, Peale).—More strongly built, and being a lighter-coloured species; only a rufous tinge on the plumage of the upper parts, with a purplish sheen on the head and neck; an obscure facial streak; the pectoral feathers with pale shaft-lines, giving a slightly streaky character to the breast. In young birds the sheen is absent and the pectoral streaks are more conspicuous. Irides red.—*Hab.* Tonga group, Savage Island, Friendly Islands, and Fiji. There is a slight difference observable in specimens from Tonga and Fiji, but nothing of any specific value.

Aplonis fuscus, Gould.—I do not think this form is separable from *A. tabuensis*. It is slightly browner on the upper parts than specimens from Tonga, but cannot be distinguished from some Fiji examples of the latter species. —*Hab.* Norfolk Island and Australia.

Aplonis brevirostris, Peale.—This species also seems to me scarcely separable from *A. tabuensis*, the only differences being in its somewhat smaller size, the darker crown, and the less streaky appearance on the underparts. In all essential respects the birds are alike. In the 'Hand-list of Birds' (vol. ii. p. 26) Mr. G. R. Gray makes *Aplonis australis*, Gould, a synonym of this species, but I have not seen this type.—*Hab.* Samoa.

Aplonis nigroviridis, Less. (= *A. pacificus*, Forst.?, = *striatus*, Gmel., = *obscurus*, Dubus, = *viridigriseus*, G. R. Gr.). —Slaty grey, with a darker head and neck, and a very perceptible gloss on the plumage, especially on the upper surface; the facial streak broader than in *A. zealandicus*. The young of this species has the entire plumage slaty grey, paler and mixed with light brown on the underparts, some specks of white on the cheeks, and the small wing-coverts narrowly margined with whitish grey; but even in the young state the facial streak is quite conspicuous, having the appearance of a dull inky stain.—*Hab.* New Caledonia and Lord Howe's Island.

Aplonis caledonicus, Bp.—Entire plumage black and glossy, with green reflections in certain lights and purplish on the head and throat. The sexes are alike, except that the female has less gloss on the plumage. Prince Bonaparte's type, marked by his own hand, is in the Museum at Paris. The British Museum contains a good number of specimens, showing very little variation, and all from New Caledonia. A specimen marked *Aplonis mavornata*, but without any reference, differs from *A. caledonicus* in having the entire plumage dingy brown, without any gloss, the feathers of the underparts narrowly margined with grey. This may prove to be the young of *A. caledonicus*, but no locality is given.

Aplonis atronitens, G. R. Gray.—This seems to be a good species, with a much more robust bill than any of the preceding, and having the entire plumage brownish black, with little or no gloss on the surface. The single specimen in the British Museum was obtained by Sir George Grey from the Loyalty Islands.

TURNAGRA HECTORI.

(NORTH-ISLAND THRUSH.)

Otagon tanagra, Schl. Ned. Tijdschr. Dierk. iii. p. 190 (1865).
Turnagra hectori, Buller, Ibis, 1869, p. 39.
Turnagra tanagra, Gray, Hand-l. of B. i. p. 284 (1869).
Keropia tanagra, Finsch, J. f. O. 1870, p. 323.

Native names.

Piopio, Koropio, Korohea, and Tiutiukata.

Ad. statura *T. crassirostris* sed rostro crassiore, suprà olivascenti-brunneus: pileo nusquam striolato: uropygio caudâque claro rufis: gutture albo: pectore superiore cinerascente: abdomine medio albo, parte imâ et subcaudalibus conspicuè flavicantibus: hypochondriis olivascentibus: rostro et pedibus saturatè brunneis: iridè flavâ.

Adult. Crown of the head, hind neck, and upper parts generally clear olive-brown; throat pure white; breast and abdomen ashy grey, darker on the former, the abdomen and the under tail-coverts tinged with yellow; sides olive-brown, washed with yellow; wing-feathers dark olive-brown, dusky on their inner webs; tail-feathers and their upper coverts bright rufous, paler on their under surface, the two middle ones tinged above with olive-brown. Irides yellow; bill and feet dark brown. Total length 11 inches; wing, from flexure, 5·25; tail 5; bill, along the ridge ·8, along the edge of lower mandible 1; tarsus 1·25; middle toe and claw 1·25; hind toe and claw 1.

Young. Birds of the first year differ in having the feathers at the base of the upper mandible, the tips of those covering the crown and sides of the head, the small feathers fringing the eyelids, and a broad zone on the upper part of the breast bright rufous; the primary and secondary wing-coverts, and sometimes the secondary quills, are also largely tipped with the same colour, and the grey of the underparts is darker, but with a tinge of orange-yellow under the wings.

In January 1869 I communicated to 'The Ibis' the description of a new species of Thrush inhabiting the North Island, and differing from the South-Island bird (*Turnagra crassirostris*) not only in plumage, but in its superior size and more strongly developed bill; and I named it in compliment to my friend Dr. (now Sir) James Hector, F.R.S., Director of the Colonial Museum and Geological Survey of New Zealand.

In an editorial footnote to my paper, Professor Newton suggested that this species might be identical with one described, in a Dutch work, by Professor Schlegel, four years before, without, however, any habitat being assigned to it. This opinion has since been verified by a careful comparison of the specimen I have figured with the type of Schlegel's *Otagon tanagra* in the Museum at Vienna; and under ordinary circumstances the name I have proposed would of course be reduced to a synonym. It will be observed, however, that Professor Schlegel has used a common generic name to distinguish the bird specifically, while he refers the form to the genus *Otagon*, established by Bonaparte in 1850. As I can see no valid reason for setting aside the generic title of *Turnagra*

proposed by Lesson as early as 1837, and as the adoption of the older specific name would, according to this view, give the confused result of *Turnagra tanagra*, I have deemed myself justified in retaining the distinctive appellation of *T. hectori*. At the same time I am anxious to give due prominence to the fact that Professor Schlegel was the first to discover the existence of this new species.

There is a peculiar charm about the New-Zealand forest in the early morning; for shortly after daylight a number of birds of various kinds join their voices in a wild jubilee of song, which, generally speaking, is of very short duration. This was the morning concert to which Captain Cook referred in such terms of enthusiasm; and the woods of Queen Charlotte's Sound, where his ship lay at anchor, are no exception to the general rule. In illustration of this, I take the following from an entry in one of my note-books:—"Tuesday, 5 A.M.—At this moment the wooded valley of the Mangaone, in which we have been camped for the night, is ringing with delightful music. It is somewhat difficult to distinguish the performers amidst the general chorus of voices. The silvery notes of the Bell-bird, the bolder song of the Tui, the loud continuous strain of the native Robin, the joyous chirping of a flock of White-heads, and the whistling cry of the Piopio—all these voices of the forest are blended together in wild harmony. And the music is occasionally varied by the harsh scream of a Kaka passing overhead, or the noisy chattering of a pair of Parrakeets on a neighbouring tree, and at regular intervals the far-off cry of the Long-tailed Cuckoo and the whistling call of its bronze-winged congener; while on every hand may be heard the soft trilling notes of *Myiomoira toitoi*." For more than an hour after this concert had ceased, and the sylvan choristers had dispersed in search of their daily food, one species continued to enliven the valley with his musical notes. This bird was the Piopio, or New-Zealand Thrush, the subject of the present article, and unquestionably the best of our native songsters. His song consists of five distinct bars, each of which is repeated six or seven times in succession; but he often stops abruptly in his overture to introduce a variety of other notes, one of which is a peculiar rattling sound, accompanied by a spreading of the tail, and apparently expressive of ecstacy. Some of the notes are scarcely distinguishable from those of the Yellow-head; and I am inclined to think that the bird is endowed with mocking-powers. The ordinary note, however, of the Piopio, whence it derives its name, is a short, sharp, whistling cry, quickly repeated.

It was when I obtained a caged Piopio that I first became acquainted with its superior vocal powers. In 1866 I purchased one for a guinea from a settler in Wellington, in whose possession it had been for a whole year. Although an adult bird when taken, it appeared to have become perfectly reconciled to confinement; but on being placed in a new cage it made strenuous assaults on the wire bars, and persevered till the feathers surrounding its beak were rubbed off and a raw wound exposed. It then desisted for several days; but when the abraded part had fairly healed, it renewed the attempt, and with such determined effort that the fore part of the head was completely disfigured, and the life of the bird endangered. On being removed, however, to a spacious compartment of the aviary, it immediately became reconciled to its condition, made no further efforts to escape, and for a period of fifteen months (when it came to an untimely end) it continued to exhibit the contentment and sprightliness of a bird in a state of nature.

I observed that this bird was always most lively during or immediately preceding a shower of rain. He often astonished me with the power and variety of his notes. Commencing sometimes with the loud strains of the Thrush, he would suddenly change his song to a low flute-note of exquisite sweetness; and then abruptly stopping, would give vent to a loud rasping cry, as if mimicking a pair of Australian Magpies confined in the same aviary. During the early morning he emitted at intervals a short flute-note, and when alarmed or startled uttered a sharp repeated whistle.

This caged bird was generally fed on dry pulse or grain; but he also evinced a great liking for cooked potato and raw meat of all kinds; in fact he appeared to be omnivorous, readily devouring earthworms, insects of all kinds, fruits, berries, green herbs, &c. He was supplied daily with a dish of fresh water, and was accustomed to bathe in it with evident delight. At one time he occupied the same division of the aviary with a pair of Australian Ring-Doves which had commenced to breed. The Doves were allowed to bring up their first brood in peace; but when the hen bird began to build a second time, she was closely watched by the Piopio, and immediately the first egg was deposited he darted upon the nest and devoured it. The innocent little Ring-Dove continued to lay on in spite of repeated robbery, and had at length to be placed beyond the reach of her persecutor. During the day the Piopio was unceasingly active and lively; at night he slept on a perch, resting on one leg, and with the plumage puffed out into the form of a perfectly round ball, the circular outline broken only by the projecting extremities of the wings and tail. Every sound seemed to attract his notice, and he betrayed an inquisitiveness of disposition which in the end proved fatal; for having inserted his prying head through an open chink in the partition, it was seized and torn off by a vicious Sparrow-Hawk in the adjoining compartment of the aviary.

In the wild state this species subsists chiefly on insects, worms, and berries. I have shot it on the ground in the act of grubbing with its bill among the dry leaves and other forest débris. Its flight is short and rapid. It haunts the undergrowth of the forest, darting from tree to tree, and occasionally descending to the ground, but rarely performing any long passage on the wing. It is very nimble in its movements; and when attempting on one occasion to catch one of these birds with an almost invisible horsehair noose, it repeatedly darted right through the snare, and defeated every effort to entrap it.

In my former edition of this work I stated that the Piopio was at that time comparatively common in all suitable localities throughout the southern portion of the North Island, but was extremely rare in the country north of Waikato. I mentioned also that a specimen which I shot in the Kaipara district in the summer of 1852 (doubtless a straggler from the south) was quite a novelty to the natives in that part of the country; that it was recognized, however, by an old Maori, who called it a "Koroke," a name quite unknown in the south, and who stated that in former years it was very abundant in all the woods. I ventured then to express a belief that the bird whose biography I had undertaken to write would soon be equally scarce elsewhere. And so it has proved, for the North-Island Piopio is now one of our rarest species, and is certainly doomed to extinction within a very few years.

In the Bay of Plenty district it has never been heard of since the time of Hongi's famous invasion (about the year 1820). A little wooded spur near Te Puke settlement, behind Maketu, frequented by a pair of these birds at that troublous period has ever since borne the name of Piopiorua; and to the present day the old men talk of the ominous appearance in their district of this "manu aitua" at the time that the bloodthirsty warrior landed in his war-canoes and spread terror and destruction with his newly acquired firearms *.

The last accessible place in which I met with it was Horokiwi, about 25 miles from Wellington.

* Captain Mair, who took a prominent part throughout the late Maori War, and finally won the New-Zealand Cross by his gallant conduct at Ohinemutu, informs me that on one occasion, when in close pursuit of Te Kooti and his followers in the Urewera country, he unexpectedly came upon a pair of these birds in the bush, and, at the risk of scaring his nobler game, could not resist the temptation of shooting both specimens of so rare a species. This was at a place between the Whakatane and Rangitaiki rivers. During very many years spent in the Bay of Plenty he has never seen or heard of the Piopio in that district or in the Rotorua country; but he once heard its unmistakable note in some low bush at the northern end of Lake Taupo. In February 1880 he shot a pair at Taumarunui, near the junction of the Wanganui river with the Ongarue in the Tuhua country, at a point about 250 miles by the river from the town of Wanganui. This pair had been known to the natives as inhabiting that particular locality for several years. In the hope of securing them he travelled more than fifteen miles through the bush. He found them perfectly tame, answering his call and hopping round him, apparently quite heedless of his presence; but his efforts to ensnare them were all in vain, the bird always darting through the loop and escaping.

This was some twenty years ago—when riding through this lovely wooded valley—at a time when the road passed through the primitive forest, all untouched by the hand of man, disclosing to the eye new beauties at every turn as it followed the course of a tortuous mountain-stream. From the time of my first visit up to the present (and I have passed through the valley hundreds of times) I have never tired of this beautiful sylvan scenery; but at the period I speak of the bush was an almost impervious tangle, the lower tree-tops bound together with kareao and other creeping plants, and the trees themselves laden with a rich epiphytic growth. Even now it is a delightfully refreshing resort. The tawa rears its feathery branches of soft pale green, and beside it rises, like a sentinel, the cone-shaped top of the darker *Knightia excelsa*; the bright green of the rimu with its graceful, drooping boughs, is everywhere present; and, as the eye scans the scene more closely, almost every tree common to the New-Zealand bush may be readily distinguished, all growing in rank profusion, plentifully sprinkled with the star-like crowns of giant tree-ferns, varied here and there with the bending palm-like top of the nikau (*Areca sapida*), its huge stem springing up from the shady depths of the uneven forest—the whole presenting a beautiful picture, in ever varying tints, and almost sub-tropical in the luxuriance of its growth. In this valley there are yet some matchless groups of *Cyathea medullaris* and other tree-ferns; but the hand of civilization is upon the wilderness, the virgin forest is receding more and more, the axe of the woodman is incessant, and the bushman's fire is doing every season its further work of devastation. A few years hence, and the sylvan beauty of Horokiwi with all its sweet memories will have passed away for ever!

One peculiarity about this species is its devotion to some particular locality, beyond which it never wanders very far. Mr. C. Field, a Government surveyor, who has spent the best part of his life in the woods, writes to me:—"I have seen the bird in the same spot year after year, and generally in pairs, except when the hen is nesting. To my certain knowledge a pair of them have kept to the same locality, on a valley flat by the side of a stream, for a period of seven or eight years." My last fresh specimens (two males and one female preserved in spirit) were received in January 1884, from this gentleman, who obtained them far up the wooded valley of the Pourewa on the west coast, where he was conducting a trigonometrical survey. A year later a skin was sent in by Mr. Tone, another Government surveyor, who was employed on the east coast, and who informed me that the bird was still to be met with in the woods at Akitio.

A pair has been known to frequent for several seasons a spot on the western side of the Rangataua lake, near the source of the Mangawhero river, at the foot of the Ruapehu mountain. A correspondent who visited the place in the summer of 1880 was informed by the resident natives that the birds had always nested there. He could hear their musical song from his camp across the lake, and on going over he found the old birds in a maire tree, but could see nothing of the young brood. They were very tame and fearless, and on his simulating their notes they readily came to the ground and hopped about, scratching the surface and turning over the leaves as if in search of insects.

It shows how rare the bird has become when its habitat is thus localized. Indeed, it has already entirely disappeared from a tract of country where in former years it was specially abundant. In proof of this, I may mention the experience of Mr. Morgan Carkeek, who in 1884, at the instance of the Public Works Department, made a careful exploration of the Mokau-Wanganui district. Starting from the foot of Mount Egmont he followed down the Patea river, then up the north-east branches of this and the Wanganui rivers, crossed the watershed, and followed up the north-west branches of the latter into the Tuhua country; and then returned, by a route lying between the White Cliffs and Mokau, to the sea-coast. All the country thus traversed is heavy bush-land and, for the most part, excessively rough and broken. During the whole journey, which occupied about two months, he never once saw or heard a Piopio!

As to its nidification, I may mention that in the Ruahine ranges I met with a breeding-pair of these birds late in December. The sudden disappearance of the female and the cautious demeanour of the male satisfied me that I was in the immediate vicinity of the nest; but I nevertheless failed in my endeavours to find it. The bird resented my intrusion on its sanctum by a peculiar purr, not unlike the alarm-note of the American Cardinal (*Cardinalis virginianus*), accompanied by a sudden spreading of the tail.

A native once described the nest to me as being of large size and composed of moss, twigs, and dry leaves. He assured me that he had twice met with it in the high scrub near the Manawatu river, and that in both cases the nest contained two eggs. This was many years ago; but that the account was reliable may be inferred from the fact, since ascertained, that this description applies very well to that of a closely allied species in the South Island.

Although *Turnagra* has hitherto been placed among the Turdidæ, the form is admittedly an aberrant one. Dr. Finsch has suggested the propriety of uniting it to *Glaucopis*, but I do not think this view has met with any acceptance or support. Fortunately I was able to bring with me to England a specimen in alcohol, which I forwarded to Dr. Gadow, of Cambridge, for anatomical study. After making an autopsy, with his accustomed care, he writes to me as follows:—"I am sorry to say that the outcome of my investigation regarding *Turnagra* is not very striking. After all, you are quite right in your suggestions as to its position and affinities. The fact is, we know so little of the anatomy of the many birds belonging to the Timeliidæ that comparison with these forms is almost out of the question. At any rate, it is satisfactory to know that there are not present any known characters to indicate other affinities, or to negative your suggestions."

Mr. Sharpe has placed *Turnagra* among his Timeliidæ; but I have decided to make it the type of a new family, Turnagridæ, because the form seems to differ quite as much from typical *Timelia* as it does from *Turdus*.

As it is important to place on permanent record the results of Dr. Gadow's patient study of the subject, I shall here append his report in full, together with his detailed remarks on *Glaucopis* (referred to on page 4), in order to show that there is no relationship between these two forms, notwithstanding the similarity of some of the external characters:—

"TURNAGRA.—Stomach quadrangular, flattened, very muscular. Crop absent. Tongue fleshy, with a few short bristles on the sides near the tip. Intestinal convolutions Thrush-like, certainly not Corvine, with decided graminivorous adaptation. Syrinx muscles acromyodean. Pterylosis agrees with Nitzsch's Subulirostres s. Canoræ. Ten primaries; terminal (or first) long; tip of wing formed by third to seventh; sixth longest. Nine secondaries. Twelve tail-feathers. Metatarsus like that of Thrushes or Sylviæ. Sternum and shoulder-girdle agree with many birds: *Struthidea*, *Graucalus*, *Strepera*, *Ptilonorhynchus*, *Turdus* (*i. e.* all alike). *Conclusion*: After examination of the digestive apparatus, the pelvic nerve-plexus, the skeleton, and the pterylosis, I feel inclined to put *Turnagra* with the wide and ill-defined group of Timeliidæ. *Turnagra* is certainly neither Corvine nor Fringilline, and it is in fact a member of the Southern (Indian-Australian) mass of Thrush-like birds. Its bill and certain modifications of its digestive apparatus seem to show that this bird is a *Thrush with graminivorous propensities*. I would put it into Sharpe's subfamily Ptilonorhynchinæ, to which *Ælurœdus* belongs, but unfortunately *Ptilonorhynchus* itself is very different from Timeliidæ in its pterylosis."

"GLAUCOPIS.—After examination of the skeleton I am satisfied that *Glaucopis* comes nearest to the Corvidæ. The skull, although in general configuration and beak very similar to that of *Struthidea*, differs from the latter. Barring the peculiar lacrymals, it agrees with *Ptilonorhynchus*, also with *Strepera*, and, more remotely, with *Paradisea*. No agreement with *Graucalus*. Comparison with *Heteralocha* and *Sturnus* is not possible. Skull, consequently, agrees with *Ptilonorhynchus* and *Strepera*. *Sternum*: agrees most with that of *Strepera*, far less with *Graucalus*, *Struthidea*, *Paradisea*. *Ptilonorhynchus* disagrees in clavicles, like *Heteralocha* and *Sturnus*. *Pelvis and sacrum*: agrees with *Graucalus*, *Heteralocha*, and *Ptilonorhynchus*, also with *Strepera*, *Paradisea*, and *Struthidea*. *Metatarsal scutes*: agree most with *Heteralocha*; through the fusing condition in which the scutes are, very much with *Ptilonorhynchus* and *Gymnorhina*. *Hyoid bones*: Corvidæ. *Pterylosis*: *Strepera* and *Ptilonorhynchus*, but the latter has considerably more remiges. *Conclusion*: *Glaucopis* is nearly allied to the Austrocoraces. It agrees best with *Strepera* (Gymnorhininæ in general), and shows some considerable similarity in structure with *Ptilonorhynchinæ*. *Struthidea* agrees with *Glaucopis* by far less than you might perhaps suppose, and *Graucalus* is still further removed. *Heteralocha* is an unmistakable Starling form, and has little of importance in common with *Glaucopis*."

TURNAGRA CRASSIROSTRIS.

(SOUTH-ISLAND THRUSH.)

Thick-billed Thrush, Lath. Gen. Syn. ii. pt. 1, p. 34, pl. xxxvii. (1783).
Tanagra capensis, Sparrm. Mus. Carls. pl. 45 (1787).
Turdus crassirostris, Gm. Syst. Nat. i. p. 815 (1788, ex Lath.).
Lanius crassirostris, Cuv. Règn. Anim. p. 338 (1817).
Campephaga ferruginea, Vieill. Nouv. Dict. d'Hist. Nat. x. p. 48 (1817).
Tanagra macularia, Quoy et Gaim. Voy. de l'Astr. i. p. 186, pl. vii. fig. 1 (1830).
Keropia crassirostris, Gray, List of Gen. of B. p. 28 (1840).
Turnagra crassirostris, id. op. cit. p. 38 (1841).
Loxia turdus, Forst. Descr. Anim. p. 85 (1844).
Otagon turdus, Bonap. Consp. Gen. Av. i. p. 374 (1850).
Ceropia crassirostris, Sundev. Krit. Framst. Mus. Carls. p. 9 (1857).
Turnagra turdus, Gray, Hand-l. of B. i. p. 284 (1869).

Ad. suprà olivaceo-brunneus, pileo vix cinerascente irregulariter fulvo striato: tectricibus alarum dorso concoloribus, rufo terminatis, fasciam duplicem alarem exhibentibus: remigibus brunneis, extùs dorsi colore marginatis, primariis ad basin rufo lavatis: supracaudalibus rufo tinctis, imis omninò rufis: caudà letè rufâ, rectricibus duabus mediis et reliquarum apicibus olivaceo-brunneis: loris cum regione oculari genisque brunneis pallidè rufo maculatis: regione paroticà pileo concolore, angustè fulvo striatâ: subtùs olivaceus, gutture toto rufescente lavato, plumis medialiter fulvescentibus: pectoris plumis medialiter albidis, utrinque olivaceo marginatis, quasi striatis: pectore superiore vix rufescente lavato: hypochondriis magis olivascentibus; abdomine imo et subcaudalibus flavo lavatis: subalaribus rufis: rostro pedibusque saturatè brunneis: iride flavâ.

Adult. General plumage olive-brown, darker on the upper parts; forehead, lores, throat, and sides of neck largely marked with rufous; breast, abdomen, and under tail-coverts covered with broad longitudinal spots of yellowish white, narrower towards the sides of the body; on the abdomen and under tail-coverts less of the olive-brown, with a strong tinge of yellow; wing-feathers dark olive-brown, dusky on their inner webs, the superior and lesser wing-coverts largely tipped with rufous, forming two broad transverse bars; lining of wings pale rufous; tail, for the most part, with the upper coverts bright rufous, the two middle feathers and the apical margins of the rest olive-brown, only slightly tinged with rufous. Irides yellow; bill and feet dark brown. Total length 11 inches; wing, from flexure, 5; tail 5; bill, along the ridge ·7, along the edge of lower mandible ·8; tarsus 1·25; middle toe and claw 1·15; hind toe and claw 1.

Young. May be distinguished from the adult by the larger amount of rufous colouring on the forehead, sides of the head, throat, and upper wing-coverts.

Obs. In some specimens the bend of the wing and the exterior edges of the outer primaries are also marked with rufous. The colour of the bill likewise varies, in different examples, from a light brown to dusky black.

This fine species is confined to the South Island. Formerly it was excessively abundant in all the elevated wooded country; but of late years it has become comparatively scarce, while in some districts it has disappeared altogether. This result is attributable, in a great measure, to the ravages of cats and dogs, to which this species, from its ground-feeding habits, falls an easy prey.

Sir James Hector informs me that, during his exploration of the West Coast in the years 1862-63, he found it very abundant, and on one occasion counted no less than forty in the immediate vicinity of his camp. They were very tame, sometimes hopping up to the very door of his tent to pick up crumbs; and he noticed that the camp-dogs were making sad havoc among them. He is of opinion that in a few years this species also will be numbered among the extinct ones.

Mr. Buchanan, of the Geological Survey, assures me that in the woods in the neighbourhood of Dunedin, where it was formerly very common, it has been quite exterminated by the wild cats. It may be here observed that there is no indigenous cat in our country; but ill-fed or ill-used members of the race, in the struggle for existence, frequently quit the settlers' houses and betake themselves to the woods, where they, in course of time, produce a purely wild breed. To this cause is partly owing the almost entire extermination of the Quail and other ground species.

It is worthy of remark that Mr. Burton obtained a specimen on Stephens Island on the south side of Cook's Strait.

The habits of this bird differ in no respect, so far as I am aware, from those of its congener in the North Island. The following incident is illustrative of its predaceous nature:—My brother, Mr. Fletcher Buller, while residing in Canterbury, obtained a live one from the woods, and placed it in a cage with a pair of tame Parrakeets (*Platycercus novæ-zealandiæ*). On the following morning he found, to his dismay, that the newly introduced bird had slain both of his fellow prisoners, and was actually engaged in eating off the head of one of them!

There is a nest of this bird in the Canterbury Museum, obtained from the River Waio, County of Westland. It is a round nest, somewhat loosely constructed, composed of small, dry twigs, shreds of bark, fragments of moss, &c., with a rather large cup-shaped cavity, lined with dry grasses and other fibres. To all appearance it is carelessly, but nevertheless firmly, fixed in the forked twigs of a small upright branch. In the same collection there is another nest from Lake Mapourika, which is formed of soft green moss on a tapering foundation of small twigs, completely filling the crutch of a manuka fork and being fully a foot in depth. Another, formed externally of dry twigs, is of more irregular shape, but is likewise built in a forked branch as a means of support. The circular cup is neatly lined with dry bents. Mr. Potts, who studied this bird pretty closely in Westland, states that the nest is generally found among the thick foliage of the tutu (*Coriaria ruscifolia*), but sometimes in karamu or manuka, that it is sometimes finished off with soft tree-fern down as a lining, and that it usually contains two eggs; and he is of opinion that the bird breeds twice in the season. The Museum collection contains four specimens of the egg, which exhibit considerable difference in form. Two of them—probably from one nest—are very ovoido-conical; one of these measures 1·3 inch by 1·05 inch, and is pure white, marked at irregular distances over the entire surface with specks and roundish spots of blackish brown. The other is slightly narrower in form, the white is not so pure, and the markings are less diffuse, being collected into reddish-brown blotches towards the larger end. The other two eggs (apparently also from one nest) are of a long ovoido-elliptical form, and of equal size; the one I tested measuring 1·6 inch in length by ·95 of an inch in its widest part. The shell is pure white, with widely-scattered irregular spots of blackish brown, less numerous and of smaller size in one than in the other. Both eggs have a rather glossy surface.

MIRO AUSTRALIS.

(NORTH-ISLAND ROBIN.)

Turdus australis, Sparrm. Mus. Carls. iii. pl. 69 (1788).
Muscicapa longipes, Garnot, Voy. Coq. i. p. 594, pl. xix. fig. 1 (1826).
Myiothera novæ-zealandiæ, Less. Man. d'Orn. i. p. 248 (1828).
Miro longipes, Less. Tr. d'Orn. p. 389 (1831).
Petroica australis, Gray, Voy. Ereb. and Terror, p. 7 (1844).
Myioscopus longipes, Reich. Syst. Av. Taf. lxvii. (1850).
Petroica longipes, Gray, Ibis, 1862, p. 223.
Miro longipes, Buller, Birds of New Zealand, 1st ed. p. 119 (1873).

Native names.

Pitoitoi, Toutou, Toutouwai, and Totoara.

♂ saturatè cinereus, scapis plumarum albidis: maculâ frontali albâ: tectricibus alarum dorso concoloribus: remigibus brunneis, extùs cinereo lavatis: caudâ nigricante: facie laterali cinereâ, albido magis distinctè striolatâ: abdomine medio albicante: corporis lateribus cinereis: subcaudalibus albidis: cruribus cinereis albido terminatis: subalaribus pallidè cinereis: primariis intùs ad basin albidis: rostro nigricanti-brunneo, mandibulâ brunnescentiore: pedibus pallidè brunneis: iride nigrâ.

♀ pallidior: remigibus brunnescentibus: facie laterali cinerascente, albo striolatâ: pectore superiore pallidè cinerascente, plumis medialiter albido striatis: abdomine albido.

Adult male. Head, neck, and all the upper surface dark slaty grey, plumbeous beneath; the shafts of the feathers greyish white, forming rather conspicuous lines on the crown and nape; a frontal spot at the base of the upper mandible pure white; rictal bristles black; throat, fore neck, and sides of the body paler slaty grey; the lower part of the breast, the middle of the abdomen, the vent, and the under tail-coverts white, blending on the edges with the darker plumage of the surrounding parts; wing-feathers dull smoky brown, with lighter shafts; lining of wings and a broad oblique bar on the under surface of all the quills except the first three primaries pure white; tail-feathers dull smoky brown, the shafts light brown on their upper and white on their under aspect. Irides black; bill blackish brown; tarsi and toes pale yellowish brown; soles dull yellow. Total length 6 inches; extent of wings 9·25; wing, from flexure, 3·5; tail 2·65; bill, along the ridge ·6, along the edge of lower mandible ·8; tarsus 1·35; middle toe and claw ·95; hind toe and claw ·8.

Female. Slightly smaller than the male and with duller plumage; the upper parts tinged with smoky brown; the throat, fore neck, and sides of the body lighter, the centre of each feather inclining to greyish white.

Young. The young of both sexes resemble the female in the comparative brownness of the plumage of the upper parts; the rictal membrane is largely developed and of a rich orange-colour.

Obs. In this and the other closely allied species the feathers of the body have loose or disunited filamentous barbs, and are very soft in texture, especially on the upper parts.

Note. I entirely agree with Dr. Finsch that this form should be separated from *Petræca* (erroneously called *Petroica*); but I am unable to follow him in adopting the genus *Myioscopus* of Reichenbach, the name of

Miro proposed by Lesson having a prior claim in regard to date. The long legs, shorter wings, and stouter bill distinguish this genus from *Petraeca* and bring it nearer to *Erythacus*.

This species is confined to the North Island, where, till within the last ten years, it was very common in all the wooded parts of the country; but it is represented in the South Island by a closely allied and still common species, the *Miro albifrons*. There is a specimen of the North-Island Robin in the Auckland Museum said to have been obtained at Nelson; but I have never found this bird south of Cook's Strait, and *vice versâ* as regards the South-Island Robin. The two species may therefore be regarded as true representatives of each other in the North and South Islands respectively.

Generally speaking, in New Zealand it is only on the outskirts of the woods that we meet with insessorial birds in any number. As we penetrate into the heart of the forest, the birds become fewer, till at length they almost entirely disappear. But there is one species whose range seems to be quite without restraint: common enough in the open coppice, it is to be found also in the gloomiest and most secluded parts of the forest. This bird is the subject of our article—the Pitoitoi or Toutouwai * of the natives and the "Robin" of the colonists.

I have been assured by officers who accompanied the celebrated Tarnnaki Expedition under Major-General Sir Trevor Chute, in 1866, that during that long and irksome march the Robin was the only bird that gave any sign of life to those interminable and gloomy forests through which the army passed. The lively twitter and song of the smaller birds had ended with the first day's march, the harsh cry of the Kaka (*Nestor meridionalis*), which had attended them far into the bush, had gradually ceased to be heard, and the Wood-Pigeon (*Carpophaga novæ-zealandiæ*), whose range extends to the summits of the low wooded ridges of the interior, was no longer to be met with. An oppressive silence reigned around them, broken only by the shrill chirp of the startled Robin as the advanced guard cut a path for the troops through the hitherto untrodden woods. Indeed the presence of this little bird was the only exception to the utter absence of animal life, and almost the only relief to the monotony of the march. Perched on a low branch, it might frequently be seen looking gravely down, as if in silent wonderment, on the weary ranks, as they toiled their way through this virgin forest in the very heart of the enemy's country!

As the popular name implies, it is naturally a tame bird; and in little-frequented parts of the country it is so fearless and unsuspicious of man that it will approach to within a yard of the traveller, and sometimes will even perch on his head or shoulder. It is the favourite companion of the lonesome wood-cutter, enlivening him with its cheerful notes; and when, sitting on a log, he partakes of his humble meal, it hops about at his feet, like the traditional Robin, to pick up the crumbs.

Like its namesake in the old country, moreover, it is noisy, active, and cheerful. Its note is generally the first to herald the dawn, while it is the last to be hushed when the evening shades bring gloom into the forest. But there is this noticeable difference between the morning and the evening performance: the former consists of a scale of notes commencing very high and running down to a low key, uttered in quick succession, and with all the energy of a challenge to the rest of the feathered tribe; and I have sometimes heard a native, when listening to this strain, exclaim "Ka kanga te manu ra!" (How that bird swears!). The evening performance is merely a short chirping note, quickly repeated, and with a rather melancholy sound. Three or four of them will sometimes join in a chirping chorus, and continue it till the shades of advancing twilight have deepened into night.

* There are some curious coincidences with Maori names, of which this is an instance. The Robin is called "Toutouwai" by the Ngapuhi tribe at the far north. The small European Owl, *Athene noctua*, has "Koukouwai" as its Greek name. Drop the final syllable, and we have the Maori name for the New-Zealand Owl, "Koukou."

It lives almost entirely on small insects and the worms and grubs which are to be found among decaying leaves and other vegetable matter on the surface of the ground in every part of the woods. Its nature is pugnacious and, in the pairing-season, the male birds often engage in sharp encounters with each other.

It generally breeds in the months of October and November. It constructs a large and compact nest, composed externally of coarse moss firmly interwoven and thickly lined inside with the soft hair-like substance which covers the young stems of the tree-fern. It is usually built against the bole of a tree, at a moderate elevation from the ground, being often found attached to and supported by the wiry stems of the kiekie (*Freycinetia banksii*), a climbing parasitical plant which is everywhere abundant. I have found scores of the nests of this species, and almost invariably in the situation described. I found one, however, placed in the fork of a tree at some elevation, and another in the truncated stem of a tree-fern (*Cyathea dealbata*). The eggs are usually three in number, broadly ovoido-conical, and measuring ·95 of an inch in length by ·70 in breadth; they are of a creamy white colour, thickly freckled and speckled with purple and brown, these markings being denser at the thick end, where they form an indistinct purplish zone.

Should the nest happen to be molested after the young are hatched, the parent birds manifest the utmost solicitude, hopping about near the intruder with outspread and quivering wings, uttering a low piping note, and showing every symptom of real distress.

The last example of the nest I examined was obtained recently on the Little Barrier Island, where it was found supported against the bole of a tree about five feet from the ground. It is not so massive as many I have seen, and is composed chiefly (and probably for protective purposes) of the green moss which clings to the trunks of old trees, mixed with dry leaves and little twigs of wood; the cup, which is rather shallow, measures three and a half inches in diameter and is deeply lined with fern-hair and vegetable fibres. It was found about the middle of December, just after the young birds had quitted it.

But for the fact that much of the foregoing article applies equally to the South-Island species, it would have been almost necessary to expunge it from the present edition; for, alas! its subject, instead of being, as formerly, the commonest of our native birds, is now one of the rarest. It is still comparatively plentiful on the Island of Kapiti, and on some of the wooded islets in the Hauraki Gulf; but it is seldom met with on the mainland, and, in common with many other native forms, its doom is sealed.

Ornithologists everywhere must regret this, because the genus to which it belongs has no representative in any other part of the world; and those who are at all familiar with the bird itself will assuredly grieve over its threatened extirpation. Personally I regard this gentle Robin with a strong sentiment of affection. In the days of my boyhood it was one of the dominant species, and some of my earliest memories are associated with it. The first nest I ever found in my juvenile excursions through the bush near the parental home—the dear old Church Mission station of forty years ago—was naturally that of the Robin. It was the first bird of which I ever prepared a specimen; and having, while yet at school, conceived the idea of writing a history of our native birds, I well remember that the first species whose biography I essayed to sketch was this everyday companion of my holiday rambles. Its presence therefore never fails to awaken reminiscences of the past; but unfortunately ere long the bird itself will be but a memory of by-gone years. Either on account of its being an easy prey to wild cats and rats, or else in obedience to some inexplicable law of nature, the species is rapidly dying out; and it requires no prophetic vision to foresee its utter extinction within a very short period. Well may the Maori say, as he laments over the decadence of his own race—"Even as the Pitoitoi has vanished from the woods, so will the Maori pass away from the land and be forgotten!"

MIRO ALBIFRONS.

(SOUTH-ISLAND ROBIN.)

White-fronted Thrush, Lath. Gen. Syn. ii. pt. 1, p. 71 (1783).
Turdus albifrons, Gm. Syst. Nat. i. p. 822 (1788).
Miro albifrons, Gray, in Dieff. Trav. ii., App. p. 190 (1843).
Petroica albifrons, Gray, Voy. Ereb. and Terror, p. 7, pl. 6. fig. 2 (1844).
Turdus ochrotarsus, Forster, Descr. Anim. p. 82 (1844).
Muscicapa albifrons, Ellman, Zool. 1861, p. 7465.

♂ similis *M. australi*, sed multo major et magis fuliginosus, scapis plumarum minùs distinctè albis: sed præcipuè pectore et abdomine medio ochrascenti-albis distinguendus.

Adult male. Head, neck, sides of the body, and all the upper surface dark sooty grey, the base of the feathers plumbeous; at the root of the upper mandible a small spot of yellowish white; breast, abdomen, and vent yellowish white, tinged with lemon-yellow on the breast, and forming a tolerably well-defined line against the dark plumage of the fore neck; inner lining of wings, flanks, and under tail-coverts greyish white; quills and tail-feathers smoky black; an oblique bar of white on the inner face of the wings, as in *M. australis*. Irides black; bill brownish black; palate and soft parts of the mouth yellow; tarsi, toes, and claws brownish black; soles of the feet dull yellow. Total length 7·25 inches, extent of wings 10·5; wing, from flexure, 4; tail 3; bill, along the ridge ·65, along the edge of lower mandible ·85; tarsus 1·5; middle toe and claw 1·05; hind toe and claw ·9.

Female. Somewhat smaller than the male, and having the plumage of the upper parts tinged with brown; there is less yellow on the breast, and the grey of the underparts is lighter.

Obs. This species may readily be distinguished from *M. australis* by its appreciably larger size, its black legs and darker coloration, and the more defined patch of yellowish white on the under surface. The white shaft-lines are not so distinct on the crown and nape, but are far more so on the throat and fore neck, owing to the ground-colour of these parts being darker than in *M. australis*. The frontal spot is smaller and less conspicuous.

Dr. Finsch has expressed an opinion in favour of uniting *M. australis* and *M. albifrons*; but a glance at the Plate will show how decidedly the two species differ from each other in their external characters*.

Var. My collection contains a specimen received from Christchurch in which the whole plumage is suffused with brown, and the underparts are smoky grey instead of being white †.

Note. The figure of this species in the 'Voyage of the Erebus and Terror' is incorrect, on account of the exaggerated extent of white on the underparts; but the attitude is a very characteristic one.

The habits of this bird differ in no respect from those of its near congener *Miro australis*; and the account given in the foregoing pages may be considered equally applicable to both species.

* "These birds seem to be scarcely distinct" (Finsch, Trans. N.-Z. Inst. vol. v. p. 207).

† In the 'Journal of Science,' vol. ii. p. 170, a full description is given of a pied example, or partial albino, the white preponderating over the normal colour, and the breast being creamy yellow.

It has large lustrous eyes, and the feathered fringe to the eyelid imparts to them an unusually prominent appearance. When its attention is excited, it assumes a very erect position, and flips its wings and tail, often uttering a short chirp between each operation. It is popularly said to have the power of expanding and contracting the small white spot on its forehead, but the explanation is a simple one: when the bird is at rest, it habitually raises the frontal feathers, making the head look large and rounded and rendering the white spot almost invisible; when excited or alarmed the feathers are immediately depressed and the frontal spot is at once conspicuous.

Under the head of *Eudynamis taitensis*, mention will be made of its services as foster-parent to the young of that Cuckoo, of which we have at least one undoubted instance.

On comparing the eggs of this species with those of *M. australis*, there is a manifest difference. They are slightly larger and more ovoido-conical in form, measuring 1·05 inch in length by ·7 in breadth. They present also more individual variation than do those of the North-Island bird, which are all marked on the same pattern. In two eggs of *M. albifrons* in my son's collection one has the entire surface minutely and indistinctly freckled with grey, whilst the other has the larger end splashed all over with confluent spots of purplish brown, with a few widely scattered specks over the rest of the surface. Another (taken from the nest in February) is somewhat pyriform in shape, measuring ·9 of an inch in length by ·7 in breadth; the obtuse end thickly smudged with dull brick-red, washed over with brown, and a few sprinkles of the same colour on other portions of the shell.

Some doubts having existed as to the true position of the genera *Miro* and *Myiomoira*, I furnished Dr. Gadow with specimens in spirit of *Miro albifrons* and *Myiomoira toitoi* to enable him to study their internal characters, and he reports, as the result of his investigations, that both forms are true Singing-birds, and that the place I had already assigned them, in my former edition, among the Sylviidæ is undoubtedly the right one *. This fact is of some importance from a systematic point of view, because of the relation of this group to others about whose location in the system there is much difference of opinion.

In the British Museum Catalogue (Birds, vol. iv.) Mr. Sharpe places both these forms among the Muscicapidæ, associating *M. toitoi* and *M. macrocephala* with twelve other species in the genus *Petrœca*, with a range extending over Australia, New Zealand, and the Pacific Islands.

Professor Newton, in his able article on "Ornithology" (Encycl. Brit.), has the following remarks:—"There is no doubt whatever as to the intimate relationship of the Thrushes (Turdidæ) to the Chats (Saxicolinæ), for that is admitted by nearly every systematizer. Now most authorities on classification are agreed in associating with the latter group the Birds of the Australian genus *Petrœca* and its allies—the so-called 'Robins' of the English-speaking part of the great southern communities. But it so happens that, from the inferior type of the osteological characters of this very group of birds, Prof. Parker has called them (Trans. Zool. Society, v. p. 152), 'Struthious Warblers' †. Now, if the *Petrœca* group be, as most allow, allied to the Saxicolinæ, they must also be allied, only rather more remotely, to the Turdidæ—for Thrushes and Chats are inseparable, and therefore this connexion must drag down the Thrushes in the scale. Let it be granted that the more highly developed Thrushes have got rid of the low Struthious features which characterize their Australian relatives, the unbroken series of connecting forms chains them to the inferior position, and of itself disqualifies them from the rank so fallaciously assigned to them."

* "MIRO ALBIFRONS (Passeres, Acromyodi, Turdiformes).—Stands very well with the Sylviidæ, where you have already put it. Tail-feathers twelve. Primary remiges ten, the terminal one being long, more than half the length of the next. Secondary remiges nine. Pterylosis typically Sylviine and Turdine. Metatarsus completely encased by three long scutes or shields, one anterior and two lateral, the latter forming a sharp posterior prominent keel; truly Sylviine. Intestines agree with Sylviine birds likewise. Stomach contained insects. Nothing peculiar about *Miro* at all. The same applies to *Myiomoira*."—H. GADOW.

† "*Petrœca* has been stated by Professor Parker to be a 'Tracheophone' (*i. e.* Mesomyodian), having 'the muscles of lower larynx quite indistinct.' In three specimens, however, of that genus examined by me I find a perfectly Oscinine syrinx with its muscles as well developed as in other birds of the same size" (Forbes, P. Z. S. 1882, p. 545).

MIRO TRAVERSI.

(CHATHAM-ISLAND ROBIN.)

Miro traversi, Buller, Birds of New Zealand, 1st ed. p. 123 (part ii.), June 1872.
Petroica traversi, Hutton, Ibis, July 1872, p. 245.

Ad. omninò niger, remigibus et rectricibus paullò brunnescentioribus: rostro nigro: pedibus nigris, plantis flavis: iride saturatè brunneâ.

Adult male. The whole of the plumage black, the base of the feathers dark plumbeous; wing-feathers and their coverts tinged with brown, the former greyish on their inner surface; tail-feathers black, very slightly tinged with brown. Irides dark brown; bill black; tarsi and toes blackish brown, the soles of the feet dull yellow. Total length 6 inches; wing, from flexure, 3·4; tail 2·6; bill, along the ridge ·5, along the edge of lower mandible ·7; tarsus 1·1; middle toe and claw 1; hind toe and claw ·8.

Female. Slightly smaller than the male, and without the brown tinge on the wings and tail.

This species was discovered by Mr. Henry Travers during an exploratory visit to the Chatham Islands in 1871. Through the courtesy of His Excellency Sir George Bowen (who forwarded them in his despatch-box to the Colonial Office), I received specimens of the male and female in time to include this bird, in its systematic order, in my former edition. In dedicating the species to the enterprising naturalist who had discovered it, I thus unavoidably anticipated Professor Hutton, who had proposed the same name for it, but did not publish his description till after mine had appeared.

Mr. Travers supplies the following note respecting it:—"I only found this bird at Mangare, where it is not uncommon. It is very fearless, possessing in other respects the habits of *Petroica albifrons* and *P. longipes*. Its ordinary note is also the same, but I did not hear it sing. It appears to be specially obnoxious to *Anthornis melanocephala*, which always attacks it most savagely when they meet."

This form appears to be the small degenerate representative of the New-Zealand Robin, which, strangely enough, does not occur in the Chatham Islands; but it is even more remarkable still that, so far as our information goes, the present bird is not found either on the main island or on its satellite, Pitt Island, being confined exclusively to Mangare, which is described as a mere rocky islet covered with low rigid scrub.

The antipathy, mentioned by Mr. Travers, on the part of *Anthornis melanocephala* towards this species is quite unaccountable, because the ordinary habits of the two birds do not conflict in any way, whilst between *Anthornis melanura* and the Robin in New Zealand the most perfect amity exists. Possibly the pugnacious habit has been developed by the insular nature of its environment, and the more severe conditions of life in the struggle for existence.

It may be here stated that the Chatham Islands, to which frequent reference will be made in the course of this work, are situated about 450 miles eastward of New Zealand, in lat. 42° S. The group consists of Wharekauri, about seventy miles in extent, shaped like an isosceles triangle, and presenting a low diversified surface of bush, lake, and open land, with much fertile soil: Pitt Island, about ten miles in circumference, separated from the main island by a deep channel, and now occupied by a sheep-farmer: Mangare, the home of *Miro traversi*, as mentioned above; South-east Island and other small rocky satellites, which are uninhabited.

MYIOMOIRA TOITOI.

(NORTH-ISLAND TOMTIT.)

Muscicapa toitoi, Garnot, Voy. Coq. i. p. 590, t. xv. fig. 3 (1826).
Miro toitoi, Gray, in Dieff. Trav. ii., App. p. 191 (1843).
Petroica toitoi, Gray, Voy. Ereb. and Terror, Birds, p. 6 (1844).
Myiomoira toitoi, Reich. Syst. Av. Taf. lxvii. (1850).
Muscicapa albopectus, Ellman, Zool. 1861, p. 7465.

Native names.

Miromiro, Komiromiro, Pimiromiro, Ngirungiru, Pingirungiru, and Pipitori.

♂ suprà sericeo-niger: maculâ frontali conspicuâ albâ: tectricibus alarum plerumque nigris, medianis brunnescentibus: remigibus brunneis, primariis interioribus ad basin albo maculatis, secundariis magis conspicuè notatis, plagam albam exhibentibus: caudâ nigrâ, rectricibus tribus exterioribus ferè omninò albis, basi pogonii interni et apice pogonii externi exceptis nigris: facie laterali, gutture toto et pectore superiore nigris, gulâ vix brunnescente: corpore reliquo subtùs albo, basi plumarum nigricante: rostro et pedibus nigricanti-brunneis: plantis pedum flavicantibus: iride nigrâ.

♀ mari dissimilis: brunnea, subtùs albida, hypochondriis brunnescente lavatis: loris et facie laterali brunneis, fulvescente variis.

Adult male. Head, neck all round, and all the upper parts black; frontal spot, at the base of the upper mandible, white; breast and underparts pure white, the black of the fore neck having a sharply defined lower edge; wing-feathers crossed near their base by an angular patch of white, which is narrow and interrupted on the primaries, broad and continuous on the secondaries, the black shafts, however, forming fine intersecting lines; tail black, the three outer feathers on each side crossed obliquely upwards by a broad bar of white, which covers more than a third of their surface. Irides and rictal bristles black; bill and tarsi blackish brown; toes paler, yellow on their inner surface. Total length 5 inches; wing, from flexure, 3; tail 2·25; bill, along the ridge ·4, along the edge of the lower mandible ·5; tarsus ·75; middle toe and claw ·8; hind toe and claw ·65.

Adult female. Upper surface smoky brown, with a minute frontal spot of white; throat, fore neck, and all the underparts greyish white, more or less clouded with dull smoky brown; wing-feathers blackish brown, a bar across the base of the secondaries and some indistinct marks on the webs of the outer primaries fulvous white; tail black, the three outer feathers on each side barred obliquely with white, as in the male.

Young. In the young male the colours are much duller and browner, and the sharply defined pectoral line is wanting; but the plumage is sufficiently different from that of the female to distinguish the sexes.

Obs. The sexes do not present any perceptible difference in size. Individuals, however, vary perceptibly. The measurements of an ordinary bird are given above; but a smaller example of the adult male which I shot in the Forty-Mile Bush gave the following results:—Length 4·75 inches; extent of wings 8; wing, from flexure, 2·75; tail 2; bill, along the ridge ·25, along the edge of lower mandible ·5; tarsus ·75; middle toe and claw ·75.

In both sexes the tongue, palate, and interior of the mouth, as well as the angle, are orange-yellow; differing in this respect from *Clitonyx*, in which the male bird has a black mouth and the female a flesh-coloured one.

This elegant little bird belongs to the North Island, where it has a pretty general distribution, being met with in all localities suited to its habits*. It is a familiar species, seeking the habitations of man, and taking up its abode in his gardens and orchards. It is always to be seen in the clearings and cultivated grounds near the bush, moving about in a peculiar fitful manner, and in the early morn may be heard uttering a prolonged trilling note, very sweet and plaintive. Its usual attitude is with the wings slightly lowered and the tail perfectly erect, almost at a right angle with the body. It has a sparkling black eye, and all its actions are lively and sprightly. The strongly contrasted plumage of the male bird renders it a conspicuous object; but the female, owing to her sombre colours and less obtrusive habits, is rarely seen. Its note in the early morning is like the Maori syllables *ngi-i-ru*, *ngiru-ngiru*, from which it derives its native name, the first syllable being somewhat prolonged. Throughout the day, and often till late in the evening, it utters, at frequent intervals, a soft note like the words "Willoughby-willoughby," repeated several times. This is often heard in association with the musical trill of *Gerygone*, the two birds warbling, as it were responsively, from the same bush.

It is very tenacious of life, and I have found it difficult to kill, even with dust-shot, the bird often flying some distance after being mortally wounded. On examining it after death, one is struck with the disproportionately large size of the head, which is kept drawn in upon the body during life, as shown in the figure. The plumage, which is peculiarly soft and yielding, is distributed in well-defined tracts or areas, as in all other Carinate birds; but the intervening spaces are unusually wide, being perfectly smooth and bare, and the skin on the hind neck rises in a peculiar, naked fold, with a narrow line of feathers on the top like a mane.

It is interesting to watch this active little creature as it flits about the fences and fallen timber in the bush-clearings, where it is to be found at all hours of the day. It rests for a moment on its perch, flirting its wings and tail in a rapid manner, then darts to the ground to pick up a grub or earthworm, and, flying upwards again almost immediately, clings by its tiny feet to the upright bole of a tree or some other perpendicular surface, a peculiar attitude which it appears to delight in. Its food consists of small insects and their larvæ; and it proves itself useful by devouring a destructive little aphide which infests our fruit-trees. I have opened many and in every instance found its stomach full of minute insect remains, proving how serviceable it must be to the husbandman.

Like its allies, *Erythacus* for instance, this bird has a pugnacious spirit, and during the pairing-season the males meet and fight on the slightest provocation, whether real or imaginary.

I have noticed that it often manifests an attachment for a particular locality, resorting to the same perch day after day. The Maoris, too, have observed this; and at Otaki they passed their title to a plot of ground through the Native Land Court under the name of "Te-tau-a-te-Miromiro" (the perching-place of the Miromiro).

It is far less plentiful than it formerly was in our fields and gardens. There seems no reason to fear, however, that the species is dying out, for in the *Fagus* forests of the interior I have found it extremely plentiful. In the woods at the foot of Ruapehu and neighbouring high lands, where, save the occasional twitter of small birds in the branches, all is silent as the grave, this pretty little creature is always to be met with. It flits noiselessly from one tree to another, then descends to the ground, and in a few instants reappears on its perch, flirting its tail upwards, and emitting at intervals a soft call-note of peculiar sweetness. Destitute of animal life as these sub-alpine woods undoubtedly are, they are not without their attractions. Owing to their high elevation vapour-clouds are continually hanging over them, causing a perpetual moisture. In consequence of this the

* Mr. Sharpe says (Cat. Birds Brit. Mus.) that *M. toitoi* is found in the Chatham Islands. But this is obviously a mistake, the only species at present known from these islands being *M. dieffenbachii*, which, as explained above, is identical with *M. macrocephala* of the South Island.

trees on their outer facies are more or less covered with kohukohu, a feathery fungus of a pale green colour, hanging like drapery from the branches, while their trunks and limbs are clad to their very tops with the richest profusion of lichens and mosses. No idea can be formed of the quasi-tropical richness of these woods in this respect by any one who has not actually visited them.

Its ally, *Myiomoira macrocephala*, in the South Island, has the same habit of frequenting high altitudes; for not only is this bird met with among the high tussock-grass on the plains, but likewise on the summits of the ranges, flitting about among the snow-grass and other stunted vegetation, at an elevation of 5000 feet or more, and subsisting on the small alpine lepidoptera and their larvæ, or such diptera and other minute insects as inhabit these mountain heights.

Common as this species is, I have found it difficult to study its breeding-habits, and have never succeeded in finding more than one nest. I met with this in the Upper Hutt valley, in the neighbourhood of Wellington, as late as the 3rd of December. It was placed in the cavity of a tree a few feet from the ground, and contained four young birds apparently about a week old. The nest was composed entirely of dry moss, shallow in its construction, but with a neatly finished rim or outer edge. The parent birds manifested some solicitude for the safety of their offspring while I was handling them. After I had replaced the young birds and retired a few steps from the spot, the female squatted upon the nest, which was sufficiently near the entrance of the cavity to be distinctly visible; and on being disturbed she fluttered away with wings outstretched and quivering, as if unable to fly, and apparently to divert attention from the nest.

Mr. Weston Brown, a bird-collector at Wellington, showed me a pair of newly fledged young birds of this species which he had taken himself. He informed me that he had found them in a rudely constructed nest in the hollow of a whitewood tree, and about 9 inches from the entrance. There were only two young birds in the nest, and these were male and female. The plumage of the former was strongly suffused with brown; but the colours were sufficiently distinct to indicate the sex.

During the early part of the breeding-season the female is never visible, and I think it is probable that while engaged in the task of incubation she is attended and fed by the male, for I have seen the latter carrying food in his bill. As late as September 30, I have seen as many as ten males in an afternoon's ramble, without catching a glimpse of the other sex. The young birds do not seem to pair till the second year; for in the breeding-season I have, on dissection, found well-plumaged birds with microscopic testes, whilst in others these organs were developed to the size of buck-shot, being conspicuously large for so small a creature.

There is every reason to believe that this species breeds twice in the season, because it is a common thing to find nests containing fresh eggs in October and again in December. The usual complement of eggs is four, but sometimes there are only three. Mr. Reischek told me that, on the Little Barrier, he came upon a nest, containing three eggs, which through some misadventure had got filled with rain-water. The birds seemed fully aware of the gravity of the situation, and were flitting around it in a very excited and distressed manner; but when he proceeded to take possession of both nest and eggs they sat perfectly quiet and did not utter a sound.

The nest is a compact round structure, with a thick foundation, and composed of dry moss, grass, and vegetable fibres, felted together; the cup, which is comparatively large, measuring 2·25 inches in diameter, is often lined with the inner bark of the ribbon-wood (*Hoheria populnea*), and the outer rim is well pressed together just as if bound by some invisible thread. The eggs are of large size in proportion to the bird, measuring ·85 inch in length by ·80 in breadth; they are in form broadly ovoido-conical and are creamy white, freckled all over with yellowish brown, the markings running together and forming a clouded zone near the larger end. Sometimes the zone is absent and the freckled appearance less pronounced. A specimen taken from a nest in the hole of a dry stump differs in being of a pale reddish tint, thickly speckled and freckled with light brown.

MYIOMOIRA MACROCEPHALA.

(SOUTH-ISLAND TOMTIT.)

Great-headed Titmouse, Lath. Gen. Syn. ii. pt. 2. p. 557, pl. lv. (1783).
Parus macrocephalus, Gm. Syst. Nat. i. p. 1012 (1788, ex Lath.).
Pachycephala? macrocephala, Steph. Gen. Zool. xiii. p. 267 (1826).
Rhipidura macrocephala, Gray, in Dieff. Trav. ii., App. p. 190 (1843).
Miro forsterorum, Gray, op. cit. ii. p. 191 (1843).
Miro dieffenbachii, Gray, op. cit. ii. p. 191 (1843).
Petroica macrocephala, Gray, Voy. Ereb. and Terror, Birds, p. 6 (1844).
Petroica dieffenbachii, id. op. cit. p. 6, pl. 6. fig. 1 (1844).
Turdus minutus, Forst. Descr. Anim. p. 83 (1844).
Miro macrocephala, Bonap. Consp. Gen. Av. i. p. 299 (1850).
Muscicapa macrocephala, Ellman, Zool. 1861, p. 7465.
Muscicapa minuta, Ellman, tom. cit. p. 7465.
Myiomoira dieffenbachii, Gray, Hand-l. of B. i. p. 229 (1869).
Myiomoira macrocephala, id. op. cit. p. 229 (1869).

Native names.

The same as those applied to the preceding species.

♂ similis *M. toitoi*, sed maculâ frontali albâ minore et pectore flavido distinguendus.

♀ similis feminæ *M. toitoi*, sed pectore flavido lavato.

Adult male. Similar to *M. toitoi*, except in the colour of the under surface, which is pale lemon-yellow instead of being white, deepening to orange where it meets the black of the fore neck, and fading away into yellowish white on the vent and under tail-coverts; the white frontal spot, moreover, is somewhat less distinct than in the former bird. Irides lustrous black. Legs and feet blackish brown, the under surface and sides of the toes orange-yellow. Total length 5·4 inches; extent of wings 8·5; wing, from flexure, 3·2; tail 2·2; bill, along the ridge ·35, along the edge of lower mandible ·55; tarsus ·7; middle toe and claw ·8; hind toe and claw ·7.

Female. Similar to the female of *M. toitoi*, but having the breast and abdomen washed with very pale lemon-yellow, and the wing-bar tinged with fulvous.

Young. In the young of both sexes the yellow is reduced to a scarcely perceptible tinge, and in some examples is altogether wanting. In the young male the breast is obscurely mottled with dusky black, and in the young female these markings are brown and extend to the flanks.

Varieties. A very pretty albino specimen, received from Otago, has nearly the whole of the body white, with a wash of bright yellow on the head, breast, and abdomen; on the fore part of the breast there is a broad mark of velvety black, and on the upper surface there are a few scattered feathers of the same; some of the wing-feathers are pure white, the rest are black; the two middle tail-feathers are white, the outer ones black, obliquely crossed with a bar of white; bill and legs as in ordinary specimens.

Another albino, in the Otago Museum, has the general plumage white, with a faint tinge of brown on the head and of yellow on the body, there being a bright wash of canary-yellow on the breast. Wings and tail parti-coloured, several of the tail-feathers being entirely black; bill and feet white.

Obs. Individuals vary much both in size and in the tone of their colouring, some males having the underparts of a uniform pale lemon-yellow, others rich canary-yellow, deepening into orange on the breast. The one figured is a highly coloured specimen in my own collection. A specimen in the Canterbury Museum measures only 4·75 inches in length, corresponding, both in size and plumage, with the type of Mr. G. R. Gray's *M. dieffenbachii*; and I have received equally small examples from the Chatham Islands; but, after a very careful comparison, I am unable to admit the validity of the supposed new species *.

THIS Tomtit is the South-Island representative of the preceding species, which is only found north of Cook's Strait. It appears, however, to enjoy a wider geographical range; for I obtained specimens at the Chatham Islands, and the Antarctic Expedition brought some from the Auckland Islands.

The stomachs of all those I opened were crammed with small diptera, coleoptera, and caterpillars, showing the strictly insectivorous character of this species.

The habits of this bird are similar to those of its northern ally (*M. toitoi*), except that it appears to be less recluse in its nidification; for it is a common thing to find its somewhat elaborate nest, and often in exposed situations, a favourite location being under the head of the ti (*Cordyline australis*).

There is much variation in nests from different localities, but a very typical example in my collection is of a rounded basket-shape, with a thick foundation, measuring four inches across the top, with a maximum depth of a little over three inches. It is composed of moss, dry leaves, roots of umbelliferous plants, minute fragments of bark and other vegetable substances, compactly bound together; and the cup, which is fully an inch and a half in depth, is thickly lined with soft tree-moss. Mixed with the building-materials I have enumerated are some small tufts of sheep's wool; and passing right through the wall of the nest, apparently to serve as a support, there is a bent fern-stalk nearly six inches long †.

The eggs, which are generally three in number, but occasionally four, are ovoido-conical, measuring ·75 inch in length by ·6 in breadth; they are white, with a broad freckled zone of purplish brown at the larger end, and with the whole surface dusted or minutely freckled with paler brown; sometimes without the zone, and beautifully speckled all over with various shades of brown.

* Professor Hutton, I believe, still recognizes two species, both of them found in the South Island. In the critical notes appended to his 'Catalogue' (1871) he remarked:—"Mr. G. R. Gray describes *Petroica dieffenbachii* as being smaller than *P. macrocephala*, and with the yellow on the chest darker; but of the two species that are found in the South Island it is the larger one that has the darker colour on the chest. It is therefore doubtful which of the two is the true *macrocephala*." The answer to the above is that I have in my possession a series of specimens showing every gradation of colour between the two extremes, and that the darkest is likewise the smallest of them all.

† Mr. W. W. Smith sends me the following note:—"I have found the nest many times under the head of the cabbage-tree, and occasionally in a suspended clump of roots on a clay bank. I have also met with it in thick masses of 'bush lawyer.' In 1880 I discovered one in a matipo tree fully nine feet from the ground. I have observed considerable difference in the size and shape of the nests, some being large and very roughly constructed, others small and highly finished." Mr. Potts writes:—"Two nests which we presented to the Canterbury Museum were of remarkable shape: one, a firm compact structure, placed in the forked head of a ti tree, resembled a very neat moss basket with a handle across the top; the second, also from a ti tree, owing, perhaps, to the foundation slipping between the leaves, was built up till it reached the great height of sixteen inches. We have found others placed on a rock; and one, now in the Colonial Museum, was built between the boards and shingles in the roof of an empty cottage" (Trans. New-Zealand Inst. 1869, vol. ii. p. 59). In a letter to myself, he adds the following interesting particulars of two other nests found by him:—"No. 1 was built chiefly of sprays of climbing plants, strengthened with grass-bents and a few pieces of split ti-palm leaf, lined with moss, as usual. The whole fabric appeared much rougher and more loosely put together than is usually the case with the nest of this bird. It was placed in a ti-palm, and contained two well-fledged young birds and three bad eggs. No. 2: this nest was composed almost entirely of moss, with a few slender strips of bark fixed to the outside, and ornamented inside with a few Parrakeet-feathers; it was placed on a ledge in a mossy recess among the rocks in dense bush, and contained four eggs."

GERYGONE FLAVIVENTRIS.

(GREY WARBLER.)

Curruca igata, Quoy et Gaim. Voy. de l'Astrol., Zool. i. p. 201, pl. xi. fig. 2 (1830).
Acanthiza igata, Gray, in Dieff. Trav. ii., App. p. 189 (1843).
Gerygone flaviventris, Gray, Voy. Ereb. and Terror, p. 5, pl. 4. fig. 1 (1844).
Gerygone igata, id. op. cit. p. 5 (1844).
Gerygone assimilis, Buller, Essay on Orn. N. Z. p. 9 (1865).
Acanthiza flaviventris, Gray, Hand-l. of B. i. p. 219 (1869).

Native names.—Riroriro and Koriroriro.

Ad. suprà grisescenti-brunneus, dorso et uropygio cum supracaudalibus olivaceo lavatis, his lætiùs tinctis: tectricibus alarum remigibusque brunneis, extùs angustè olivaceo limbatis: rectricibus cinerascenti-brunneis versùs apicem conspicuè nigricantibus, duabus externis maculâ anteapicali albâ notatis, reliquis ad apicem pogonii interni albo maculatis: facie laterali gutturéque toto sordidè cinereis: corpore reliquo subtùs albicante, abdomine imo et hypochondriis flavido tinctis, his etiam paullò olivascentibus: rostro et pedibus saturatè brunneis: iride rubrâ.

Juv. similis adultis, sed coloribus dilutioribus.

Adult male. Upper parts brownish grey, tinged on the back with olivaceous brown; throat, fore part of neck, breast, and sides cinereous grey; abdomen and under tail-coverts white, the former slightly tinged with yellow; primaries dark brown, paler on the inner webs; tail-feathers dark brown in their basal, almost black in their apical portion, and, with the exception of the two median ones, having an angular white spot near the tip on their inner webs. The plumage is sooty black at the base, but this is only observable on moving the feathers. Irides red; bill, tarsi, and toes dark brown. Total length 4·5 inches; extent of wings 6; wing, from flexure, 2·12; tail 2; culmen ·3; tarsus ·75; middle toe and claw ·5; hind toe and claw ·75.

Female. Similar in plumage, but of smaller size.

Young. In the young bird the tints of the plumage generally are paler and there is an entire absence of the yellow tinge. Irides brown.

Obs. In some adult examples the measurements are slightly larger than those given above, there is an absence of the yellow tinge on the abdomen, and the white spot on the lateral tail-feathers is terminal.

Note. A figure of this bird in the act of feeding a young Cuckoo will be found on the Plate representing *Eudynamis taitensis*. The illustration given in the 'Voy. de l'Astrolabe' is scarcely recognizable.

IN the warm sunlight of advancing summer, when the manuka-scrub is covered with its snow-white bloom and the air is laden with the fragrance of forest flowers, amidst the hum of happy insect-life, a soft trill of peculiar sweetness—like the chirping of a merry cricket—falls upon the ear, and presently a tiny bird appears for an instant on the topmost twigs of some low bush, hovers for a few moments, like a moth before a flower, or turns a somersault in the air, and then drops out of sight again. This is the Grey Warbler, the well-known Riroriro of Maori history and song.

This little bird, of sombre plumage and unobtrusive habits, is an interesting species, whether we regard it merely as the familiar frequenter of our gardens and hedgerows, or, more especially, as the builder of a beautiful pensile nest and the foster-parent of our two parasitical Cuckoos (*Eudynamis taitensis* and *Chrysococcyx lucidus*). It belongs to a group of which there are numerous representatives in Australia, and its habits are in no way different from those of its relations.

It is plentiful in every part of New Zealand, and appears to be as much at home in the woods as in the open scrub. I have seen it hunting for its minute prey in the leafy tops of forest trees, the tawa being its favourite resort, probably on account of some special kind of insect food. On one occasion, after very cold weather, I picked up a dead one at the foot of an aged kauri tree, with a smooth trunk fully seventy feet in height. In the Hot Lakes district I have found it flitting round the steaming geysers, apparently unaffected by the sulphur fumes, and catching the minute flies that are attracted thither by the humid warmth. Down by the sea-shore its note may be heard in the low vegetation that fringes the ocean beach; whilst far up the mountain-side, where the scrub is scarce and stunted, it shares the dominion with the ever-present *Zosterops*. Its sweet trilling warble is always pleasant to the ear, being naturally associated in the mind with the hum of bees among the flowers, and the drumming of locusts in the sunshine. It becomes louder and more persistent in the spring-time; and "Kua tangi te riroriro" has become a sort of watchword among the Maoris, signifying "Planting-time has commenced: let us be up and doing." I remember the late Sir Donald McLean commencing one of his most successful Maori speeches with those figurative words, using them of course in a political sense *.

Its food consists of minute flies and insects and their larvæ, in the eager pursuit of which it appears to spend its whole time, moving about with great agility and uttering at short intervals a note of much sweetness, though of little variety. The bird is easily attracted by an imitation of this note, however rudely attempted, and may be induced to fly into the open hand by quickly revolving a leaf or small fern-frond, so as to represent the fluttering of a captive bird. Layard compares the note to the creaking sound of a wheel-barrow; and I have sometimes heard it so subdued and regular, as to be scarcely distinguishable from the musical chirping of the pihareinga or native cricket.

When resting on a twig, it has a habit of flipping its wings after the manner of a Goldfinch. Its ordinary flight is in short undulations with the tail outspread, showing the markings on the lateral feathers.

Where the rank growth of bracken covers the open land, mixed here and there with the flowering *Leptospermum* and blending its sombre tints with the dark-green clumps of tupakihi—forming together a close thicket over which the wild convolvulus twines itself and exhibits its pendent flowers of pink and white—here the Grey Warbler has its home in absolute security, and here in some shady recess it hangs its pear-shaped nest and rears its little brood. It builds a large and remarkably ingenious nest, in which it lays from three to six eggs, and, as I am inclined to think, breeds twice in the season. The construction of the nest, which is of great size as compared with the bird, occupies of necessity a considerable time. In one instance noted, I observed the birds collecting materials for their work towards the end of August, and the young did not quit the clump of climbing-rose in which the nest was placed till the first week in October.

Selected on account of its unwearied industry, or because of the peculiar fitness of its warm domed nest for the nurture of a semitropical species, this little bird is the willing victim of our two migratory Cuckoos, the Warauroa and Koheperoa—the former of which, at any rate, deposits its egg in the nest of this species, while both of them delegate to this tiny creature the task of rearing their young.

* "I hea koe i te tangihanga o te riroriro"? (Where were you when the Riroriro began to sing?): a proverb applied to a lazy man who neglects his planting.

I have found the intrusive egg of the former in the nest with those of the Grey Warbler, and I have frequently observed the voracious young Cuckoo being attended and fed by the foster-parent, but I have never seen the young of these birds together. Either the parasitic egg being the first hatched, the others are neglected and allowed to perish, or the intruder, finding the accommodation insufficient, by virtue of his superior size and strength casts out the rightful occupants and usurps entire possession of the nest.

Although, as already mentioned, the Grey Warbler appears to lay twice in the season *, it would seem that one nest serves the purpose of rearing two broods; for, allowing that the family would require the attention of the old birds up to the middle of October (though probably it would be later), there would not be time to build another nest before the arrival of the Cuckoos to spend the summer with us and to deposit their eggs for incubation. The production of double broods in this case would seem to be a provision of nature to enable this species to maintain its ground, seeing that the demands of the parasitical Cuckoos involve in many cases the loss or destruction of the legitimate offspring. Instead of being scarce, the Grey Warbler continues to be one of our commonest species—a circumstance owing, no doubt, in some measure, to its being a pensile-nest builder, and thus escaping the ravages of the Norway rat, the great enemy to the increase or perpetuation of our indigenous birds.

The young on leaving the nest are extremely nimble and somewhat shy. For several days after quitting their domed cradle they remain in its vicinity, following the old birds about in a restless manner and emitting incessantly a scarcely audible piping note. On these occasions I have noticed that the birds hunt all day long in a wide circle, with the nest-home as a centre; and they probably take their young family back to it at night for shelter and warmth. The nests of most birds, when the young have flown, are polluted and unserviceable, being easily distinguishable as "old nests;" but this is not the case with the nest of the species under consideration. The cavity or chamber is deeply lined with soft feathers; and to keep the interior clean and pure, the young birds may be seen elevating their bodies to the edge of the orifice on the side of the nest and ejecting the alvine discharge to some distance. Thus the nest is kept in perfect condition for continued use, in the manner suggested, for the rearing of a second brood. At the close of the breeding-season it may be observed that this bird has the shafts of the tail-feathers denuded, often to the extent of a quarter of an inch, the result, no doubt, of its laborious building-operations.

I have examined a large number of their nests in various parts of the country and in almost every variety of situation; and while invariably exhibiting the pensile character, they are, as a rule, referable to one or the other of two distinct types—the bottle-shaped nest with the porch or vestibule, and the pear-shaped form without the porch. This peculiarity, coupled with the significant fact that in some instances the eggs were pure white, in others speckled or spotted with red, led me at one time to suspect the existence of two distinct but closely allied species; and a manifest difference in the size of some examples tended to strengthen that view. In my 'Essay on the Ornithology of New-Zealand' (1865) I described the two forms of nest, and proposed to distinguish the builder of the large pear-

* In further support of my view as to a double brood, I am glad to find room for the following valuable note received from Mr. J. Brough, of Nelson:—"It may interest you to know exactly the time it takes the Grey Warbler to construct its nest. On November 29 I took a nest with five eggs which I had found close to my camp. On December 1 the birds commenced a fresh nest near the site of the old one. I watched them carefully, and will give you the result as entered every night in my diary.—Dec. 2. Showery day; warblers hard at work. Dec. 3. Snow showers; but no interruption in the work. Dec. 4 & 5. Snowing all day, but warblers hard at work from morning till night. Dec. 5. Fine day; birds working diligently. Dec. 6. Another fine day; warblers completed their nest. Dec. 8. First egg laid. Dec. 11. Another egg. Dec. 12. Third egg. Dec. 13. Fourth egg laid, and hen commenced to sit. Whilst the building of the nest was proceeding, I noticed that the male bird undertook the chief part of the labour in collecting and carrying materials, and that the weaving of these materials together and building of the nest was performed almost entirely by the female."

shaped structure as *Gerygone assimilis*. It may yet be necessary to recognize the existence of a larger and a smaller race, although the subject requires further investigation. My present belief is that the difference in size is only sexual. It may be considered settled, however, that the ascertained difference is not such as to justify a specific separation.

Since the foregoing was written, Mr. R. B. Sharpe has expressed his belief that the bird brought from New Zealand by the 'Astrolabe' Expedition in 1829 (*Gerygone igata*, Quoy et Gaim.) is a distinct species. Being anxious to determine the point for myself, I lately paid a visit to Paris and examined the type. I was unable to find any character by which it could be distinguished from the common species. It is apparently a young bird with soft plumage; there is no tinge of yellow on the underparts, and the dark grey of the upper surface is somewhat suffused with brown *.

Strictly speaking, according to this view, *Gerygone igata* ought to take the place of *Gerygone flaviventris*, owing to its priority over the latter; but, in the first place, the name is a barbarous one and objectionable on that account, and, secondly, I am unwilling to disturb a name that has been in general currency for close upon fifty years.

The two forms of nest above alluded to were thus described in my 'Essay' (p. 9):—"'That of the smaller species is a compact little nest, measuring about 6 inches by 3·5. It is 'bottle-shaped,' full and rounded at the base, and tapering upwards to a point, by which it is suspended. It is composed of a variety of soft materials—spiders' nests, dry moss, grass, vegetable fibres, &c. The spiders' nests consist of a soft silky substance, by the aid of which the materials composing the nest are woven into a compact wall, with a smooth and finished exterior. The entrance, which is situated on the side of the nest, is so small as barely to admit the finger, and it is protected from the weather by a very ingenious contrivance. It is surrounded by a protecting rim or ledge, composed of extremely fine roots interlaced or loosely woven together and firmly secured to the groundwork of the nest. This facing is arched at the top so as to form a vestibule or porch, while at the base it stands out boldly from the wall, and is nearly an inch in depth, thus furnishing a firm and secure threshold for the bird in its passage to and from the cell. The interior apartment or cavity is about two inches deep, and is thickly lined with soft feathers; and the nest forms altogether a well-proportioned and symmetrical structure, testifying alike to the skill and industry of the modest little builder. The nest of the other species is of a somewhat similar size; but it is fuller in the middle than the one described, and is pear-shaped towards the apex instead of tapering. The materials composing it are of coarser texture, there is less execution or finish about it, and the ingenious porch, the peculiar feature of the one, is altogether wanting in the other."

A specimen of the nest, with a porch entrance, in Dr. Sisson's possession, measures nine inches, and is produced downwards to a point, instead of being rounded as in the typical examples.

* Having given the result of my own examination of the type of *Gerygone igata*, I think it is only right to quote, in full, the conclusion in an opposite direction arrived at by Mr. Sharpe, in his notes to the 'Voyage of the Erebus and Terror,' pp. 25, 26:—"During a recent visit to Paris I examined, in company with Dr. Oustalet, the type of this species, which still exists in the Jardin des Plantes. . . . . . We compared the type with Dr. Buller's figure and with the specimens of *Gerygone flaviventris* and we could not believe that the two species were identical. I take the following observations from my note-book:—'It is very close to *G. flaviventris*, but instead of being grey on the throat, the latter is whitish washed with yellow, a shade of which is also apparent on the cheeks; sides of the breast washed with brown; abdomen white, the flanks washed with yellow. Wing 1·05 inch, tarsus ·75.' The tail is imperfect, but on the feathers which remain the white spot is decidedly more correctly described as terminal instead of subterminal. I mention this latter observation *à propos* of the following remarks made by Dr. Buller in his great work: 'In some examples the measurements are slightly larger, there is an absence of the yellow tinge on the abdomen, and the white spot on the lateral tail-feathers is terminal.' The last-named author does not seem to allow these differences to be specific; but I think that further investigation by the field-observers in New Zealand may prove *G. igata* to be a good species, and I leave the matter in their hands." On the other hand, Dr. Finsch, in a letter to myself, stated, as the result of an independent examination:—"It will interest you to hear that the specimen of the so-called *Gerygone igata* in the Museum at Paris is positively *Gerygone flaviventris*."

As I have previously pointed out, in a communication to the Wellington Philosophical Society (November 12, 1870), among the substances used as building-materials by this bird, spiders' nests are always conspicuous; indeed, in some specimens, the whole exterior surface is covered with them. The particular web chosen for this purpose is an adhesive cocoon of loose texture and of dull green colour. These spiders' nests contain a cluster of flesh-coloured eggs or young; and in tearing them off the bird necessarily exposes the contents, which it eagerly devours. Thus, while engaged in collecting the requisite building-material, it finds also a plentiful supply of food—an economy of time and labour very necessary to a bird that requires to build a nest fully ten times its own size, and to rear the Cuckoo's offspring in addition to its own. Curiously enough, the bird uses only the green-coloured nests of *Epeira verrucosa*, and rejects the orange-coloured nests of *E. antipodiana*. I think this may be explained on the principle of assimilative or protective colouring. Dry freshwater algæ are sometimes used for binding the exterior and giving additional firmness to the structure.

In the Canterbury Museum there is a beautiful nest of this species, composed almost wholly of sheeps' wool intermixed with soft dry leaves. It is almost globular in shape, with the entrance near the top, and is lightly suspended from a branch of *Leptospermum*. There is also another of much larger size, composed of wool and spiders' nests, with fragments of cotton and twine carefully interwoven, and furnished with a hoodless vestibule or porch, composed of fibrous rootlets; the threshold is unusually deep and firm, probably because of the very yielding materials of which the nest is built.

Another series presents some curious departures from the normal type, showing that the exact form of the nest is often the result of accident, the structure being adapted to the materials of which it happens to be composed and to the circumstances of its location. The subjoined woodcuts may help to illustrate the subject. Fig. 1 represents a nest of larger size than usual, and of a long elliptical shape, which exhibits the uncommon feature of several soft Emu-feathers, worked into the felting among the other building-materials. Fig. 2 shows a nest of the ordinary form, ornamented with the long dry leaves of the red gum (*Eucalyptus rostrata*), around and among which the neat structure is most cleverly built. In fig. 3 there is a manifest departure from the typical character exhibited in fig. 4. Lastly, fig. 5 shows the condition of the nest after the young Cuckoo usurper has pulled it out of shape and symmetry. Four is the normal number of eggs, although there are sometimes six. They differ somewhat in size; and in shape are ovoido-conical or slightly pyriform. They are sometimes pure white, but more generally freckled and marked with purplish brown, and are so fragile in texture as to bear only the most delicate handling. Ordinary specimens measure ·7 of an inch in length by ·5 in breadth. I have remarked that among the highly variable eggs of this species several distinct types may be recognized, and that all the eggs in one nest are invariably alike. Thus there is the spotted variety, in which the whole surface is studded with scattered dots of purplish brown; secondly, the freckled variety, in which the coloration is more diffuse; and, thirdly, the zoned variety, presenting a broad zone of colour near the thick end. Two examples, taken from a nest which contained also an egg of the Shining Cuckoo, had the thick end broadly capped with reddish brown.

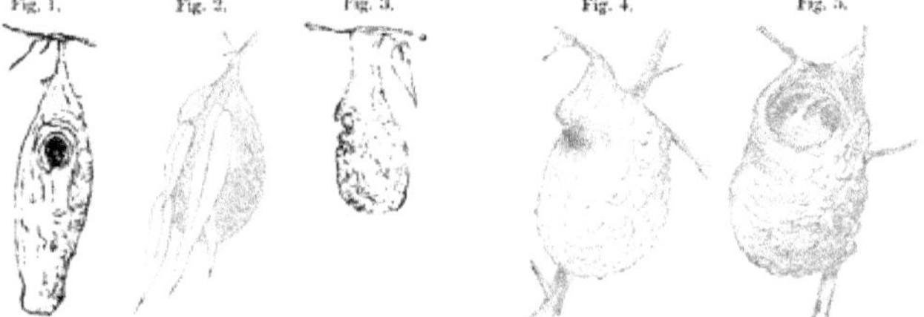

GERYGONE ALBOFRONTATA.

(CHATHAM-ISLAND WARBLER.)

Gerygone? albofrontata, Gray, Voy. Ereb. and Terror, p. 5, pl. 4. fig. 2 (1844).
Acanthiza albofrontata, Gray, Hand-l. of B. i. p. 219 (1869).

Ad. suprà olivascenti-brunneus, pileo obscuriore, uropygio et supracaudalibus latè et conspicuè rufescenti-fulvis: tectricibus alarum et remigibus cinerascenti-brunneis, dorsi colore limbatis: rectricibus cinerascenti-brunneis, versus apicem purpurascenti-nigris et fasciâ fulvescente transversim notatâ, pennis duabus centralibus reliquarumque apicibus omninò cinerascenti-brunneis: fronte, superciliis et facie laterali albidis, loris et regione parotica brunneo notatis: subtùs albicans, abdomine imo et hypochondriis flavicantibus: subcaudalibus et tibiis fulvis: subalaribus albicantibus flavido lavatis: iride coccineâ: rostro brunneo, gonyde pallidiore: pedibus saturatè brunneis.

Adult male. Upper surface rusty brown, lighter on the wings and rump; the whole of the plumage plumbeous beneath; forehead, sides of the head, fore neck, breast, and the underparts generally greyish white, tinged with yellow on the flanks and abdomen; an obscure streak of dusky brown passes through the eyes; wing-feathers dusky brown, with lighter shafts, margined on their outer webs with yellowish brown; inner lining of wings yellowish white; tail-feathers rusty brown, tinged with rufous towards the base, darker brown in their apical portion, with the tips paler; the two outermost feathers on each side with a broad subterminal bar of fulvous white, and the two succeeding ones with an obscure triangular spot of fulvous white on the inner webs; upper tail-coverts rufous-brown. Irides blood-red; bill and feet blackish brown. Total length 5·75 inches; wing, from flexure, 2·6; tail 2·5; bill, along the ridge ·4, along the edge of lower mandible ·5; tarsus ·9; middle toe and claw ·65; hind toe and claw ·65.

Female. Similar to the male, but slightly smaller, and without the yellow tinge on the underparts.

Obs. In my former edition, under the head of *Gerygone albofrontata*, I observed:—"I have never met with this bird in New Zealand; but it is highly probable that the supposed new species of *Gerygone* lately observed by Mr. Potts and his son in Westland, of which an account will shortly appear in 'The Ibis,' will prove to be the same." As Professor Hutton, however, has since pointed out (Trans. N.-Z. Inst. vol. v. p. 222), "Mr. Potts's specimen, as he describes it, differs from *G. albofrontata*, not only in the absence of the white forehead, but also in the dark colour of the wings, in having the two centre tail-feathers black, and in the chin, cheeks, and breast being grey; in all which respects it agrees with *Gerygone flaviventris*."

This fine species was originally described and figured by Mr. G. R. Gray, in the 'Voyage of the Erebus and Terror,' from a specimen alleged to have been "brought by Dr. Dieffenbach from New Zealand." The specimen itself, however, which is now in the British Museum, is labelled as from the Chatham Islands, whence other examples have since been obtained. Mr. Henry Travers reports that he met with it on all the islands of the group, although it is by no means common. He observed that it had much the same habits as the New-Zealand species. The nest of this bird is similar to that of *Gerygone flaviventris*; but with a larger aperture, and without any threshold projection, although the upper edge is overhanging. The green-coloured nests of the meadow-spider (*Epeira*) are used among the building-materials, and likewise the white cocoons of some ground species, which I have not been able to identify. The eggs (of which I have three specimens) are slightly ovoido-conical, measuring ·75 inch in length by ·55 in breadth; pinkish white, marked over the entire surface with minute specks and linear freckles of reddish brown, which coalesce and form a cap at the larger end.

GERYGONE SYLVESTRIS.

(BUSH-WARBLER.)

Gerygone sylvestris, Potts, Trans. N.-Z. Inst. 1872, vol. v. p. 177.

Ad. ♂ similis *G. flaviventri*, sed suprà saturatior: tectricibus alarum nigris, extùs flavido lavatis: remigibus brunneis, extùs flavido lavatis: subalaribus albidis: caudâ brunneâ, nigro conspicuè transfasciatâ, rectricibus duabus mediis nigris brunneo terminatis, duabus externis albidis conspicuè transfasciatis et brunneo terminatis: supracaudalibus schistaceo-nigris, flavido terminatis: rostro nigro, versus apicem flavicante: pedibus nigris, plantis flavicantibus: iride cruentatâ.

Adult male. "Upper surface dark olivaceous; wings smoky black, except first two feathers, outer webs fringed with yellow; cheek dark grey, darkest in a line from the gape through the eye; chin grey; neck and breast pale grey; abdomen white; under wing-coverts white; upper wing-coverts brown, margined with yellow; upper tail-coverts slaty black, tipped with yellow; tail brown, with a broad band of black, two centre feathers black, tipped with brown, four feathers on each side tipped with white on inner webs, pale brown on outer web, two outer feathers broadly barred with white, tipped with brown. Bill black; both mandibles horn-colour at the point; legs and feet black; inside of feet yellowish flesh; irides bright blood-red. Bill, from gape, 6 lines; wing from flexure 2 inches; tail 2 inches 2 lines; tarsus 9 lines; middle toe and claw 5 lines; total length 4 inches 5 lines. Male bird killed in full song, Dec. 20." (*Potts.*)

THE above Warbler, of which unfortunately no specimen exists in any collection, by which to test its value as a species, was first made known by Mr. T. H. Potts, who described it as above in a communication he made to 'The Ibis' in 1872 (p. 325), but without then proposing any name for it. He afterwards characterized it, under the name of *Gerygone flaviventris*, in the 'Transactions of the New-Zealand Institute'; and, never having seen the bird, I have quoted his description of it.

The following is the account he gives:—"Whilst journeying in the dense bush which clothes the western slopes of the Middle Island, making acquaintance with the Kiwi and Kakapo, the note of a bird was heard that was new to us; it was evidently that of a *Gerygone*, but differed much from that of our familiar gully-haunting Warbler. The habitat was unusual, *in the thick bush*, between the bluff of Okarito and Lake Mapourika; whereas our little Riroriro delights in trilling from the shrubs on the creek-side or more open country, or in flitting about the bushy vegetation of the gullies that fringe or form the outskirts of a forest. Neither my son, who accompanied me, nor myself had ever heard a similar note. For the next few days, whilst rambling in that locality, we heard the same note repeatedly, and saw the birds, but we never observed one of them on the outside of the bush."

Possibly to this species belongs the alpine bird mentioned by Mr. Reischek in a letter to myself, as having been met with by him during his trip to the west-coast sounds in search of *Notornis*:—"At Dusky Sound (on the 2nd July, 1884) I ascended one of the heavily wooded ridges, and on arriving at the top I heard a new bird. It was out of sight in the foliage of a tree, and got away before I could get a glimpse of it. Its call consists of three notes, like *di-di-di*, repeated several times. I went in search of it again but without success. I have been exploring in the New-Zealand forests for the last eight years and am familiar with all the birds' notes; but this one was quite new to me, and was evidently produced by some small bird which I have not yet seen."

CERTHIPARUS NOVÆ ZEALANDIÆ.

(NEW-ZEALAND CREEPER.)

New-Zealand Titmouse, Lath. Gen. Syn. ii. pt. 2, p. 558 (1783).
Parus novæ seelandiæ, Gm. Syst. Nat. i. p. 1013 (1788, ex Lath.).
Parus novæ zealandiæ, Lath. Ind. Orn. ii. p. 571 (1790).
Parus zelandicus, Quoy & Gaim. Voy. de l'Astrol. i. p. 210, pl. xi. fig. 3 * (1830).
Certhiparus novæ zelandiæ, Lafr. Rev. Zool. 1842, p. 69.
Certhiparus novæ zealandiæ, Gray, in Dieff. Trav. ii., App. p. 189 (1843).
Certhiparus maculicaudus, Gray, op. cit. ii. p. 189 (1843).
Parus urostigma, Forst. Descr. Anim. p. 90 (1844).
Certhiparus novæ zealandiæ, Finsch, J. f. O. 1870, p. 254.

Native names.—Pipipi and Toitoi.

♂ suprà chocolatino-brunneus, pileo paullulum obscuriore: facie laterali nuchâque cinerascentibus: tectricibus alarum dorso concoloribus: remigibus brunneis, primariis extùs angustè fulvescente limbatis, secundariis latiùs dorsi colore lavatis: caudâ rufescenti-chocolatinâ, rectricibus (duabus mediis exceptis) fasciâ nigrâ transnotatis; subtùs rufescenti-albus, corporis lateribus et tectricibus subcaudalibus chocolatino lavatis: rostro et pedibus pallidè brunneis, unguibus fulvescenti-brunneis: iride saturatè brunneâ.

♀ mari omninò similis.

Juv. vix ab adultis distinguendus, sed magis vinaceo tinctus.

Adult. Fore part of head, crown, back, rump, and upper surface of wings bright cinereous brown, inclining to rufous; quills light brown, the outer webs tinged towards their base with rufous; tail-feathers pale rufous, and, with the exception of the two middle ones, crossed on their inner web, about half an inch from the tip, with a broad band of black; sides of head and nape cinereous grey; throat, breast, and abdomen rufous-white. Irides grey; bill, tarsi, and toes pale brown; claws lighter brown. Total length 5·25 inches; extent of wings 6·75; wing, from flexure, 2·5; tail 2·6; tarsus ·75; bill, along the ridge ·5, along the edge of lower mandible ·6; middle toe and claw ·6; hind toe and claw ·6.

Young. Plumage as in the adult, but suffused with vinous brown.

Obs. The sexes are alike, both as to size and colouring.

Remarks. I carefully examined, with the late Mr. G. R. Gray, the examples in the British Museum on which he had founded his distinction between *Certhiparus novæ zealandiæ* and *C. maculicaudus*. The individual differences were trivial, and I felt perfectly satisfied that the two species could not be maintained—a conclusion in which Mr. Gray concurred.

* In the 'Voyage de l'Astrolabe' there is a figure intended to represent this bird, under the title of "Mésange de la Nouvelle Zélande;" but without the descriptive text it would be quite impossible to identify the species, the drawing being very defective and the colouring incorrect.

This lively little species is confined to the wooded parts of the South Island *. I met with it in Nelson and in Otago, but more abundantly in the Canterbury provincial district. On Banks Peninsula I found it particularly numerous, but I was never able to discover its nest.

Like the other members of the group to which it belongs, it is a gregarious species, associating together in small flocks, and hunting diligently for its insect food among the branches and dense foliage of the forest undergrowth. On being disturbed or alarmed they quickly assemble and chirp round the intruder for a few minutes; and on being reassured they disperse again in search of food.

One of their ordinary notes is not unlike the cry of *Creadion carunculatus*, although, of course, much feebler.

I have seen them consorting with the Yellow-head in the low underwood, owing doubtless to a community of interest, their habits of feeding being very much the same. They seem to prefer the outskirts of the bush, where insect-life is more abundant; but they are also to be met with in the thick forest.

During severe seasons it has been known to leave the shelter of the bush and to frequent the sheep-stations, flitting about the meat gallows and picking off morsels of fat from the bones and skins of the butchered animals, exactly after the manner of *Zosterops* under similar circumstances.

In the stomachs of those I examined I found the scale-insect, with minute coleoptera, diptera, and their larvæ, all testifying to the strictly insectivorous character of the bird. The ovary of one which I opened on November 3 contained a small cluster of eggs, the largest being of the size of buck-shot, indicating a late nesting-season.

A nest of this species in the Canterbury Museum is of a rounded form, with a slightly tapering apex, and not unlike a large pear in shape. The structure is composed of dry vegetable fibres, fragments of wood, moss, spiders' nests, and other soft materials closely felted together. The entrance is placed on the side, about one third distant from the top, and is perfectly round, with smoothened edges. The interior cavity is deeply lined with soft, white, pigeon feathers †. It will be seen, therefore, that the nest of this species shows its affinity to *Gerygone* rather than to *Clitonyx*, with which it is associated in the British Museum Catalogue (vol. viii.). I have grouped the birds together on one Plate merely for the sake of artistic convenience.

This bird breeds late in the year, for the nest just mentioned was found far above the Rangitata gorge, in the month of December, and contained three nestlings. Mr. Potts reports that it was "placed in a black-birch between the trunk and a spur, from whence sprouted out a thick tuft of dwarfed sprays, about seven feet from the ground." He says that it usually lays three eggs and that he has a note of finding the young in the nest as late as December 25th.

There are two eggs of this bird in the Otago Museum. They are broadly ovoido-conical, measuring ·7 of an inch in length by ·6 in breadth, and white with small purplish and brown spots, which run together and form a zone round the larger end.

* Captain Hutton, writing from Auckland, in the North Island, stated, in a letter to 'The Ibis' (1867, p. 379), that *Certhiparus novæ zealandiæ* is "one of the commonest birds in the bush about here;" but he was evidently confounding this bird with some other species, probably *Clitonyx albicapilla*, at that time common enough. He repeated, in his 'Catalogue of the Birds of New Zealand' (published in 1871) that *Certhiparus novæ zealandiæ* inhabits "both islands;" but this is undoubtedly an error. I have never heard of the occurrence of this bird, even as a straggler, in any part of the North Island.

† *Cf.* Trans. N.-Z. Instit. 1872, vol. v. pl. 37.

CLITONYX ALBICAPILLA*.

(THE WHITE-HEAD.)

Fringilla albicilla, Less. Voy. Coq. i. p. 662 (1826).
Parus senilis, Dubus, Bull. Acad. Roy. Brux. vi. pt. 1, p. 297 (1839).
Certhiparus senilis, Lafr. Rev. Zool. v. p. 69 (1842).
Certhiparus albicillus, Gray, Voy. Ereb. and Terror, p. 6 (1844).
Certhiparus cinereus, Ellman, Zool. 1861, p. 7465.
Mohoua? albicilla, Gray, Ibis, 1862, p. 220.
Orthonyx albicilla, Finsch, J. f. O. 1870, p. 253.
Orthonyx albicilla, Buller, Birds of New Zealand, 1st ed. p. 101 (1873).
Certhiparus albicillus, Gadow, Cat. Birds Brit. Mus. vol. viii. p. 75 (1883).

Native names.

Popotea, Poupoutea, Popokotea, and Upokotea.

Ad. pileo nuchâque et pectore superiore albis: dorso toto brunneo, supracaudalibus pallidioribus: tectricibus alarum dorso concoloribus: remigibus saturatè brunneis, extùs dorsi colore lavatis, primariis paullò pallidiùs limbatis: pogonio interno flavicanti-albo marginato: caudâ flavicanti-brunneâ: pectore medio fulvescenti-albo: corporis lateribus brunneis, dorso concoloribus: subalaribus albis, brunneo lavatis: rostro nigro: tarso et pedibus plumbescenti-nigris, plantis pallidioribus, unguibus brunneis: iride nigrâ.

Juv. vix ab adultis distinguendus, sed coloribus dilutioribus et pileo brunneo lavato.

Adult male. Head and neck all round, breast, inner face of the wings, and middle of the abdomen white, slightly tinged with brown; sides of the body and flanks pale vinous brown; the whole of the back, rump, and upper surface of wings vinous brown, paler on the upper wing-coverts; quills blackish brown, the primaries narrowly margined on their outer webs with grey, and more broadly on their inner webs with yellowish white; tail-feathers and their coverts pale yellowish brown on their upper aspect, sometimes tinged with rufous, the shafts darker; paler on the under surface, with white shafts. Irides black; bill and rictal bristles black; tarsi and toes bluish black, with paler soles and brown claws. Total length 6·5 inches; extent of wings, 8·4; wing, from flexure, 2·9; tail 2·75; bill, along the ridge ·4, along the edge of lower mandible ·5; tarsus 1; middle toe and claw ·8; hind toe and claw ·6.

Female. Similar to the male but somewhat smaller. Total length 6 inches; extent of wings 7·75; wing, from flexure, 2·6; tarsus ·9; middle toe and claw ·6.

Obs. In the male bird the palate and soft parts of the mouth are black, and in the female flesh-coloured.

Young. Upper parts pale vinous brown, whitish on the head; throat and underparts greyish white, shading into brown on the sides; wings tinged with yellow on their inner edges.

* Acting on Professor Newton's suggestion, I have substituted *albicapilla* for *albicilla*; for the bird is white-headed and not white-tailed, and I cannot believe that Lesson ever intended to apply the latter name to it. Although it has hitherto been the practice to use it, I think I am justified in rectifying what was obviously a *lapsus calami*.

My account of this species in the former edition of this work commenced thus:—"This interesting little bird is distributed all over the North Island, but is replaced in the South by a representative species, the *Orthonyx ochrocephala* or Yellow-head. It frequents all wooded localities, but seems to prefer the outskirts of the forest and the low bush fringing the banks of rivers and streams. It is gregarious in its nature; and the report of a gun, the cry of a Hawk, or any other exciting cause will instantly bring a flock of them together, producing a perfect din with their loud chirping notes. It is a curious or inquisitive bird, following the intruder as he passes through the bush, and watching all his movements in a very intelligent manner. If he remains stationary for a few moments, it will peer at him through the leaves with evident curiosity, and will hop gradually downwards from twig to twig, stretching out its neck and calling to its fellows in a loud chirp, and approaching the object of this scrutiny till almost within reach of his hand."

But alas! what of the Popokotea in this year of grace 1887? In the interesting account which Mr. Reischek has furnished me of a collecting tour he made through almost every part of the island lying to the north of Hawke's Bay, he says:—"I found one pair of *Orthonyx albicilla* on Castle Hill, Coromandel, one pair in the Pirongia ranges, Waikato, and one pair in the Tuhua ranges, near Mokau; that was all." So this is the rapid fate of the pretty, noisy, little White-head, once the commonest bird in all our northern forests!

Even five years ago it was quite plentiful on Te Iwitaaroa, at the north-east extremity of the Kuranui-whaiti range in the Waikato district; but now it has disappeared entirely. It is still numerous on the island of Kapiti in Cook's Strait, and on the Little Barrier; but, strange to say, it no longer exists on the Great Barrier, Kawau, the Hen and Chickens, or indeed, so far as I am aware, on any of the other islands in the Hauraki Gulf. The only localities on the mainland in which I have met with it of late are the wooded hill-tops in the Upper Wairarapa district, and a clump of bush near the Owhaoko station in the Patea country, at a considerable elevation above the sea. It has a simple but very melodious song, some bars of it reminding one of the musical notes of English birds. Its loud chirp is not unlike that of the House-Sparrow, but sharper.

Its food consists of insects and minute seeds. It is very active in all its movements, flitting about among the leafy branches and often ascending to the lofty tree-tops; clinging by the feet head downwards, and assuming every variety of attitude as it prosecutes its diligent search for the small insects on which it principally subsists. I have frequently observed it inserting its beak into the flower of the *Metrosideros*, either for the purpose of extracting honey, or, as is more likely, to prey on the insects that are attracted by it. I have also known them occasionally caught on the *tuke* baited with these flowers to allure the Tui and Korimako, which are genuine honey-eaters.

I have found scores of nests of this species, and have made frequent but ineffectual attempts to rear the young in a cage. The nest is usually fixed in the fork of a low shrubby tree, frequently that of the Ramarama (*Myrtus bullata*), and is always so placed as to be well concealed from observation. It is a round, compact, and well-constructed nest, being composed of soft materials, such as moss, dry leaves, spiders' nests, shreds of native flax, and sometimes wool, all firmly knit together. The cavity is deep and well rounded, the walls being formed of dry bents and vegetable fibres, and thickly lined with soft feathers. The lip or outer edge of the nest is carefully bound in with these fibres, sometimes mixed with spiders' webs, and often presenting a high degree of finish. The eggs are usually three in number, but sometimes four; they are of proportionate size, measuring ·8 of an inch in length by ·6 in breadth, rather rounded in form, and with a shell of very delicate texture. They are creamy white, minutely speckled or marbled over the entire surface with reddish brown, the markings being denser towards the thick end, where they sometimes form an irregular zone. During incubation the hen bird sits closely, and leaves the nest with reluctance, almost permitting herself to be touched by the hand before quitting it.

I have before me now a beautiful nest of this species, which was taken on the Little Barrier in December, and contained three young birds. It is almost spherical, except at the top, which is flattened, measuring in its largest part 4 inches by 3; and its structure is very close and compact, all the materials composing it being well felted together; the cup or cavity is rather deep and rounded with an overhanging lip, the edges being very closely bound and interlaced; and the opening measures just two inches in diameter. The nest is composed of many coloured mosses and lichens, dry leaves, grasses, vegetable fibres, and here and there a feather closely interwoven with the web; and the interior is lined with fine grass-bents and a few feathers.

For the rapid disappearance of our indigenous birds it is hard to assign any special cause. The introduced rat is undoubtedly an important factor in the business by preying on the eggs and young of such species as habitually nest in places accessible to them; but we can hardly account in this way for the almost total disappearance of the pert little White-head, once the commonest denizen of our woods. The introduced bee gets a share of the blame in the case of honey-eating and tree-hole nesting birds, like the Korimako and Stitch-bird on the one hand, and the Kaka and Parrakeet on the other; but with even less probability than the Norwegian rat can this agent be credited with the destruction of the White-head. The disappearance of the Quail we are accustomed to attribute to the introduction of sheep and the prevalence of tussock fires; the diminution of the Wild Duck to the extensive draining-operations of the farmer; and the thinning of the Wood-Pigeon to the wholesale slaughter of these birds by both Europeans and natives, and in some districts without cessation all the year through. But we find it extremely difficult to discover any sufficient reason for the wonderfully rapid extinction of the White-head, or Popokatea, in most parts of the island. No doubt it is due to a variety of causes, operating with more or less force, all round, and thus furnishing another illustration of what appears to be an almost universal natural law—that indigenous forms of animal and vegetable life sooner or later succumb to, and are displaced by, more vigorous types from without. As the Maori is being rapidly supplanted by his Anglo-Saxon neighbour, as the rat has exterminated and replaced the *Kiore maori*, as the native fern and other herbaceous vegetation disappears in all directions before the spreading grass and clover of the colonist, so in like manner the native birds, or at any rate many of the well-known species, are giving place to the ever-increasing numbers of Sparrows, Linnets, Greenfinches, Yellowhammers, Starlings, and other introduced birds that are now to be met with in every part of the country.

On the other hand, how are we to account for the almost total disappearance of the introduced Pheasant from the Waikato and other districts, where a few short years ago they were excessively abundant, proving almost a plague to the farmers and Maori cultivators? Some ascribe it to the Hawks, but these were always as numerous as they are now; some to poisoned wheat laid for rabbits, but the Pheasant has disappeared from districts where there are no rabbits, and consequently no poisoned wheat. Others believe that the native Woodhen is responsible for the change; but the habit of feasting on Pheasants' eggs, whenever it gets the chance, is by no means a newly acquired one with this bird. Doubtless there are agencies at work of which at present we have no knowledge. The fact nevertheless remains, and is quite as inexplicable as in the case of some of our indigenous birds.

For my own part, I deplore very much this displacement of the natural Avifauna, which appears to be almost inevitable, because many interesting types will disappear for ever. Efforts are being made to save some of them by means of island reserves, but I fear the task is a hopeless one. All therefore that remains to us now is to record their history as fully and minutely as possible for the benefit of science. This I shall endeavour to accomplish in the present work, describing faithfully their habits of life, and omitting nothing that may seem likely to prove of interest or value to the student of the future.

CLITONYX OCHROCEPHALA.

(THE YELLOW-HEAD.)

Yellow-headed Flycatcher, Lath. Gen. Syn. ii. p. 342 (1783).
Muscicapa ochrocephala, Gm. Syst. Nat. i. p. 944 (1788, ex Lath.).
Certhia heteroclites, Quoy & Gaim. Voy. Astrol. i. pl. 17. fig. 1 (1830).
Mohoua hua, Less. Compl. Buff. ix. p. 139 (1837).
Orthonyx icterocephalus, Lafr. Rev. Zool. 1839, p. 257.
Orthonyx heteroclitus, Lafr. Mag. de Zool. 1839, pl. 8.
Mohoua ochrocephala, Gray, List of Gen. of B. p. 25 (1841).
Muscicapa chloris, Forst. Descr. Anim. p. 87 (1844).
Orthonyx ochrocephala, Gray & Mitch. Gen. of B. i. p. 151, pl. 46 (1847).
Orthonyx ochrocephala, Buller, Birds of New Zealand, 1st ed. p. 103 (1873).
Certhiparus ochrocephalus, Gadow, Cat. Birds Brit. Mus. vol. viii. p. 76 (1883).

Native names.

The same as those applied to the preceding species; "Canary" of the colonists.

Ad. pileo undique et corpore subtùs lætè citrinis, nuchâ vix olivascente, abdomine imo cum cruribus crissoque cinerascis; dorso toto olivascente-brunneo, flavido lavato, uropygio conspicuè lætiore flavo; tectricibus alarum et supracaudalibus olivaceo-flavis, illarum majoribus saturatioribus, potiùs olivaceo-viridibus; remigibus brunneis, extùs dorsi colore lavatis, primariis cano limbatis, pogonii interni margine lætè flavicante: caudâ olivaceo-flavâ, subcaudalibus et subalaribus olivaceo-flavis, his albido lavatis: rostro nigro: pedibus nigris, unguibus saturatè brunneis: iride nigrâ.

♀ mari similis, sed coloribus obscurioribus.

Juv. similis adulto, sed pileo et nuchâ olivascente lavatis.

Adult male. Head and breast, sides of the body, and upper part of the abdomen bright canary-yellow; shoulders, back, and upper surface of wings yellowish brown, with an olivaceous tinge; upper surface of tail and the outer margins of the secondary quills dark olivaceous yellow; the colours are blended where they meet, the nape being more or less mottled with yellowish brown; lower part of abdomen greyish white; thighs and flanks pale brown; upper and lower tail-coverts yellow; the whole of the plumage dark plumbeous at the base. Irides black; bill and feet black; claws dark brown. Total length 6·75 inches; extent of wings 9·5; wing, from flexure, 3·25; tail 2·75; bill, along the ridge ·5, along the edge of lower mandible ·7; tarsus 1; middle toe and claw ·87; hind toe and claw ·75.

Female. Similar to the male, but with the tints of the plumage generally duller.

Young. The young bird differs from the adult in having the yellow plumage tinged with olivaceous, especially on the crown and nape, where the latter colour predominates; rictal membrane yellow.

Obs. The shafts of the tail-feathers are often found denuded at the tips. During the breeding-season the testes are enormously developed, attaining to the size of small marbles.

This bright-coloured bird is the southern representative of *Clitonyx albicapilla*. Its range is confined to the South Island, where it is quite as common as the preceding species formerly was in the North. A narrow neck of sea completely divides their natural habitat—a very curious and suggestive fact, inasmuch as this rule applies equally to several other representative species treated of in the present work.

The habits of this bird are precisely similar to those of its northern ally; but it is superior to the latter in size and in the richer colour of its plumage, while its notes are louder and its song more varied and musical. A flock of these Canary-like birds alarmed or excited, flitting about among the branches with much chirping clamour, and exhibiting the bright tints of their plumage, has a very pretty effect in the woods. Even under ordinary conditions it is very pleasing to watch their movements. Hopping from twig to twig, and calling to each other almost continuously in a short clear note, they pass quickly through the branches, moving the body deftly, first to one side then to the other, as they pry into every crevice for the insect food on which they live; then, after remaining stationary a few seconds, they utter a louder and more plaintive note and fly a few yards further to repeat these movements; and so on, all through the day, with never tiring persistence. Sometimes they may be seen hunting among the mosses and lichens that grow on the bark of old forest trees, on which occasions they will ascend the trunks in company, clinging to the hanging vines or any other projecting point, as they make their rapid search, and finally consorting together in the topmost branches. Their black eyes, in a setting of yellow plumage, have a pretty effect, and nothing seems to escape their close scrutiny. They love to move about in the thick foliage, indicating their presence when not chirping by an audible rustling of the green leaves.

The discharge of a collector's gun, the snapping of a stick under foot, or the cry of a wounded bird, will sometimes bring a flock of forty or fifty of these bright-coloured creatures into the branches overhead, where they move restlessly about, peering down and chirping with noisy din, as if in eager consultation.

In all the specimens opened by me the stomach contained comminuted insect remains, chiefly those of minute coleoptera, and larvæ of various kinds.

A life-size drawing of this species, by Mitchell, appeared long ago in the 'Genera of Birds;' but the attitude is unnatural, the bird being placed on the ground instead of a tree. The attitude in which Mr. Keulemans has depicted the bird is a highly characteristic one.

On comparing the nest of this species with that of *Clitonyx albicapilla*, it appears to exhibit more care and finish in its general construction, although composed of the same materials. It is a round and compactly built structure, composed chiefly of mosses, felted together with spiders' webs, and having the cup lined with fine grasses. In the specimen under examination there are a few feathers of the Tui and Parrakeet intermixed with the other materials. Mr. Potts has "sometimes found it placed in the hollow trunk of a broad-leaf." His son found a nest containing two young birds. It was built of moss, grass, and sheep's wool, with a few feathers intermixed, and was placed in a cluster of young shoots on the side of a black birch, near a shepherd's homestead.

The eggs differ in colour from those of *C. albicapilla*, but the type is the same. They are ovoido-elliptical in form, measuring ·9 inch by ·7 inch, although some specimens which I have examined were slightly smaller. They are of a uniform reddish cream-colour, minutely and faintly freckled over the entire surface with a darker tint, approaching to pale brown. In one of my specimens the entire surface is of a warm salmon-colour, without any freckled markings; and another is minutely freckled and dotted with reddish brown, of which colour there are also some irregular smeared markings towards the smaller end. The last-mentioned specimen differs also from the typical form in being almost pear-shaped, with the thick end rather flattened, and measuring only ·75 of an inch in length by ·65 in breadth.

As to the systematic position of this form, much doubt and uncertainty existed till the appearance of a paper "On the Structure of the genus *Orthonyx*," by the late Mr. Forbes, the Prosector to the Zoological Society, in which he gave the results of a careful dissection and comparison of the typical *Orthonyx spinicauda* of Australia with the so-called *Orthonyx ochrocephala* from New Zealand *. This examination convinced him that the two forms are not really congeneric, the New-Zealand bird, apart from its entirely dissimilar coloration, differing from the Australian in its more slender bill, less development of the nasal operculum, less spiny tail, and more slender claws. He states further that internally the skull and the syrinx exhibit differences, slight in amount, but greater than those usually found in birds of the same genus; and he concludes thus:—"Under these circumstances it seems that *Clitonyx* of Reichenbach † will be the correct generic term for the New-Zealand birds, as Lesson's name *Mohoua*, though of prior application, is not only barbarous but, what is more important, liable to be confounded with *Mohoa*, also a genus of Passeres from the Pacific Subregion. In the present unsatisfactory condition of the systematic grouping of the Oscinine Passeres, it is impossible for me to point out clearly any definite position either for *Orthonyx* or *Clitonyx*, though both forms might, I apprehend, be safely placed in Mr. Sharpe's somewhat vaguely defined 'Timeliidæ.'"

The above conclusions were based upon an examination of *C. ochrocephala* only from New Zealand. It will be seen that I have placed the North-Island form (*C. albicapilla*) in the same genus. I am aware that Dr. Finsch has proposed to separate these birds generically, and that his views have been adopted by one or two of our local naturalists. It appears to me, however, quite impossible to find any sufficient distinguishing characters. It will be seen, on comparison, that the wing-feathers present the same proportional arrangement in both species, and that the bill and feet of *C. albicapilla*, although somewhat more slender, are formed on exactly the same model as in *C. ochrocephala*. Apart from these external characters, the two forms agree in other essential respects. The peculiar feature of a black mouth (in the male) is common to both; their style of song is the same; the sexes are alike in both, and their habits of nidification are very similar. It is true that the colour of the plumage is different, and that there is some dissimilarity in the coloration of the eggs, but these differences have no generic value. On these grounds I adhere to my old contention ‡ that the two species belong to the same genus.

Dr. Gadow, in the 'Catalogue of the British Museum' (*l. c.*), while accepting this relationship of the two forms to each other, has grouped them together with the New-Zealand Creeper in the genus *Certhiparus*. So far, however, from adopting this arrangement, I have deemed it necessary not only to separate these birds generically but to place them in different Families.

* "Both forms are *typical* Singing-birds ('Oscines Normales'), with a well-developed Oscinine syrinx with its normal complement of *four* pairs of muscles. Of these the *short anterior* muscle runs to the anterior end of the third bronchial semi-ring alone in *O. spinicauda*; whilst in *O. ochrocephala* this ring receives its muscular supply from a fasciculus of the *long* anterior muscle. They thus differ essentially from *Menura*, with which they have been associated, that bird having but three pairs of muscles peculiarly arranged. In this, as in all other points examined—with one exception in the case of *Orthonyx spinicauda*—these birds quite resemble the normal Passeres, as they do in having the bilaminate tarsus and reduced 'first' (tenth) primary nearly always associated with the normal Acromyodian syrinx. *Orthonyx spinicauda*, however, has a peculiarity quite unknown to me in any other bird, inasmuch as its carotid artery, the left alone of these vessels (as in all Passeres) being developed, is not contained anywhere in the subvertebral canal, but *runs up superficially in company with the left vagus* nerve to near the head, where it bifurcates in the usual manner. . . . . In *Orthonyx ochrocephala* the left carotid retains its normal situation, though the point of entrance into the canal is somewhat higher up than is usual in other Passeres." (P. Z. S. 1882, pp. 544, 545.)

† Handb. Spec. Ornith. p. 167 (1854).

‡ Trans. N.-Z. Inst. vol. vii. p. 204.

SPHENŒACUS PUNCTATUS.

(FERN-BIRD.)

Synallaxis punctata, Quoy & Gaim. Voy. de l'Astrol. i. p. 255, t. 18. fig. 2 (1830).
Sphenœacus punctatus, Gray, Voy. Ereb. and Terror, p. 5 (1844).
Megalurus punctatus, Gray, Gen. of B. i. p. 169 (1848).

Native names.

Mata, Matata, Kotata, Nako, and Koroatito.

Ad. suprà ochrascenti-fulvus, dorsi plumis medialiter nigris, lineas latas longitudinales formantibus: pileo rufescente, fronte immaculatâ, vertice angustiùs nigro striolato: loris et regione oculari albidis: facie laterali albidâ, brunneo maculatâ, regione paroticâ brunnescente: tectricibus alarum dorso concoloribus et eodem modo medialiter nigris: remigibus rectricibusque nigricanti-brunneis, ochrascenti-fulvo limbatis, his acuminatis, scapis versùs apicem nudis: subtùs albescens, hypochondriis et subcaudalibus ochrascenti-fulvis, latè nigro striolatis: gutture indistinctè, pectore superiore magis distinctè, brunneo punctatis et pectore laterali nigro lineato: rostro brunnescente, mandibulâ flavicante: pedibus flavidis: iride nigrâ.

Adult. Upper parts dark brown, each feather margined with fulvous, shading into rufous-brown on the forehead and crown; streak over the eyes white; throat, fore neck, breast, and abdomen fulvous-white, each feather with a central streak of black, giving to the underparts a spotted appearance; wing-feathers and their coverts blackish brown, edged with bright fulvous; tail-feathers dark brown, with black shafts. Irides black; bill and feet pale brown. Total length 6·5 inches; wing, from flexure, 2·25; tail 3·25; bill, along the ridge ·4, along the edge of lower mandible ·6; tarsus ·75; middle toe and claw ·7; hind toe and claw ·6.

Young. The young assume the adult plumage on quitting the nest.

Obs. The tail-feathers have the barbs disunited in their whole extent.

THIS recluse little species is one of our commonest birds, but is oftener heard than seen. It frequents the dense fern (*Pteris aquilina*) of the open country, and the beds of raupo (*Typha angustifolia*) and other tall vegetation that cover our swamps and low-lying flats. In these localities it may constantly be heard uttering, at regular intervals, its sharp melancholy call of two notes, *u-tick, u-tick*, and responsively when there are two or more. When the shades of evening are closing in, this call is emitted with greater frequency and energy, and in some dreary solitudes it is almost the only sound that breaks the oppressive stillness. In the Manawatu district, where there are continuous raupo-swamps, covering an area of 50,000 acres or more, I have particularly remarked this; for, save the peevish cry of the Pukeko, occasionally heard, and the boom of the lonely Bittern, the only animate sound I could detect was the monotonous cry of this little bird calling to its fellows as it threaded its way among the tangled growth of reeds.

Large portions of the North Island consist of rolling land covered with stunted brown fern, this being the characteristic feature for twenty miles at a stretch, broken only by little patches of bush in the gullies. Intersecting these fern-ridges are narrow belts of wiwi-swamp, of a dark green colour from the character of the vegetation. These beds of rushes, which form blind watercourses during the winter season, are dry in summer and are then a favourite resort for the "Swamp-Sparrow," as this bird is sometimes called. In these localities it may always be found, sometimes in pairs but usually singly, the habits of the species being solitary, except of course in the breeding-season. But other places also are frequented by it. As already mentioned, it inhabits the raupo-swamps; and in the tangled vegetation which fringes our low-lying rivers, under a thick screen of native bramble and convolvulus, its melancholy note may frequently be heard, particularly towards nightfall. But it is never met with in the forest, or at any great elevation from the sea.

During my last visit to the Hot Lakes district, I found it still plentiful in all suitable localities. There are marshy tracts occurring at intervals along the road from Taupo to Ohinemutu, and the familiar note of this little bird was the only relief to those quiet solitudes. The pairing-season had commenced, and it was most pleasant to hear the couples singing their simple duet, the notes being always in harmony and responsive. When excited or alarmed its cry becomes sharper, being not unlike the call "Philip, Philip!" with a short pause between.

Like the other members of the group to which it belongs, it is a lively creature, active in all its movements, and easily attracted by an imitation of its note; but, when alarmed, shy and wary. Its tail, which is long and composed of ten graduated feathers, with disunited filaments, appears to subserve some useful purpose in the daily economy of the bird; for it is often found very much denuded or worn. When the bird is flying, the tail hangs downwards. Its wings are very feebly developed, and its powers of flight so weak that, in open land where the fern is stunted, it may easily be run down and caught with the hand; but in the swamps it threads its way through the dense reed-beds with wonderful celerity, and eludes the most careful pursuit. When surprised or hard pressed in its more exposed haunts, it takes wing, but never rises high, and, after a laboured flight of from fifteen to twenty yards in a direct line, drops under cover again. Its food consists of small insects and their larvæ and the minute seeds of various grasses and other plants.

Major Jackson, of Kihikihi, who is a keen sportsman, assures me that this bird has a very strong scent, so much so that when he has been out pheasant-shooting his pointer has "stood" to it quite as staunchly as if it had been a game bird.

This pretty little creature is not exempt from the common ills that "flesh is heir to." A specimen brought to me on the 8th March presented a remarkable diseased swelling, larger than a pea, at the root of the beak. After carefully examining it, I turned the little sufferer free, leaving Dame Nature, in this case as in others, to work out her own cure.

It is a matter of extreme difficulty to study the breeding-habits of species that resort to the dense vegetation of the swamps. Even a systematic search for the nests, in such localities, is of very little use, and the collector must trust to the chapter of accidents for opportunities of examining them. Although so common a bird, I have only once succeeded in finding the nest. This discovery was made many years ago, on the edge of a raupo-swamp, near the old Mission Station on the Wairoa river. The nest was a small cup-shaped structure, composed of bents and dry grass-leaves, not very compact, but with a smooth and carefully lined interior. It was attached to reed-stems standing together, and contained four young birds, which showed remarkable nimbleness, darting out of the nest and disappearing in the long grass on the first moment of my approach. I have, however, heard of others, containing sometimes four eggs, sometimes three. The eggs are ovoido-conical in form, measuring ·8 of an inch in length by ·6 in breadth, and are creamy white, thickly speckled over the entire surface with purplish brown.

Mr. Potts describes the nest as being composed of grass-leaves, with generally a few feathers of the Swamp-hen, and sometimes a small tuft of wool. The breeding-season appears to embrace the months of October and November; for on November 4 he found a nest containing three young birds, and three days later, but in another locality, a nest with four eggs in it.

SPHENŒACUS FULVUS.

(FULVOUS FERN-BIRD.)

Sphenœacus fulvus, Gray, Ibis, 1862, p. 221.
Megalurus fulvus, Gray, Hand-l. of B. i. p. 206 (1869).

Ad. similis *S. punctato*, sed paullò major: ubique lætiùs fulvescens, plumis vix ita distinctè medialiter lineatis: pectore etiam minùs distinctè maculato: caudâ minùs acuminatâ, scapis plumarum haud nudis, sed ad apicem ipsum plumiferis.

Adult. Upper parts dark fulvous, each feather centred with black; forehead and crown slightly stained with rufous; line over the eyes, throat, fore neck, breast, and upper part of abdomen fulvous-white, obscurely spotted on the breast with brown; sides of the body, flanks, thighs, and lower part of abdomen bright fulvous; primaries and secondaries blackish brown, margined on their outer webs, and the three innermost secondaries broadly margined all round, with bright fulvous; tail-feathers fulvous, with a dark shaft-line, and lighter on the edges. Total length 7·5 inches; wing, from flexure, 2·5; tail 4; bill, along the ridge ·4, along the edge of lower mandible ·6; tarsus ·75; middle toe and claw ·7; hind toe and claw ·6.

Young. An example in the Canterbury Museum, so immature that the tail-feathers are only two inches long, has more fulvous in the plumage and no indication whatever of a superciliary streak.

Obs. Mr. Sharpe says of the type in the British Museum:—"Similar to *S. punctatus*, but rather larger, and very much lighter and more ochraceous in colour. Both on the upper and under surface the black centres to the feathers are not so broad, and thus the plumage appears more distinctly streaked" (Cat. Birds B. M. vii. p. 98).

THIS species, as distinguished by Mr. G. R. Gray, bears a general resemblance to *Sphenœacus punctatus*; but, on comparing them, the following differences are manifest:—The present bird is larger and has the whole of the plumage lighter; the upper parts have the central marks much narrower, and on the hind neck and rump they are entirely absent; the white superciliary streak is less distinctly defined, the spots on the under surface are less conspicuous, and the tail-feathers, which are much paler than in *S. punctatus*, differ likewise in their structure, the webs being closely set, instead of having loose disunited barbs.

Several specimens have passed through my hands, all of which have been obtained in the South Island.

Mr. Potts distinguishes the eggs of this bird from those of *S. punctatus* as being "slightly larger and white, marked with reddish-purple freckles."

Whilst, however, keeping the form distinct for the present, I am far from being satisfied that it can be separated from *S. punctatus*. I am more inclined to regard it as a somewhat larger local race, with a corresponding modification of plumage. But for the fact that the latter species is as common in the South Island as in the North, this might be treated as the representative form.

SPHENŒACUS RUFESCENS.

(CHATHAM-ISLAND FERN-BIRD.)

Sphenœacus rufescens, Buller, Ibis, 1869, p. 38.
Megalurus rufescens, Gray, Hand-l. of B. i. p. 206 (1869).

Ad. suprà saturatè castaneus, pileo concolore: dorso paullò fulvescente, plumis latè medialiter nigris: tectricibus alarum medialiter nigris, dorso concoloribus: remigibus nigris, rufescente limbatis: caudâ rufescente, subtùs fulvescentiore, scapis pennarum nigris: loris et supercilio distincto fulvescenti-albis: regione paroticâ saturatè castaneâ, nigro notatâ: genis fulvescentibus, nigro maculatis: subtùs fulvescenti-albus, corporis lateribus castaneis nigro striolatis, dorso concoloribus: subalaribus stramineis, rufescente lavatis: rostro corneo, mandibulâ flavicante: pedibus flavicanti-brunneis: iride nigrâ.

Adult male. Upper parts dark rufous-brown, brightest on the crown and hind neck; streak over the eyes, throat, breast, and abdomen dull rufous-white, slightly tinged with yellow on the throat; sides of the head, ear-coverts, and a series of spots from the base of the lower mandible brownish black; sides of the body and the flanks bright rufous-brown, each feather with a central streak of black; wing-feathers dusky black, margined on both webs with rufous-brown; the wing-coverts and the scapularies broadly centred with brownish black; tail-feathers clear rufous-brown, with glossy black shafts, paler on their under surface. Irides black; bill and feet yellowish brown. Total length 7·25 inches; extent of wings 7; wing, from flexure, 2·25; tail 4·25; bill, along the ridge ·5, along the edge of lower mandible ·7; tarsus 1; middle toe and claw ·85; hind toe and claw ·75.

Female. Similar to the male, but somewhat smaller in size and with rather duller plumage.

Obs. Prof. Hutton states that two of the specimens collected by Mr. Travers are "variegated with white feathers, principally on the wings."

THIS well-marked species is confined to the Chatham Islands, where it was first discovered, in 1868, by Mr. Charles Traill, a gentleman greatly devoted to conchology, who visited that group for the purpose of collecting its marine shells. He obtained it on a small rocky isle, lying off the coast of the main island, during one of his dredging-expeditions; but he was unable to give me much information respecting its habits or economy, merely stating that he observed it flitting about among the grass and stunted vegetation, and succeeded in knocking it over with a stone.

Mr. Henry Travers says:—"I only found this bird on Mangare, where it is not uncommon. Its peculiar habit of hopping from one point of concealment to another renders it difficult to secure. It has a peculiar whistle, very like that which a man would use in order to attract the attention of another at some distance; and although I knew that I was alone on the island, I frequently stopped mechanically on hearing the note of this bird, under the momentary impression that some other person was whistling to me. It also has the same cry as *Sphenœacus punctatus*. It is solitary in its habits and appears to live exclusively on insects."

I am indebted to Mr. Walter Shrimpton for an egg obtained on Pitt Island, and assigned, I believe correctly, to this species. It is broadly ovoido-conical, measuring ·80 of an inch in length by ·65 in breadth, and has the entire surface covered with a speckled or marbled graining of reddish brown on a creamy-white ground.

ANTHUS NOVÆ ZEALANDIÆ.

(NEW-ZEALAND PIPIT.)

New-Zealand Lark, Lath. Gen. Syn. ii. pt. 2, p. 384, pl. 21 (1783).
Alauda novæ seelandiæ, Gm. Syst. Nat. i. p. 799 (1788).
Alauda littorea, Forst. Descr. Anim. p. 90 (1844).
Anthus novæ zealandiæ, Gray, Voy. Ereb. and Terror, Birds, p. 4 (1844).
Anthus grayi, Bonap. Consp. Gen. Av. i. p. 249 (1850).
Anthus aucklandica, Gray, Ibis, 1862, p. 254.
Corydalla aucklandica, id. Hand-l. of B. i. p. 253 (1869).
Corydalla novæ zealandiæ, id. op. cit. i. p. 253 (1869).

Native names.

Pihoihoi and Whioi; "Ground-Lark" of the colonists.

Ad. brunneus, fulvescente lavatus, plumis medialiter paullò saturatioribus, uropygio unicolore fulvescenti-brunneo: loris et supercilio lato fulvescenti-albis: lineâ brunneâ per oculum ab ortu rostri ductâ: genis et regione paroticâ albidis, hîc paullò brunneo maculatâ: fasciâ mystacali irregulari brunneâ: colli lateribus dorso concoloribus et eodem modo notatis: tectricibus alarum brunneis, minimis latè et conspicuè aurantiaco-fulvo lavatis, majoribus angustè fulvido marginatis: remigibus brunneis, primariis angustissimè, secundariis latiùs fulvo marginatis: caudâ brunneâ, fulvo marginatâ, rectrice extimâ ferè omninò albâ, pogonio interno versùs basin brunneo, proximâ versùs apicem obliquè albâ, tertiâ extùs angustè albo limbatâ: subtùs fulvescenti-albus, hypochondriis brunneis: pectore superiore brunneo longitudinaliter maculato: rostro corneo, mandibulâ flavicante: pedibus flavicanti-brunneis: iride saturatè brunneâ.

Juv. similis adultis, sed pallidior, plumis indistinctè fulvo marginatis: collo postico conspicuè fulvescente: tectricibus alarum, remigibus et rectricibus latiùs fulvo marginatis: subtùs sordidè albus, pectore superiore vix distinctè brunneo striolato.

Adult. Upper parts brownish grey, darker on the rump and upper tail-coverts; on the back, each feather centred with brown; from the base of the bill a broad line of white passes above, and an irregular band of black extends across the eyes; cheeks greyish white, minutely spotted with black; chin, or interramal space, white; throat, fore neck, and upper part of breast fulvous, with numerous broad dashes of brown; under-parts white, tinged on the flanks and under tail-coverts with fulvous; sides of the body greyish white, with longitudinal streaks of brown; all the plumage of the underparts plumbeous at the base; wing-feathers and their coverts dark brown, margined on their outer webs with fulvous-grey, broadest on the tertiaries, and reduced to a mere line on the primaries; the marginal colour changes to fulvous-white on the secondary-coverts, presenting, when the wings are closed, a series of small crescentic bands; tail-feathers dark brown, with paler edges, except the two outermost ones on each side, which are white, the inner one crossed by an oblique band of dusky brown, and the outer one with a mere streak of the same colour near the root. Irides very dark brown, almost black; bill and feet yellowish brown. Total length 8 inches; extent of wings 12; wing, from flexure, 3·75; tail 3; bill, along the ridge ·5, along the edge of lower mandible ·75; tarsus 1; middle toe and claw ·85; hind toe and claw ·75.

Young. The young has the breast more spotted, and the feathers of the upper parts narrowly margined with pale rust-colour.

Obs. The sexes are alike. In some examples the under tail-coverts are pure white, while in others the upper wing-coverts are broadly margined with light rufous-brown. Allowing for this variation, I cannot see the propriety of admitting the supposed new species from Queen Charlotte's Sound (*Anthus grayi*, Bonap.), which I have accordingly expunged from our list.

Varieties. Albinos, more or less pure, are of common occurrence. The following is the description of an example in the Canterbury Museum:—General plumage pure white, varied on the back and wings with brownish grey; some of the quills and tail-feathers pure white, the others dark brown, as in ordinary specimens; bill and feet white horn-colour; the hind claw conspicuously long, measuring ·55 of an inch. Another specimen, in Mr. J. C. Firth's fine collection at Mount Eden, has the whole plumage dull creamy white, stained and washed on the upper parts of the body with yellowish brown. Captain Mair writes to me:—"I saw a pure white Lark, two days in succession, at Sulphur Point, but could not find it when I went with a gun"; and several other correspondents refer to similar occurrences in various parts of the country.

Of this bird I may remark that it is a true Pipit both in structure and in its habits of life. It bears a general resemblance to an Australian species (*Anthus australis*); but the specific differences are sufficiently manifest on an actual comparison of the two birds.

It is common throughout the country, frequenting the open land, and sometimes resorting to the dry sands along the sea-shore. During the autumn months it is gregarious, and may then be observed in flocks varying in number from twenty to fifty or more, alternately collecting and mounting in the air with a loud cheerful note, and scattering themselves again on the open ground to search for their food, which consists of insects and their larvæ, small earthworms, and occasionally minute seeds as well. At sundown the flocks break up, each bird seeking a convenient resting-place for the night; and with the first streak of daylight they begin to reassemble. On the approach of winter the flocks disperse *, and the birds appear to pair off at once, and remain so till the breeding-season arrives. They are always plentiful on the settlers' farms, and may be seen during the summer months perched in large parties on the roofs of the country houses or on the surrounding fences and outbuildings. They may sometimes be observed in similar situations within the towns, and notably on the roofs of churches and other lofty edifices. They love to resort to the roads and beaten paths, where they amuse the traveller by their playfulness, running before him as he advances, then rising in the air with a sharp but pleasant chirp, settling down again and running forward as before, and continually flirting the tail upwards. During the heat of the day they may often be seen sitting on the logs or fences with their beaks wide open as if gasping for air. They repose at night on the ground, finding shelter among the grass or fern on the open ridges or on the wayside, where the benighted traveller, as he plods along, may often disturb them and hear the sharp rustling of their wings as they rise startled at his very feet.

When searching for food, a flock of these birds will spread themselves out in all directions; but the instant a Hawk appears in sight, or some other common danger threatens, they will rise into the

* The accuracy of the above statement, in my former edition, having been called in question (Ibis, 1874, p. 38), I made careful observations over a continuous period of ten years, during which time I was constantly moving from one part of the colony to the other. From the notes in my diary I have abstracted the following particulars:—Autumn months (March, April, and May), numerous flocks, and often of considerable size, all over the country; winter months (June, July, and August), always in pairs; spring and summer (September to February, inclusive), still in pairs, but sometimes congregating. I have seen a flock numbering upwards of fifty as early as September 4. In the months of November and December it is a common thing to see parties of five or six, consisting probably of early broods of the year; and I find a note of one party of five on the 23rd October. The autumnal gathering commences about the second week in March, at the close of a prolonged breeding-season, with probably two broods; and I have no record of any flock after the beginning of June. Professor Hutton's statement that "they congregate in the autumn after the breeding-season is over and disperse to breed in spring" would seem to imply that the flocks keep together during the winter; but this is certainly not the case.

air together with much clamour, and sometimes mount to a considerable height. I have frequently seen a number of them pursue and harass the Bush-Hawk, which is doubtless their worst natural enemy. Their ordinary flight is rapid and undulating, being performed, as it were, by a succession of jerks. During the breeding-season the male bird frequently soars, mounting to a height in the air, and then descends with tremulous wings and outspread tail, uttering a prolonged trilling note, very pleasant to the ear.

This is one of the few species that appear to thrive and increase in the cultivated districts; and in localities where formerly it was only tolerably plentiful it has kept pace with the progress of colonization, becoming every year more abundant. It frequents the mountain-tops, being often met with above the snow-line. Mr. Ernest Bell observed one on the very summit of Mount Egmont.

It is never met with in the woods; and I have observed that in the open country it is rarely seen to alight on a green tree or shrub, although often poising itself on the slender stalks of the *Phormium tenax* or on a bunch of fern. I have occasionally seen it dusting itself after the manner of some gallinaceous birds, rolling in the dust with evident delight, and then shaking its feathers, probably in order to free the body of parasitic insects.

It is amusing to watch a pair of them chasing and making love to each other at the commencement of the breeding-season, each one alternately springing up in the air, with expanded wings and tail, and curvetting over the other in the most playful manner. The call of the young resembles the sharp note of the Silver-eye (*Zosterops cærulescens*); and when engaged in feeding them, the parent bird displays an unusual degree of caution in the presence of an intruder, alighting ten or fifteen yards from the nest, and loitering about for a considerable time with the food in its bill before attempting to deliver it. I have seen a pair skimming playfully together over the ground but close to the surface, when one would suddenly drop out of sight in the vicinity of the nest, leaving the other to pursue its wayward flight, as if to divert attention.

The natives catch this bird by means of a running-noose at the end of a long stick; and there are various modes of trapping it, very generally known and appreciated among colonial school-boys.

I have noticed that it is very subject to a disease of the foot, which takes the form of a large irregular swelling. This may probably result from accidental burns; for I have often observed these birds alight on ground over which a fire had recently passed, leaving a light surface of smouldering ashes, and rise again immediately in evident pain.

On the Hastings-Napier railway line and elsewhere I have observed a peculiar habit which this species has developed of following the train. I have seen, in autumn, a flight of a hundred birds keeping abreast or a little ahead of the train in rapid motion for two or three miles at a stretch, picking up stragglers *en route*, and to all appearance thoroughly enjoying the excitement.

The breeding-season of the New-Zealand Pipit extends from October to February or March, and, like the other members of the same group, it appears to rear two broods; for I have seen well-fledged young ones in November, while nests containing eggs are often met with as late in the season as January or the early part of February. The nest is composed of dry grass and other fibrous substances loosely put together, and is always placed on the ground, generally in a horse's footprint or in some natural depression, and under shelter of a tussock or clump of rushes. The eggs are usually four in number, rather ovoido-conical in shape, measuring, as a rule, ·9 of an inch by ·65, and marked over the entire surface with numerous spots or freckles of dark grey on a paler or ashy ground. A fine series of eight in my son's collection exhibit a considerable amount of individual variation both in form and colouring. The smallest of these measures ·85 inch in length by ·65 in breadth, and is almost a perfect oval; and the largest 1 inch in length by ·70 in breadth. The ground-colour varies from pale stone-grey to a warm creamy-grey, and the markings pass through every gradation, from a covering of uniform speckles and freckles of greyish brown to a much darker character, blotched and mottled with purplish brown of different shades.

GRAUCALUS MELANOPS.

(AUSTRALIAN SHRIKE.)

Black-faced Crow, Lath. Gen. Syn. Suppl. ii. p. 116 (1801).
Corvus melanops, Lath. Suppl. Ind. Orn. p. xxiv (1801).
Rollier à masque noir, Levaill. Ois. de Paradis, pl. 30 (1806).
Ceblepyris melanops, Temm. Man. d'Orn. i. pl. lxii. (1820).
Graucalus melanops, Vig. & Horsf. Tr. Linn. Soc. xv. p. 216 (1826).
Graucalus melanotis, Gould, P. Z. S. 1837, p. 143.
Campephaga melanops, Gray, Cat. B. N. Guin. p. 32 (1859).
Colluricincla concinna, Hutton, Cat. B. New Zealand, p. 15 (1871).
Graucalus concinnus, Hutton, Trans. N.-Z. Inst. vol. v. p. 225 (1872).
Graucalus melanops, Buller, Birds of New Zealand, 1st ed. p. 148 (1873).

Descr. exempl. ex N.Z. Suprà cinereus: tectricibus alarum dorso concoloribus: remigibus nigricanti-brunneis, primariis angustè, secundariis latiùs albido marginatis: rectricibus nigricanti-brunneis, parte basali cinereâ, pennis externis ad apicem albis, duabus exterioribus gradatim obliquè albis, rectrice extimâ etiam albo marginatâ: facie laterali totâ nigrâ: gutture et pectore superiore cinereis dorso concoloribus: corpore reliquo subtùs albo: rostro nigro versùs basin mandibulæ brunnescente: pedibus saturatè brunneis.

New-Zealand example (young). General plumage light cinereous or ashy grey; a patch of black fills the lores, crosses the eyes, and covers the cheeks and ear-coverts; on the upper part of the breast the grey fades into white, with a purplish tinge; lower part of breast, lining of wings, flanks, abdomen, and under tail-coverts pure white; wing-feathers dark brown, the primaries narrowly and the secondaries broadly margined with greyish white; tail-feathers dark brown, the two middle ones tinged with ashy grey, especially in their basal portion; the lateral ones tipped progressively outwards with white, the outermost one on each side having an inch at the extremity and a narrow line along the apical portion of its outer web pure white. Bill black, changing to brown at the base of the lower mandible; legs blackish brown. Total length 13 inches; wing, from flexure, 8; tail 5·5; bill, along the ridge ·9, along the edge of lower mandible 1·25; tarsus 1·12; middle toe and claw 1·2; hind toe and claw 1.

Adult male (from Australia). General colour above light French grey, the wing-coverts like the back, with edgings of still lighter grey; primary-coverts and primaries black, externally edged with grey, inclining to white towards the tips of the quills; secondaries black, the outer aspect of the feathers light grey on the innermost, with the outer web grey and the inner one black; two centre tail-feathers ashy grey, blacker towards the tip, which is white, all the other feathers black, washed with grey towards the base and tipped with white, which increases in extent towards the outermost feather, which is also edged with white along the outer web; entire forehead, feathers above the eye, ear-coverts, sides of face, sides of neck, entire throat, and fore neck black, with a greenish gloss, fading off paler towards the chest, which is iron-grey, becoming gradually lighter and more delicate grey on the sides of the body, so as to leave only the lower abdomen and under tail-coverts pure white; thighs grey; under wing-coverts and axillaries pure white, as also the inner lining of the quills, which are otherwise ashy grey below; bill black; feet dull ashy; iris black. Total length 12·5 inches, culmen 1·05, wing 7·65, tail 5·75, tarsus 1·05. (Cat. Birds Brit. Mus. vol. iv. p. 31.)

Obs. "♂, Louisiade Islands specimen, wing 8·1 inches; ♀, N.W. Australia, wing 7·1 inches. These two seem to be the extremes, and every intermediate link between them can be found." (*Id. l. c.*)

The example from which the above description is taken was shot by Mr. Giblin at Motueka, in the Provincial district of Nelson, and now forms part of the public collection in the Nelson Museum. Mr. Huddleston informs me that he saw the bird in the flesh, and knows the precise locality in which it was shot. There can be no doubt, therefore, as to the authenticity of the specimen as a New-Zealand bird; but as it appears to be quite unknown to the natives of the country, it may, I think, be safely assumed that this was an accidental visitant from Australia, where the species is very plentiful. Another example was shot at Invercargill in April 1870, and forwarded to the Colonial Museum. Of this Professor Hutton writes (*l. c.*):—"Like the bird shot in Nelson province, this one also has the general plumage of the young of *G. melanops*; but the feathers of the chin and forehead are similar to those on the throat and top of the head, and not lighter as in *G. melanops*; there is also no indication of any black feathers coming on the chin or upper part of the head. It differs from the Australian bird in having a more slender bill, a rather longer tail, the feathers of which are acutely pointed at the tip instead of being rounded, and in having much more white on the wings. These differences are, I think, quite sufficient to warrant its being kept as a distinct species"*. He adds:—"Mr. Mantell has informed me that he saw this bird many years ago at Port Chalmers in Otago; Mr. W. Travers says that he has seen it at Nelson, and Captain Fraser says that he saw it near Hawea Lake in Otago."

This species is liable to so much variation, both in plumage and size, that I am unable at present to consider the form which has thus occurred at such rare intervals in New Zealand as distinct from the Australian one. Of the latter Mr. Gould says that the "infinite changes of plumage which these birds undergo from youth to maturity render their investigation very perplexing."

Dr. Finsch expresses his belief that the bird which has occurred in New Zealand is *G. parvirostris*, Gould; but Mr. Sharpe, in his account of *G. melanops* (Cat. of Birds Brit. Mus. iv. p. 31) says:—"This species varies much in size, but it is impossible to believe in the existence of more than one species; and *G. parvirostris* is little more than a race of the present bird."

I have gone carefully over the whole series of skins in the British Museum, and am confirmed in my original conclusion that our bird is the young or immature state of *G. melanops*. I attach no value to the two characters on which Professor Hutton appears mainly to rely, namely, the white margins to the greater wing-coverts and the more acutely pointed tail-feathers. In a large series, of all ages, I find the extent of white on the wings very variable, and in the younger birds the tail-feathers are undoubtedly narrower at the points than in fully adult specimens. In the Nelson bird, of which a full description is given above, it will be seen that the former of Professor Hutton's distinguishing characters is absent. I should be inclined to give more weight to the colour of the primaries, as described by him, because in every specimen of *G. melanops* examined by me the first five primaries are uniform brownish black, or with only a very narrow greyish-white margin on the outer web, there being no sign of any white tips. This difference, however, appears to me too trivial to separate the species, the more so as it is wanting in the Nelson example. The "white circular

* Graucalus concinnus, Hutton (*l. c.*):—"The whole of the upper surface uniform pale grey, the feathers of the forehead with the shafts darker; feathers of the throat and breast pale grey, slightly tipped with white; those of the upper abdomen and thighs pale grey, with white circular bands; lower abdomen, vent, and under tail-coverts pure white; a broad band of black passes from the nostrils and gape through and below the eye to the region of the ears; primaries brownish black, the first slightly tipped with white, the second, third, fourth, and fifth margined outwardly and slightly tipped with white, the remainder margined all round with a white band, which is broader on the tip and inner web; secondaries greyish black, with more or less grey on the outer web near the base, and with a rather broad white margin on the outer web and tip; *greater wing-coverts margined externally with white; tail-feathers acutely pointed at the tip*, the two middle ones brownish grey, laterals brownish black tipped with white, the white decreasing inwards; shafts of the tail-feathers greyish black above and pure white below; bill (dry) brownish black, paler at the base; legs and feet (dry) black. Wing 8 inches; tail 7; tarsus 1·1; hind toe ·5; middle toe 1·1; bill, culmen, ·86, breadth at nostrils ·4, height at nostrils ·35."

bands" afford to my mind further evidence of immaturity. If, however, it were the young of *G. parvirostris* (as suggested by Dr. Finsch), it ought to present other markings, for the young of this form exhibits numerous arrow-heads of brownish black on its chin and throat.

Assuming, therefore, the species to be the same, this bird is very common in New South Wales, especially in the summer months, frequenting "plains thinly covered with large trees," rather than the thick brushes. It is said to be also abundantly dispersed over the plains of the interior, such as the Liverpool, and those which stretch away to the northward and eastward of New South Wales.

"It breeds in October and the three following months. The nest is often of a triangular form, in consequence of its being made to fit the angle of the fork of the horizontal branch in which it is placed; it is entirely composed of small dead twigs, firmly matted together with a very fine, white, downy substance like cobwebs and a species of lichen, giving the nest the same appearance as the branch upon which it is placed, and rendering it most difficult of detection. The ground-colour of the eggs, which are usually two in number, varies from wood-brown to asparagus-green, the blotches and spots, which are very generally dispersed over their surface, varying from dull chestnut-brown to light yellowish brown; in some instances they are also sparingly dotted with deep umber-brown; their medium length is thirteen lines, and breadth ten lines. Its note, which is seldom uttered, is a peculiar single purring or jarring sound, repeated several times in succession." (Gould, Handb. Birds Austr. i. pp. 193, 194.)

The ornithology of New Zealand has now been so thoroughly explored that we cannot hope to make any further additions to our list of species, except by recording accidental visitants like the above at long intervals of time—such birds, for example, as *Acanthochæra carunculata* and *Eurystomus pacificus*; or the occurrence of foreign Waders, such stragglers from the flock as may occasionally pass out of their course to New Zealand during their seasonal migration—as, for instance, *Charadrius fulvus* and *Phalaropus ruficapillus*; or of oceanic species whose home is on the rolling sea and whose habitual range, within uncertain degrees of latitude and longitude, is often extended almost indefinitely by the terrific and long-continued storms that sweep over the face of the great Pacific Ocean—such as the beautiful red-tailed Tropic-bird (*Phaëton rubricauda*), or that noble "Vulture of the sea," *Tachypetes aquila*, and the rarer kinds of Petrel. The opportunities, however, of recording such occurrences are becoming every year more difficult for the practical ornithologist, owing to the number and variety of foreign birds that are being introduced into the country through the efforts of Acclimatization Societies and other local agencies. In the early days of the colony nothing that was new escaped the vigilant eye of the Maori, and the appearance of a strange bird, whether on the sea-shore, in the lagoons, or on the land, was immediately noticed, and the fact sooner or later reported to the colonists. But nowadays the country teems with imported birds of every kind—Thrushes and Blackbirds, Greenfinches and Linnets in the woods and shrubberies; Pheasants, Partridges, and Quail in the open, with Sky-Larks and Starlings on the meadows; Black Swan and Egyptian Geese on the lagoons, and the ubiquitous Sparrow in every street and hedgerow, besides numberless other introduced species of more or less importance. Consequently, when a Maori sees a bird hitherto unknown to him he puts it down in his mind as a "manu pakeha," and pays no further heed to it.

The occurrence in New Zealand, from time to time, of Australian and Polynesian forms, without any suspicion of human intervention or of artificial assistance such as that afforded by ships' rigging, is a matter of extreme interest to the philosophic naturalist, because these cases serve to illustrate the manner in which the avifauna of oceanic islands lying far apart from one another or from any continental area—as, for example, Norfolk Island and Lord Howe's Island—may undergo, in process of time and by insensible degrees, important changes of feature through the accidental intrusion of foreign types. For this reason, I have been very careful to notice in the present work every instance of the kind that has come to my knowledge.

RHIPIDURA FLABELLIFERA.

(PIED FANTAIL.)

Fan-tailed Flycatcher, Lath. Gen. Syn. ii. pt. 1, p. 340, pl. xlix. (1783).
Muscicapa flabellifera, Gm. Syst. Nat. i. p. 943 (1788, ex Lath.).
Rhipidura flabellifera, Gray, in Dieff. Trav. ii., App. p. 190 (1843).
Muscicapa ventilabrum, Forst. Descr. Anim. p. 86 (1844).
Rhipidura albiscapa, Cass. U. S. Expl. Exp. p. 150 (1858, nec Gould).

Native names.

Piwaiwaka, Tiwaiwaka, Piwakawaka, Tiraираka, Pirairaka, Tiwakawaka, and Pitakataka.

Ad. suprà olivascenti-brunneus, pileo nigricante: lineâ supraoculari albidâ: tectricibus alarum brunneis, olivaceo lavatis, albido terminatis: remigibus nigricanti-brunneis extùs dorsi colore lavatis: caudâ sordidè albâ, scapis purè albis, rectricibus duabus centralibus nigricantibus ad apicem albidis, reliquis extùs brunnescenti-nigris, pennâ extimâ omninò albidâ: facie laterali pileo concolore: gulâ albidâ: torque pectorali nigrâ: subtùs aurantiaco-fulvus, pectore superiore et subcaudalibus pallidioribus: cruribus nigricantibus: rostro nigro: pedibus brunnescenti-nigris: iride nigrâ.

Juv. similis adulto, sed suprà magis brunnescens: gutture griscescenti-albo: corpore reliquo subtùs sordidè fulvescente: torque pectorali absente: tectricibus alarum fulvido apicatis, et secundariis extùs eodem colore marginatis.

Adult male. Crown, nape, and sides of the head sooty black; the whole of the back, rump, and upper surface of wings dark olive-brown; the small wing-coverts tipped with fulvous white; rictal bristles black; throat and mark over the eyes greyish white; across the fore neck and upper part of breast a broad band of sooty black; lower part of breast and all the under surface fulvous, tinged with cinnamon, the base of the feathers plumbeous; quills dark olive-brown, with paler shafts, the inner secondaries edged with fulvous white; the two middle tail-feathers brownish black, with pure white shafts, and tipped with greyish white; the lateral feathers greyish white and, with the exception of the outermost one on each side, margined on their outer webs with brownish black, all having pure white shafts. Irides and bill black; feet blackish brown; soles greyish. Total length 6·5 inches; extent of wings 8; wing, from flexure, 2·75; tail 4; bill, along the ridge ·3, along the edge of lower mandible ·4; tarsus ·7; middle toe and claw ·6; hind toe and claw ·5.

Female. Similar in plumage to the male, but slightly smaller.

Young. The young bird has the throat greyish white; the breast and all the under surface dark fulvous brown; the small wing-coverts are largely tipped and the secondaries narrowly edged with fulvous brown, and the plumage of the back is more or less tinged with the same colour.

Obs. I have observed birds in the young plumage as late as the middle of March; but the adult livery is certainly assumed at the first moult.

THE Pied Fantail, ever flitting about with broadly expanded tail, and performing all manner of fantastic evolutions, in its diligent pursuit of gnats and flies, is one of the most pleasing and attractive objects in the New-Zealand forest.

It is very tame and familiar, allowing a person to approach within a few feet of it without evincing any alarm, sometimes, indeed, perching for an instant on his head or shoulders. It will often enter the settler's house in the bush, and remain there for days together, clearing the window-panes of sand-flies, fluttering about the open rooms with an incessant lively twitter during the day, and roosting at night under the friendly roof*. It is found, generally in pairs, on the outskirts of the forest, in the open glades, and in all similar localities adapted to its habits of life. It loves to frequent the wooded banks of mountain-streams and rivulets, where it may be seen hovering over the surface of the water collecting gnats; and I have counted as many as ten of them at one time so engaged. It affects low shrubby bushes and the branches of fallen trees; but it may often be seen catering for its insect-food among the topmost branches of the high timber.

You may always make sure of finding it flitting noiselessly about the bushes at the edges of the little mountain-stream which

"Chatters over stony ways,
In little sharps and trebles,
And bubbles into eddying bays,
And babbles on the pebbles."

These localities often swarm with minute diptera, on which the bird subsists. And I have seen five or six of them together displaying their fans, and hawking, as it were, for these invisible flies above the surface of the water.

In winter it generally frequents the darker parts of the forest, where insect-life is more abundant at that season; but it is nevertheless to be met with, wherever there is any bush, all the year round. It is a true Flycatcher, subsisting entirely by the chase: darting forth from its perch, it performs a number of aërial evolutions in pursuit of invisible flies—the snapping of its mandibles as it catches its prey being distinctly audible—and generally returns to the twig from which it started. It hops about along the dry branches of a prostrate tree, or upwards along the tangled vines of the kareao (*Rhipogonum scandens*), with its tail half expanded and its wings drooping, seizing a little victim at almost every turn, and all the while uttering a pleasant twitter. When hurt or alarmed it immediately closes its pretty fan, and silently flies off in a direct course, disappearing in the denser foliage.

It breeds twice in the season, producing four young ones at each sitting. It generally commences to build in September, and brings out its first brood about the last week in October. The second brood appears to leave the nest about the beginning of January.

The nest is a beautiful little structure, compact and symmetrical. A forked twig is the site usually selected; and the nest, instead of being placed within the fork for support, is built around it, the branchlets being thus made to serve the purpose of braces and stays to strengthen the work and to hold it together. It is therefore generally impossible to remove or detach the nest from the branch without tearing it to pieces. In form it is cup-shaped, the upper part towards the rim being closely interwoven and securely bound, while the base is left unfinished or loosely constructed. The materials composing the foundation are light fragments of decayed wood, coarse mosses, and the skeletons of dead leaves. The centre and upper portion of the nest consist principally of the tough and elastic seed-stems of various mosses finely interwoven. There is an exterior wall composed of

* Major Jackson told me a romantic story about this bird. A friend of his met with an accident and got his leg broken. He was carried into a little country house, and could not safely be moved for some time. During his detention he suffered very much from the heat and the swarms of small flies that invaded his improvised hospital. On one occasion, however, the window being open, a Fantail came in from the adjoining garden, took up its station on a peg in the wall, and soon cleared the room of flies, flitting airily about, and snapping its mandibles so long as a single fly remained. After this and as long as the invalid remained, the bird was a daily visitor, ministering in the manner described to the peace and comfort of the fly-pestered inmate.

cow-hair, the downy seed-vessels of plants, and other soft materials, and the whole is admirably bound together with fine spiders' webs. The interior cavity, which is rather large in proportion to the nest, is closely lined with fibrous grasses, or bents, disposed in a circular form. I have examined numbers of nests, and I have observed that the materials employed vary slightly, according to the locality, specimens collected in the vicinity of farmhouses disclosing tufts of wool, fragments of cloth, remnants of cotton-thread, &c. among the building-materials; nevertheless, in every instance that has come under my notice, the use of spiders' webs for binding the walls has been adhered to, thus manifesting a very decided instinct. The eggs are usually four in number, slightly ovoido-conical, and measuring ·7 of an inch in length by ·5 in breadth; they are white, with numerous purplish-brown freckles, denser and forming an obscure zone towards the larger end.

Mr. J. H. Gurney ('Ibis,' 1860, p. 212), in his account of the Red-throated Widow-bird (*Vidua rubritorques*, Swains.), says:—"These birds build amongst the grass in the open country. The nest is curiously built; they select a convenient tuft of grass, and interlace the blades as they stand, without breaking them off; so that the nest is green during the whole time of incubation, and is very beautiful when thus seen." This brings to my recollection a very pretty nest of the Pied Fantail which I found in the Kaipara woods many years ago. It was smaller and more cup-shaped than the generality of these nests, and was composed chiefly of moss firmly bound together with spiders' webs; but it was an "old nest," and the winter rains had soaked it, causing the moss to vegetate afresh; and when it came into my hands it was covered on the outer surface with a luxuriant growth of stunted moss of the brightest green, and presented a very beautiful appearance.

To any one having any experience of bird-craft, it is very easy to discover the nest of this species. The movements of the old birds, properly interpreted, are a very sure index. As you approach the nest, the Fantails, which follow your steps with an incessant twitter, become ominously silent. If you fail immediately to discover the object of your search, and chance to wander away from it, the anxious little birds give vent to their joy by an exuberant strain of notes, which, as I have often thought, might be appropriately compared to the supposed merry laugh of one of Gulliver's Liliputians*. On one occasion I succeeded in capturing the old bird on the nest, which was found to contain four unfledged young ones. I placed my captive in a cage, together with the nest and young: she refused food, and vented her rage by pecking her young ones to death. On the following morning I liberated the parent, regretting much that I had invaded her domestic happiness.

The multiplication of numbers by second broods, in the proportion of four to one, as already noticed, appears to me a wise provision of Nature to save the species from extinction. At the close of the breeding-season the Fantails, principally in the immature plumage, are excessively abundant; by the end of the year their numbers have been considerably thinned, owing to the joint ravages of the wild cat, the Bush-Hawk, and Morepork, to all of which this defenceless little creature falls an easy prey. The reproduction by each pair of eight young ones every season seems, therefore, almost necessary to preserve the very existence of this species in the balance of life.

Long may the Pied Fantail thrive and prosper, in the face of cats, owls, naturalists, and the whole race of depredators; for without it our woods would lack one of their prettiest attractions, and our fauna its gentlest representative!

* In one of the Maori legends we are told that the great ancestor Maui-Potaka, whose ordinary companions were a flock of Piwaiwaka, was betrayed by this "laugh" when eating up the body of Hinenuitepo and was forthwith killed. The myth relates how these little birds could contain themselves no longer, and when Hinenuitepo's head and shoulders had disappeared down Maui-Potaka's throat "they danced about and laughed," a pretty allusion to the habits of the Fantail.

RHIPIDURA FULIGINOSA.

(BLACK FANTAIL.)

Muscicapa fuliginosa, Sparrm. Mus. Carls. pl. 47 (1787).
Muscicapa deserti, Gm. Syst. Nat. i. p. 949 (1788, ex Sparrm.).
Rhipidura melanura, Gray, in Dieff. Trav., ii. App. p. 191 (1843).
Leucocerca melanura, Bonap. Consp. Gen. Av. i. p. 324 (1850).
Rhipidura tristis, Hombr. et Jacq. Voy. Pôle Sud, Ois. iii. p. 76, pl. xi. fig. 5 (1853).
Rhipidura sombre, iid. op. cit., Atlas, pl. xi. fig. 4 (1853).

Ad. nigricans, dorso alisque brunneo tinctis: maculâ postauriculari parvâ albâ: subtùs dilutiùs brunneus: rostro nigro, mandibulâ versùs basin albicante: pedibus nigricanti-brunneis: iride nigrâ.

Adult. Entire plumage black, tinged on the back and wings with rusty brown, and on the under surface with paler brown; behind each ear a small spot of white. Irides black; bill black, white at the base of the lower mandible; tarsi and toes blackish brown. Total length 6·5 inches; extent of wings 8; wing, from flexure, 2·75; tail 4; bill, along the ridge ·3, along the edge of lower mandible ·4; tarsus ·7; middle toe and claw ·6; hind toe and claw ·5.

Female. Similar to the male, but with the white spots behind the ears much reduced.

Obs. In the full-plumaged male the white mark described above usually consists of twelve diminutive feathers. In an example which came under my notice at Kaiapoi this feature was exaggerated, the white spreading entirely over the ear-coverts and surrounding feathers. In some it is scarcely visible, while in others (probably young birds) it is altogether wanting.

THIS dark-coloured species is, generally speaking, restricted to the South Island, where it is far more common than the preceding one.

Its life-history differs in no respect from that of its congener, as described in the foregoing pages. The stomachs of two which I dissected contained, in addition to the remains of small dipterous insects, the minute seeds of some wild berry.

Mr. G. R. Gray gives Cook's Strait as its habitat; but although common enough on the Nelson side, at the date of my former edition I knew of only one instance of its occurrence on the northern shore of the strait, or in any part of the North Island. After very stormy weather in May 1864, I shot a specimen in a flax-field near the mouth of the Manawatu river, on the south-west coast of the Wellington Province. It was evidently a straggler from the opposite mainland, and having by some means been deprived of its ample tail, which serves to balance the body, it had probably lost command of itself, and thus been borne across the sea by the prevailing gales. That the Flycatcher does sometimes indulge voluntarily in a water excursion, I have myself had proof; for in April 1869, when entering the Whangarei Heads, a Pied Fantail (*Rhipidura flabellifera*) flew off from the shore, and after making a circuit of our little steamer, apparently to satisfy its curiosity, returned to the land.

Ten years later another specimen was killed near a streamlet in the Pirongia Ranges, Waikato; and a third was obtained by my son in a shrubbery near Wellington on the 2nd April, 1876. Again, a pair of these Black Fantails visited my garden on Wellington Terrace on the 15th of the same month, and, as I would not allow them to be molested, returned on several successive days. They disappeared together, and I did not see them afterwards, although fondly hoping that they would breed with us, and that this pretty bird might become at length fairly acclimatized in the North Island.

Several more instances of its occurrence in the North Island, in the year following, have come to my knowledge. Major Muir recorded a second example from the Pirongia ranges in the Waikato; another was seen by Mrs. Howard Jackson in the shrubbery at Major Marshall's, near Rangitikei; and another was reported from Auckland. Of the last-mentioned Mr. T. F. Cheeseman, the Curator of the Auckland Museum, writes to me:—"You will be interested to hear that a solitary individual of the Black Fantail has been repeatedly seen near Auckland this winter. It was first noticed by Mr. James Baber in his garden at Remuera; afterwards it visited Mr. Hay's nursery-garden, where it remained for some weeks; and it has since been noticed about several of the residences at Remuera. I was fortunate enough to see it one evening when walking home, and can consequently vouch for its being the South Island species. Its occurrence so far to the north is certainly very remarkable."

Mr. Colenso, F.R.S., informs me that he met with one, in February 1882, at Napier; and to Mr. Leonard Reid, the Assistant Law Officer at Wellington, I am indebted for the following note:— "It may interest you to know that I met with a specimen on the Pukerua Range near Pauatahanui, when out shooting there in May 1883, in company with two residents of the district who had never seen a Black Fantail before. We tried hard to secure it alive, but though, like its northern congener, it was remarkably tame and fearless, our efforts were unsuccessful. We observed none others of either species in the same locality, and though a frequent visitor to the bush in various parts of this district I have never observed its occurrence on any other occasion" *.

A very interesting phase of character exhibited by this species is that, in its wild state, it associates and interbreeds with the Pied Fantail (*Rhipidura flabellifera*), as represented in the Plate.

There is a nest of this sort in the Canterbury Museum, containing three eggs. It was taken, in October 1870, by Mr. Potts, who informed me that the female was a dark bird and the male a pied one †. In another case of intercrossing which came under his notice the relative position of the sexes was reversed, the female being *R. flabellifera*: the eggs proved to be fertile, and the young assumed the plumage of the female parent.

On the nesting-habits generally he has furnished me with the following interesting note:—"To my view, the most remarkable feature in the breeding-habits of our Flycatchers is the situation usually selected for rearing their young. Security does not appear to be the first consideration; security by concealment seems the leading feature which guides most arboreal birds in choosing the site of their home, and it is one in which the most admirable displays of instinct may be frequently observed. The Flycatchers rather appear to be led by the same consideration which actuates many sea-birds in selecting the position of their breeding-place—proximity to the food supply. Stroll carefully along the rocky bed of a creek which rambles through some bushy gully, and you may perchance see the beautiful nest perched on some slender bough, in so delicate a manner that it appears scarcely so much to be fixed as to rest balanced there, and without any attempt at concealment."

The eggs of this species are of similar size and shape to those of the Pied Fantail, but I have remarked that they usually have a darker zone of purple and brown spots.

* Mr. Hamilton writes:—"I obtained a specimen of this bird in the Pohue Bush, about 20 miles north of Napier, July 7, 1865. I have seen it occasionally nearer Napier. In 1876 I got two or three in the Horokiwi district, near Wellington." (Trans. N.-Z. Inst. vol. xviii. p. 125.)

† Writing of another, exhibited at a meeting of the Zoological Society in November 1884, he says:—"Before I removed it, I saw both parent birds undertake the duties of incubation in turn, relieving each other at brief intervals. The cock bird was *R. fuliginosa*, with the aural plumes very small but quite distinct; the hen, *R. flabellifera*, occupied the nest till gently pushed off with the finger."

PETROCHELIDON NIGRICANS.

(AUSTRALIAN TREE-SWALLOW.)

Hirundo nigricans, Vieill. N. Dict. d'Hist. Nat. xiv. p. 523 (1817).
Dun-rumped Swallow, Lath. Gen. Hist. of B. vii. p. 309 (1823).
Hirundo pyrrhonota, Vig. & Horsf. Trans. Linn. Soc. xv. p. 190 (1826).
Herse nigricans, Less. Compl. Buff. viii. p. 497 (1837).
Herse pyrrhonota, id. tom. cit. p. 497 (1837).
Cecropis nigricans, Boie, Isis, 1844, p. 175.
Collocalia arborea, Gould, B. of Austr. ii. pl. 14 (c. 1845).
Chelidon arborea, id. op. cit. i. Intr. p. xxix (1848).
Petrochelidon nigricans, Cab. Mus. Hein. i. p. 47 (1850).
Hylochelidon nigricans, Gould, Handb. B. of Austr. i. p. 111 (1865); Buller, Birds of New Zealand, 1st ed. p. 141 (1873).

Ad. suprà purpurascenti-niger: fronte conspicuè ferrugineâ indistinctè nigro maculatâ: uropygio rufescenti-fulvo, scapis plumarum brunneo indicatis: supracaudalibus brunneis uropygii colore lavatis, scapis eodem modo indicatis: tectricibus alarum minimis dorso concoloribus, majoribus et remigibus brunneis, concoloribus: caudâ brunneâ, rectrice extimâ pogonio interno albo notatâ: remigum rectricumque scapis suprà brunneis, subtùs albidis: loris cum regione oculari et paroticâ nigricantibus: genis et colli lateribus sordidè fulvis brunnescente variis: subtùs fulvescens, corporis lateribus et subalaribus ferrugineis: guttare lineis longitudinalibus parvissimis, pectore et hypochondriis lineis angustioribus et longioribus striatis: rostro brunneo: pedibus brunneis: iride nigrâ.

Juv. similis adulto, sed suprà magis brunnescens: uropygio fulvescenti-albido: subtùs albicans, corporis lateribus vix rufescente tinctis.

Adult male. Forehead chestnut-brown; crown of the head, hind neck, the whole of the back, and the small wing-coverts glossy steel-blue; rump and inferior upper tail-coverts yellowish buff mixed with pale rufous, each feather with a narrow shaft-line of dark brown; longer upper tail-coverts dark brown with paler edges; throat, fore part and sides of neck, and all the under surface pale yellowish buff, marked on the throat with numerous touches of brown, stained on the sides of the body, inner linings of wings, and under tail-coverts with chestnut-brown; quills and tail-feathers dark brown, with paler shafts, greyish on their under surface and slightly glossed above. Irides black; bill, tarsi, and toes light brown. Total length 5·25 inches; wing, from flexure, 4·5; tail, to extremity of lateral feathers, 2·25 (middle feathers ·4 shorter); bill, along the ridge ·25, along the edge of lower mandible ·5, breadth at the gape ·4; tarsus ·4; middle toe and claw ·55; hind toe and claw ·45.

Female. Slightly smaller than the male, with the colours somewhat duller and the markings on the throat less distinct; but, as a matter of fact, the sexes are scarcely distinguishable from each other.

Young. Plumage of the upper parts duller, the head and back being dark umber-brown with only a slight steel gloss; the rump and tail-coverts yellowish brown, with darker shafts; the underparts altogether lighter, the abdomen and under tail-coverts being fulvous white, and the throat more distinctly spotted with brown.

THE Tree-Swallow, which is a native of Australia, was first admitted into our list of birds on the authority of a specimen shot by Mr. Lea at Taupata, near Cape Farewell, on the 14th of March, 1856, and still preserved in the Otago Museum.

In the summer of 1851, Mr. F. Jollie observed a flight of Swallows at Wakapuaka, in the vicinity of Nelson, and succeeded in shooting one, the description of which, as given by him, left no doubt in my mind that it was of the same species. According to a statement made by the late Sir David Monro at a meeting of the Wellington Philosophical Society in February 1875 *, it would appear that about the same period there were other appearances of this Swallow in the vicinity of Nelson.

At a later period, again, the bird appeared at Blenheim, in the provincial district of Marlborough, the fact being announced to me in a letter † from Mr. J. R. W. Cook, dated June 22, 1878, from which I quote the following:—

"On Sunday, the 9th instant, about two miles from Blenheim, on the bank of the Opawa river, I saw the first Martin I have met with in New Zealand. The bird was hawking after insects close to the ground in a ploughed field. I was accompanied by two residents in the town of Blenheim, and we watched it closely for some time. It passed us at one time within a few yards. There was no mistaking either the appearance or the flight of the bird. It seemed to me more like the English House-Martin than the common Australian Martin. It seemed, however, dingier in the black than the English bird, and rather smaller—more like the Sand-Martin, in fact. Unfortunately I was absent from the district for some days after seeing it, but since returning I have carefully watched for its reappearance. I have not again seen the bird, so presume it has shifted its quarters."

I had a further communication from Mr. Cook on August 23rd in which he said:—"I saw what I believe to be the same bird, about half a mile from where I saw it before, a month after its first appearance."

In April of the following year I had the pleasure of receiving from him a freshly skinned specimen of this bird, accompanied by the following letter:—

"Since writing to you last winter, reporting the occurrence here of the Australian Swallow, I did not again notice the bird until the 16th of February last, when I saw another hawking over one of my stubble paddocks. I watched it for some time, and had good opportunities of remarking its plumage. The bird appeared to me either immature or weary, the flight being weak and uncertain. I found, too, that the white on the rump was dingy, and the chestnut on the breast faded-looking. There was a stiffish nor'-west breeze blowing at the time, and the bird tried in vain to get past a belt of willow and poplar so long as I was watching.

"On the 20th of last month (March) when duck-shooting, I mentioned the occurrence to a party of sportsmen, when one remarked 'Oh! there have been some birds answering to your description flying about Grovetown for some time back.' Grovetown, I may remark, is situated about four miles from this, and nearly in the centre of the Wairau valley. After a little talk on the subject it struck me that possibly the birds had been bred there. I said—'The next time you see them, shoot one and send to me.' Yesterday morning one was handed in, but unfortunately I did not see the man who brought it. Fearing that the weather might not allow me to send it to you in the flesh, I have skinned the bird and now send it to you."

Mr. Cook having very thoughtfully sent me also the body, preserved in spirit, I was able to dissect it. It proved to be an adult female, and the stomach contained four large blue-bottle flies, almost uninjured, and the remains of others in black comminuted matter.

As bearing on this point, he remarks:—"Certainly the condition of the specimen is not that of one which has lately made a long aerial trip. In skinning it, although I freely used cotton wool and kept the pepper-castor going, I could not help getting the plumage saturated with oil, owing to the excessive fatness of the body."

* Trans. New-Zealand Instit. vol. vii. p. 510.

† *Op. cit.* vol. xi. p. 360.

Writing to me again, under date of June 11th, Mr. Cook says:—

"Since I wrote I have seen no further specimens, but note a paragraph in the 'Kaikoura Star' newspaper, stating that two Swallows had been seen at Kaikoura about the same time as the birds appeared here.

"I have since seen Mr. Cheeseman, who shot the specimen I sent. He tells me there were some six or seven birds in all; that they had been hanging about Grovetown for some weeks before he shot the one; and that he fancied they were young birds, or, at least, that some of them were. He could not, however, say that the party consisted of a pair of old birds with their brood. . . . . The one interesting question possibly may be why the first notice of occurrence of the Swallow is on our east coast. If the 'drift' (from Australia or Tasmania) is to and through Cook Straits, I can understand it. Otherwise we should expect notice of arrivals on the west coasts of both islands." Commenting on the fact that this bird appears in our country only at long intervals and as a stray migrant from a warmer clime, he makes the following very pertinent remarks:—"Is our New-Zealand winter too rigorous for this family of birds? I scarcely fancy so. Even here there are few winter days when an occasional blink of sunshine does not fetch out dancing myriads of *Ephemeridæ* on the river-banks. In olden days I fancy this was not so much the case. The rapid growth of willows now overhanging the water must afford protection to delicate new-born insects such as mosquito and other gnats which the old fringe of flax and toetoe never could have given. The temperature of the water in which the larvæ reach their fullest development is scarcely affected by the season. Indeed, in many snow-fed rivers the temperature, far from the source, when the water is at its lowest, must often be higher in winter than in summer, when the melting snows are in full swing and the river body too great to be affected materially by sun-heat. I hope you will agree with me that the natural acclimatization of the Australian Swallow is not impossible."

Mr. J. D. Enys states that he observed this Swallow skimming over the Avon, near Christchurch, in the year 1861 (Journ. of Science, ii. p. 274).

On another occasion (as reported in the 'Otago Daily News') a flight of five was seen at Moeraki, still further south, by Mr. Bills, who was then engaged catching native birds for the Acclimatization Society, and got near enough to the Swallows to be sure of their identification.

There can be no doubt that these occasional visitants are stragglers from the Australian continent, and that to reach our country they perform a pilgrimage on the wing of upwards of a thousand miles!

In its own country it is a migratory species, visiting the southern portions of Australia and Tasmania, arriving in August and retiring northwards as autumn advances.

It visits the towns, in company with the Common Swallow (*Hirundo frontalis*); and I remember seeing it comparatively numerous in and about Sydney, during a visit there in August 1871.

Mr. A. R. Wallace brought specimens from the Aru Islands; so also did the 'Challenger' Expedition; and it is likewise recorded from New Guinea, New Britain, and the Ké Islands.

According to Gould it breeds during the month of October, nesting in the holes of trees, and depositing its eggs (three to five in number) on the soft, pulverized wood. The eggs are pinky white, freckled at the larger end with fine spots of light reddish brown, and measure eight lines in length by six in breadth.

ZOSTEROPS CÆRULESCENS.

(THE SILVER-EYE.)

Cærulean Warbler, Lath. Gen. Syn. Suppl. ii. p. 169 (1801).
Zosterops cærulescens, Lath. Ind. Orn. Suppl. p. xxxviii (1801).
Sylvia lateralis, Lath. Ind. Orn. Suppl. p. lv (1801, nec Sund.).
Zosterops dorsalis, Vig. & Horsf. Trans. Linn. Soc. xv. p. 235 (1826).
Zosterops lateralis, Reich. Handb. Meropinæ, p. 94, t. cccclxiii. (1852).
Zosterops cærulescens, Gould, Handb. B. of Austr. i. p. 587 (1865).
Zosterops lateralis, Buller, Birds of New Zealand, 1st ed. p. 80 (1873).

Native names.

Tau-hou, Whiorangi, Hiraka, Motingitingi, Kanohi-mowhiti, Karu-patene, Karu-ringi, Karu-hiriwha, Poporohe, and Iringatau.

Ad. pileo et facie laterali, dorso postico et uropygio, cum tectricibus alarum lætè flavicanti-olivaceis: interscapulio scapularibusque sordidè cinereis: remigibus et rectricibus brunneis, extùs dorsi colore limbatis: regione orbitali anticè nigricante, annulo ophthalmico albo: gulâ albidâ vix flavicante tinctâ: gutture imo cinereo: abdomine medio et subcaudalibus albidis, his flavicante lavatis: corporis lateribus conspicuè badiis: rostro saturatè brunneo, mandibulâ ad basin albicante: pedibus et iride pallidè brunneis.

Adult. Crown, sides of the head, nape, upper surface of wings, rump, and upper tail-coverts bright yellowish olive; back and scapularies cinereous tinged with green; eyes surrounded by a narrow circlet of silvery-white feathers, with a line of black in front and below; quills and tail-feathers dusky brown, margined with yellowish olive; throat, fore neck, and breast greyish white, tinged more or less with yellow towards the angle of the lower mandible; abdomen and under tail-coverts fulvous white; sides pale chocolate-brown; lining of wings white, the edges tinged with yellow. Bill dark brown; under mandible whitish at the base; irides clear reddish brown; tarsi and toes light brown. Total length 5 inches; extent of wings 7·5; wing, from flexure, 2·5; tail 2; tarsus ·6; middle toe and claw ·6; hind toe and claw ·5; bill, along the ridge ·4, along the edge of lower mandible ·5.

Fledgling. Colours paler than in the adult; the throat and breast pale cinereous grey; the sides of the body fulvous brown; the white eye-circle absent, the orbits being still destitute of feathers; irides hazel-brown; tarsi and toes light flesh-colour; bill pale brown; rictal membrane yellow.

Obs. The sexes are precisely alike, the plumage of the female being in no way inferior to that of the male. Although I have examined a great number, I have only detected very slight variation in the adult birds. But Archdeacon Stock, of Wellington, who is a good practical ornithologist, has favoured me with the following note on this subject:—"I saw on Friday last, November 11, at Wilkinson's 'tea-gardens' (Wellington), what appeared to be a new variety of the Blight-bird. The white circle around the eye was not so distinct; and the head and throat were orange-coloured."

THE story of the irregular appearance of this little bird in New Zealand has for many years past been a fruitful topic of discussion among those who take an interest in our local natural history. Whether it came over to us originally from Australia, or whether it is only a species from the extreme south of New Zealand, which has of late years perceptibly increased, and has migrated northwards, is still a matter of conjecture *. The evidence which, with Sir James Hector's assistance, I have been able to collect on this subject is somewhat conflicting; but I have myself arrived at the conclusion that the Silver-eye, although identical with the Australian bird, is in reality an indigenous species. The history of the bird, however, from a North-Island point of view is very interesting and suggestive. It appeared on the north side of Cook's Strait, for the first time within the memory of the oldest native inhabitants, in the winter of 1856. In the early part of June of that year I first heard of its occurrence at Waikanae, a native settlement on the west coast, about forty miles from Wellington. The native mailman brought in word that a new bird had been seen, and that it was a visitor from some other land. A week later he brought intelligence that large flocks had appeared, and that the "tau-hou" (stranger) swarmed in the brushwood near the coast; reporting further that they seemed weary after their journey, and that the natives had caught many of them alive. Simultaneously with this intelligence, I observed a pair of them in a garden hedge, in Wellington, and a fortnight later they appeared in large numbers, frequenting the gardens and shrubberies both in and around the town. They were to be seen daily in considerable flocks, hurrying forwards from tree to tree, and from one garden to another, with a continuous, noisy twitter. In the early morning, a flock of them might be seen clustering together on the topmost twigs of a leafless willow, uttering short plaintive notes, and if disturbed, suddenly rising in the air and wheeling off with a confused and rapid twittering. When the flock had dispersed in the shrubbery, I always observed that two or more birds remained as sentinels or call-birds, stationed on the highest twigs, and that on the slightest alarm, the sharp signal-note of these watchers would instantly bring the whole fraternity together. The number of individuals in a flock, at that time, never exceeded forty or fifty; but of late years the number has sensibly increased, it being a common thing now to see a hundred or more consorting together at one time. They appeared to be uneasy during, or immediately preceding, a shower of rain, becoming more noisy and more restless in their movements. They proclaimed themselves a blessing by preying on and arresting the progress of that noxious aphis known as "American blight" (*Schizoneura lanigera*). They remained with us for three months, and then departed as suddenly as they had come. They left before the orchard-fruits, of which they are also fond, had ripened; and having proved themselves real benefactors they earned the gratitude of the settlers, while all the local newspapers sounded their well-deserved praises.

During the two years that followed, the *Zosterops* was never heard of again in any part of the North Island; but in the winter of 1858 it again crossed the strait, and appeared in Wellington and its environs in greater numbers than before. During the four succeeding years it regularly wintered with us, recrossing the strait on the approach of spring. Since the year 1862, when it commenced to breed with us, it has been a permanent resident in the North Island, and from that time it continued to advance northwards. Mr. Colenso, of Napier, reports that it was first seen at Ahuriri in 1862. On his journey to Te Wairoa, in that year, he saw it at Aropaunui, and found its nest containing four fledglings. The natives of that place told him that it was a new bird to them, they having first observed it there in the preceding year, 1861. The Hon. Major Atkinson, on the occasion of a visit, as Defence Minister, to the native tribes of the Upper Wanganui, in April 1864,

* The substance of the above article on *Zosterops* was read by the author before a meeting of the Wellington Philosophical Society on November 12th, 1870, and led to a discussion, in the course of which Dr. Hector made the following remarks:—"He said that on the south-west coast of Otago the bird was numerous, and there was very good evidence to show that this region was its native habitat. While exploring there, some years ago, he had remarked that the whole country was covered with forest, which extended down to the sea, and that the whole of the vegetation, both trees and shrubs, especially those near the sea-shore, seemed to have a coating of scaly insects, the entire bush being, in fact, covered with blight. He therefore thought it probable that as these birds increased from the superabundance of their particular food, they in course of time sent out migratory flocks, which worked their way up the coast, and at length spread over the country."—*Trans. N.-Z. Inst.* 1870, vol. iii. p. 79.

made inquiries on the subject, and was informed by the natives that the *Zosterops* had appeared in their district for the first time in 1863.

As far as I can ascertain, they penetrated to Waikato in the following year, and pushed their way as far as Auckland in 1865. Major Mair, R.M., writing to me from Taupo in 1866, said:—"It is now to be seen, in flocks of from 10 to 30, all over the Taupo and Rotoiti districts; and all the natives agree that it is a recent arrival in these parts." Professor Hutton reports that in the winter of 1867 they had spread all over the province, as far north as the Bay of Islands, and in 1868 he writes:—"They are now in the most northerly parts of this island." That they have continued to move on still further northward would appear to be the case from the following suggestive notes by Mr. G. B. Owen:—"On my passage from Tahiti to Auckland, per brig 'Rita,' about 300 miles north of the North Cape of New Zealand, I saw one morning several little birds flying about the ship. From their twittering and manner of flying I concluded that they were land-birds, and they were easily caught. They were of a brownish-grey and yellowish colour, with a little white mark round the eye. I saw several pass over the ship during the day, travelling northwards. I arrived in Auckland a few days afterwards, on the 20th of May, when the so-called Blight-birds appeared here in such numbers, and I at once recognized them as the same." Mr. Seed, the Inspector of Customs, has furnished me with the following interesting particulars bearing on the same point. When on an official visit to the lighthouse on Dog Island, situated about seven miles eastward of the Bluff, he was informed by the keeper that on one occasion a great number of these birds had killed themselves by striking against the lighthouse, either during the night or before the lights were put out in the morning, as he found them in scores lying dead in the gallery *. Mr. Seed could not ascertain positively the direction whence they came, but he understood that it was from the southward; and other inquiries at the time led him to conclude that they had come from Stewart's Island, the extreme southern limit of New Zealand.

This tendency of migration *northwards* appears to me quite inconsistent with the idea of the species having come to us from Australia.

Now let us ascertain something of its recorded history in the South Island. Mr. Potts says, in a letter to me:—"I first observed it (in Canterbury) after some rough weather, July 25, 1856. I saw about half a dozen specimens on some isolated black birch trees in the Rockwood valley in the Malvern Hills." In the Auckland Museum there is a specimen of this bird, sent from Nelson by Mr. St. John (an industrious bird-collector) in 1856. The skin was labelled "Stranger," and in the letter accompanying it Mr. St. John states that these birds had made their first appearance in Nelson *that winter* (the same in which they crossed to the North Island), and that "no one, not even the natives, had ever seen them before."

On a visit to Nelson in the winter of 1860, I saw numerous flights of them in the gardens and shrubberies. The result of very careful inquiries on the spot satisfied me that since their first appearance there, in 1856, they had continued to visit Nelson every year, arriving at the commencement of winter, and vanishing on the approach of warmer days as suddenly as they had come. On every hand the settlers bore testimony to their good services in destroying the cabbage-blight and other insect pests.

About the middle of June 1861, I met with small flocks of this bird on the Canterbury Plains, evidently on their passage northward. I first observed them in the low scrub on the broad shingle-

* The fact that they continue their flight at night is very curious. I may mention that on a dark evening in August, about 8 p.m., I observed what seemed to be a large moth fluttering against the glass of a lamp-post on Wellington Terrace. Apparently stunned, or wearied out, it fell to the ground, and on picking it up I found it to be a *Zosterops*, which had evidently been attracted by the gas-light. Its poor condition indicated that it was a migrant, doubtless a straggler from one of the flocks, large numbers of these birds having about that time made their appearance on the northern side of Cook's Strait.

beds of the Rakaia, advancing in a very hurried manner, not high in the air, as migrations are usually performed, but close to the ground, and occasionally resting. But that this bird is capable of protracted flight is evidenced by the form of its wings, which are of the lengthened, acuminate character common to most birds of passage.

During a visit to Dunedin, in the summer of 1860, the Rev. Mr. Stack observed numerous flocks in the gardens and thickets in the environs of the town. At this season they had disappeared from the Province of Canterbury and all the country further north. In the following summer (1861) I met with numerous stragglers in the northern parts of the Canterbury Province, and I understand from Mr. Potts that since that time it has been a permanent resident there, increasing in numbers every year. Mr. Buchanan, late artist to the Geological Survey Department, informs me that he observed the *Zosterops* at Otago, on his first arrival there in 1851, five years previous to its appearance in the North Island; and the following letters from correspondents go still further to prove that the species is an indigenous one there, and is only new to the country lying further north.

Mr. Newton Watt, R.M., of Campbell Town (Southland), writes as follows:—"Paitu, a chief here, and I believe the oldest man in the tribe, says it was always here. Howell says that he first noticed them on the west coast, about Milford Sound, in the year 1832, in flocks of thirty or forty, but never noticed them here (Riverton) till about 1863, when he saw them inland and in smaller flocks. On my way back from Riverton, I was mentioning it at the Club at Invercargill, and a gentleman present told me he had first noticed them, about eighty miles inland, about the year 1861, and that his attention was first called to them from the circumstance that they were gregarious,—a habit not common with New-Zealand birds. At Campbell Town it appeared to be more scarce, being seen only in small flocks, varying in number from six to twelve. In 1866 my sons noticed numbers of them among my cabbages, and observed that the cats caught many of them; and, further, that whilst my cabbages in the three preceding years were infested with blight, in that year there was little or no blight upon them till very late in the season. They appear to migrate from this locality in the winter, or at any rate to be *scarce*."

Mr. James P. Maitland, R.M., of Molyneux, writes:—"From what I hear from old settlers of seventeen or eighteen years' standing (whom I can trust as men of observation), I am convinced we have had the birds here for that time at any rate, although all agree that they have become much more numerous everywhere during the last seven years; and this year (1867) in particular I observe them in larger flocks than ever. I confess I do not recollect noticing the bird until about six years ago; but the smallness of their number at that time, and the smallness of the bird itself, may easily account for its being unnoticed in the bush. The gardens seem to be the great attraction here, and they are the best hands I know at picking a cherry- or plum-stone clean!"

All my own personal inquiries at Otago, during my first visit there in February 1865, led me to the same conclusion.

Referring again to the migration of *Zosterops* from the South Island in 1856, it may, I think, be assumed that the large flights which came across Cook's Strait made the island of Kapiti in their passage, and tarried there for a time before they reached the North Island. It will be remembered that the flocks which afterwards spread over the province appeared first at Waikanae and Paekakariki, on the lee shore from that island. I found *Zosterops* excessively abundant at Kapiti during a visit there in April 1875. Every bush swarmed with them, and sometimes fifty or more would crowd together in the leafy top of a stunted karaka, warbling and piping in chorus, producing sylvan music of a very sweet description. They appeared to be feeding on a species of *Coccus* that afflicts that tree.

The large numbers of these birds that appeared in flocks at Waikanae and Otaki in the early part of June of the same year would seem to indicate another incursion from the South Island at that date *.

The bird whose history has been so fully recorded in these pages being once fairly established among us, it has continued to increase and multiply, and now it disputes possession of our gardens and hedgerows with the introduced Sparrows and Finches, and indeed swarms all over the country. On my last visit to the Hot Lakes I found it extremely abundant everywhere; even amid the noxious fumes at Sulphur Point I met with small flocks flitting about in the stunted manuka scrub, and apparently quite at home in sulphuretted hydrogen! In the Bay of Plenty district it is particularly plentiful, so much so as to form an article of food to the natives. They are in season in the months of March and April, and are then collected in large numbers, singed on a bush fire to take the feathers off, and forthwith converted into *huahua* and potted in calabashes. The catching is effected in a very primitive way. The birds have their favourite trees upon which they are accustomed to congregate. Selecting one of these, the bird-catcher clears an open space in the boughs and puts up several straight horizontal perches, under which he sits with a long supple wand in his hand. He emits a low twittering note in imitation of the birds', and, responding to the call, they cluster on the perches, filling them from end to end. The wand is switched along the perch, bringing dozens down together, and a boy on the ground below picks up the stunned birds as they fall. Captain Mair, when visiting Ruatahuna on one occasion, had brought to him, by two Urewera lads, a basket containing some five or six hundred of these little birds which had been killed in the manner described.

In front of the Rev. Mr. Spencer's house at Tarawera, in a hedge of *Laurustinus*, scarcely six yards from the door, upwards of twenty nests of *Zosterops* were found at one time, each containing from three to five eggs (generally the former) of a lovely blue colour. Usually, however, these birds do not breed in communities, but scatter themselves in the nesting-season.

My son discovered a nest, containing three eggs, attached to a fern-stalk at the very edge of a boiling and steaming fumarole, near the White Terrace of Rotomahana, and suspended as it were in the midst of a perpetual vapour-bath.

In the selection of its breeding-home, this bird has manifested with us somewhat erratic tendencies: thus, for the first three or four years after its permanent location in the North Island, it wintered in the low lands and the districts bordering on the sea-coast, and retired in summer to the higher forest-lands of the interior to breed and rear its young. In the summer of 1865 a few stragglers were observed to remain behind all through the season, and in the following year they sojourned in flocks and freely built their nests in our shrubberies and thickets, and even among the stunted fern and tea-tree (*Leptospermum*) near the sea-shore. From that time to the present it has ranked as one of our commonest birds all the year round; and, what is even more remarkable, it has very perceptibly increased in numbers, whilst most of our other insectivorous birds are rapidly declining, and threaten ere long to be extinct.

To the philosophical naturalist the history of the *Zosterops* in New Zealand is pregnant with interest, and I feel that no apology is needed for my having thus minutely recorded it.

The natives distinguish the bird as Tau-hou (which means a stranger), or Kanohi-mowhiti (which may be interpreted spectacle-eye or ring-eye). It is also called Poporohe and Iringatau, names suggested by its accidental or periodical occurrence.

* Six years later, about the month of July, there was another irruption of the kind, the gardens and shrubberies in and around Wellington swarming with them, many hundreds often consorting together in one flock. On this occasion, they freely visited the poultry enclosures and back-yards in their search for food, and I have counted as many as thirty at one time exploring a drain-trap or clustering together on a discarded bone at the dog-kennel, and eagerly tearing off the particles of meat adhering to it. As a rule, they seemed to be unusually tame, as if weary after their long flight; and some of them, emboldened by hunger, entered the houses and outbuildings, whilst numbers fell victims to the remorseless cat.

By the settlers it has been variously designated as Ring-eye, Wax-eye, White-eye, or Silver-eye, in allusion to the beautiful circlet of satiny-white feathers which surrounds the eyes; and quite as commonly the "Blight-bird," or "Winter-migrant."

I have frequently watched the habits of this little bird, and with much interest. As already stated, it is gregarious, flying and consorting in flocks, except in the breeding-season, when they are to be observed singly or in pairs. As soon as a flock of them alights on a tree, or clump of brush-wood, they immediately disperse in quest of food; and, on a cautious approach, may be seen prosecuting a very diligent search among the leaves and flowers, and in the crevices of the bark, for the small insects and aphides on which they principally subsist. I have opened many specimens, at all seasons, and I have invariably found their stomachs crammed with minute insects and their larvæ. In some I have found the large pulpy scale-insect (*Coccus*, sp.), of a dull green colour, which is commonly found adhering to the leaves of the ramarama (*Myrtus bullata*); also small caterpillars, grasshoppers, and coleoptera, and occasionally the small fruity seeds of *Rubus australis* and other native plants. In our orchards and gardens it regales itself freely on plums, cherries, figs, goose-berries, and other soft fruits; but it far more than compensates for this petty pilfering by the whole-sale war it carries on against the various species of insects that affect our fruit-trees and vegetables. It feeds on that disgusting little aphis known as American blight, which so rapidly covers with a fatal cloak of white the stems and branches of our best apple-trees; it clears our early cabbages of a pestilent little insect that, left unchecked, would utterly destroy the crop; it visits our gardens and devours another swarming parasite that covers our roses and other flowering plants, to say nothing of its general services as an insectivorous bird. Surely, in return for these important benefits, to both orchard and garden, the flocks of *Zosterops* may justly be held entitled to an occasional feed of cherries, or to a small tithe of the ripe fruits which they have done so much to defend and cherish!

It is very pretty to see a pair of them feeding together on a single berry of the poroporo (*Solanum nigrum*) or diligently scooping out the centre of a ripe fig, their ever-changing positions being very artistic.

A favourite resort of this bird in the early part of November is the kohia creeper (*Passiflora tetrandra*), which covers much of the low scrub on the outskirts of the forest, and is at this time a mass of white bloom. The little bell-shaped flowers, which diffuse so much fragrance through the woods, being full of nectar, attract the little golden butterfly (*Chrysophanus enysi*) and swarms of gaily-coloured diptera. Here the *Zosterops*, in addition to the sip of honey, finds an abundance of its favourite insect food. When thus engaged, it emits a soft plaintive cry, repeated at short intervals; but on the wing, and especially when consorting in a flock, it utters a rapid twittering note. During the breeding-season the male indulges in a low musical strain of exquisite sweetness, but very subdued, as if singing to himself or performing for the exclusive benefit of his partner. This song is something like the subdued strain of the Korimako (*Anthornis melanura*), but much softer.

I have already mentioned the circumstance of a flock of these birds being generally attended by two or more sentinels or call-birds, who take their station on the topmost twigs, as a post of observa-tion, and whose sharp signal-note instantly brings the whole fraternity together. On one occasion, while out pheasant-shooting at Wangaehu, the sound of my companion's whistle, although more than 200 yards away, attracted the notice of a flock of *Zosterops* consorting together in the top of a lofty kahikatea tree. The call-birds gave the alarm, and the whole flock, amidst much clamour, ascended high in the air and disappeared behind a neighbouring hill. The sentinels appear to be always on the alert; and I have seen the same effect produced on a flock of these birds by the cry of a hawk, or any other suspicious sound, although there was no appearance of immediate danger.

If shot at and wounded it generally manages to escape capture by scrambling nimbly off into the thicket, hiding itself and remaining perfectly silent till the danger has passed. Frequent attempts have been made to keep it caged; but although it will readily feed, it seldom survives confinement many weeks. Only one instance of complete success has come to my knowledge. Mrs. Fereday, residing near Christchurch, kept several of them caged for upwards of two years; and I am indebted to that lady for the following amusing account of these captives:—They were adult birds when taken, but soon became reconciled to the restraints of a canary-cage, and partook readily of bread soaked in milk. They were interesting objects on account of their extreme display of mutual affection, as they were always caressing one another and preening each other's feathers. This demonstration of affection, however, was at length carried too far, as one of them contracted a habit of pulling out his neighbour's feathers, in order to suck the oily matter from the roots of the quills. The practice was commenced during the seasonal moult, when the pen-feathers were present, but was continued afterwards, till it became necessary to turn out the offender and introduce a wild bird in its place. But the practice soon became general, each bird plucking and submitting to be plucked in the most business-like manner. The operation was usually commenced on the neck, and it was very droll, said my informant, to see the bird holding its head up, as a man would sit to be shaved, while the feathers were plucked out one by one. The birds were then separated, but they manifested the utmost distress, crying plaintively and refusing their food. On the first opportunity they resumed their old habit, and at length one of them was plucked completely bare! Finding the case hopeless, Mrs. Fereday then liberated the birds in the garden, where they seemed to suffer from the colder temperature of the open air, and shortly disappeared altogether, probably falling victims to some predatory cat.

At the period when they were most plentiful at Wellington, an unaccountable mortality manifested itself; and in one particular locality, near Te Aro, sometimes as many as twenty dead ones were found in the morning under the *Eucalyptus* tree in which the flock had roosted for the night.

Mr. Colenso observes that "when they retire to roost they sleep in pairs, cuddling quite close together, like love-parrots; and before they fold their heads under their wings they bill and preen each other's head and neck most lovingly, uttering at the same time a gentle twittering note."

Mr. Potts informs me that, in Canterbury, this species begins nesting early in October. In one instance, within his own observation, the birds commenced incubation on October 10, the young were hatched on October 25, and left the nest on November 4. In the North Island the breeding-season is somewhat later. As late as the 24th of December I met with a nest in the Taupo-Patea country, containing two perfectly fresh eggs. The nest is a slight cup-shaped structure, with a rather large cavity for the size of the bird, and is generally found suspended by side-fastenings to hanging vines, or to the slender twigs of *Leptospermum*, *Olearia*, and other shrubs, and sometimes to the common fern (*Pteris aquilina*). The eggs are generally three in number (sometimes four), ovoido-conical in form, measuring ·7 of an inch in length by ·5 in breadth, and of a beautiful, uniform pale blue colour.

Nests of this species exhibit some variety, both as to structure and the materials of which they are composed. Of three specimens now before me, one is of slight construction and shallow in its cavity, composed externally of green-coloured lichen, spiders' nests, the downy seed-vessels of the pikiarero (or flowering clematis), and a few dry leaves, lined internally with long horse-hair disposed in a circular form; another is of smaller size, more compact, composed externally of crisp dry moss, and internally of grass-bents with a few long hairs interlaced; while the third has the exterior walls constructed entirely of spiders' nests and stiff fibrous mosses, the former predominating, and the interior lining composed wholly of long horse-hair.

At Akitio (in the North Island), where wild pigs are very plentiful, the Blight-birds habitually line their nests with pigs' bristles, as a substitute for horse-hair, which is generally used by them in other parts of the country. In a multitude of cases I have found the cavity of the nest lined with long horse-hair intermixed with dry bents, all carefully twined together. An example in the Canterbury Museum has the cavity lined entirely with long horse-hair, and two other specimens in the same collection have a lining composed exclusively of fine grass-stems, carefully bent.

A specimen which I found suspended in a clump of creeping kohia was composed externally of the pale green and rust-coloured lichen so abundant on the branches of dead timber, intermixed with spiders' webs, and lined inside with dry fibrous grasses, the whole being laced together with hair, the long straggling ends of which projected from every part of the nest; and another, which was obtained from the low brushwood bordering on the sea-shore, was built of sheep's wool, spiders' nests, pellets of cow-hair, and fine seaweed firmly bound together with long thread-like fibres, apparently the rootlets of some aquatic plant, and lined internally with fine grass-bents and soft feathers. Sometimes the nest is constructed wholly of bents and dry grass.

I have lately had an opportunity of examining a beautiful series of the nests of this species, and I remarked that through all the varieties of individual form and structure they presented these two essential features—the large cup-like cavity with thin walls, and the admixture of long hairs in the lining material. In one of the nests forming this series the proximity to civilization was proclaimed by a lining consisting of the flaxen hair from a child's doll!

Zosterops is not, strictly speaking, a suctorial bird; but it is closely allied to the Tubilingues. The tongue has no brush, but ends in two short filaments; and, as shown by Dr. Gadow, in his 'Account of the Suctorial Apparatus of the Tenuirostres,' is far from being the complicated and elaborate organ generally exhibited in the tubular tongues of the Nectariniidæ and Meliphagidæ.

The genus has an extensive range, for, according to the British Museum Catalogue, its members are spread all over South Africa south of the Sahara, Madagascar and the Comoro Islands, the entire Indian Peninsula and Ceylon, Burmese countries, the whole of China (extending into Amoor Land), Japan, Formosa, Hainan, Malay Peninsula, all the Indo-Malayan islands, Moluccas, New Guinea and the adjacent Papuan group, and (with few exceptions) throughout the islands of the great Pacific Ocean.

Mr. Gould states that *Zosterops cærulescens* "is stationary in all parts of Tasmania, New South Wales, and South Australia, where it is not only to be met with in the forests and thickets, but also in nearly every garden."

At the Chatham Islands, where it is now very abundant, it is said to have made its first appearance shortly after the great fire in Australia known as Black Thursday.

ANTHORNIS MELANURA.

(BELL-BIRD.)

Mocking-Creeper, Lath. Gen. Syn. ii. p. 735 (1782).
Certhia melanura, Sparrm. Mus. Carls. pl. v. (1786).
Certhia sannio, Gm. Syst. Nat. i. p. 471 (1788).
Philedon dumerilii, Less. Voy. Coq. Zool. i. p. 644, t. 21. fig. 2 (1826).
Anthomiza cæruleocephala, Swains. Classif. of B. ii. p. 327 (1837).
Philedon sannio, Less. Compl. Buff. xi. p. 165 (1838).
Anthornis melanura, Gray, List of Gen. of B. p. 15 (1840).
Certhia olivacea, Forst. Descr. Anim. p. 79 (1844).
Anthornis ruficeps, Von Pelz. Verh. zool.-bot. Gesellsch. Wien, 1867, p. 316.

Native names.

Mako, Makomako, Komako, Kokomako, Korimako, Kohimako, Kokorimako, Kohorimako, Titimako, and Kopara. Of the above names, Korimako is most generally used by the northern and Makomako by the southern tribes. The Ngatiawa call this bird Rearea; and the natives of the Bay of Plenty distinguish the male and female as Kokorohimako and Titapu.

♂ suprà flavicanti-olivaceus, uropygio vix lætiore: pileo undique metallicè violaceo nitente: loris et mento ipso nigricantibus: tectricibus alarum nigricantibus dorsi colore lavatis: remigibus nigricantibus vix sub certâ luce indigotico nitentibus, extùs angustè olivaceo limbatis, scapis suprà nigricantibus, subtùs brunnescentibus: caudâ nigrâ, subtùs pallidiore, rectricibus extùs sordidè indigotico lavatis: subtùs flavicanti-olivaceus, hypochondriis paulò lætioribus: crisso et subcaudalibus flavicanti-albis, olivaceo-brunneo variis: subalaribus cinerascentibus, olivaceo lavatis: fasciis axillaribus flavidis: rostro nigro: pedibus plumbeis, unguibus brunneis: iride rubrâ.

♀ mari similis, sed magis olivaceo-brunnescens, et ubique sordidior: pileo dorso concolore, metallicè viridi obscurè nitente: alis et caudâ brunnescentibus, secundariis fulvo terminatis et rectricibus olivaceo-viridi limbatis: fasciâ mystacali parvâ albidâ: subtùs brunnescens, pectore pallidè ferrugineo lavato, abdomine magis olivascente: subalaribus et fasciis axillaribus sordidè flavidis.

♂ *juv.* similis mari adulto, sed pallidior: fasciâ mystacali indistinctâ.

Adult male. The whole of the plumage olive-green, changing to yellowish-olive on the sides of the body and abdomen; beneath plumbeous; forehead, crown, and sides of the head glossed with deep purple; primary quills and tail-feathers dusky black, darker and having a steel gloss on the outer webs; the secondary quills narrowly margined outwardly with olive-green, which colour spreads on the inner ones till it nearly covers the entire web; inner lining of wings, as well as the soft ventral feathers and under tail-coverts, pale fulvous yellow. Irides cherry-red; bill black; tarsi and toes dark leaden grey; the claws brown. Total length 7·75 inches; wing, from flexure, 3·4; tail, to the extremity of lateral feathers, 3·6; bill, along the ridge ·6, along the edge of lower mandible ·75; tarsus 1; middle toe and claw ·6; hind toe and claw ·75.

Adult female. Smaller than the male, with little or no purple gloss on the head, and readily distinguished by a narrow streak of white, which extends downwards from the angles of the mouth, fading off in a line with the ear-coverts. Upper parts dull olivaceous; throat, breast, and underparts generally yellowish brown, strongly tinged with olive; quills and tail-feathers dusky black, margined on their outer webs with olivaceous; linings of wings, vent, and under tail-coverts fulvous white, washed with yellow.

Young male. Plumage lighter than in the adult bird, with a narrow indistinct line of yellowish white from the angles of the mouth.

Nestling. Plumage fluffy and colours dull. The membrane at the corners of the mouth strongly developed and of a bright yellow colour.

Obs. The bird described by Herr von Pelzeln (*l. c.*) under the name of *Anthornis ruficeps* was, what I had always contended for *, nothing but a flower-stained example of the present species. In acknowledgment of this I have received the following note from my friend Dr. Finsch, of Bremen:—"You are quite right in respect to *A. ruficeps*. The red colour on the face is caused by external influences; for my friend Von Pelzeln has washed the type in the Vienna Museum, and the red tinge has partially disappeared." But, even as far back as 1782, Latham mentions (*l. c.*) the existence of a red stain in some specimens, and ascribes it to the true cause, adding "this in time rubs off, and the colour of the head appears the same as the rest of the plumage."

Varieties. On the 10th October, 1874, a partial albino was brought to the Canterbury Museum, and I had an opportunity of examining it in the flesh. Although I had seen probably some thousands of this species, this was the first instance I could remember of any departure from the normal colour, unless it were an occasional very slight tendency to melanism. This specimen, which is still in the collection, is a fine male bird, with the body-plumage as in ordinary specimens, but having the whole of the quills and tail-feathers ashy white, the edges of the outer webs slightly tinged with yellow. The shafts of the quills are dark brown, those of the tail-feathers white in their greater portion, becoming brown towards the base; the bastard-quills and tertiary coverts are ashy white; the large secondary coverts dark grey tipped with whitish and margined with dull olive; the axillary tufts, lower part of abdomen, flanks, and under tail-coverts pale lemon-yellow. Irides, bill, and feet as in ordinary examples.

Much more recently, however (in April 1885), a more perfect albino was brought to the Museum from Akaroa. The whole plumage is white, washed with pale yellow on the back, upper surface of wings, rump, and underparts, the basal portion of each feather being pale plumbeous; under surface of wings and tail-feathers pale slaty grey; bill and feet as in the normal condition.

THE praises of the Bell-bird were sung, a hundred years ago, by the illustrious navigator Cook, whose 'Voyages' contain the following record:—"The ship lay at the distance of somewhat less than a quarter of a mile from the shore †; and in the morning we were awakened by the singing of the birds: the number was incredible, and they seemed to strain their throats in emulation of each other. This wild melody was infinitely superior to any that we had ever heard of the same kind; it seemed to be like small bells most exquisitely tuned, and perhaps the distance and the water between might be no small advantage to the sound." One has but to read this early tribute to realize how great a loss we have suffered from the almost total disappearance of this bird from the North Island. Even when writing its biography for my former edition, I had to make the following discouraging statement:—

This species, formerly very plentiful in every part of the country, appears to be rapidly dying out. From some districts, where a few years ago it was the commonest bird, it has now entirely vanished. In the Waikato it is comparatively scarce, on the East Coast it is only rarely met with,

* Trans. New-Zealand Inst. 1868, vol. i. p. 108.

† Queen Charlotte's Sound.

and from the woods north of Auckland it has disappeared altogether. In my journeys through the Kaipara district, eighteen years ago, I found this bird excessively abundant everywhere; and on the banks of the Wairoa the bush fairly swarmed with them. Dr. Hector, who passed over the same ground in 1866, assures me that he scarcely ever met with it; and a valued correspondent, writing from Whangarei (about 80 miles north of Auckland), says:—"In 1859 this bird was very abundant here, in 1860 it was less numerous, in 1862 it was extremely rare, and from 1863 to 1866 I never saw but one individual. It now seems to be entirely extinct in this district."

The above remarks were intended to refer principally to the North Island; but even in the South, as I have elsewhere pointed out *, it is far less plentiful than it formerly was. Doubtless it is only a question of a few years, and the sweet notes of this native songster will cease to be heard in the grove; and naturalists, when compelled to admit the fact, will be left to speculate and argue as to the causes of its extinction.

My observations as to the extreme rarity of this species in the North Island, where in former years it was the commonest of the perchers, are confirmed by Captain G. Mair, who informs me that during the last ten years he has never met with it at all, except on the Island of Mokoia (a place of some historic interest in the Rotorua Lake, about 600 acres in extent), in a tract of manuka bush covering about a thousand acres of land at the foot of Mount Edgecumbe, and in the high scrub at Waitahanui about ten miles from Taupo. In the first named of these localities it is still very plentiful †.

In 1868, Professor Hutton found the Korimako abundant on Great Barrier Island, although even then scarce on the mainland ‡; and in 1871 Major Mair met with it on the Rurima Rocks and on Whale Island, in the Bay of Plenty, places about five miles apart. He records the delight with which he again listened to its sweet note, and adds, "the Maoris think there is only one, that it is the sole survivor of the race, and that it flies backwards and forwards between these islands."

Although I travelled a good deal through the forests of the interior during the ten years after my return from Europe in 1874, on one occasion only did I ever meet with this species on the mainland, and then only with a solitary bird; but during a storm-bound visit to the island of Kapiti (Cook's Strait) in April 1877, I was charmed immediately on landing to hear the musical notes of the Bell-bird again, and to meet with it in every direction among the stunted karaka groves that clothe the western slopes of that island. In the course of an afternoon I saw a score or more of them within a very limited area, and on a second and more extended visit on the following day I found them equally numerous. I met with another bird also, which has likewise become well-nigh extinct on the mainland (*Miro australis*), although not in such numbers as the former.

Several years later I met with the Korimako again, in sufficient abundance, on a wooded islet called Motu-taiko in the very centre of the Taupo Lake, having put in there for shelter.

The facts I have mentioned are interesting, as furnishing another illustration of the observed natural law, that expiring races of animals and plants linger longest and find their last refuge on sea-girt islands of limited extent.

* Trans. New-Zealand Instit. vol. ix. p. 209.

† Captain Mair informs me also that on the small island of Motiti, in the Bay of Plenty, the Bell-bird is very numerous, although it is never seen or heard on the mainland opposite. He adds:—"On Whale Island also, although there are no Tuis, Korimakos are very plentiful. It was really delightful to see and hear them again. They abound in numbers in the shrubbery, and hearing them sing at daylight carried me back in spirit to my boyhood, at the North, thirty years ago!" My son having gone to this island, to indulge in deep-sea fishing, had to camp there for the night. He and his party found shelter in a little rocky cavern, and being the last day of the old year, the new year's morn was ushered in by a delightful chorus from the Bell-birds in the pohutukawa trees above them.

‡ Ten years later, Reischek could not find one on the Great Barrier, although the bird was still to be heard and seen on the Little Barrier.

The cause of the rapid disappearance in New Zealand of some species of birds, and absolute extinction of others, is a very interesting question, and I have already called attention to it in various published papers. In a newly colonized country, where the old fauna and flora are being invaded by a host of foreign immigrants, various natural agencies are brought into play to check the progress of the indigenous species, and to supplant them by new and more enduring forms, more especially in the case of insular areas of comparatively small extent. These agencies are often too subtle in their operation to arrest the notice of the ordinary observer; and it is only the ultimate results that command his attention and wonder. But in New Zealand some special cause, apart from this general law, must be assigned for the alarmingly rapid decrease of many of the indigenous birds: in the course of a very few years, species formerly common in every grove have become so scarce throughout the country as to threaten to become extinct at no very distant date.

Various reasons have been suggested to account for this. The natives believe that the imported bee, which has become naturalized in the woods, is displacing the Korimako, Tui, and other honey-eating birds. One of the oldest settlers in the Hokianga district (the late Judge Maning), speaking to me on this subject, said:—"I remember the time, not very long ago, when the Maori lads would come out of the woods with hundreds of Korimakos hung around them in strings; now one scarcely ever hears the bird: formerly they swarmed in the northern woods by thousands; now they are well nigh extinct." On asking him his opinion as to the cause of this, he told me that he agreed with the Maoris, that the bee, having taken possession of the woods, had driven the honey-eating birds away from the flowers, and practically starved them out; and he referred to the scarcity of the Tui, another honey-eater, in support of this view *. But it must be remembered that both of these species subsist largely on berries and insects, and that the comparative failure of their honey-food, even if granted, will not of itself account for the rapid decrease of these birds; while, on the other hand, the Totoara (*Miro australis*) and other species which do not sip flowers are becoming equally scarce. It appears to me that the honey-bee theory is quite insufficient to meet the case, and that we must look further for the real cause. As the result of long observation, I have come to the conclusion that, apart from the effects produced by a gradual change in the physical conditions of the country, the chief agent in this rapid destruction of certain species of native birds is the introduced rat. This cosmopolitan pest swarms through every part of the country, and nothing escapes its voracity †. It is very abundant in all our woods, and the wonder rather is that any of our insessorial birds are able to rear their broods in safety. Species that nest in hollow trees, or in other situations accessible to the ravages of this little thief, are found to be decreasing, while other species whose nests are, as a rule, more favourably placed, continue to exist in undiminished numbers. As examples of this latter class, I may instance the Kingfisher, which usually scoops out a hole for

* In this connection it is worth mentioning that on the Great Barrier and Island of Kawau, from both of which the Korimako has now disappeared, bees are plentiful; whereas on the Little Barrier and the Chickens, where the bird still lingers, there are no bees.

† In a letter which I had the pleasure of receiving from the Rev. T. Chapman, of Rotorua, some years ago, that gentleman states:—"Wild Ducks were particularly numerous in this district on my arrival here; you saw them by dozens; you hardly see them now by twos. I have no doubt we owe this to the Norway rat. There is a place on the Waikato river, some twenty miles below Taupo, where the chiefs occasionally assembled to set out two important matters,—to discuss politics and eat kouras (crayfish). A few years after the Norway rat fully appeared, the kouras were no longer plentiful; and as the New Testament made Maori politics rather unnecessary, the usage of meeting no longer exists. The natives assured me that the Norway rat caught the crayfish by diving. Rowing up the rivers you see little deposits of shells: upon inquiry I found they were the selections of the Norway rats, who, by diving for these freshwater pipis, provide a *kinaki* (relish) for their vegetable suppers."

Herr F. von Fischer (Zool. Gart. 1872, p. 125) calculates that a single pair of these rats might have, after ten years, a progeny of 48,319,698,843,030,344,720 individuals.

its nest in the upright bole of a dead tree, quite beyond the reach of rats, and appears to be more abundant now than ever; also the *Rhipidura*, *Zosterops*, *Gerygone*, and other small birds, whose delicate nests are secured to slender twigs or suspended among vines and creepers. And the Ground-Lark, again, which nests in open grass or fern land, where the Harrier keeps the rat well under control, has of late years sensibly increased, being now very common. As a matter of fact, I have known a case in which half a dozen nests of the Tui, within a radius of a hundred yards, were robbed by rats of both eggs and young *.

But to resume our history of the "Bell-bird"—so-called from the fanciful resemblance of one of its notes to the distant tolling of a bell. Its ordinary song is not unlike that of the Tui or Parson-bird, but is more mellifluous. Its notes though simple are varied and sweetly chimed; and as the bird is of social habits, the morning anthem, in which scores of these sylvan choristers perform together, is a concert of eccentric parts, producing a wild but pleasing melody. When singing it arches its back and puffs out the feathers of the body. I have occasionally heard a solitary Bell-bird pouring forth its liquid notes after the darkness of advancing night had silenced all the other denizens of the grove. It ought to be mentioned, moreover, that both sexes sing. When alarmed or excited they utter a strain of notes which I can only compare to the sound produced by a policeman's rattle quickly revolved. This cry, or the bird-catcher's imitation of it, never fails to attract to the spot all the Bell-birds within hearing. The Maoris are accustomed to snare them by means of a *tuke* baited with the crimson flowers of the climbing *Metrosideros*. The same device is adopted for catching the Tui.

This snare, of which a figure is here given, is formed of a carefully selected piece of kareao vine, having the necessary curve upwards. The lower part of this is fastened to the thick end of a bush-rod, eight or ten feet in length, through a small hole in which a looped flax line is passed, a crook, to serve as a support, being placed on the opposite side. At the upper extremity of the artificial perch thus produced a circular flower-holder, made of split vine, is fixed, and a string connects it with the stem of the *tuke*, whilst the attachment of the lower end to the support is concealed by a covering of soft moss, carefully tied round with a strip of green flax, every precaution being taken to give it a natural appearance. Having baited and set his snare, the bird-catcher hitches it by the crook to a branch in some favourable position and prepares for action. Concealing himself in a shelter of fronds, torn from a tree-fern and hastily stuck into the ground with the tops overlapping, he imitates the alarm-cry of the bird by means of a nikau leaf placed between his lips. The call is soon responded to, and birds from far and near hurry to the fatal spot. The artful Maori then stops calling, and the birds, as soon as their excitement has subsided, begin to look about them and are

attracted by the flowers. The instant one touches the treacherous perch, a pull on the string, bringing the loop home, secures it firmly by the leg. The *tuke* is then gently unhitched and lowered from the branch, cleared of its victim, and quickly reset.

* Mr. W. T. L. Travers, in an interesting article on the subject, says:—"The rat and the bee may each have played a part in bringing about its disappearance from the North Island, as both of these swarm all through the forest there, whilst in the South Island the rat has been nearly extirpated from the great *Fagus* forests by the Woodhen (*Ocydromus*), and the bee is limited in its range to the cultivated districts. But the cause of the disappearance of this bird is more matter of speculation, and I have only cited the case in order to show how little we really know of the circumstances which may govern or limit the distribution of any particular species." (Trans. N.-Z. Inst. 1882, vol. xv. p. 182.)

In former times, when this species was abundant throughout the whole country, certain forest-ranges were famed as Korimako preserves, and were highly prized on that account by the natives owning them. At the present day, in the investigation of native titles to land, the "snaring of Korimakos" by their ancestors is an act of ownership frequently pleaded in support of the tribal claim.

The flight of this bird is undulating, but very rapid, the wings and tail being alternately opened to their full extent and sharply closed. It sometimes mounts to a considerable height in the air, and I have occasionally observed large parties of them indulging in a playful flight far above the tree-tops.

Its food consists of minute flies and insects, as well as small berries, such as those of the karamu (*Coprosma lucida*) and other shrubs, and the honey of various kinds of bush-flowers. When feeding on the latter, it may be seen hanging by the feet in all positions from the slight flower-bearing twigs, while the slender bill, with the pencilled tongue protruded, is thrust into the corolla of each flower in quick succession.

In the gardens of the South Island it is still daily to be seen, moving actively about and collecting honey from various flowers. It is specially fond of the common black wattle; and it is a pretty sight to watch the bird clinging to the flower-stems in the manner described and assuming every variety of attitude as it sips the nectar from the golden tassels that cover the tree in such thick profusion *. It also attacks the full-blown flower of the common foxglove, which now grows wild in some parts of the country, piercing or tearing open the corolla with its bill in order to get at the honeyed juice.

When the korari (*Phormium tenax*) is in full bloom, the horn-shaped flowers are filled with delicious nectar, which the natives are accustomed to collect in calabashes, to be used as a drinking-beverage for visitors. The Bell-bird, too, loves to regale itself on this saccharine production; and while the season lasts its forehead is often stained red from the colouring-matter that adheres to the feathers. When the bird, with the change of season again, is feasting itself from the smaller cups of the pretty native fuchsia (*F. excorticata*), the stain on the forehead changes to a very bright purple or blue.

Its ordinary chime consists of the following four notes (as set by Dr. Shortland):—

No one who has not actually listened to the melody can form any idea of the effect produced by these high notes coming from a hundred throats independently, and blending together in the richest harmony of song.

The Bell-bird commences breeding towards the end of September or early in October, and sometimes even as late as November and December. I have met with a brood of fully-fledged young birds as early as October 28; while, on the other hand, Mr. Potts informs me that he has observed it building its nest at the end of January or beginning of February. It seems probable, therefore, that this species rears two broods in the year. Its nest may be looked for in deep wooded gullies and in the low brushwood along the outskirts of the forest. It is usually placed in the fork of a low branch, and the bird in selecting a site seems generally to prefer those bushes over which the native bramble (*Rubus australis*) has thrown a protecting mantle. It is a common thing to find four or five old nests of former years in the immediate vicinity of the occupied one, as if the birds formed an attachment for a locality once chosen as a breeding-place. The nest is a rather loose structure, composed externally of small dry twigs, sometimes interlaced with the wiry stems of the bush convolvulus, over

* Forty years ago literally thousands of these birds annually frequented the groves of wattle around the old mission-station at Tangiteroria (on the northern Wairoa). The wattles still are there, grown to the size of forest trees, with many generations of younger ones; but, alas! the "chime of silver bells" is no longer to be heard: the Korimakos have gone, and the groves are silent!

which there is a layer of fine grass disposed in a concave form, and then deeply lined with feathers. The eggs are usually three in number, but sometimes four, broadly elliptical or slightly ovoido-conical in shape, and measuring ·88 inch in length by ·65 in breadth. They are pure white, creamy, or pinkish white with a broad zone of reddish-brown spots towards the larger end, besides a few widely scattered dots of red over the general surface. In some specimens the ground-colour exhibits a delicate pinkish tinge, and the reddish markings are more numerous and distinct, often deepening to a dark chestnut-red. Among the examples in the Canterbury Museum, some are pinkish white, blotched at the larger end and densely freckled all over with pale reddish brown, whilst one of them presents delicate pencilled markings or veins towards the smaller end.

My son's collection contains a beautiful series of thirteen, presenting a considerable amount of individual variation, not only in the surface tint, but in the extent and character of the markings. In some the reddish spots coalesce at the large end, forming a sort of cap, in others they present distinct blots and smudges; some have a polar zone of confluent freckles, while others are studded with roundish rust-spots; in some the markings are sharp and distinct, in others smeared or blurred; one is of a pinkish cream-colour, clouded over its major portion with reddish brown, and another is perfectly white, with a cluster of reddish dots on its larger pole and a few scattered specks below. In form, too, they vary from the perfect ovoid to the types mentioned above.

In the selection of feathers for the lining of its nest this bird shows an extraordinary love of decoration, the preference being given to those of striking colours. The scarlet feathers of the Kaka, the bright green of the Parrakeet, and the ultramarine of the Kingfisher are sometimes found intermixed; the shining breast-feathers of the Wood-Pigeon are invariably used; and in the vicinity of habitations (as a correspondent informs me) the nest is occasionally found supplied from a neighbouring poultry-yard, the spotted plumes of the Guinea-fowl being most conspicuous*.

A nest from the Little Barrier is composed entirely of small black twigs carefully worked together and deeply lined with dark Pigeon's feathers, the cup being very wide, having a diameter at its rim of 3·25 inches. It was found at an elevation of thirty feet from the ground, under shelter of a clump of parasitic *Astelia*, and contained four young birds.

During the breeding-season the parent birds evince much tender solicitude for the safety of their offspring. On leaving the nest, the young have the rictal membrane (at the angles of the mouth) very large and of a bright yellow colour. The old birds hunt for them with untiring industry; and the young brood may be seen perched side by side on a branch patiently waiting for their food, and on the approach of their parents, quivering their wings with excitement, and eagerly gaping their throats, all of them together, to receive the coveted morsel.

I have made frequent attempts to rear the young, but have never succeeded. I have known instances of the adult birds being caged with success; but, like the Tui, they are liable to sudden convulsive fits, and seldom survive their confinement very long.

The Korimako from the south-west region appears to be a somewhat larger race. An egg of this bird taken at Preservation Inlet, in the month of January, is ovoido-conical, measuring ·9 of an inch in length by ·7 in breadth. It is of a delicate pinky white, with irregular stained markings of reddish brown, chiefly towards the larger end, and particularly on one side of the egg, without any appearance of a zone; the other end towards the pole being quite free from markings of any kind.

* This statement in my former edition having been questioned by Prof. Hutton ('Ibis,' 1874, p. 36), I may quote the following observations since recorded by Mr. Potts ('Journal of Science,' vol. ii. p. 278):—"Keeping several kinds of choice poultry not far from the bush afforded me special opportunities of observing this fact. I noticed nests lined with coloured feathers as follows: red from the Kakas, green from the Parrakeets, black from the Norfolk Turkeys, buff from Cochin fowls, speckled from the Pintadoes, and white from the Geese. I have not seen a red- or green-lined nest for years, as the destruction of the woods about here (Ohinitahi) has made both Kakas and Parrakeets rare visitors."

ANTHORNIS MELANOCEPHALA.

(CHATHAM-ISLAND BELL-BIRD.)

Anthornis melanocephala, Gray, in Dieff. Trav. ii., App. p. 188 (1843).
Anthornis auriocula, Buller, Essay on the Orn. of N. Z. p. 8 (1865).

♂ similis *A. melanuræ*, sed conspicuè major: pileo undique chalybeo, indigotico vel purpureo nitente.

♂ *juv.* similis adulto, sed pallidior: abdomine imo cum crisso et hypochondriis imis fulvescentibus: fronte vix chalybeo nitente: filamentis pilei gulæque chalybeo-nigris: fasciâ mystacali indistinctâ, pallidè flavâ: tectricibus alarum, remigibus et rectricibus brunnescenti-nigris, paullò chalybeo lavatis, extùs angustè flavicanti-olivaceo limbatis: rostro nigro: pedibus brunneis, plantis pallidioribus, unguibus saturatè brunneis: iride aureâ.

Adult male. The whole of the plumage olive-green, lighter on the sides of the body and lower part of abdomen; beneath dark plumbeous, this being observable only on raising the feathers; forehead and crown steel-blue, changing to a purplish-blue gloss on the sides of the head, nape, throat, and fore part of the breast, these parts appearing shot with purple and blue in certain lights; quills dusky brown, with yellowish-brown shafts, margined on the outer webs with yellow; the small wing-coverts steel-blue, margined with olive-green; tail-feathers dusky black, with steel-black margins; the soft ventral feathers and under tail-coverts fulvous yellow, the latter with an olivaceous tinge. Irides golden yellow (?); bill black; tarsi, toes, and claws dark brown. Total length 10 inches; wing, from flexure, 4·25; tail 4·5; bill, along the ridge ·7, along the edge of lower mandible ·9; tarsus 1·5; middle toe and claw 1·05; hind toe and claw 1·15.

Female. Although, since the publication of my former edition, I have received four or five examples of this bird from the Chatham Islands, I have never yet had an opportunity of comparing the female. Prof. Hutton says that "it is similar to the female of *A. melanura* except as to size."

Young. An examination of the type of *Anthornis melanocephala* in the British Museum satisfied me that the bird named by me (*l. c.*) was only the young of this species. The following is a description of this specimen, which is now in the Colonial Museum at Wellington:—The whole of the plumage yellowish olive, paler on the underparts, and tinged with fulvous on the abdomen, flanks, and under tail-coverts; faint steel gloss on the forehead; produced filaments on the crown, sides of the head, and throat steel-black; from the angle of the mouth a narrow indistinct streak of pale yellow; wing-feathers and their coverts, also tail-feathers, blackish brown, with a faint steel gloss, their outer webs narrowly margined with yellowish olive; inner lining of wings pale yellow. Irides golden yellow; bill black; tarsi and toes brown, with paler soles; claws umber-brown. Total length 9·5 inches; wing, from flexure, 4·4; tail 4·5; tarsus 1·5. (On a close inspection of this specimen two minute feathers of steel-blue on the side of the head give indication of a change of plumage.)

Obs. Gray's type was obtained by Dr. Dieffenbach, the naturalist to the New-Zealand Company, who visited the Chatham Islands in 1839. I may mention that it is not in the fully-matured plumage. Three of the tail-feathers on one side are dusky black, deepening to glossy steel-black on the outer webs; the rest are, like the wing-feathers, dusky brown, margined with olivaceous green. In the adult male the primaries and secondaries, as well as the tail, assume the dark colour.

THIS species, which is a native of the Chatham Islands, is very similar to the well-known *Anthornis melanura*; but, as will be seen on referring to the measurements given above, it is considerably larger. It differs, moreover, in having the whole of the head and neck brightly glossed with purplish or steel-blue.

During a visit to the Chatham Islands in 1855, I observed this *Anthornis* in the woods near Waitangi, and procured a specimen, although, as already mentioned, I was unable at the time to identify it. In giving it a provisional name, I selected the beautiful golden irides as presenting a good distinguishable feature, those of *A. melanura* being bright cherry-red. I observed that its habits were precisely similar to those of the common Bell-bird, but that its notes appeared to be louder and somewhat less musical. Its gregarious instincts are the same; for, on imitating the alarm-cry, I was immediately surrounded by a number of these birds in a high state of excitement.

Mr. Henry Travers, from whom I have received several specimens, states that he found it in great numbers on Mangare, less frequent on the main island, and rare on Pitt Island. It had commenced to breed in October, and its nest, which he describes as being "composed of grass and feathers, large and coarsely constructed," contained as a rule three eggs. He considers its song richer and fuller than that of its New-Zealand congener. It seemed to me very much the same, but louder.

It is said that of late years this bird has deserted the neighbourhood of the native villages and settlers' homesteads, and retired to the southern portion of Wharekauri (as the main island is called), where the woods have not yet been destroyed *.

It is a remarkable fact that whereas the New-Zealand bird is common enough at the Chatham Islands, this larger form has never been found in any part of New Zealand. The two species subsist on the same kind of food; and it is difficult to account for this peculiarity of range on any principle of geographical distribution. Where species are representative of each other in neighbouring islands, as is the case with several birds inhabiting the North and South Islands respectively, this differentiation of character, with the necessary lapse of time, is intelligible enough; but the present case is entirely different. If the long-continued separation had affected the New-Zealand bird to any appreciable degree, the same result must presumably have happened to the same bird in the Chatham Islands, four hundred miles distant; we find, however, the same type common to both places, which in itself would occasion no surprise but for the singular fact that the larger and stronger form, associated with it, is confined strictly to the smaller area, and preserves its distinctive character.

It seems to me probable that in former times both species inhabited New Zealand, and that, as *Anthornis melanura* is now rapidly disappearing from the mainland, so in like manner the other species may have died out before we became acquainted with the country. In that case, however, it would be necessary to discover some other factor than the Norwegian rat, which, as explained on a former page, is suspected of the principal mischief now. The survival of the extirpated race in the Chatham Islands is consistent with this supposition, because it is an observed law of nature that expiring races of animals and plants linger to the last in such insular areas.

The nest of this species is very much larger than that of the *Anthornis melanura*. A specimen in the Canterbury Museum measures in its largest diameter about 8 inches by 7 inches. It is composed chiefly of dry narrow flags or grasses bent in a circular form, the outer wall being strengthened with an admixture of fibrous twigs. The cavity, which is rather loosely formed, as compared with that of the latter, is roughly lined with sheep's wool, with a few small feathers intermixed. It contained two eggs, which differ somewhat from each other, both in form and colour. One of them is of a warm salmon-pink, thickly blotched at the larger end, and spotted at irregular intervals on the general surface with reddish brown, ovoido-elliptical in form, and measuring 1·05 inch by ·75 inch. The other egg is more oval in form, paler in colour, and less marked with reddish brown, the spots being much smaller and more scattered over the surface.

* Zoologist, 1885, vol. xliii. p. 422.

PROSTHEMADERA NOVÆ ZEALANDIÆ.

(TUI OR PARSON BIRD.)

New-Zealand Creeper, Brown, Illustr. Zool. pl. ix. (1776).
Poë Bee-eater, Lath. Gen. Syn. ii. p. 682 (1782).
Merops novæ seelandiæ, Gm. Syst. Nat. i. p. 464 (1788, ex Lath.).
Merops cincinnatus, Lath. Ind. Orn. i. p. 275 (1790).
La Cravate Frisée, Levaill. Ois. d'Afr. ii. pl. 92 (1800).
Sturnus crispicollis, Daud. Traité d'Orn. ii. p. 314 (1800, ex Levaill.).
Philemon cincinnatus, Bonn. et Vieill. Enc. Méth. p. 613 (1823).
Prosthemadera concinnata, Gray, List Gen. of B. 1840, p. 3.
Certhia cincinnata, Forst. Descr. Anim. p. 78 (1844).
Prosthemadera circinata, Reich. Handb. Merop. p. 127, t. cccexcii. fig. 3466 (1852).
Meliphaga novæ zealandiæ, Ellman, Zool. 1861, p. 7466.

Native names.

Tui and Koko; the young bird distinguished as Pi-tui or Pikari.

♂ pileo toto metallicè viridi, collo postico, uropygio et supracaudalibus purpurascentibus: collo undique filamentis albis ornato: dorso reliquo et scapularibus cuprescenti-brunneis: alis supernè metallicè viridi, tectricibus alarum paullò purpurascentibus, medianis albo terminatis, fasciam alarem distinctam formantibus: remigibus nigris, extùs viridi metallico lavatis, secundariis latiùs: caudâ nigrâ, suprà purpurascenti-viridi nitente: subtùs metallicè viridis, versus pectus imum purpurascens: abdomine toto cuprescenti-brunneo: hypochondriis elongatis latè brunneis: gutture imo fasciculis duobus albis globosis ornato: subalaribus nigris: subcaudalibus metallicè viridibus: rostro et pedibus nigricanti-brunneis: iride saturatè brunneâ.

♀ mari similis, sed paullò minor: coloribus sordidioribus: hypochondriis fulvescentioribus.

Juv. schistaceo-niger: tectricibus alarum medianis ut in adultis albis: collo plus minusve albicante: rictu flavo: iride nigrâ.

Male. General plumage shining metallic green, with bluish-purple reflections on the shoulders, rump, and upper tail-coverts; the hind neck ornamented with a collar of soft filamentous plumes, curving outwards and with a white line down the centre; the middle of the back and the scapulars bronzy brown, the latter with blue reflections; the greater wing-coverts are metallic green, those near the arm of the wing shining blackish purple, and the intermediate ones white in their apical portion, forming a conspicuous alar bar; the remiges are black, the primaries having an outer margin of metallic green in their basal portion, this colour spreading on the secondaries till it covers the whole of the web; tail-feathers metallic green on their upper surface, with purplish reflections; lower part of breast metallic green changing into purplish blue; sides and abdomen blackish brown, the long flank-feathers shading into pale brown; under surface of wings and tail black; the under tail-coverts metallic green. The throat is ornamented with two tufts of white filamentous feathers, which curl in upon each other in a globose form. Irides dark brown; bill and feet blackish brown. Total length 12·75 inches; extent of wings 18·5; wing from flexure 6; tail 5; culmen 1; tarsus 1·35; middle toe and claw 1·55; hind toe and claw 1·25.

Female. The female is somewhat smaller than the male; but the plumage differs in no essential respect. The

metallic tints are not so bright, and there is more brown in the plumage of the underparts. The throat is adorned with white tufts as in the other sex, but they are usually smaller.

Young. Uniform slaty black, with a broad undefined patch or circlet of greyish white on the throat, varying in extent, more conspicuous in the female, and sometimes spreading all round the neck; median wing-coverts white, as in the adult; irides black; rictal membrane yellow.

Obs. In the young bird the plumage is soft and fluffy, and entirely wants the metallic lustre. In the adult state examples vary in the brilliancy of their tints, and some have a bright coppery bronze on their upper parts.

Progress towards maturity. About the first week of November I obtained from the nest a fledgling, in which the membrane at the angles of the mouth was very conspicuous and the plumage partly undeveloped; by the second week in December it had assumed the full juvenile dress, with a faint greyish collar, the rictal membrane had disappeared, and the throat-tufts had commenced to sprout; at the end of another month the lappets had formed but were very small; two weeks later, the new metallic plumage had begun to supplant the adolescent growth, appearing at first in tracts, or irregular strips, on the breast and sides of the body, and then spreading outwards; and by the end of February the bird had acquired the full adult livery, although the tints of the plumage were not so brilliant as in the more matured condition.

Varieties. Uniform brown-coloured varieties have been occasionally met with; and it is not an unusual thing to find specimens with a single white quill or tail-feather, or marked about the throat and face with scattered white feathers. In the Christchurch Acclimatization Gardens I observed a caged one with a broad patch of white covering the outer webs of the secondaries on both wings. A beautiful albino was obtained some years ago in the Wanganui district, and now forms part of my collection in the Colonial Museum: the general plumage is pure white; a shining black band fills the lores, crosses the forehead, and spreads down each side of the neck in an irregular patch of sooty black; lower part of back, rump, and thighs sooty black, with white feathers interspersed; wings pure white, excepting the outer secondaries and the long primary coverts, which are glossy black; bill white; tarsi and toes yellowish white.

There is another abnormally coloured bird in the Colonial Museum: head, neck all round, breast, and fore part of abdomen smoky brown; the rest of the plumage pale creamy brown, darker on the quills and tail-feathers. The throat-tufts are as in ordinary examples, and there is a broad bar of white across the smaller wing-coverts; the frilled collar is rather inconspicuous, although the central line of white is present, and there is a narrow streak of the same from the angles of the mouth; the feathers of the breast have likewise fine white shaft-lines; bill and feet white horn-colour.

Sir William Fox informs me that at Porirua Harbour (near Wellington) he once observed a bird of this species with the entire plumage of a delicate fawn-colour.

This bird is one of our most common species, and on that account generally receives less attention in its own country than its singular beauty merits. It was described and figured, as early as the year 1776, in Brown's 'Illustrations of Zoology,' and has since been mentioned by nearly every writer on general ornithology. In 1840 Mr. G. R. Gray made it the type of a new genus, in which, up to the present time, it stands quite alone.

The early colonists named it the "Parson bird," in allusion to the peculiar tufts of white feathers that adorn its throat, and their fancied resemblance to the clerical bands. To those who are familiar with the bird in its native woods, this name is certainly appropriate; for when indulging in its strain of wild notes it displays these "bands," and gesticulates in a manner forcibly suggestive of the declamatory style of preaching, or, as Dr. Thompson graphically expresses it, "sitting on the branch of a tree, as a *pro tempore* pulpit, he shakes his head, bending to one side and then to another, as if he remarked to this one and to that one; and once and again, with pent-up vehemence, contracting his muscles and drawing himself together, his voice waxes loud, in a manner to waken sleepers to their senses!"

Owing to its excellent powers of mimicry, and the facility of rearing it in confinement, it is a favourite cage-bird, both with the natives and the colonists. Although of very delicate constitution, it has been known to live in confinement for upwards of ten years. More frequently, however, it becomes subject, after the first year, to convulsive fits, under which it ultimately succumbs. Cleanliness, a well-regulated diet, and protection from extremes of temperature are the proper safeguards. I had as many as ten of them caged at one time; but they died off one by one, and invariably in the manner indicated. Naturally of a sprightly disposition, it is cheerful and playful in captivity, incessantly flitting about in its cage and mimicking every sound within hearing. It will learn to articulate sentences of several words with clearness, to crow like a cock, and to imitate the barking of a dog to perfection. One, which I had kept caged in the same room with a Parrakeet (*Platycercus auriceps*), acquired the rapid chattering note of that species; and another, in the possession of a friend, could whistle several bars of a familiar tune in excellent time. Another, which I kept for two years, although a female bird, proved to be a good mimic. I first taught it to imitate the soft whistling note of the Huia, in repetition. When perfect in that, I gave it lessons in the long plaintive whistling-cry of the Shining Cuckoo, thrice repeated; and, strange to say, after the bird had acquired that, and was accustomed to practise it a hundred times over during the day, I taught it to add, or interject, the sharp four-times-repeated note which precedes the final strain. The bird learnt all this to perfection, and never mixed the parts, exhibiting in this respect a remarkable exercise of memory *.

It has several times been brought alive to this country; and there is now to be seen in the Zoological Society's Gardens, at Regent's Park, a very healthy one which I succeeded in bringing to England last year, and had the pleasure of presenting to the Society. It was one of three scarcely-fledged nestlings brought to me by a Maori shortly before I embarked on my trip home; and although all of them survived the sea-voyage, the others soon succumbed to the severity of the English climate.

The Maoris fully appreciate the mocking-powers of this bird, and often devote much time and patience to its instruction. There are some wonderful stories current among them of the proficiency it sometimes acquires; and I may mention an amusing incident that came under my own notice at Rangitikei some years ago. I had been addressing a large meeting of natives in the Whare-runanga, or Council-house, on a matter of considerable political importance, and had been urging my views with all the earnestness that the subject demanded: immediately on the conclusion of my speech, and before the old chief, to whom my arguments were chiefly addressed, had time to reply, a Tui, whose netted cage hung to a rafter overhead, responded, in a clear emphatic way, "Tito!" (false). The circumstance naturally caused much merriment among my audience, and quite upset the gravity of the venerable old chief Nepia Taratoa. "Friend," said he, laughing, "your arguments are very good: but my *mokai* is a very wise bird, and he is not yet convinced!"

In a state of nature the Tui is even more lively and active than in captivity. It is incessantly on the move, pausing only to utter its joyous notes. The early morning is the period devoted to melody, and the Tuis then perform in concert, gladdening the woods with their wild ecstacy. Besides their chime of five notes (always preceded by a key-note of preparation), they indulge in a peculiar outburst which has been facetiously described as "a cough, a laugh, and a sneeze," and a variety of other notes, fully entitling it to be ranked as a songster.

* The Tui, as a caged bird, is apt to become excessively fat, through overfeeding and the want of proper muscular exercise; and this may account for its tendency to fits. The intelligent bird mentioned above, without any apparent cause, began to mope and refused its food. After a day or two it became subject to epileptic fits, falling suddenly from its perch, screaming in its convulsions, and then lying perfectly inert for several minutes. These fits continued to increase in frequency and severity, till finally it succumbed to one of them, and died in my hand. On dissecting it, I found the cavity of the stomach choked up with an accumulation of yellow fat, and the vital organs completely enveloped in fat. This excessive fatness had no doubt interfered with the performance of life's regular functions and had caused the fits, which in the end proved fatal.

When engaged in song, the Tui puffs out the feathers of his body, distends his throat, opens wide his beak, with the tongue raised and slightly protruded, and gesticulates with his head, as he pours forth the wild harmony of his soul. A pair may often be observed, scarcely a foot apart, on the same branch, performing in concert, for (as with the Korimako also) both sexes sing. The notes are rich and varied—now resembling the striking together of hollow metallic rods, then a long-drawn sigh, a warble, and a sob, followed by a note of great sweetness, like a touch on the high key of an organ. The last time I listened to the wild music of this bird, in all its depth and richness, was from the pew of a little country chapel, where a Maori deacon of the Church of England was delivering a sensible discourse and drawing his illustrations from surrounding objects. The chapel was overshadowed by tall *Eucalyptus* trees, amongst the flowers of which the Tuis were regaling themselves on their viscid nectar, and stopping at intervals to pour forth their full volume of song, thus giving emphasis to the preacher's appeal to nature.

One of its finest notes is a clear, silvery toll, followed by a pause and then another toll, the performance lasting sometimes an hour or more. This is generally heard at the close of the day, or just before the bird betakes itself to its roost for the night. I have, however, on one or two occasions, heard the Tui's sweet toll long after the shadow of darkness had settled down upon the forests and all other sounds were hushed.

At other times it may be heard uttering a sweet warbling note, followed by a sneeze, after that a pause, then a sharp cry of *tu-whit, tu-whit, o-o-o*, a pause again, and then its warbling note with variations, very soft and liquid, but ending abruptly in a sound like the breaking of a pane of glass. It has indeed such an endless variety of notes that it is impossible to convey in writing any adequate idea of its vocal powers *.

Its flight is rapid, graceful, and slightly undulating, the rustling of the wings as they are alternately opened and closed being distinctly audible. Layard mentions ('Ibis,' 1863, p. 243) the peculiar habit which this bird has of mounting high in the air during fine weather, in parties of six or more, and performing wide aërial circles or indulging in a sportive flight, "turning, twisting, throwing somersaults, dropping from a height with expanded wings and tails, and performing other antics, till, as if guided by some preconcerted signal, they suddenly dive into the forest and are lost to view." High in the air it may sometimes be seen closing its wings and supporting its body for a few moments by a rapid perpendicular movement of the expanded tail; and slowly descending in this

* The compass and variety of the Tui's song may be judged by the following Maori paraphrase, as reduced to writing by Sir George Grey ('Poetry of the New-Zealanders'):—

Ko tu koe.
Ko rongo koe.
Ko te manuwhiri.
Nau mai.
Moemoe hia mai te kuri.
Haere mai te manuwhiri.
No runga te manuwhiri.
No raro te manuwhiri.
No te ti.
No te ta.
No waka i oio.
Tupu kere kere.
Tupu a nanga.
Ka hea e wa.
I ki e rero.
Ki tahi ka tu ke he.

Ko wai wai.
Korero rero.
Ka koro kore te toki.
Te whare pa tahi.
Te whare pa rua.
Te ui te rangi ora.
E roro ki waho.
Ko tu koe.
Ko rongo koe.
Ko tenei te manuwhiri.
Nau mai.
Kaore te kai i te kainga.
E ronga.
E ronga.
E ronga maru awa.
Ka ha te tai.

Ka timo te tai.
Nga tai o te tu.
Ko waka rara na tauna.
Ma nga wai.
E tai taua?
E tai, homai te wai.
Ka hi te kai.
Ka kawa te kai.
Ka whakarere te kai.
E kai.
Ari nui.
Ari roa.
Ari ma noa noa.
E titi rau ma hewa.
E to kai moana.
E roro ki waho.

manner to a lower level, it speeds forwards with half-closed wings and tail, and then rises high in the air again by a rapid vibration of those members.

It is a pretty sight to watch a pair of them mount together in playful flight, high above the tree-tops; then, by a simultaneous movement, they descend in company and alight on the topmost twigs of some tall forest tree, where they puff out their plumage, giving a very exaggerated appearance to their bodies, and gesticulate as if in angry altercation with each other; and then, as if by mutual agreement, they rise together in the air, and disappear in opposite directions.

There is, I believe, a popular notion in the Colony that the plumage of the Tui is black, and even some old settlers, familiar enough with the bird, considered the plate in my first edition too highly coloured. But this is entirely a mistake, as may be easily proved by holding the bird against the light, at the angle of incidence. It will then be seen that the plumage presents beautiful steel-blue and purple colours, with high metallic reflections, particularly on the breast, wings, rump, and upper tail-coverts. On the shoulders and mantle, also, there are bronze reflections which no artist could ever do justice to. In the sunny glades of the forest the glancing of the light on its burnished plumage and the gleaming of its pure white epaulettes renders the Tui a very attractive object, as it glides rapidly from one tree to another, or darts into the sunshine to capture a vagrant butterfly.

The food of the Tui consists of ripe berries of various kinds, flies and other insects, and the honey of certain wild flowers. To enable it to collect the latter, the tongue is furnished at its termination with a brush of extreme fineness—a characteristic common to all the true honey-eaters—the nectar ascending to the tubular portion of the tongue, apparently by capillary attraction *. When the functions of life are suspended or interfered with, this little brush protrudes from the bill. This occurs not only after death, but in the case of the sickly Tui; and the involuntary protrusion of the tongue may generally be accepted as a fatal symptom. It also feeds with avidity on the sugary bract-like spadices of the kiekie (*Freycinetia banksii*). In the months of October and November, when the kowhai (*Sophora grandiflora*), which grows so luxuriantly on the river-flats, has cast its leaves and is covered with a beautiful mantle of yellow flowers, its branches are alive with Tuis; and in December and January, when the *Phormium tenax* is in full bloom, they leave the forest and repair to the flax-fields to feast on the korari honey. At these times large numbers are caught in snares or speared by the natives, who thus supply themselves with a delicious article of food.

On these occasions the best-conditioned birds are preserved in their own fat, and potted in calabashes, "hua-hua koko" being esteemed a great delicacy. At the periodical festivals one or two of these pots, decorated with Pigeons' feathers, are placed on top of the great pile of food which is presented to the visitors at these ceremonials. Calabashes of kaka, titi, and kereru are plentiful enough, but one of "tui" gives the finishing touch to the *menu* at a Maori feast of the kind I have indicated.

Among introduced trees, the Tui is particularly partial to the Australian blue-gum (*Eucalyptus globosa*) and the common black-wattle. When these trees are in full bloom, this bird holds high carnival among the flowers, making playful sallies into the air from time to time, and uttering its mellifluous notes, as if in the highest ecstasy.

At certain seasons of the year, when its favourite berries have fully ripened, the Tui becomes

* Dr. Gadow, after describing fully the muscular apparatus in *Prosthemadera*, thus explains the suctorial process:—"The contraction of the mylo- and serpi-hyoid muscles presses the whole tongue and larynx upwards against the palatal roof of the mouth-cavity. The mouth is thus wholly filled up. Through the contraction of the genio-hyoid muscles the tongue will be protruded from the mouth. Now, if the serpi-hyoid muscles relax, and the tracheo-laryngeus and tracheo-hyoideus, on the other hand, by their contraction depress the larynx and at the same time depress the posterior part of the tongue, a vacuum will be produced between tongue and palate. This space, again, is in connexion with the tubes of the tongue, and therefore will be filled by the fluid into which the tips of these tubes may be inserted. In the birds in question the fluid is honey or nectar. Consequently sucking is accomplished automatically through the mere protrusion of the tongue." (Proc. Z. S. 1883, pp. 68, 69.)

exceedingly fat—so much so as very much to embarrass the operations of the taxidermist, who finds it almost impossible to keep the feathers free from the oily matter that exudes under the operator's knife. But I am unable to endorse the statement made by the reverend author of 'New Zealand and its Inhabitants' (probably on the authority of a native), that on these occasions the Tui relieves itself of its exuberant fat by pecking its breast!

The Tui is still very plentiful over both Islands. It has apparently been driven away from some districts where formerly it was abundant; but this is hardly to be wondered at when I state that (in spite of the wise protective legislation) I was assured by a dealer in Wellington that he had sent as many as five hundred skins to London for the ornamentation of ladies' hats!

It is easily approached and shot; but I have often remarked its extreme tenacity of life, reminding one of Mr. Gosse's charming account of *Conurus flaviventer* in his 'Birds of Jamaica.' Sometimes, when mortally wounded, the grasp of the feet by which the bird was clinging to the twigs or vines becomes convulsively tightened, and the falling body is seen suspended, head downward, for several minutes, the wings now and then giving an ineffectual flutter, till at last one foot relaxes its hold and then the other, and the quivering body falls heavily to the ground.

There can be little doubt that the Tui breeds twice in the year. I have found birds nesting as early as August, the young being abroad in October and November; and I have received from the Maoris nestlings, not more than ten days old, as late as May 12th, although young birds can always be obtained in March and April.

Under the head of "progress towards maturity," I have described (at page 95), from personal observation, the successive development of plumage in a young bird, taken from the nest in November, and presenting an adult appearance at the end of February. As late as the 23rd October, I saw a young bird at Atiamuri, on the Waikato river, in which this change had scarcely been completed, much of the body-plumage being adolescent, with only vestiges of the frill and lappets. This fact tends strongly to support the view of there being two broods in the year.

The nest of this species is usually placed in the fork of a bushy shrub, only a few feet from the ground; but I have also found it at a considerable elevation, hidden among the leafy top of a forest tree. It is a rather large structure, composed chiefly of sprays or dry twigs, intermixed with coarse green moss, the cavity being lined with fibrous grasses, very carefully bent and adjusted. Sometimes the interior is composed of the black hair-like substance from the young shoots of the tree-fern, the cavity being sparingly lined with dry bents. One which I examined at Rangitikei was composed almost entirely of dry *Leptospermum* twigs, with a little green moss intermixed, the ends of the twigs projecting more or less, so that the exterior of the nest measured nearly 12 inches across; the twigs were largest at the foundation and got smaller upwards; the cavity was large, but somewhat shallow or saucer-shaped, and the interior thickly lined with brown fern-hair, with a few long grass-leaves carefully interlaced; thus giving the nest a neatly finished appearance.

The eggs are generally three or four in number and present some variety both in form and colour. There are some good examples in the Nelson Museum: the eggs (numbering three) in one of the nests are of a pyriform character, being blunt and rounded at the large end and tapering upwards to a point, measuring 1·3 inch in length by ·75 in their widest part; they are white, with a faint rosy blush, stained and mottled at the larger end and lightly freckled or dusted all over with pale reddish brown. Those contained in another nest (also numbering three) are ovoido-conical, measuring 1·05 in length by ·75; these are of a delicate rosy tint, obscurely freckled, darker and more or less speckled with brown at the large end. A third nest contains two almost pure white eggs, intermediate in form between those described above, stained and freckled, at the larger end only, with brick-red. There is likewise an interesting series of these eggs in the Canterbury Museum, varying in character from the true ovato-pyriform to a fusiform outline, something like a skittle-head.

The former measure 1·5 inch in length by ·9 in breadth, and are of a pinky-white colour, freckled and spotted at the larger end with reddish brown, and with marbled markings of the same colour at the smaller end: the other extreme form measures 1·7 in length by ·8 in its widest part, and the whole surface is white with scattered specks of rust-red at the large end, each surrounded by a light stain or halo, as if the colour had run; there are also two or three of these specks with the same stained circumference in the anterior or produced portion of the egg. Sir James Hector informs me that Tui's eggs in his possession vary from a decidedly elliptical shape to a narrow oval, and that both forms are "spotted with round dabs of red." One of my specimens from the South Island is ovoido-elliptical or slightly pyriform, measuring 1·25 inch in length by ·10 in breadth, and is creamy white, much smeared and blotted with pale lake-red towards the smaller end.

The newly-hatched Tui is almost entirely bare, there being mere indications of linear tracts on the upper surface, with light woolly filaments adhering. The feathers, however, soon begin to appear, and the growth of the nestling is rapid; but the gradation in size of the three or four occupants of the same nest is very noticeable. Till about three weeks old, they have a very feeble *cheep*; but it is curious to see them, in their eagerness to be fed, stretch up their bodies and necks, four inches or more above the nest, with widely-gaping mouths bordered with a membrane of vivid yellow. As their development proceeds, their cry strengthens; and when they are fully fledged it becomes an almost incessant plaintive note, which changes to an impatient scream on the approach of the parent bird with food, all the nestlings craning their necks together for the first attention. After it quits the nest, and before it has attempted any song, it acquires the peculiar alarm-cry, *ke-e-e-e*, so familiar to the ear.

Head and neck of the Tui, showing arrangement of feathers.

POGONORNIS CINCTA.

(STITCH-BIRD.)

Meliphaga cincta, Dubus, Bull. Acad. Sc. Brux. vi. pt. 1, p. 295 (1839).
Ptilotis auritus, Lafr. Rev. Zool. 1839, p. 257.
Ptilotis cincta, Gray, Voy. Ereb. and Terror, Birds, p. 4 (1844).
Pogonornis cincta, Gray, Gen. of B. i. p. 123 (1846).

Native names.

Hihi, Tihe, Kotihe, Kotihewera, Tiora, and Tiheora; male and female sometimes distinguished as Hihi-paka and Hihi-matakiore or Tihe-kiore.

♂ suprà nigerrimus: fasciis duabus conspicuis postocularibus albis: dorso imo et uropygio cinerascenti-brunneis, vix olivaceo tinctis: dorsi plumis quibusdam lateralibus latè aurantiaco terminatis: tectricibus alarum minimis latè aurantiacis, plagam magnam formantibus, majoribus nigris, extùs aurantiaco marginatis: alæ spuriæ plumis ad basin albis speculum exhibentibus: remigibus nigris, primariis versus apicem albido, secundariis aurantiaco marginatis: tectricibus alarum majoribus intimis et secundariis dorsalibus purè albis, plagam distinctam formantibus, dorso proximis medialiter nigris: gutture toto et collo laterali nigerrimis: torque pectorali angustâ aurantiacâ: corpore reliquo subtùs cinerascente, hypochondriis et subcaudalibus saturatioribus, illis brunneo striatis: subalaribus cinerascentibus, margine alarum aurantiaco: rostro brunnescenti-nigro: pedibus brunneis: setis rictalibus nigris: iride nigrâ.

♀ mari omninò dissimilis: suprà brunnea olivaceo lavata, pileo obscuriore: maculâ parvâ postoculari albâ: tectricibus alarum olivaceo-fulvo lavatis, minimis aurantiaco nitentibus, majoribus intimis et secundariis dorsalibus albis, brunneo medialiter lineatis et marginatis, plagam magnam albam formantibus: remigibus et rectricibus cinerascenti-brunneis, extùs latè fulvescente lavatis, primariis ad basin pogonii interni albis: subtùs obscurè brunnea, pectore et abdomine fulvescentibus, obscurè brunneo striatis.

Adult male. Head, neck, and upper part of the back velvety black; on each side of the head there is a tuft of snow-white feathers which the bird has the power of erecting. A band of rich canary-yellow encircles the breast, contrasting finely with the dark plumage immediately above it; narrow in the centre, it widens on both sides and expands on the wings, covering the small coverts and the margins of the scapularies, and becomes very conspicuous when the wings are spread. Underparts light greyish brown, inclining to olivaceous brown on the sides of the body. Primaries and tail-feathers black, margined outwardly with olivaceous brown; the secondaries in their basal portion and their coverts white; the upper tail-coverts olivaceous brown. Irides and rictal bristles black; bill brownish black; tarsi and toes pale brown. Total length 8 inches; extent of wings 12·5; wing, from flexure, 4; tail 3; bill, along the ridge ·60, along the edge of the lower mandible ·75; tarsus 1; middle toe and claw 1; hind toe and claw ·75.

Female. Obscure olivaceous brown, darker on the upper parts, and changing to pale brown on the abdomen and under tail-coverts. The primaries and outer tail-feathers have their external webs narrowly margined with very pale brown; the rest of the quills and tail-feathers are dusky black, edged externally with olivaceous brown. There is a large spot of white on the secondaries corresponding to that in the male, with faint indications of yellow towards the root of the wing; but this is only apparent when the wings are spread. There are a few minute touches of white on each side of the head, corresponding in position to the tufts in the male bird; but these adornments are wanting in this sex. Total length 7·25 inches; wing, from flexure, 3·75; tail 2·73; culmen ·55; tarsus 1.

Young male. In the Auckland Museum there is a young male, in transitional plumage, which is very interesting, as showing that in the young state both sexes have the colours of the female. At the first moult the male bird puts on the adult livery, although the tints of the plumage are less bright than in the fully matured bird. In the present example the plumage of the head, neck all round, and shoulders is changing from dull olive-brown to black, the new feathers being very conspicuous and predominating. The white tufts on the head have appeared, but have not attained their full development, being only about one third of the usual size; the canary-yellow band on the upper edge of the wings is well defined, but the pectoral zone is narrow and indistinct; on the breast the old plumage has almost entirely disappeared, being replaced by the black, but there are enough remnants to show what it was originally; the rest of the body-plumage the same as in ordinary examples, being alike in both sexes.

Young female. The specimen in the Auckland Museum has no appearance of the white marks on the head; the spots covering the base of the secondaries are yellowish white or very pale fawn-colour, becoming pure white at the roots of the feathers; the small feathers at the carpal flexure are pale yellow; quills blackish brown, the primaries very narrowly, and the secondaries broadly, margined with pale olive; tail-feathers blackish brown margined on their outer webs with dull olive.

Obs. In some examples of the male the colours are brighter, the pectoral zone being wider and deepening to a clear orange-yellow, while the quills and larger wing-coverts have a narrow external margin of yellowish olive.

Remarks. This species is furnished with hair-like bristles at the angles of the mouth measuring half an inch in length. The tongue has a pencilled or brush-like termination; the hind claw is almost twice the length of those of the fore toes, which are about equal, measuring ·25 of an inch in their curvature; the tail is of medium length and slightly cuneiform. The plumage, especially that of the female, is soft to the touch, and, in the adult, has a peculiar silky gloss.

This New-Zealand form approaches closely to a numerous group of Australian birds comprehended under the generic name of *Ptilotis*, among which it originally was placed. It has since, however, been recognized as the type of a distinct genus.

In my former edition of this work I wrote thus:—"This handsome species has only a limited range. It is comparatively common in the southern parts of the North Island, and may be met with as far north as the wooded ranges between Waikato Heads and Raglan, beyond which it is extremely rare. It is never found in the country north of Auckland, with the exception of one locality, the Barrier Islands, where Captain Hutton records it 'not uncommon' in December 1868. I have never heard of its occurrence anywhere in the South Island. It affects deep wooded gullies, and is seldom found on the summits of the ranges. In the dense timber covering old river-bottoms or low-lying flats it may be sought for; but it rarely frequents the light open bush or the outskirts of the forest. It is, moreover, a very shy bird; and being most active in all its movements, it is not easily shot. Its food consists of insects, the honey of various bush-flowers, and the smaller kinds of berries. It often frequents the topmost branches of the high timber, where it may be seen flitting about in search of insects. If disturbed by the report of a gun, it will fly off to a neighbouring tree with a light and graceful movement of the wings; but when descending to a lower station, it adopts a different manner of flight, elevating the tail almost to a right angle with the body, and scarcely moving the wings at all. The male bird erects the tail and spreads the ear-tufts when excited or alarmed; but the female habitually carries the tail perfectly erect and the wings drooping. The sexes vary so much in appearance that many of the natives regard them as distinct species, and call them by different names. The male bird utters at short intervals and with startling energy a melodious whistling call of three notes. At other times he produces a sharp clicking sound like the striking of two quartz stones together: the sound has a fanciful resemblance to the word 'stitch,' whence the popular name of the bird is derived. The female also utters this note, but not the former one; and being recluse in her habits as well as silent, she is seldom seen."

Although only fifteen years have elapsed since the above was penned, the Hihi has become the rarest of our existing native birds. To show how rapidly it has disappeared, I may mention that after my return to the colony, in 1874, I met with it only twice—the first time in a sunny glade in the Forty-mile Bush, near Eketahuna, and two years later in a strip of forest at Tarawera, midway between Napier and Taupo. I know of one other instance of its being seen of late years on the mainland. Mr. Tone, a Government Surveyor who has been working for many years past in the bush, and is familiar with all the native birds, met with it in February 1883 on the summit of one of the wooded spurs of the Tararua range, leading down into the Wairarapa valley. He saw the bird several times during the day, heard its note and carefully observed its habits.

The Maoris allege that it still lingers in the Kauwhanga range, above the famous gorge of the Manawatu, but the report lacks confirmation. A more likely refuge is another they assign to it on the Island of Kapiti, where Korimako, Popokatea, and Toutouwai (all absent from the mainland) are still to be found.

In 1880, the indefatigable Austrian collector, Herr Reischek, determined to visit the Little Barrier in quest of this bird *. He remained on the island three weeks without any sign of it. Two years afterwards he sent down his assistant who, after a sojourn of three months, succeeded in shooting a pair, but unfortunately knocked them to pieces with heavy shot. In October 1882, he went down again himself, determined to remain till he had secured good specimens. After five weeks' continuous search, traversing every part of this rugged island and climbing over ranges some 2000 feet above the level of the sea, he was at length rewarded by the sight of *Pogonornis*. A beautiful male bird was disporting himself in the sunlight, erecting his snow-white tufts and hopping about in a very excited manner. Suddenly the bird disappeared as by magic; and the discovery immediately afterwards of an unfinished nest explained the singular performance he had witnessed. This structure was composed of small twigs, partially lined with fine native grasses, and was placed in a bunch of mangimangi creeper hanging from a low tree, about eight feet from the ground. Frequently after this he heard the sharp call of the male bird in the vicinity of the nest and at length, on November 8, succeeded in shooting both male and female. He had now discovered that the favourite haunt of the Stitch-bird was a deep ravine near the top of the range, where the rocks formed steep precipices and the low scrub was covered over with a mass of creeping mangimangi so rank and thick in its growth as to be almost impenetrable. Some idea of the inaccessible nature of the place may be gathered from the fact that it took Reischek two whole days tramping, climbing, and scaling precipices to get back to his landing-place; but he had visited the last home of the Hihi and had obtained, besides several specimens of the male bird, a female in perfect plumage.

The nature of the ground often prevented his using his gun, even with dust-shot, but he was able to make some interesting observations on the habits of the bird.

He often observed it using its brush tongue among the wild flowers, and in the stomachs of those he skinned he found some minute seeds as well as insect-remains.

On one occasion he noticed a female performing very singular antics, hopping round and round within a restricted circle, with her wings drooped and the tail slightly elevated. She kept up this

* Mr. Reischek has communicated to the New-Zealand Institute (Trans. vol. xviii. pp. 84, 87) a short account of his expedition in search of *Pogonornis cincta*; but I prefer to give, in my own words, the more detailed information obtained from him immediately after his return. On the general habits of this species he says:—"I have only once seen these birds sitting still and that was near the nest. They appear always on the move, carrying their heads proudly, their wings drooped, and their tails spread and raised; and, at each successive movement, they utter that peculiar whistle from which the natives have named them *Tiora*. The female has a different note, sounding like *tee, tee, tee*, repeated several times."

performance for fully twenty minutes and apparently for mere sport. Then some movement alarmed the bird, and in an instant she had disappeared amongst the mangimangi.

It is somewhat curious that whereas the male bird never descended to the ground, his mate seemed to delight in doing so, hopping about with outstretched wings, and uttering every now and then her peculiar note. On the slightest alarm, however, she would hide herself and remain perfectly quiet. The male appeared to be always on the alert, keeping a strict guard, and giving the signal at the least sign of danger. The instinct of caution must be strongly developed in this bird, to manifest itself thus in the most secluded part of a lonely island, where probably the face of man had never appeared before.

I have already remarked upon the shy and retired habits of the female. Twenty years ago, when the bird was comparatively common in the valley of the Hutt, and at Makara, near Wellington, although frequently out with the gun I never succeeded in shooting more than two of this sex; and whilst the bright-plumaged male bird was being constantly brought in to the local birdstuffers I never saw a female in their hands. One of those shot by me was too much shattered to be of any use; the other is in my old type collection in the Colonial Museum at Wellington. There is a specimen from the Little Barrier in the Auckland Museum; but this sex is a desideratum in all the other local museums. There is one old and dingy skin in the British Museum (obtained by Percy Earl in 1842), and another (from Sir William Jardine's collection) in the new University Museum at Cambridge, but no other English or foreign museum can boast a specimen of the female Hihi.

My own private collection was equally deficient till I induced Mr. Reischek, in 1884, to make another visit to the Little Barrier in quest of it. In this further search he succeeded, although the rarity of the bird may be inferred from the fact that he was fifteen days on the island and did not even hear the Hihi till within the last three days of his stay. As already stated, the bird frequents the deep wooded ravines in the highest part of the Barrier, and to reach this ground he had to perform a toilsome journey of two days, on foot, being accompanied all through by his trusty dog, who had in places to be hoisted up with a rope. In the end his efforts were rewarded by his finding a family party of five—an adult male and female with three birds of the year, curiously enough, all males. At first the male birds alone were visible. They seemed much interested in the movements of the dog, and hopped about in the branches above him, peering down in a very inquisitive manner. The female bird had secreted herself on the ground and kept perfectly silent. Once or twice she left her place of concealment, and darted off uttering on the wing her peculiar rapid snapping note. For two hours this watch was continued before there was an opportunity of shooting her.

The Maoris state that formerly this bird was very plentiful in the Rotorua district, where it was known under the name of Kotihe; and that, at a certain season of the year, it was accustomed to come out of the woods to feast on the berries of the tupakihi (*Coriaria sarmentosa*), on which occasions numbers were killed for the oven, sometimes as many as a hundred being taken in a day. They were caught in the same manner as the Korimako, by means of a tuke or pewa, baited with flowers, as described at page 89. If the birds proved to be *matakana*, or shy, the hunter would at once move his snare to another place, it being perfectly well recognized that these birds were often fastidious and had to be humoured.

The fine old Wanganui chief, Topine Te Mamaku, who was almost a centenarian when I last saw him, told me that in his young days this bird was very plentiful in the Upper Wanganui district—so much so that one of the chiefs of that period always appeared on public occasions in a gorgeous feather robe which was largely ornamented with the canary-yellow feathers from the wing of the Hihi. Considering how very minute these feathers are, it may be imagined how many were sacrificed in order to make this colour conspicuous in the historic mantle, which Topine called, by way of distinction, the *kahu-hihi*.

It may be here mentioned that the Maoris excel in the manufacture of feather robes, many of which are very beautiful. The robe itself is formed of hand-prepared *Phormium*-fibre, soft and silky, which is woven and plaited into a thick fabric, over which the feathers are tastefully laid, with the webs overlapping, the shaft of each being doubled back and tied, thus imparting strength and durability to the garment. The pattern varies according to the kind of feather used, and sometimes much artistic skill is displayed in the grouping and arrangement of the colours. The *kahu-kereru* is composed of the bronzy-green feathers of the Wood-Pigeon, quite resplendent in the sunshine, and relieved with stars and stripes of snow-white tufts taken from the breast of that bird; the *kahu-kaka* is a mass of scarlet of different shades, from under the wings of *Nestor meridionalis*; the *kahu-kakariki* displays the brilliant green plumage of the Parrakeet, with which is usually mixed the feathers of the Tui and other birds, in squares or crosses or other fanciful designs; but far more valuable than any of these is the *kahu-kiwi*, covered entirely with the soft back-feathers of the *Apteryx*, and having a peculiarly rich effect when held against the light. One of special beauty, for which some forty adult birds were placed under contribution, was presented many years ago by the loyal Wanganui tribes to Her Majesty the Queen. There is a very fine one in the ethnological collection at the British Museum, and another in my own collection, in both of which there is a broad margin of bright Tui and Pigeon feathers, to heighten the effect. Far and away more precious than any of these must have been that mantle of golden yellow of which old Topine had so vivid a recollection; and one can only compare it, in imagination, with that gorgeous coronation-robe of costly yellow plumes worn by the kings and queens of Hawaii, of which mention is made by the early writers on Polynesia *.

Mr. F. H. Meinertzhagen informs me that in the spring of 1871 he observed a pair of Hihi nesting in a clump of blue-gums at Waimarama, near his own pretty homestead (Paparewa) about thirty miles from Napier. He at first mistook the female bird for a Green Linnet; but discovered his error the moment he saw the male and observed its peculiar flight. Hoping to retain these rare visitants, he allowed them to hatch out and rear their brood without molestation; but he never saw anything of them afterwards. The nest, which Mrs. Meinertzhagen fortunately preserved, is now in the Canterbury Museum.

A nest was discovered many years ago in the bush above the Kai-warawara stream, in the vicinity of Wellington, and is still preserved in the Colonial Museum. It is a shallow structure, with thin walls, and measures 4·75 inches across the top, with a cavity of 2·35 by 1·35. It is built of sprays, above which are laid fibres and dry rootlets of tree-fern; and the cavity is formed of fine grass, lined with cow-hair. This nest contained a single egg, of a narrow ovoid form, measuring ·75 inch in length by ·6 in breadth, of a yellowish-white colour, thickly spotted and clouded with pale rufous.

The assumption of the female plumage by the young of both sexes, as described above, is very singular, being the only instance of the kind we have among the Passeres in New Zealand. According to Reischek's observations on the Little Barrier, the brood generally numbers three, and the young birds keep together till the change of plumage has been effected. He met with four broods in all, and out of those shot, no less than five young males were in transitional plumage. He generally shot the adult male first, then attracted the female by imitating the cry of the young birds, and after securing her, the rest of the family fell an easy prey to this insatiable collector.

* These Hawaiian robes are made with the beautiful red feathers of the Hohiwi (*Vestiaria coccinea*) mixed with the golden-yellow plumes of a rare species of *Nectarinea*. "A cloak of yellow feathers could only be worn by the king" (see Lord Byron's 'Voyage of H.M.S. Blonde' in 1826). "A feathered cloak which in point of beauty and magnificence is perhaps nearly equal to that of any nation in the world" (Cook's 'Third Voyage,' 1785, p. 127). Mr. Fornander states that such cloaks, "irrespective of their value as insignia of the highest nobility in the land," represent, in feathers alone, at their present price, apart from the cost of manufacture, from five to ten thousand dollars each ('Polynesia,' vol. ii. p. 186).

ACANTHOCHÆRA CARUNCULATA.

(AUSTRALIAN HONEY-EATER.)

Merops carunculatus, Lath. Ind. Orn. i. p. 276 (1790).
Pie à pendeloques, Daud. Orn. ii. p. 246, pl. 16 (1800).
Corvus paradoxus, Lath. Ind. Orn. Suppl. p. 26 (1801).
Wattled Crow, Lath. Gen. Syn. Suppl. ii. p. 119 (1801).
Wattled Bee-eater, Lath. ibid. p. 150 (1801).
Corvus carunculatus, Shaw, Gen. Zool. vii. p. 378 (1809).
Creadion carunculatus, Bonn. et Vieill. Encycl. Méth. ii. p. 874 (1823); Vieill. Gal. Ois. i. pl. 94 (1825); Lesson, Traité d'Orn. i. p. 359 (1831).
Anthochæra lewini, Vig. & Horsf. Trans. Linn. Soc. xv. p. 322 (1827), note.
Anthochæra carunculata, Gould, Birds of Australia, iv. pl. 55 (1848).
Anthochæra bulleri, Finsch, Journ. f. Orn. 1867, p. 342.
Anthochæra carunculata, Buller, Trans. N.-Z. Instit. vol. xvi. p. 313 (1884).
Acanthochæra carunculata, Gadow, Cat. Birds Brit. Mus. vol. ix. p. 263 (1884).

Descr. exempl. ex N. Z. Brunnea, plumis omnibus longitudinaliter albo striatis: supracaudalibus brunneis latè albo marginatis: tectricibus alarum magis nigricantibus, albo marginatis, minimis et medianis albo medialiter lineatis: remigibus et rectricibus nigricantibus cinerascenti-brunneo marginatis et albo angustiùs fimbriatis, primariis et rectricibus latè albo terminatis: pileo nigricante, vix albido lineato: collo postico dorso concolore et in eodem modo lineato: loris quoque nigricantibus: fasciâ latâ suboculari argentescenti-albâ: regione parotica auriculari nigricante, albo minutè lineatâ: plagâ pone regionem paroticam fulvescente: genis nigricanti-brunneis, vix albo minutè striatis: gutture et præpectore brunneis albo lineatis: pectore albido, brunneo striato, plumis lateraliter brunneo marginatis: pectore imo et abdomine flavis: corporis lateribus albidis, brunneo marginatis, vix flavo lavatis: tibiis brunneis albo striatis: subcaudalibus albidis, medialiter brunneis: subalaribus et axillaribus nigricanti-brunneis, albo latè marginatis: remigibus infrà nigricantibus, intùs versus basin rufescentibus: rostro nigricante: pedibus pallidè brunneis.

New-Zealand example. Crown and hind-head glossy brownish black, the feathers of the nape with narrow shaft-lines of white; upper surface generally blackish brown, the feathers of the hind neck and shoulders with a broad central streak of greyish white; quills and tail-feathers darker brown; the primaries and all the rectrices, except the two middle ones, largely tipped with white; the secondaries and the two middle tail-feathers largely margined with grey, and the latter minutely tipped with white; all the wing-coverts and the upper tail-coverts broadly margined with fulvous white; all the primaries except the first largely margined on their inner vanes with pale rufous. From the angle of the mouth a widening patch of short silvery-white feathers, speckled with brown, extends below the eye and is bounded by a small bare space, below which is a minute caruncle (now dried and colourless); on the sides of the neck, behind the ear-coverts, a broad rounded patch of pale fulvous; sides of the head, behind the eyes, conspicuously speckled with silvery white; fore neck and upper part of the breast dark fulvous brown with a more or less distinct streak of white down the shafts; lower part of breast greyish brown, each feather largely centred and also margined with fulvous white; the whole of the abdomen canary-yellow, with a silky sheen, the underpart of the plumage plumbeous; sides of the body faintly washed with yellow; vent yellowish white; under tail-coverts fulvous white, the centre of each feather, except the shaft, dull brown; thighs darker brown with light shaft-lines. Bill brownish black; legs pale reddish brown. Total length 15 inches; wing, from flexure, 6·5; tail 6·75; bill, along the ridge ·9, along the edge of lower mandible 1·25; tarsus 1; middle toe and claw 1·5; hind toe and claw 1·2.

Obs. Mr. Gould, describing the freshly killed bird in Australia, states that the oblong naked spot on the sides of the head is flesh-coloured, the pendulous wattle is of a pinky blood-red colour, the irides hazel-red, and the inside of the mouth yellow.

In my 'Essay on the Ornithology of New Zealand,' 1865, I included the above species among our birds, on the authority of a specimen in the Auckland Museum, preserved by Mr. St. John, and said to have been obtained at Matakana, to the north of Auckland. The bird was retained in our lists for many years, but no fresh examples having been heard of, and St. John's specimen being of doubtful authenticity, its name was ultimately expunged *.

After a lapse of nearly twenty years I had again the pleasure of recording it (*l. c.*) as a New-Zealand bird.

During a visit to Marton, I was invited by Mr. Avery, the local bird-stuffer, to examine his novelties. Among these was a bird which he had himself obtained when serving with the volunteers in Mr. Bryce's expedition against Parihaka; he met with it in some high scrub at the rear of the camp at Rahotu, when on fatigue-duty, and was fortunate enough to shoot it. The bird was new to him and he therefore skinned it, performing the operation very successfully. The skin was in a fresh condition when it came into my hands, and proved on examination to be a well-plumaged example of *Acanthochæra carunculata*.

This specimen, which Mr. Avery was generous enough to give me, is now in my collection; and the claim of this species to a place in the New-Zealand avifauna is undoubted. Its occurrence in our country, as a straggler from Australia, may only happen at long intervals; but the rule in such cases is to admit every species of which even a single individual has been met with in a wild state, unless there is a suspicion of its having been introduced by man.

Its habits are thus described in Gould's 'Birds of Australia':—

"It is a showy, active bird, constantly engaged in flying from tree to tree and searching among the flowers for its food, which consists of honey, insects, and occasionally berries. In disposition it is generally shy and wary, but at times is confident and bold. It is usually seen in pairs, and the males are very pugnacious.

"It breeds in September and October. The nests observed by myself in the Upper Hunter district were placed on the horizontal branches of the *Angophoræ*, and were of a large rounded form, composed of small sticks, and lined with fine grasses; those found by Gilbert in Western Australia were formed of dried sticks, without any kind of lining, and were placed in the open bushes. The eggs are two or three in number, one inch and three lines long by ten lines and a half broad; their ground-colour is reddish buff, very thickly dotted with distinct markings of deep chestnut, umber, and reddish brown, interspersed with a number of indistinct marks of blackish grey, which appear as if beneath the surface of the shell. Eggs taken in New South Wales are somewhat larger than those from Western Australia, and have markings of a blotched rather than of a dotted form, and principally at the larger end."

* Dr. Otto Finsch (*l. c.*) proposed to distinguish the example described by me, as a new species, under the name of *Anthochæra bulleri*; but it is well known that this bird is subject to a considerable amount of variation, and I do not consider the differences relied upon as having any specific value. The same remarks apply to the form described by Messrs. Vigors and Horsfield (*l. c.*) under the distinctive name of *A. lewini*.

Curiously enough, when hunting up the synonymy and bibliography of this species, I found that it had been described by M. [illegible] (*l. c.*) as far back as 1860, as a New-Zealand bird. His words are:—"Cet oiseau de la nouvelle Zélande est dans la galerie du Muséum de Paris."

XENICUS LONGIPES.

(BUSH-WREN.)

Long-legged Warbler, Lath. Gen. Syn. ii. pt. 2, p. 465 (1783).
Motacilla longipes, Gm. Syst. Nat. i. p. 979 (1788, ex Lath.).
Sylvia longipes, Lath. Ind. Orn. ii. p. 529 (1790).
Acanthisitta longipes, Gray, List of Gen. of Birds, App. p. 6 (1841).
Xenicus longipes, Gray, Ibis, 1862, p. 218.
Xenicus stokesii, id. tom. cit. p. 219.

Native names.

Matuhituhi, Piwauwau, Puano, and Huru-pounamu.

♂ pileo umbrino: dorso toto viridi, uropygio lætiore: supercilio distincto albo: plumis anteocularibus nigris: regione paroticâ brunneâ vix viridi lavatâ: tectricibus alarum dorso concoloribus, vix flavido tinctis: alâ spuriâ nigrâ: remigibus brunneis, extùs olivaceo-viridi lavatis; caudâ suprà olivaceo-viridi, subtùs flavicante: mento albido: corpore reliquo subtùs pulchrè cinereo, pectore vix argenteo-nitente: abdomine imo et subcaudalibus viridibus, hypochondriis olivaceo-flavis: cruribus brunneis: subalaribus et margine alari pallidè citrinis: rostro saturatè brunneo: pedibus flavicantibus.

♀ dissimilis: suprà ferrugineo-brunnea: uropygio vix olivaceo tincto: supercilio lato albo: subtùs pallidè chocolatina, hypochondriis et abdomine imo sordidè flavis.

Adult male. Upper parts dark green, tinged with yellow, shading into dark brown on the forehead and crown; sides of the head black, with a broad superciliary streak of white extending beyond the ears, and then changing to yellow; sides, thighs, and rump bright greenish yellow; fore neck, breast, and abdomen cinereous grey, with a beautiful gloss (sometimes tinged with cobalt), and softening into greyish white on the throat; lining of wings pale yellow; quills, on their outer webs, and the tail-feathers olivaceous green. Irides and bill brownish black; tarsi and toes pale brown. Extreme length 4 inches; wing, from flexure, 2·25; tail 1 (more than half of it concealed by the soft coverts); bill, along the ridge ·5, along the edge of lower mandible ·7; tarsus 1; middle toe and claw ·9; hind toe and claw ·8.

Adult female. Upper parts umber-brown, tinged with yellowish green, especially on the rump; crown shaded with purplish brown; superciliary streak white; throat, sides of the neck, breast, and upper part of abdomen delicate vinous brown; sides of the body, flanks, and thighs dull lemon-yellow; inner lining of wings pure yellow.

Young male. Plumage generally as in the adult, but with the green tints of the upper parts paler, and the silky grey of the breast tinged with purple; crown of the head and hind part of neck chocolate-brown, blending into the olivaceous green of the upper parts; superciliary streak broad and conspicuous.

Obs. The figure of *X. longipes* in the 'Voyage of the Erebus and Terror,' which represents a bird with a white eye-circlet and an upturned bill like that of *Acanthidositta*, is copied from a rough half-finished drawing of Forster's (1777) and is strikingly incorrect. Professor Hutton, whose views are entitled to respect, wrote

* Strictly speaking Acanthidositidæ is the right name for this family, *Acanthidositta* being an older genus than *Xenicus*, but I am unwilling to disturb a name that has already obtained currency.

thus, after the appearance of my former edition:—"Dr. Buller obtained specimens of *X. stokesii* which he wrongly determined as *X. longipes*; in fact all the specimens of *X. longipes* in his collection were *X. stokesii*; these he compared with *X. stokesii* in the British Museum, and naturally found them identical. But until it is explained how it is that the figure and description of *X. longipes* in the 'Voyage of the Erebus and Terror' differ so much from specimens of *X. stokesii*, I must continue to regard them as two species" ('Ibis,' 1874, p. 37). A specimen, however, labelled by Prof. Hutton "*Xenicus stokesii*, female," and sent to Dr. Finsch for examination, was referred by this naturalist, without hesitation, to *X. longipes*, Gmelin. It is perfectly clear that *X. stokesii* has no existence as a species.

This species is confined to the *Fagus*-forests which clothe the sides of our subalpine ranges in the South Island, never being met with in the low country. In many parts of the Nelson provincial district it is quite abundant, but only in the dense bush. In the dark forest lying between Wallis-head and Tophouse, also along the wooded banks of the Pelorous river, it is said to be very plentiful, and even in the *Fagus*-covered hills in the vicinity of Nelson it is a comparatively common bird, although less numerous now than formerly. Mr. Travers found it numerous in the Spencer ranges (Nelson) at an elevation of 3000 feet; Sir J. Hector obtained specimens in the high wooded lands of Otago, where, as he informs me, it was a very rare bird; Sir J. von Haast met with it frequently during his exploration of the interior of the Canterbury district; and I observed it in the high wooded ranges forming the inland boundary of Westland. The localities I have enumerated are all in the South Island. There are specimens, however, in the British Museum which are said to have been obtained by Captain Stokes in the Rimutaka ranges (in the provincial district of Wellington); and although I never met with the species in that district, or, indeed, in any part of the North Island, an intelligent Maori, to whom I showed a coloured drawing of the bird, appeared at once to recognize it. He said that he had often seen it in the Ruahine mountains, and that during severe winters it sometimes appeared in the low country; and he further spoke of the plumage as being "like silk," an expression so aptly descriptive of its peculiar softness, that I believe the man was quite familiar with the bird.

The Maoris have a saying that if you kill this bird "ka panga te huka" (or, "snow will fall").

It is generally met with singly or in pairs, but sometimes several are associated, attracting notice by the sprightliness of their movements. They run along the boles and branches of the trees with restless activity, peering into every crevice and searching the bark for the small insects, chrysalids, and larvæ on which they feed. It is generally arboreal in its habits, seldom being seen on the ground, in which respect it differs conspicuously from the closely allied species, *Xenicus gilviventris*. It has a weak but lively note, the sexes always calling to each other with a subdued trill, and its powers of flight are very limited.

On comparing my specimens of this bird with the type of Mr. G. R. Gray's *Xenicus stokesii* in the British Museum, I feel satisfied that they are referable to one and the same species, the difference of plumage being only sexual.

In June 1882 the late Mr. W. A. Forbes made a communication to the Zoological Society, showing, from an investigation of their anatomy, that *Xenicus* and *Acanthisitta* (rectius *Acanthidositta*), hitherto supposed to be allied to *Certhia* and *Sitta*, were in reality mesomyodian forms, the discovery of such low types among the Passerine birds in New Zealand being a fact of considerable interest in zoo-geographic distribution. The characters pointed out by this able investigator * went

* The following is extracted from the paper referred to:—

"The subjoined drawings of the syrinx of *Xenicus*—with which in all points *Acanthisitta* appears to agree in every essential respect—will show that it has none of the complex nature of that organ in the Oscines, the thin lateral tracheal muscle terminating on the upper edge of a somewhat osseous box formed by the consolidation of the last few tracheal rings, and there being no other intrinsic syringeal muscle whatsoever. The box has a well-developed antero-posterior pessular piece. The bronchial rings are to show that the affinities of the Xenicidæ (as he proposed to call the family) could only make it compare with Pipridæ, Tyrannidæ, Pittidæ, and Philepittidæ. From all these groups, however, it differs widely in the number of rectrices, the character of the tarsus, and the non-oscine syrinx.

The nest (of which there is a specimen in the Canterbury Museum) is usually placed among the upturned roots of a fallen tree, or in the fork of a double trunk, at a low elevation from the ground. It is a compact structure, composed entirely of green moss, oval in form, measuring about eight inches in length by about five inches in breadth, with a small entrance on the side not far from the top, and so small as scarcely to admit the tip of the finger *.

Mr. W. W. Smith has furnished me with the following note:—"The Bush-Wren nests in small holes and forks of trees and builds a very comfortable nest. When found in a hole they are generally open at the top or cup-shaped; when built in a fork they are slightly hooded. The eggs, which are usually five or six in number, are small, roundish in shape, and white with irregular pink blotches on the thick end."

---

throughout of quite simple form, and are separated by but narrow intervals. None are modified in form to serve for the insertion of a vocal muscle, as the latter terminates higher up, as already described, on the tracheal box, and therefore quite out of the region of the bronchi.

"The lateral position of the single syringeal muscle is that characteristic of all the Mesomyodian Passeres, though in most of these it terminates on one of the bronchial rings, and not, as in the birds under consideration, on the sides of the trachea. This may easily be seen by comparing the accompanying figures of *Xenicus* with the beautiful series given by Johannes Müller of the

Syrinx of *Xenicus longipes*, much enlarged. A. From in front. B. From behind. m. Lateral tracheal muscle.

syrinx of many of the Neotropical Mesomyodi, with those of Garrod of *Pitta*, or my own of *Eurylæmus*, *Cymbirhynchus*, and *Philepitta*. In fact it resembles rather that of *Todus*, as lately described and figured by myself. Externally the non-oscine nature of *Xenicus* and *Acanthisitta* is at once proclaimed by the structure of their wings, which have a 'first' (tenth) primary nearly as long as the preceding one, and by the non-bilaminate tarsus. The latter is covered almost completely by a single large scute, with only some very obsolete traces of transverse division below, whilst behind its edges are contiguous for the greater length of the tarsus, leaving only small areas at each end of that bone, which are covered by very small scutellæ of irregular form. The digits are slender and compressed, the foot being slightly syndactyle by the union of the fourth toe to the third for the greater part of its two most basal joints. The tail is short and weak; and there are only ten rectrices in each of my specimens. [This is the normal number.] In all other points, *Xenicus* and *Acanthisitta* conform to the general Passerine type. There is no trace of a plantar *vinculum*. The *tensor patagii brevis* has the peculiar arrangement characterizing the Passeres, only slightly masked by the muscular fibres somewhat concealing the two superimposed tendons, as is frequently the case in the short-and-rounded-winged forms of the group. The *gluteus primus* is well-developed. The tongue is lanceolate and horny, with its apex somewhat frayed out and its base spiny. The main artery of the leg is the sciatic. The sternum has a single pair of posterior notches and a bifid manubrium. In the skull the nostrils are holorhinal, the vomer broad and deeply emarginate anteriorly, the maxillo-palatines slender and recurved." (Proc. Zool. Soc. 1882, pp. 569–571.)

* One which was found in a birch-forest far up the Havelock, in the month of December, was so admirably hidden amongst the surrounding moss that its detection was quite accidental. It was situated beneath the moss-covered roots of a ribbon-wood tree, and was pouch-shaped in form, with the opening near the top, and composed almost entirely of minute fern-roots, carefully interwoven, especially at the entrance. It measured about 3·5 inches in height by 3 in breadth; and the cavity, which was profusely lined with feathers, extended to a depth of 2·5 inches, with an opening one inch and a half across. ('Journal of Science,' vol. ii. p. 281.)

XENICUS GILVIVENTRIS.

(ROCK-WREN.)

Xenicus gilviventris, Von Pelz. Verh. k.-k. zool.-bot. Ges. Wien, 1867, p. 316.
Xenicus haasti, Buller, Ibis, 1869, p. 37.
Acanthisitta gilviventris, Gray, Hand-l. of B. i. p. 183 (1869).
Acanthisitta haastii, id. tom. cit. p. 183 (1869).

♂ staturâ X. longipedis, sed halluxis ungue maximo distinguendus: suprà pallidè viridis, pileo et dorso superiore brunnescentioribus concoloribus: superciliis albo, loral flavo tincto: subtùs dilutè chocolatino-brunneus, crisso cum cruribus viridescentibus, hypochondriis lætè flavis: subalaribus pallidè flavis.

♀ femina X. longipedis dissimilis et hujus mari magis assimilata: suprà ochrascenti-brunnea, uropygio vix viridescente: tectricibus alarum conspicuè nigris: remigibus brunneis, extùs dorsi colore lavatis: subtùs pallidè isabellina, hypochondriis vinafescentibus.

Adult male. Upper parts dull olive-brown, with a greyish gloss, darker on the forehead and crown, and tinged on the back, wing-coverts, and rump with yellowish green; sides of the head dark brown, with a narrow superciliary streak of fulvous white, widening above the ears; underparts delicate purplish brown, with a silky appearance, and fading into fulvous white at the base of the lower mandible; the sides of the body lemon-yellow; wing-feathers brown, the primaries margined on their outer webs with dull olive, the secondaries with an apical spot of fulvous on their outer webs; tertials and the lesser wing-coverts black, forming a conspicuous triangular spot; inner lining of wings pale yellow; tail-feathers dull olive. Irides and bill blackish brown; tarsi and toes pale brown; claws darker. Total length 3·75 inches; wing, from flexure, 2·1; tail ·75 (nearly two thirds of it being concealed by the coverts); bill, along the ridge ·4, along the edge of lower mandible ·5; tarsus 1; middle toe and claw ·9; hind toe and claw ·9.

Adult female. Differs from the male in having the plumage of the upper parts dull yellowish brown, shaded with umber on the crown, and tinged with yellowish olive on the wings and rump; the superciliary streak less distinct; and the underparts pale fulvous, stained on the sides of the body with lemon-yellow.

Obs. It will be necessary to obtain a larger series of specimens than is at present available, and to make a closer investigation of the subject, before the differences supposed to be characteristic of the sexes (both of this and the preceding species) can be considered finally determined. It is probable that the colours undergo some change in the progress of the bird towards maturity; and there is likewise reason to suspect that a seasonal change takes place in the plumage of the male.

MY first specimens of this bird were received from Dr. (now Sir Julius von) Haast, F.R.S., who discovered it in the Southern Alps, during a topographical survey of the Canterbury Province. In a notice which I communicated to 'The Ibis' (*l. c.*), I described the species as new, and named it *Xenicus haasti*, in compliment to the discoverer. This name, however, must, in obedience to the inflexible law of priority, give place to *Xenicus gilviventris*, a description of the species under that title having previously been published by Von Pelzeln, although it had not then reached the colony. Nevertheless I am glad to be able to quote Haast's account of the bird's habits as communicated to me at the time:—"It lives exclusively amongst the large taluses of débris high on the mountain-sides. Instead of flying away when frightened or when stones are thrown at it, or even when shot at, it hides itself among the angular débris of which these large taluses are composed. We tried several times in vain to catch one alive by surrounding it and removing these blocks. It reminded me strongly of the habits and movements of the lizards which live in the same regions and in similar localities."

Another correspondent says that "they move about so nimbly that to procure specimens was like shooting at mice."

This species is confined to the South Island, being met with in the mountains, at an elevation of 3000 feet and upwards, their range appearing to commence with the snow-line, below which I have never heard of their being found.

Mr. Brough, who sent me a specimen from Nelson, says it was one of five which he met with in February on the very summit of a barren mountain. "They were dodging about among the angular rocks right on the top of the peaks, where there was no vegetation except the so-called 'vegetable sheep' (*Raoulia eximia*), which grows freely enough among the débris or shale." These birds were, at that time, catching a bright-coloured alpine butterfly, which I have since identified as *Phaos huttoni*.

Mr. Reischek writes that he found it very plentiful on the top of Mount Alexander, near Lake Brunner, also on Mount Alcidus, near Rakaia forks, "hopping about among the débris grown over with alpine vegetation." On the heights overlooking Dusky Sound, he found it extremely rare, a circumstance which he attributes to the thousands of rats infesting that region.

Sir J. Hector found it frequenting the stunted vegetation growing on the mountain-sides in the Otago Province; and Mr. John Buchanan, the artist attached to the Geological Survey, met with it on the Black Peak, at an elevation of 8000 feet. There, where the vegetation is reduced to a height of only a few inches, it was constantly to be seen, fluttering over the loose rocks or upon the ground in its assiduous search for minute insects and their larvæ.

It is worthy of remark that in this species the claw of the hind toe is considerably more developed than in the tree-frequenting *X. longipes*—even exceeding the toe in length—a modification of structure specially adapted to the peculiar habits of the bird, which differ from those of the former species consistently with its habitat. They hunt much on the ground, particularly in wet weather; and will freely visit the explorer's camp, hopping about on the ground, picking at mutton-fat or anything of the kind lying outside. The young are fed on insects; and it is amusing to see the old birds coming to the nest, sometimes with a dragon-fly almost as large as its captor, and dividing it among the brood.

Mr. W. W. Smith informs me that on one occasion he collected twelve of these birds, and that the stomachs of all of them contained the minute chrysalids commonly found among fallen leaves and other decaying vegetation.

The nest, which is a more finished construction than that of the Bush-Wren, is placed in a sheltered crevice among the loose rocks or débris of the mountain. One found under these circumstances by Mr. Brough in the Nelson district, on Sept. 24, contained five eggs. This nest, which is now in my collection, is of a rounded form, laterally compressed, and measuring five inches in its widest diameter. It is composed externally of wiry rootlets, intermixed with very small twigs and dry leaves. The entrance is on the side, being a perfectly round aperture about an inch in diameter. The interior of the nest is lined with soft feathers.

The egg (of which I have a single damaged specimen) is ovoido-elliptical in form, measuring ·7 of an inch in length by ·5 in breadth, and is perfectly white, with a slightly polished surface.

ACANTHIDOSITTA* CHLORIS.

(RIFLEMAN.)

Citrine Warbler, Lath. Gen. Syn. ii. pt. 2, p. 464 (1783).
Sitta chloris, Sparrm. Mus. Carls. pl. 33 (1787).
Motacilla citrina, Gm. Syst. Nat. i. p. 979 (1788, ex Lath.).
Sylvia citrina, Lath. Ind. Orn. ii. p. 529 (1790).
Sitta punctata, Quoy et Gaim. Voy. de l'Astrol. i. p. 221, pl. 18. fig. 1 (1830).
Acanthiza tenuirostris, Lafr. Rev. Zool. 1841, p. 242.
Acanthisitta tenuirostris, Lafr. Mag. de Zool. 1842, pl. 27.
Motacilla citrinella, Forst. Descr. Anim. p. 89 (1844).
Acanthisitta tenuirostris, Ellman, Zoologist, 1861, p. 7466.
Acanthisitta punctata, Ellman, tom. cit. p. 7466.

Native names.

Tititipounamu, Kikimutu, Kikirimutu, Pihipihi, Piripiri, Tokepiripiri, and Moutuutu.

♂ *ad.* suprà viridis, uropygio lætiore, pileo brunneo lavato: tectricibus alarum nigris, extùs viridi lavatis: alâ spuriâ nigrâ, extùs albicante: remigibus nigricanti-brunneis, extùs viridi (ad basin pennarum lætiore) limbatis, secundariis dorsalibus pogonio externo albo conspicuè maculatis: caudâ nigrâ, ad apicem albo viridi lavato maculatâ: loris, supercilio et facie laterali albidis, strigâ per oculum eunte fuscâ: subtùs albus, vix fulvo tinctus, corporis lateribus flavo lavatis: rostro saturatè brunneo: pedibus pallidè brunneis: iride saturatè brunneâ.

♀ mari omninò similis, sed saturatior: pileo magis brunnescente.

Juv. suprà cinerascenti-brunneus, plumis utrinque nigro marginatis, uropygio olivascente: alâ ut in adultis coloratâ, sed extùs ad basin secundariarum conspicuè flavâ: facie laterali cinerascente, nigricante variâ: subtùs albescens, hypochondriis flavicantibus, gutture et pectore superiore maculis triquetris nigricantibus notatis.

Male. Upper parts dull green, tinged with yellow on the wings and rump; throat, breast, and underparts generally fulvous white, with a tinge of yellow on the sides of the body and abdomen; a streak over and beyond the eyes and a lower-eyelid fringe of fulvous white; wing-feathers black, edged on their outer webs with green, and crossed with a band of dull yellow immediately below the coverts, which are black; the first tertial white on its outer web; tail-feathers black, tipped with fuscous. Irides and bill dark brown; legs and feet paler brown, changing to yellow on the under surface of the toes. Total length 3 inches; extent of wings 5·25; wing, from flexure, 1·5; tail ·25; bill, along the ridge ·4, along the edge of lower mandible ·53; tarsus ·75; middle toe and claw ·6; hind toe and claw ·55.

Female. Crown, hind neck, and upper part of back olivaceous yellow, each feather margined with brown; lower part of back and rump olivaceous yellow, tinged with green; tail-coverts dull green; underparts buffy

* This has hitherto been written *Acanthisitta*; but Professor Newton has drawn my attention to the fact of its being erroneous. I have therefore adopted the more classic form of *Acanthidositta*, the etymology of which is ἀκανθίς (ἀκανθιδ-), a crude form of *Acanthis* = *Carduelis*, and σίττα = sitta.

white, washed on the sides with yellow; wing-feathers dusky, margined on the outer web and marked at the base with olivaceous yellow; superior wing-coverts black; outer tertials margined with white; innermost secondary with an oblong spot of yellowish white on the outer vane; tail-feathers black, tipped with fulvous.

Young. Plumage generally duller and suffused with yellowish brown; marked on the breast with numerous small longitudinal spots of brown.

Obs. As will be seen above, the plumage of *A. chloris* differs in the male, female, and young. Examples vary in the tone of their colouring; and a specimen in my collection (received from the South Island) has the rump and upper tail-coverts almost orange-coloured, without any mixture of green. I do not believe in the existence of *Acanthidositta citrina*, Gmelin, although recognized as a distinct species by Dr. Finsch *.

THE Rifleman is the smallest of our New-Zealand birds. It is very generally distributed over the middle and southern portions of the North Island, in all suitable localities, and throughout the whole extent of the South Island. It is to be met with generally on the sides and summits of the wooded ranges, seldom or never in the low gullies. Professor Hutton found it on the Great Barrier, and was assured by the native residents of that island that it was a migratory bird, coming and going with the Cuckoo! Mr. Reischek met with it also on the Little Barrier, but not on the other islands in the Hauraki Gulf.

In the hilly pine-forests at the head of the Wairarapa valley I found this bird comparatively plentiful in the summer of 1883. This was the more noticeable on account of the general absence of bird-life in these dark woods at all seasons of the year. On the outskirts small flocks of *Zosterops* consort together in the underwood, and a few Flycatchers and White-heads share the solitude with the sober Tomtit; but as we enter the woods the stillness becomes oppressive, unbroken even by the chirp of a cricket or the drumming of a locust, and, apart from the active little Rifleman, which seems perfectly at home under all conditions, the only sign of animation is an occasional night-moth lazily flapping its wings in the gloomy shade of the forest.

In its habits it is lively and active, being incessantly on the move, uttering a low feeble *cheep* (like the cry of a young bird), accompanied by a constant quivering of the wings. I have noticed that this cry becomes louder and more continuous towards evening. It is generally to be seen running up the boles of the larger trees, often ascending spirally, prying into every chink and crevice, and moving about with such celerity that it is rather difficult for the collector to obtain a shot. Its powers of flight are very feeble, and it simply uses its wings for short passages from one tree to another. Its tail is extremely short, and is hardly visible when the bird is in motion.

The stomachs of all that I have opened contained numerous remains of minute Coleoptera and other insects, sometimes mixed with finely comminuted vegetable matter.

It is naturally a shy bird, but of so excitable a nature that, during the breeding-season, it may be decoyed into the open hand by rapidly twirling a leaf, so as to simulate the fluttering of a bird, accompanied by an imitation of its simple note.

A bird-collector at Wellington showed me a brood of three young ones which he had taken from a nest in the cavity of a hinau, at an elevation of 20 feet or more from the ground. Finding the aperture too small to admit his hand, he cut into the tree about a foot below it, and thus disclosed the nest, which he described as being composed entirely of fern-hair, about 10 inches in length, and bottle-shaped, with a long vertical tube forming the entrance to it. In the Canterbury Museum there is a nest of this species, which appears to have been torn out of some natural cavity. It is pear-shaped, with the entrance on the side and near the bottom, and is very loosely constructed, the

* Trans. N.-Z. Inst. 1874, vol. vii. p. 227.

materials composing it being the skeletons of decayed leaves, the wiry stems of plants, rootlets, and a few feathers.

Captain Mair discovered a nest under the thatched eaves of a Maori hut; and Mr. E. Pharazyn sent me an egg taken from another nest found concealed among the dry roots of a fallen tree. Mr. Potts has found the nest "very cleverly built, in a roll of bark that hung suspended in a thicket of climbing convolvulus," and, at another time, in a small hole in the trunk of a black birch. More than once he has known the bird to occupy the mortice-hole of a stock-yard post; also to utilize the skull of a horse, and to build between the slabs of a bush hut, adapting the form of its nest to the immediate surroundings.

The Rifleman has been found breeding as early in the year as the month of August; and in a specimen which I killed in the Ruahine ranges on the 23rd of December the ovary contained an undeveloped egg of the size of buck-shot, while the bareness of the underparts bore indication that the bird had already been sitting. From these facts we may, I think, reasonably infer that this species produces two broods in the season. The companion male bird on this occasion also had the abdomen bare, thus affording presumptive evidence that the sexes share the labour of incubation. The eggs vary in number from three to five; they are very fragile, broadly ovoid, or inclined to a spherical form, measuring ·6 of an inch in length by ·5 in breadth, and perfectly white, with a slightly glossy surface.

Before leaving the great Order of Passeres and passing on to the next, the Picariæ, it may be useful to note that most of the Passerine genera found in New Zealand are strictly endemic or peculiar to the country. Without of course taking into account the undoubted stragglers from abroad, the only exceptions to this rule are *Sphenœacus*, which occurs also in Australia; *Gerygone* and *Rhipidura*, of which there are representatives in Australia, Tasmania, Norfolk Island, New Guinea, and many of the Indo-Malayan Islands; *Zosterops*, whose range extends over the entire southern hemisphere; and *Anthus*, which occurs in most parts of the world.

I have already explained in my account of *Xenicus* why it became necessary to remove that form and *Acanthidositta* from their old position among the Certhiidæ and to place them in a new family at the end of the Passeres. Both these forms are, in fact, dwarf Pittas of a degenerate type. They have no relations in New Zealand, and their nearest allies in Australia are the true Pittidæ, a highly specialized group extending to New Guinea and, through the entire Malay Archipelago, to India and China. One species occurs in West Africa; but in all the other zoological regions of the earth, so far as we at present know, this type is absent.

CYPSELUS PACIFICUS.

(AUSTRALIAN SWIFT.)

Hirundo pacifica, Lath. Ind. Orn. Suppl. ii. p. lviii (1801).
New-Holland Swallow, Lath. Syn. Suppl. ii. p. 259 (1801).
Cypselus pacificus, Steph. Cont. of Shaw's Gen. Zool. vol. x. p. 132 (1817).
Hirundo apus, *var.* β, Pall. Zool. Ross.-Asiat. tom. i. p. 540 (1831).
Cypselus australis, Gould in Proc. Zool. Soc. part vii. 1839, p. 141.
Cypselus vittata, Jard. Ill. Orn. ser. 2, pl. 39 (1840).
Micropus australis, Boie, Isis, 1844, p. 165.
Micropus vittata, id. tom. cit.
Cypselus australis, Gould, Birds of Austr. fol. vol. ii. pl. xi. (1848).
Cypselus pacificus, Gould, Handb. Birds of Austr. vol. i. p. 105 (1865).

Descr. exempl. ex N. Z. Suprà nigricanti-brunneus: dorso metallicè nitente: uropygio albo: subtùs intensè fusco: gutture cinerascenti-albo: plumis pectoris abdominisque angustè albo marginatis: remigibus caudâque nigricantibus: rostro nigro: pedibus nigris.

New-Zealand example. Crown of the head and general upper surface blackish brown, with a metallic lustre on the back and upper surface of wings and tail; rump pure white; throat and upper part of fore neck greyish white; the rest of the under surface blackish brown, but paler on the lower fore neck and under tail-coverts; the feathers of the breast and abdomen narrowly tipped with white; quills and tail-feathers brownish black, the shafts greyish towards the base on their under aspect; the inferior primaries, the whole of the secondaries, and the inner lining of the wings minutely margined with greyish white. Bill and feet black. Total length 7·75 inches; wing, from flexure, 7·2; tail 3; bill, along the ridge ·3, along the edge of lower mandible ·8; tarsus ·5; middle toe and claw ·55.

Young (Australian specimen in British Museum). Has the plumage of the head, shoulders, and back very narrowly margined with paler brown; in front of each eye an angular spot of black and above that a line of greyish white; throat greyish white, with indistinct shaft-lines of brown; the plumage of the underparts conspicuously marked in crescents, each feather becoming black in its apical portion and then broadly tipped with greyish white; the lining of wings uniform dark brown; the whole of the rump white with fine black shaft-lines; under tail-coverts broadly tipped with white.

Obs. The only sexual difference is that the female has somewhat duller plumage than the male. The amount of white on the throat is very variable, being reduced in some specimens to a mere wash of fulvous-white. The extent of white also on the uropygium varies much in individual examples, sometimes spreading down to the thighs.

ONE of the most recent cases, and perhaps the most interesting, of the occurrence of common Australian forms in New Zealand is that of the Swift, which made its appearance for the first time, so far as we know, in the history of the colony, in December 1884.

On seeing the newspaper accounts of the flight which had visited the White Cliffs (near the town of New Plymouth) I naturally concluded that this was another instance of the Tree-Swallow visitant from Australia, with which we had already become familiar.

Fortunately, however, one of the birds had been shot, and the skin having been forwarded to me in a fresh condition, I saw at a glance that we had now to add another bird to our list of species.

Major W. B. Messenger, to whom I am indebted for this unique specimen, sent me also the following notes:—

"Respecting the Swift I shot here, I am glad to be able to furnish you with particulars. One evening, at about 6 P.M., four strange birds were flying about the camp, evidently in pursuit of insects. Their flight so reminded me of that of the Swift, that to make sure I shot one and took it to the office of the 'Taranaki Herald.' I believed it to be an English Swift, but from what I have since heard, I conclude that it is an Australian bird. I did not know until I received your letter that Swallows had ever been seen in New Zealand."

In bringing this Australian bird before the Zoological Society in October 1839, Mr. Gould wrote:—"This species is about the size of *Cypselus murarius*. I first met with it on the 8th March, 1839. They were in considerable abundance, but flying very high. I succeeded in killing one, which was immediately pronounced by Mr. Coxen to be new to the colony. On the 22nd I again saw a number of these birds hawking over a piece of cleared land at Yarrondi, on the Upper Hunter; upon this occasion I obtained six specimens, but have not met with it since."

In his account of the species in his 'Birds of Australia' he adds:—"Those I then observed were flying high in the air, and performing immense sweeps and circles, while engaged in the capture of insects. I succeeded in killing six or eight individuals, among which were adult examples of both sexes; but I was unable to obtain any particulars as to their habits and economy. It would be highly interesting to know whether this bird, like the Swallow, returns annually to spend the months of summer in Australia. I think it likely that this may be the case, and that it may have been frequently confounded with *Acanthylis caudacuta*, as I have more than once seen the two species united in flocks, hawking together in the cloudless skies, like the Martins and Swallows of England. It is considered by some ornithologists that this bird and the Swift with crescentic markings of white on the breast, which inhabits China and Amoorland, are the same. If this supposition be correct, this species ranges very widely over the surface of the globe."

The British-Museum collection contains specimens from N. S. Wales, Queensland, Tasmania, Cape York, Formosa, Penang, Tenasserim, Assam, Japan, China, and Siberia.

I have carefully examined all these examples and can find nothing whatever to justify specific separation, although as a rule the birds from India and China have a larger and therefore more conspicuous patch of white on the throat.

The specimen from Japan differs from typical examples in having black shaft-lines on the throat; but there is an exactly similar one from Cape York obtained during the voyage of H.M.S. 'Rattlesnake.'

There is, however, a bird in immature plumage from the Hume Collection, marked "*Cypselus pacificus*, ♂, Bankasoon," which may prove to be distinct. It is of appreciably smaller size, the wing from the flexure being fully half an inch shorter; the throat-patch is covered with linear brown markings, and the whole of the uropygium is greyish white with dark shaft-lines.

EURYSTOMUS PACIFICUS.

(AUSTRALIAN ROLLER.)

Coracias pacifica, Lath. Ind. Orn. Suppl. ii. p. xxvii (1801).
Pacific Roller, Lath. Gen. Syn. Suppl. ii. p. 371 (1801).
Galgulus pacificus, Vieill. Encycl. Méth. tom. ii. p. 870 (1823).
Eurystomus australis, Swains. Anim. in Menag. p. 326 (1838).
Eurystomus pacificus, G. R. Gray, Ann. Nat. Hist. 1843, vol. xi. p. 190.
Eurystomus australis, Gould, Birds of Austr. fol. vol. ii. pl. 17 (1848).
Eurystomus pacificus, Id. Handb. Birds of Austr. vol. i. p. 119 (1865).
Eurystomus pacificus, Buller, Manual of Birds of N. Z. p. 7 (1882).

Descr. exempl. ex N. Z. Suprà pallidè viridi-griseus: tectricibus alarum minoribus dorso concoloribus, medianis et majoribus lætiùs cyaneo-viridibus: alâ spuriâ, tectricibus primariorum remigibusque nigris, extùs purpureo-cæruleis, cyaneo marginatis, primariis basin versus plagâ magnâ pallidè argentescenti-cyaneâ notatis: secundariis intimis viridioribus: caudâ nigrâ, basaliter viridi, medialiter purpurascenti-cyaneâ: pileo toto brunneo, versus interscapulium brunnescente pallidè viridi-cyaneo lavato: loris nigricantibus: regione paroticâ brunneâ: gulâ cyanescente, plumis medialiter argentescenti-cyaneo angustè lineatis: mento et gulæ lateribus brunneis; corpore reliquo subtùs pallidè viridi-cyaneo, pectore summo obscuriore: subcaudalibus et subalaribus pallidè viridi-cyaneis: alis subtùs purpureis, versus apicem nigris, plagâ basali magnâ argentescenti-cyaneâ.

New-Zealand specimen. Head and hind neck dark brown; shoulders, back, and scapulars dull brownish green, becoming brighter on the rump and upper tail-coverts and tinged with blue in certain lights; upper surface of wings greenish blue, brighter on the large primary-coverts; lores black; throat dark purplish blue, each feather with a central streak of lighter blue; under surface generally dull greenish blue, paler on the lining of the wings and under tail-coverts, and suffused with brown on the breast and sides of the body; quills black on their upper surface, the first primary margined externally with indigo-blue and having on its inner web towards the base a broad bar of pale silvery blue, which increases on the four succeeding primaries and occupies both webs, forming in the open wing a conspicuous rounded patch, the shafts brown, and the outer webs beyond indigo-blue shading into black; the secondaries and tertials bright indigo on their outer webs, changing in the former to bluish green towards the base; tail-feathers on their upper surface bluish green at the base, changing to bright indigo in their central portion and becoming entirely black beyond; under surface of wing- and tail-feathers with a bright blue lustre, shot with green and purple in certain lights. Irides dark brown; bill orange-red, shading into black at the tip of upper mandible, and becoming yellow towards the base of under mandible; legs and feet pale reddish brown. Total length 10·5 inches; wing, from flexure, 7·75; tail 4; bill, along the ridge 1, along the edge of lower mandible 1·25; tarsus ·6; middle toe and claw 1·25.

Young (*Australian specimen in British Museum*). General plumage dull brownish grey, paler on the underparts; crown of the head, hind neck, and shoulders dark brown; lores black; the whole of the fore neck dull brown, faintly washed on the throat with metallic green; the whole of the wing-coverts pale bluish green; primaries and secondaries black, the former in their basal portion and the latter in their whole extent broadly edged on their outer webs with dark blue, fading into green; the primaries crossed in their middle portion by a band of pale silvery blue, fading into white on their inner webs; tail-feathers green towards the base, blue in the middle, and black in their apical portion. Bill brownish black; legs and feet yellowish brown.

Obs. "The sexes are alike in plumage. Irides dark brown; eyelash, bill, and feet red; inside of the mouth yellow." (*Gould.*)

The first occurrence of this bird in New Zealand was recorded by Mr. F. E. Clarke in a communication to the Westland Institute, on the 18th February, 1881 *, the author regarding it as the representative of a new genus which he characterized as *Hirundolanius*. His description of the form left no doubt on my mind that the bird was the common Australian Roller, and the subsequent receipt of the specimen itself at the Colonial Museum confirmed that view. Another example (now in my collection) was shortly afterwards obtained at Parihaka, a few miles from New Plymouth. There is a specimen in the Auckland Museum † shot by Mr. Cowan in a patch of bush near the sea, at Piha Bay, about ten miles north of Manukau harbour, towards the end of 1881; Mr. Tryon has in his possession the skin of another obtained near the Waiwakaio river, in the district of Taranaki, about a month later; and the Canterbury Museum contains a specimen received about the same period from Westland ‡.

Thus the bird has occurred almost simultaneously at no less than four places, far apart from one another, but all on the west coast; and, although of course only a visitant from Australia, the species has fully established its right to a place in the New-Zealand avifauna.

Mr. Caley, writing of this Roller in New South Wales, says §:—" It is a bird of passage. The earliest period of the year at which I have noticed it was on the 3rd of October, 1809; and I have missed it early in February. It is most plentiful about Christmas."

Mr. Gould gives us the following account of its habits:—

"In Australia the Roller would appear to be a very local species, for I have never seen it from any other part of the country than New South Wales: but the late Mr. Elsey informed me that he found it very common in the Victoria basin, and that it became very numerous about the head of the Lynd. It arrives early in spring, and, having brought forth its progeny, retires northwards on the approach of winter. It appeared to be most active about sunrise and sunset; in sultry weather it was generally perched upon some dead branch in a state of quietude. It is a very bold bird at all times, but particularly so during the breeding-season, when it attacks with the utmost fury any intruder that may venture to approach the hole in the tree in which its eggs are deposited.

"When intent upon the capture of insects it usually perches upon the dead upright branch of a tree growing beside and overhanging water, where it sits very erect, until a passing insect attracts its notice, when it suddenly darts off, secures its victim, and returns to the same branch; at other times it may constantly be seen on the wing, mostly in pairs, flying just above the tops of the trees, diving and rising again with many rapid turns. During flight the silvery-white spot on the centre of each wing shows very distinctly, and hence the name of Dollar Bird bestowed upon it by the Colonists.

"It is a very noisy bird, particularly in dull weather, when it often emits its peculiar chattering note during flight.

"It is said to take the young Parrots from their holes and kill them, but this I never witnessed; the stomachs of the many I dissected contained the remains of Coleoptera only.

"The breeding-season lasts from September to December; and the eggs, which are three and sometimes four in number, are deposited in the hole of a tree, without any nest; they are of a

* Trans. N.-Z. Instit. vol. xiii. p. 454.

† *Cf.* Cheeseman, *op. cit.* vol. xiv. pp. 264–265.

‡ "In addition to Mr. F. E. Clarke's note, it may be mentioned that an old Australian, then living at Okarito, was certain that he had seen the bird in the Queen Charlotte's Sound district." (Journ. of Science, ii. p. 275.)

§ Trans. Linn. Soc. vol. xv. p. 262.

beautiful pearly white, considerably pointed at the smaller end; their medium length is one inch and five lines, and breadth one inch and two lines."

Sir T. M. Mitchell states that on dissecting a specimen obtained in N.E. Australia he found the stomach crammed with wasps and coleopterous insects.

Dr. Ramsay writes*:—"I found this bird nesting in the hollow Eucalyptus boughs on the Richmond river in 1867. They make no nest, but lay their eggs on the dust formed by decayed wood; not unfrequently they fight with and dispossess the *Dacelo gigas*, and I have seen them take the young of this bird and throw them out of the nest. The eggs are two or three in number, of a dull white, rather glossy, and sometimes variable in form, some being oval and pointed, others almost round."

This species occurs on Lord Howe's Island, where there is a perceptible blending of Australian and New-Zealand forms. It also abounds in some parts of New Guinea; and Mr. Macleay is of opinion that the birds which spend the summer in Australia pass on to the southern coast of New Guinea for the winter.

I have examined a large series of specimens collected by Mr. A. R. Wallace and others in the Malay Archipelago and now forming part of the magnificent collection of cabinet skins in the British Museum. There are specimens from Penang, Malacca, Labuan, Celebes, the Sula Islands, Timor, Lombock, Flores, Gilolo, Matabello Island, Bouru, Ceram, and Dorey. As a rule these differ from Australian birds in having the patch of blue on the throat of a brighter colour. The brightest of them all is an adult example from the Sula Islands. Another specimen from the same locality (evidently a young bird) differs in having all the colours much duller, and instead of the throat-patch of purplish blue a mere wash of the prevailing greenish colour, with just the faintest tinge of indigo on the sides of the throat. In this bird the bill is reddish yellow with a brownish ridge and tip.

Another specimen of the young (obtained by Mr. Wallace from Matabello Island) differs in having the general plumage lighter, with a small patch of purple mixed with grey on the throat, each feather having a central streak of cobalt-blue. In this example the feathers of the head and shoulders are narrowly, and those of the breast broadly, margined with grey; bill blackish brown, the outer edge of lower mandible dull yellow.

Eurystomus orientalis, Vig. & Horsf., although generally regarded as merely a local race, appears to me to be a good species, readily distinguishable from *E. pacificus* by its brighter plumage and decidedly darker head and neck; but after carefully comparing Mr. Wallace's Javan specimen of *E. cyanicollis*, Wagl., with the former, I can find nothing to justify their separation.

Eurystomus crassirostris, Sclater, is appreciably larger and brighter than *E. orientalis*, although the colouring is the same, and its bill is conspicuously broader and more robust. This species, which comes from the Solomon group and New Guinea, is said to have dark red eyes.

Eurystomus azureus, G. R. Gray (brought by Wallace from Batchian), is a very distinct species, remarkable for its large size and uniform dark blue plumage. The young of this Roller has similar plumage to the adult, but the blue is of a duller hue, and the bill is blackish brown instead of being orange.

* Proc. Linn. Soc. N. S. W. vol. vii. p. 46.

HALCYON VAGANS.

(NEW-ZEALAND KINGFISHER.)

Alcedo sacra, var. D, Lath. Gen. Syn. Suppl. p. 114 (1790).
Alcedo sacra, var. ε, Bonn. et Vieill. Enc. Méth. i. p. 295 (1823).
Alcedo vagans, Less. Voy. Coq. i. p. 694 (1826).
Alcedo chlorocephala, var. γ, Less. Traité d'Orn. p. 246 (1831).
Halcyon vagans, Gray, Voy. Ereb. & Terror, p. 3, pl. 1 (1844).
Alcedo cyanea, Forst. Descr. Anim. p. 76 (1844).
Todiramphus vagans, Bonap. Consp. Gen. Av. i. p. 157 (1850).
Dacelo sancta (pt.), Schl. Cat. Mus. Pays-Bas, Alced. p. 37 (1863).
Halcyon sanctus, Finsch, J. f. O. 1870, p. 246; Hutton, Cat. Birds N. Z. p. 3 (1871).

Native names.

Kotare and Kotaretare; "Kingfisher" of the colonists.

Ad. suprà sordidè viridis, pileo laterali et dorso postico uropygioque cyanescentibus: loris et supercilio antico fulvis: genis, cum regione paroticâ utrâque circà collum posticum conjunctâ, nigris vix viridi tinctis: maculâ nuchali et collo toto albidis, torquem collarem latam formantibus: tectricibus alarum cyanescenti-viridibus: remigibus nigricantibus, primariis ad basin et secundariis extùs lætè cyanescentibus: caudâ suprà cyanescente, subtùs griseâ: corpore subtùs toto lætè fulvescente, gutture albicante: rostro nigro, ad basin mandibulæ albo: pedibus saturatè brunneis: iride nigricanti-brunneâ.

Juv. similis adultis, sed sordidior: tectricibus alarum fulvo marginatis: pectoris et colli postici plumis brunneo marginatis.

Adult male. Crown, shoulders, and scapulars deep sea-green, with an olive tinge; back, tail-coverts, and upper surface of wings ultramarine, changing to green in certain lights; quills and tail-feathers washed with cobalt on their outer webs. A spot of bright fulvous fills the lores, a dash of ultramarine blue, bordered above the eyes and on the occiput with white, surrounds the crown; and a broad band of black, proceeding from the angles of the mouth, completely encircles the hind head. Throat, breast, and a broad nuchal collar buffy white; the rest of the under surface delicate fawn-colour, with deepening tints. Irides black; bill black, with the basal portion of the lower mandible white, feet dark brown, with paler soles. Extreme length 9·75 inches; extent of wings 13·6; wing, from flexure, 4; tail 2·6; bill, along the ridge 1·75, along the edge of lower mandible 2·1; tarsus ·6; middle toe and claw 1·05; hind toe and claw ·6.

Female. Tints of the plumage generally duller.

Young. In the young bird the throat is pure white; the underparts fulvous-white, tinged on the sides with fawn-colour; feathers of the breast broadly margined with dusky brown, forming an irregular pectoral zone; loral spots and nuchal collar rufous, with markings of the same colour on the fore part of the crown; nuchal collar indistinct and largely marked with brown; plumage of the upper parts darker than in the adult; the wing-coverts margined with yellow, in the form of narrow crescentic bands.

Progress towards maturity. Tints of the plumage brighter; the loral spots bright fulvous; the sides, flanks, lining of wings, and under tail-coverts bright fawn-colour; pectoral zone indistinct, the dark margins being very narrow; nuchal collar well defined and almost pure white. The full adult dress is not attained till after the second or third moult.

Obs. Mr. Reischek brought from the Little Barrier a brightly coloured specimen, which comes very near to *Halcyon sancta*. He saw a pair of them together on the south-west side of the island. They appeared to be exactly alike in plumage, and the one he shot proved to be a female. In this specimen the nuchal collar is half an inch wide, quite regular, creamy white, and margined on both sides with black; throat white; underparts and flanks plain fawn-colour; hind head, wings, and rump very bright blue; mantle largely tinged with verditer-green. Extreme length 9·5 inches; wing, from flexure, 4; tail 3; bill, along the ridge 1·5, along the edge of lower mandible 2; tarsus ·3; middle toe and claw 1. The bill differs from that of ordinary examples, the lower mandible having a more upward curve, and the upper, viewed vertically, being much compressed laterally, especially towards the point. Although I have thought it right to record these differences, I do not propose at present, at any rate, to separate this bird from *H. vagans*, which as a species is not very far removed from the Australian form.

Much difference of opinion has existed as to whether this bird is really distinct from the *Halcyon sancta* of Australia. Mr. R. B. Sharpe, in his 'Monograph of the Kingfishers,' pronounces it a good species, being "always of a more robust size, and having the colours much less bright than the Australian bird." Professor Schlegel and Dr. Finsch proposed uniting it to *H. sancta*; but in a letter which I afterwards received from the latter of these experienced ornithologists he admits that the species is quite distinct, adding that his former conclusions were based on two specimens only, and not on the good series of skins since obtained. I have always contended for the recognition of *Halcyon vagans*; and the question may now be considered fairly set at rest.

In habits the two species are very much alike. The New-Zealand bird is very generally dispersed, being met with in all suitable localities. It frequents alike the sea-shore, the open country, forest-clearings, and the banks of freshwater streams. It is, moreover, one of those birds that seem instinctively to resort to the habitations of man, and instead of, like many other indigenous species, decreasing, it thrives and multiplies under the altered physical conditions resulting from the colonization of the country. It seeks out the new home of the settler, and becomes the familiar companion of his solitude. During the winter months especially, it resorts to cultivated grounds in quest of grubs and worms, which at this season constitute its principal food. In the early morn it may be seen perched on the fences, gateways, and out-buildings of the farmyard, sitting upright with contracted neck, looking stiff and rigid in the cold frosty air; and as the day advances, it enlivens the landscape by its darting flight, while it attracts notice by its shrill, quickly repeated call, which is not unlike the note of the European Kestrel. In the pairing-season this species becomes very noisy and lively, the mated birds chasing each other, in amorous play, from tree to tree or from post to post, with loud unmusical cries, something like the syllables *cree-cree-cree* uttered in quick succession. Its ordinary call-note is more like *chiu-chiu-chiu*, with a clear accent. When wounded or caught in a trap it utters a peculiar rasping cry, exactly like that of the Indian Mynah when alarmed or excited, only louder.

They breed late in the year; the brood numbers five or six; and for several weeks after quitting the nest the young family keep together. This will probably account for the abundance of Kingfishers in the autumn months, which has been regarded by some as indicating a seasonal migration.

The flight of this species is short, rapid, and direct, being performed by a quick vibration of the wings. It flies with considerable velocity; and I have known several instances of its dashing headlong through a pane of glass. On one occasion this occurred in the church at Raglan during divine service; and the Kingfisher, after recovering from the shock, remained to the last perched on the end of a pew, looking more devout, says our correspondent, than the Jackdaw of Rheims! Another instance occurred more recently at Wanganui, where, according to a local paper, the family of the Rev. C. H. S. Nicholls were startled one day at dinner by the entrance of a Kingfisher, which "flew through a pane of glass in one of the windows, scattering the fragments around," and was forthwith made prisoner by the household cat.

Its food consists of lizards, small fish, grubs, earthworms, locusts, insects of all kinds, and even mice. On examining a young Kingfisher just taken from the nest, I observed the tail of a half-grown mouse protruding from its bill; and on taking hold of it I drew the unmutilated carcass of the rodent from the throat of the bird. I was not previously aware that mice formed part of the Kingfisher's bill of fare *. I have often, however, witnessed its fondness for lizards, two species of which (*Mocoa zealandica* and *M. ornata*) are very common in all the open glades. I have seen it seize the nimble little reptile by the tail, and after battering its head against a stone or the branch of a tree, to destroy life, swallow the captive, head foremost. It has been known to attack and kill chickens in the poultry-yard. On one occasion, at Otaki, I saw one of these birds dart down into the midst of a very young clutch; but the old barn-door hen proved too active, and with one rapid stroke of her bill put the assailant *hors de combat*. The bird was picked up stunned with the blow, but soon after, recovering itself, escaped from the hands of its captor. In Wanganui it provoked the hostility of the Acclimatization Society by preying on the young of the House-Sparrow (*Passer domesticus*), which had been introduced at much expense; and the committee encouraged a crusade against the offenders by offering a premium for Kingfishers' heads. But in the present attitude of the public towards the ubiquitous Sparrow, which has become a nuisance, it would be scarcely prudent to repeat such an offer. According to the Report of the Auckland Acclimatization Society for 1868–69, it has proved very troublesome in destroying birds, and has even attacked and killed a Californian Quail. In Otago it has been accused of purloining the speckled trout; and in Canterbury it was found necessary to protect the newly hatched fish by stretching wire netting over the shallow artificial streams. A valued correspondent, and very careful observer, informs me that on one occasion he killed a blackfish about twelve feet long in Whangarei harbour, and dragged it ashore; and on visiting the place a few days later he observed an unusual number of Kingfishers present. On watching them, he found that they were preying on the swarms of flies attracted by the dead cetacean, darting after them with the swiftness of an arrow, and capturing them on the wing.

In light rainy weather the Kingfisher is in his element in the meadows. The moisture brings the grubs, earthworms, and other small animal life to the surface. From his post of observation on the fence he drops nimbly to the ground, swallows his captive and remounts to his perch, repeating the operation every few minutes, and for more than an hour at a time. It is evident, therefore, that this bird is of use to the agriculturist, and deserves protection rather than persecution.

When engaged in fishing, it does not plunge into the stream, like the common British Kingfisher, but dips into it lightly as it skims the surface of the water or darts downwards from its post of observation on a rock or overhanging branch †.

* "It may not be generally known that Kingfishers are excellent vermin destroyers, and on this account are well worthy of protection. Yesterday a number of gentlemen in the neighbourhood of the Park Hotel, Wellesley Street, observed a curious scene. A Kingfisher which was perched on one of the newly-planted trees was observed to make a sudden dart towards the high bank at the side of the street, and he speedily returned to his perch on the fence which protects the tree. It was then seen that the bird had a mouse, which was alive and struggling, in his beak. The attention of those present became concentrated on the movements of the bird, and they saw him repeatedly strike the mouse's head against the rail. As the latter became stunned, the bird removed its hold from the centre of the animal's back to his hind quarters and tail, and while so holding it beat the mouse to death against the rail, and flew off to devour its prey."—*New Zealand Herald*.

† As the fact of our Kingfisher being piscivorous has been challenged, it may be well to reprint here a note which I sent to the Wellington Philosophical Society in 1878 (Trans. N.Z. Inst. vol. xi. p. 369):—

"On driving round Porirua harbour on the 19th July last, I noticed an unusual number of Kingfishers perched on the

On the feeding-habits of this species, Mr. Henry C. Field of Wanganui has sent me the following interesting observations, which exhibit the Kingfisher in the new character of a frugivorous bird:—

"Knowing the interest you take in our New-Zealand birds, I have thought you might like to be informed of the following trait in the habits of the Kotare, which I think is not generally known. About a week before Christmas my children reported to me that in what they took to be a rat's hole in the pumice bank of the stream, just behind my garden, there was something which growled at them whenever they passed the hole or looked into it. On the matter being mentioned a second or third time the hole was examined, and proved to be a Kotare's nest, containing four young ones about half-fledged. The old birds, of course, manifested a strong objection to the nest being touched, flying round, screaming, and darting at us whenever we went close to it. I desired the children not to meddle with the young birds, but told them that if they sat a little way off and watched they would see the old ones catch fish, lizards, and insects, and bring them to the nest for the young ones to eat. The children were very pleased to do this, but quickly discovered that very few fish, and apparently very little animal food of any kind, was brought to the nest, and the young brood were being reared on the cherries out of our garden. I at first thought the children were mistaken, but as they assured me they saw the birds fly to the trees, and bring back the cherries in their bills, I examined the nest, and from the quantity of cherry stones that it contained saw that the youngsters' eyes had not deceived them. It was evident, in fact, that, up to the time they left the nest, fruit formed the chief food of the young birds. It has occurred to me that possibly the Kingfisher, from its habits, consumes a large quantity of fluid with its food, and that the juice of the fruit supplies moisture necessary to the proper growth of the young birds. At all events it is clear that young fruit forms an important article in their diet, though I never saw them eating it, or heard of their doing so at a later stage of their existence.

"I accidentally got corroborative information as to the frugivorous habits of the Kotare lately. I met Mr. Enderby, who mentioned that he had been greatly annoyed by these birds this autumn.

---

rocks along the beach, and on the telegraph wires stretched across the numerous little bays. They were evidently attracted by the shoals of little fish that were frequenting the shallow water at the time; and at one spot I had an ocular demonstration of my argument with Captain Hutton, which I should like him to have witnessed. Ten little Kingfishers sitting in a row were in possession of a short span of telegraph wire overhanging the water, and, one after the other, they were dipping into the shallow sea-water in pursuit of fish. Sometimes two or even three of them would dip at the same moment, raising a tiny splash all round, and then mount again to the wire or fly off to the shore with their finny prey. In further illustration of the piscivorous habits of the bird it may be mentioned that Mr. Brandon, of this city, has an indictment to file against the Kingfisher for robbing the fountain in his garden of goldfish."

I have frequently observed these birds fishing from the scaffolding under Queen Street Wharf, in Auckland Harbour, at a distance of fully two hundred yards from the shore; and my son, on one occasion, saw a Kingfisher capture a sea-minnow about four inches long and devour it.

The custodian of the trout-ponds at Hastings, near Napier, informed me, on my last visit there, that the Kingfisher had proved very destructive to the young trout, often attacking even good-sized fish in the ponds and picking out their eyes!

The following communication from Capt. Mair (*l. c.* vol. x. p. 202) bears on the same point:—

"The Kingfisher is found in all the mountain streams of the Urewera and Bay of Plenty districts. It subsists largely on small freshwater fish (mohiwai of the natives), also on flies, moths, and beetles."

Mr. Potts says, in his interesting little volume 'Out in the Open' (p. 150):—"It remains for me to state that these Kingfishers really do fish at times. We have watched with great pleasure and interest displays of their remarkable skill and activity. In the lovely island of Kawau, these birds are very numerous; and well they merit the protection extended to them for their useful labours in clearing off many of the crickets that are to be seen there in abundance. At ebb-tide we have noticed Kingfishers settled on the twisted trunks of pohutukawa trees that spread out their crooked limbs over wave-washed and shelly beaches. From such convenient perches the birds plunge boldly into the sea, often wholly immersed, sprinkling round showers of spray. They swiftly emerge, rarely failing to bear back with them to their standing places their finny spoil."

He said that scarcely a peach in the garden escaped having one or more large pieces pecked out of it, and that the birds did not meddle with the ripe fruit, but attacked it when it was just ripening and before it became soft. This seems to indicate that, as in my case, the fruit was wanted not for the consumption of the old birds themselves, but as food for their young, and that it was taken therefore before it was too soft to be carried in the bill, or not required after the fruit was ripe, because the young birds were then fledged. Mr. Enderby was quite positive that it was the Kotares and not Sparrows who were the depredators, as he saw them taking the fruit, and said he at first had a great mind to shoot them, till he noticed that they evidently carried it away to their nests."

I am not aware that the Kingfisher is ever nocturnal in its habits; but on one occasion, when travelling by coach along the banks of the Manawatu river, about 2.30 A.M., it being a cloudy night and quite dark, I heard the loud call-notes of this bird with startling distinctness. Probably it was a sleeper disturbed by the passing of the coach; although under these circumstances birds, as a rule, betake themselves off in silence to another roosting-place.

The New-Zealand Kingfisher commences to breed towards the end of November or early in December, usually selecting for its nesting-operations a tree denuded of its bark and decayed at heart, standing near the margin of the forest or in an old Maori clearing. By means of its powerful bill it cuts a round passage through the hard exterior surface, and then scoops out a deep cavity, proceeding in a horizontal direction for several inches, and then downwards to an extent of ten inches or more. The bird thus instinctively protects its chamber from the inclemencies of the weather. There is no further attempt at forming a nest, the eggs being deposited on a layer of pulverized decayed wood, the shavings and sawdust, so to speak, of the borer's operations in finishing the cavity.

The labour of boring a cavity is often greatly augmented by natural impediments. If, after drilling through the hard external surface, the bird finds the inner wood too hard for its tools, it at once abandons the spot and sounds the tree in another place. I have counted half a dozen or more of these abortive borings on a single tree, in addition to the finished one, affording evidence of indomitable perseverance on the part of the bird, and a determination not to forsake a tree which it had instinctively selected as a suitable one for its operations. In two instances, however, I have known the Kingfisher to adopt an existing hollow in a partially decayed kahikatea tree, dispensing altogether with the labour of boring and forming it.

The nestling of this species is a very curious object. On bursting from the shell it presents the following appearance: the abdomen, as in most young birds, is perfectly bare; on the other parts each feather is encased in a sharp-pointed sheath of a greyish colour, closely studded, and bristling like the quills of a porcupine. Before the young birds quit the nest, the sheathings gradually burst, exposing the true feathers in all their brilliancy; vestiges, however, of this spiny condition adhere to the fore part of the head for several days after the birds have quitted their cell.

On being alarmed or excited, the young Kingfisher utters a prolonged rasping cry, sounding very harsh to the ear. The parent birds are very fierce when their nest is molested, darting into the face of the intruder, and flying off again, with a loud, quickly repeated note of alarm.

Mr. Robertson, of Waireka, near Wanganui, informs me that he once saw a cat killed by a pair of these birds. The unfortunate puss had been treed by a dog and was hanging on to the bole, spread-eagle fashion, when she was fiercely attacked by a pair of Kingfishers who appeared to consider their nest in danger. After receiving repeated thrusts from the bills of her assailants the cat fell to the ground and shortly afterwards expired.

In the Canterbury district, where timber is scarce, it more frequently burrows a hole in a bank, and often near the sea-beach. On examining one of these holes, Mr. Potts observed that the bottom inclined slightly upwards from the entrance, and that the eggs were deposited on a layer of crustacean remains about a foot from the outside. The exuviæ within the nest consisted of mud, with numerous remains of crustacea and the wings of coleopterous insects *.

The eggs are generally five in number, sometimes six, broadly oval in form, and measuring 1·2 inch by ·95. They are of the purest white, with a smooth or polished surface, and very fragile in texture; sometimes the shell is marked by minute limy excrescences at the larger end.

In the British Museum collection there is a specimen from Norfolk Island (marked *Halcyon sancta*, ♂) which is undoubtedly referable to *H. vagans*.

In 'The Ibis' (1880, p. 459) Mr. Layard described a new Kingfisher from the Solomon Islands, under the name of *Halcyon tristrami*, stating that it was distinguishable from *H. juliæ*, *H. chloris*, and *H. sancta* by the well-marked supercilium and the rich colour of the underparts, in which respects, he said, it exactly accords with *H. vagans* from New Zealand. Dr. Ramsay has already pointed out (Proc. Linn. Soc. N. S. W. vol. vi. p. 833) that Layard's description does not altogether agree with the coloured figure which accompanied it. Canon Tristram states (Ibis, 1882, p. 609) that the type of *H. tristrami* has "no occipital patch whatever" and seems to be "further removed from *H. vagans* than from any other of the group." In reference to this Mr. Sharpe remarks (Gould's 'Birds of New Guinea'):—"We cannot understand why Canon Tristram should object to the close resemblance of *H. tristrami* and *H. vagans*." He expresses a doubt whether the bird exists at all in the Solomon group, all the examples in the British Museum having come from New Britain; and he adds that in all of these the nape-patch is present, being plainly discernible even in the nestling.

I have examined all these specimens and I do not hesitate to say that *H. tristrami* is readily distinguishable from *H. vagans* by its larger size, brighter blue on the upper surface, more conspicuous nuchal collar, and greater extent and depth of cinnamon hue on the underparts. But, after all, it is hardly possible to resist the conclusion that these closely allied forms are little more, if anything, than local or climatic races of one common species. For example, in the British Museum there is a Tongan specimen of *H. sacra* (from the collection of Sir E. Home, Bart.) in which the bill is quite as large, the upper surface fully as bright, and the nape-patch as distinct as in *H. tristrami*, although the underparts, as well as the nuchal collar, are perfectly white. On the other hand, a specimen of *H. vagans* (brought by H.M.S. 'Herald' from Raoul Island), albeit a comparatively young bird, is as highly coloured in every respect as ordinary examples of *H. tristrami*, although it is appreciably smaller in all its dimensions.

* "Referring to your interesting account of its nesting-habits in the 'Birds of New Zealand' (1st ed.), I may mention that I have found three or four pairs building in close association in a clay bank, and that on one occasion I counted ten pairs boring in the standing trunk of a dead and decaying rimu. I have never found more than five eggs in a nest."—Gilbert Mair.

"The Kingfisher makes its nest in our neighbourhood (Oamaru) by digging out a hole in a clay bank. A tunnel is driven horizontally into the bank for twelve inches; at the end of this a round hole, five inches in diameter and two in depth, is formed, and here the beautiful white eggs are deposited. There are usually four or five in a nest, and the incubation takes nineteen days. After the young are hatched out, a strong stench is experienced at the mouth of the nest, owing to the nature of the food supplied to them, consisting of small fish, lizards, &c. On one occasion I caught an adult bird in the vinery, where it appeared to be testing the quality of the grapes. It bit my hand savagely when captured, and uttered a loud discordant scream."—W. W. Smith.

"On a Kingfisher's nest and its contents:—October 10th, first egg laid in a nest on a cliff; second egg laid on the 12th before 10 A.M.; third egg laid on the 14th; fourth egg on the 15th; fifth egg on the 16th; sixth and last egg on the 17th. Subsequently the nesting-place was measured and gave the following dimensions: entrance rather over two inches in diameter; tunnel sixteen inches in length; egg-chamber of ovoid form, 7 inches in length, 5½ in width, with a height from the bottom of 4 inches. The size of the nest may create surprise when one thinks of the small space occupied by the eggs, but a roomy house is necessary, for, like those of most troglodytal breeders, the young remain in their hole till their wings are well grown."—T. H. Potts.

EUDYNAMIS TAITENSIS.

(LONG-TAILED CUCKOO.)

Le Coucou brun varié de noir, Montb. Ois. vi. p. 376 (1779).
Society Cuckoo, Lath. Gen. Syn. ii. p. 514 (1782).
Cuculus taitensis, Sparrm. Mus. Carls. t. 32 (1787).
Cuculus taitius, Gm. Syst. Nat. i. p. 412 (1788).
Eudynamys taitensis, Gray, Dieff. Trav. ii., App. p. 193 (1843).
Cuculus fasciatus, Forst. Descr. Anim. p. 160 (1844).
Eudynamys cuneicauda, Peale, U. S. Expl. Exp. p. 139, pl. 38. f. 2 (1848).
Eudynamys tahitius, Gray, B. Tr. Isl. Pacif. Ocean, p. 35 (1859).
Eudynamis taitiensis, Cab. & Heine, Mus. Hein. Th. iv. p. 56 (1862).
Eudynamis tahitiensis, Potts, Trans. N.-Z. Inst. vol. iii. p. 90 (1870).

Native names.

Koekoea, Kawekawea, Kohoperoa, and Kohaperoa.

♂ *ad.* brunneus, pileo longitudinaliter fulvo striato: corpore reliquo superiore brunneo et pallidè ferrugineo conspicuè et irregulariter transfasciato: tectricibus alarum fulvo maculatis: caudâ brunneâ et ferrugineo transfasciatâ alboque terminatâ: remigibus brunneis, ferrugineo maculatis, fascias irregulares formantibus: superciliis angusto fulvo: regione auriculari brunneâ angustissimè fulvo lineatâ: genis et colli lateribus albis ferrugineo lavatis et brunneo longitudinaliter striatis: subtùs albicans, plumis medialiter brunneo striatis et ferrugineo tinctis: hypochondriis brunneo transfasciatis: subalaribus fulvescenti-albis, angustè brunneo striatis: rostro pallidè brunneo, ad basin saturatiore, mandibulâ flavicante: pedibus viridi-flavis, unguibus brunneis: iride rubescente, interdum flavicante: regione ophthalmicâ nudâ sordidè viridi.

♀ vix a mari distinguenda: paullò minor: coloribus sordidioribus.

Juv. pallidior, suprà ubique albido maculatus, nec fasciatus: caudâ pallidè fulvo transfasciatâ: subtùs ochraceus, pectore abdomineque maculis elongatis triquetris notatis: rostro flavicanti-brunneo: pedibus viridi-flavis.

Adult male. Upper surface dark brown, with a purplish gloss, longitudinally streaked on the head and neck, barred and spotted on the wings and back with rufous; wing-coverts tipped with fulvous white; quills dark brown, banded with pale rufous; tail-feathers marked in their whole extent with narrow alternate bars of dark brown and rufous, tipped with white and finely glossed with purple; a broad line of yellowish white passing from the nostrils over the eyes, and another extending downwards from the angles of the mouth; lores and chin white, with numerous black hair-like filaments; sides of the neck dark brown mixed with rufous; throat, fore part of neck, breast, and sides of the body pure white, with numerous longitudinal streaks of brown, each feather having a broad mark down the centre; lining of wings fulvous white or pale fawn-colour; femoral plumes and under tail-coverts crossed with broad arrow-head marks of brown. Bill pale brown, darker at the base, and yellowish on the lower mandible; irides reddish brown, inclining in some to yellow; bare skin surrounding the eyes dull green; tarsi and toes greenish yellow; claws dark brown. Total length 16·5 inches; extent of wings 21; wing, from flexure, 7·5; tail 9·75; bill along the ridge 1, along the edge of lower mandible 1·4; tarsus 1·5; longer fore toe and claw 1·4; longer hind toe and claw 1·25.

Adult female. Slightly smaller than the male, and with the tints of the plumage duller, the purple gloss on the upper parts being scarcely perceptible.

Young. Upper surface blackish brown, marked on the crown with narrow streaks, on the hind neck with fusiform and on the back with rounded spots of fulvous yellow; quills and tail-feathers blackish brown, barred and tipped with fulvous brown. Under surface pale cinnamon-brown; on each side of the throat two longitudinal streaks, and on the breast and sides of the body broad shaft-lines of dusky black; under tail-coverts barred and tibial plumes crossed with marks of the same colour in the form of an inverted V. Bill yellowish brown; tarsi and toes greenish yellow.

Obs. In examples of the young birds much difference is observable both in the ground-tints and in the markings of the plumage. Some are much darker than others, and have the spots on the upper surface pale rufous instead of fulvous yellow; in others, again, they are yellowish white; some have the barred markings on the tail-feathers very obscure, while in others they are as distinct as in the adult, although not so regular in form.

Note.—There is a remarkable phenomenon in the animal world known to naturalists as "mimicry," or the law of protective resemblance. It is developed chiefly among insects, and particularly among the Lepidoptera. Mr. Wallace describes, at page 205 of his enchanting book on the 'Malay Archipelago,' a butterfly which, when at rest, so closely resembles a dead leaf as almost to defy detection. The varied details of colouring combine to produce a disguise that so exactly represents a slightly curved or shrivelled leaf as to render the butterfly quite safe from the attacks of insectivorous birds, except when on the wing. The flight of the insect, on the other hand, is so vigorous and rapid that it is well able then to protect itself. Mr. Wallace adds that in many specimens there occur patches and spots, formed of small black dots, so closely resembling the way in which minute fungi grow on leaves that it is impossible not to believe that fungi have grown on the butterflies themselves! This protective imitation must obviously favour the species in the general struggle for existence, and may of itself be sufficient to save it from extinction. But there is another kind of "mimicry," where one insect which would, on discovery, be eagerly devoured, assumes for similar protective purposes a close resemblance to some other insect notoriously distasteful to birds and reptiles, and often belonging to a totally different family or order. Numberless instances might be given in illustration of this singular fact, every department furnishing examples of adaptation more or less complete, and all being explainable on the principle of variation under natural selection or the "survival of the fittest." Mr. Wallace, when exploring in the Moluccas, was the first to discover similar instances of mimicry among birds, although the law of protective colouring had long been known to exist in the case of birds' eggs. He gives two very curious examples of external resemblance co-existing with very important structural differences, rendering it impossible to place the model and the copy near each other in any natural arrangement. In one of these a Honey-sucker has its colours mimicked by a species of Oriole, and the reason is thus stated:—"They must derive some advantage from the imitation, and as they are certainly weak birds, with small feet and claws, they may require it. Now, the Tropidorhynchi are very strong and active birds, having powerful grasping claws, and long, curved, sharp beaks. They assemble together in groups and small flocks, and they have a very loud, bawling note, which can be heard at a great distance, and serves to collect a number together in time of danger. They are very plentiful and very pugnacious, frequently driving away crows and even hawks, which perch on a tree where a few of them are assembled. It is very probable, therefore, that the smaller birds of prey have learnt to respect these birds, and leave them alone, and it may thus be a great advantage for the weaker and less courageous Mimetas to be mistaken for them. This being the case, the laws of Variation and Survival of the fittest will suffice to explain how the resemblance has been brought about, without supposing any voluntary action on the part of the birds themselves; and those who have read Mr. Darwin's 'Origin of Species' will have no difficulty in comprehending the whole process."

Among the many minor instances that have attracted notice, the English Cuckoo (*Cuculus canorus*) is supposed to derive protection from the resemblance of its markings to those of the Sparrow-Hawk (*Accipiter nisus*); but the resemblance is far more striking between our Long-tailed Cuckoo and a North-American species of Hawk (*Accipiter cooperi*). In fully adult specimens of the former it will be observed that the markings of the plumage are very pronounced, while the peculiar form of the bird itself distinguishes it very readily from all other New-Zealand species. Beyond the general grouping of the colours there is nothing to remind us of our own Bush-Hawk; and that there is no great protective resemblance is sufficiently manifest from the fact that our Cuckoo is persecuted on every possible occasion by the Tui, which is timorous enough in the presence of a Hawk. During a trip, however, on the Continent, in the autumn of 1871, I found in the Zoological Museum at Frankfort what appeared to be the accipitrine model, in a very striking likeness to our bird. Not only has our Cuckoo the general contour of Cooper's Sparrow-Hawk, but the tear-shaped markings on the underparts and the arrow-head bars on the femoral plumes are exactly similar in both. The resemblance is carried still further in the beautifully banded tail and marginal wing-coverts, and likewise in the distribution of colours and markings on the sides of the neck. On turning to Mr. Sharpe's description of the "young male" of this species in his Catalogue of the Accipitres in the British Museum (p. 137), it will be seen how many of the terms employed apply equally to our *Eudynamis*, even to the general words "deep brown above with a chocolate gloss, all the feathers of the upper surface broadly edged with rufous."

The coincident existence of such a remarkable resemblance to a New-World form cannot of course be any protection to an inhabitant of New Zealand, and I do not pretend in this instance to apply the rule; but in the light of natural selection, to which at present no limit can be assigned, the fact itself is a suggestive one, and sufficiently striking to call for special mention.

The illustration which accompanies this article, although it may have the appearance of an exaggeration, is in reality a true picture of bird-life. The Long-tailed Cuckoo, which is a native of the warm islands of the South Pacific, visits our country in the summer and breeds with us; but the task of rearing its young (as many witnesses can testify) is entrusted to the Grey Warbler (*Gerygone flaviventris*), figured in our Plate—a species that performs the same friendly office for the Shining Cuckoo (*Chrysococcyx lucidus*), another summer visitant.

Drs. Finsch and Hartlaub, in their valuable work on the Birds of Central Polynesia, record the occurrence of this species in Samoa, as well as in the Friendly *, the Society, the Marquesas, and the Fiji groups of islands; but although it migrates to New Zealand, there is no mention of its occurrence in any part of Australia or Tasmania.

In the still summer's evening when the landscape is wrapped in the gloom of faded twilight—when no sound meets the ear but the low musical song of the pihareinga cricket and the occasional hum of a *Prionoplus* on the wing—there comes from the thicket a long-drawn cry, shrill and clear; then a pause of five minutes or more, followed by another cry; and so on at intervals till long after the pihareingas have ceased to chirp and the nocturnal beetles have folded their wings in sluggish repose. This is the first intimation we get that the Long-tailed Cuckoo has come amongst us.

It begins to arrive about the second week in October, but is not numerous till the following month, when the pairing commences. It is, however, somewhat irregularly dispersed over the country; for in the far north it is at all times a very rare bird. In the southern portion of the North Island, and throughout the wooded parts of the South Island, it is comparatively common. It appears to be most plentiful in November and December, becoming scarcer in January and disappearing altogether by the end of February. I have a note, however, of its occurrence at Otaki (in the North Island) as late as the first week in April.

Young birds are not unfrequently met with in the month of March or even later; but it seems probable that these are only solitary individuals hatched too late to permit of their joining in the return migration, and accordingly left to perish as the cold season advances; and this is likewise the case with our Shining Cuckoo. As an illustration of this, I may mention that a young bird of this latter species, which had been picked up dead in a garden, was brought to me at the end of February (long after the old birds had quitted the country), and that I found it excessively fat, and the stomach crammed with caterpillars—strong presumptive evidence that the bird had not suffered from the neglect of its foster-parents, but had succumbed to the exigencies of its late birth.

In the early dawn and during the cool hours of the morning, the Long-tailed Cuckoo resorts to the low underwood and brushes; but although its cry may be frequently heard, it is not easy to find the bird, inasmuch as the sound, though produced within a few yards of the listener, has the effect on the ear of one coming from a remote distance. This species, in fact, appears, like some others of the same family, to be endowed with a sort of natural ventriloquism, and its apparently far-off cry is often very deceptive.

While searching for his food the Koheperoa moves about with much activity; but as soon as the sun is up he betakes himself to the top branches of a kahikatea or other lofty tree, where he remains closely concealed till sunset. He continues to utter, at intervals of ten or fifteen minutes,

* Dr. Finsch has identified a young male in the spotted dress in a collection of birds from the island of Eua.

his prolonged shrill note (quite distinct from all other sounds of the forest, and very pleasant to hear) till about noon, when he remains perfectly silent for two hours or more. As soon, however, as the heat of the day is over, he resumes his cry, and shortly afterwards leaves his retreat to hunt for food again.

It is not unusual to hear a pair of these birds answering each other for hours together from the tops of neighbouring trees. Indeed, I have observed that it is habitually stationary, for it may often be heard uttering its long, plaintive scream for a whole day in the same tree, but always quite out of view. During the quiet nights of December its piercing cry may be heard at intervals till break of day, varied only in the earlier watches by the solemn hooting of the Morepork.

This species is more predatory in its habits than is usual with the members of this group. Lizards and large insects form its principal diet; but it also plunders the nests of other birds, devouring alike the eggs and young. From the stomach of one which I shot in December 1856, I took the body of a young bird (apparently a Piopio), partly fledged and only slightly mutilated, showing the enormous capacity of the Cuckoo's throat. This interesting object, preserved in spirit, is now in the collection of the Colonial Museum at Wellington. The large nocturnal beetle (*Prionoplus reticularis*), the various species of *Deinacrida* and *Phasmidæ*, the kekereru or fetid bug, the large bush *Cicada*, and different kinds of spiders and caterpillars, all contribute to the support of this bird; for I have found their remains in abundance in the stomachs of specimens I have dissected.

As already stated, it is accustomed to rob the nests of other birds; and whether from this or some other cause, it is an object of constant persecution to the Tui or Parson bird. The instant one of these birds shows itself, the Tui commences its pursuit, chasing it from tree to tree, and fairly driving it out of the woods. I have actually seen three or four of these persecutors at one time following the unfortunate Cuckoo, with loud cries of intimidation, and, finally, compelling it to take refuge in the long grass on the banks of a stream.

During its sojourn with us it is generally met with singly or in pairs, but Captain Mair gives the following interesting particulars of a summer flight:—"Passing down the Hurukareao river, in the Urewera country, during the intensely hot weather of February 1872, I was astonished at the number of Kohoperoa that coursed about overhead. During the three days that we were making the passage, I saw some hundreds of them, swarming about in the air like large dragon-flies, as many as twenty or thirty of them being sometimes associated together. The loud clamour of their notes became at length quite oppressive. There was much dead timber on the banks of the river, and it appeared to me that the birds were feasting on the large brown cicada. This is the only occasion on which I have observed this bird consorting as it were in parties."

Very little is at present known of the breeding-habits of this species. As I have mentioned above, it is parasitical; but to what extent is not yet fully determined. The theory put forward in my former edition was, that it performs itself the duty of incubation, and then abandons its young to the Grey Warbler, which instinctively accepts the charge and caters untiringly for its support. In the first place it is difficult to conceive how a bird of the size and form of the Long-tailed Cuckoo could deposit its eggs in the domed nest of the last-named species; and even supposing that it did, it would seem almost a physical impossibility for so small a creature to hatch it; and, again, even were this feasible, it is difficult to imagine how the frail tenement of a suspension nest could support the daily increasing weight of the young Cuckoo *. Over and above all this, there was the significant

* Mr. Justice Gillies thus describes a nest of the Grey Warbler which he found depending from a manuka bush close to the roadside, and about five feet from the ground, at the Bay of Islands (it was on the 7th October, and the nest contained four eggs):—"It is of the shape of a soda-water bottle, eight inches in length by about four in diameter at its widest part. The side aperture is fully one-third way down from the twig on which it hung, and measured one and a half inches across by about one inch

fact that I had once shot an adult female in which the underparts were quite denuded of feathers, as if the bird had been long incubating. Strange as such an hypothesis may appear, we are not altogether without a parallel instance in bird-history; for in the case of the *Chrysococcyx smaragdineus* of Western Africa, it is alleged that this Cuckoo hatches its single egg and then, utterly unmindful of its parental obligations, casts the care of its offspring on a charitable public; and that almost every passing bird, attracted by the piping cry of the deserted bantling, drops a caterpillar or other sweet morsel into its imploring throat! My artist, Mr. Keulemans, assures me that he often witnessed this himself during his residence on Prince's Island. As entirely opposed to this theory, however, there is one undoubted case of an egg of this Cuckoo being hatched out by a Wood-Robin (*Miro albifrons*) in whose nest it had been deposited. The young Cuckoo was removed by the finder and soon afterwards died *. The question arises, If it had been allowed to leave the nest, would the Robin have reared it, or would she have delegated this task to the Grey Warbler?

An egg forwarded to me some years ago by the Rev. R. Taylor, of Wanganui, as belonging to this species, is almost spherical in shape, with a slightly rough or granulate surface; it is of a pale buff or yellowish-brown colour, and measures 1·25 inch in length by 1·15 in breadth. A specimen in the Canterbury Museum, taken by Mr. Smith from a Warbler's nest at Oamaru, in November 1885, corresponds exactly with mine (which is now in the Colonial Museum at Wellington) except that it is slightly narrower.

---

perpendicular. The upper portion of the nest somewhat overhangs the aperture, forming a sort of hood. The nest is composed of twigs, grass, cow-hair, and greenish spider-nests, with a white coral-like moss scattered over the outside. The eggs are ten sixteenths of an inch in length by seven sixteenths of an inch greatest diameter, ovoid, of a faint pinkish colour, with small brown spots, more numerous at the larger end of the egg." The learned author continues:—"How the Long-tailed Cuckoo (*Eudynamis taitensis*) can, as stated by Dr. Buller ('Birds of New Zealand,' p. 75), deposit its eggs in such a nest I can scarcely understand. On the 22nd instant (October) one of my children discovered, under a large *Cupressus macrocarpa* in my garden, a specimen of the *Eudynamis taitensis*, recently killed, apparently by a Hawk. It would have been impossible for the *Eudynamis* to have entered the opening in the nest of the *Gerygone*." (Trans. N.-Z. Inst. vol. vii. p. 524.)

On referring to the page of my first edition as cited above, it will be seen that, so far from making the supposed statement, I then expressed, as I now repeat, a very decided opinion to the contrary.

* Mr. W. W. Smith sends me the following particulars of this case:—

"Oct. 29th. Found Wood-Robin's nest with two eggs. Oct. 31. Visited Robin's nest; four eggs. Nov. 3. Agreeably surprised to find egg of *Eudynamis taitensis* placed among the rest; for this is the first time I have seen its egg in the nest of this species. It was almost round in shape, with a deeper shade of colour than the specimen in the Canterbury Museum. Nov. 7. Found Robin sitting, and did not disturb her. Nov. 10. Made bird fly off, in order to examine the eggs, which I found to be all right. She was very tame, and came close to my face whilst I was looking at the eggs. Nov. 15. Again found bird on the nest, and left her undisturbed. Nov. 24. Visited nest again, and found all the eggs hatched; young Cuckoo of enormous size compared to its mates; must have been hatched out later than the others, as one of the young Robins was dead. I took the former in my hand, and found it to be a very helpless creature, with the skin almost entirely naked and the eyes closed. Nov. 28. Found young Cuckoo thriving well, being kept constantly supplied with food by the Robin, whose own surviving offspring, three in number, appear likewise to be doing well. Dec. 2. Young Cuckoo growing rapidly. It will soon be too large for the nest, and already has to lie on the top of the young Robins. Dec. 6. Cuckoo still in nest, and now covered with thick blackish downy feathers. It seems very robust; and I observed it raise its body over the edge of the nest in order to void its excrement. Dec. 8. Young Cuckoo has grown so much that it quite fills the cavity of the nest. The young Robins appear instinctively to remain at the bottom for self-preservation; for if the Cuckoo could displace them, he could occupy the whole of the cavity of the nest. Dec. 9. Removed two of the young Robins, in order to make room for the increasing size of the Cuckoo. Dec. 10. Young Cuckoo and remaining Robin doing well, the latter being nearly ready to fly. Dec. 11. Placed the nest, with both occupants, inside a box with wire-netting in front—the mesh being large enough to admit the head of the parent—and left it there. Dec. 15. Found young birds quite active, having been fed by the old ones through the netting. Liberated the Robin and brought the Cuckoo home. It is now in fine plumage, spotted with white or greyish white on a brown ground. Dec. 17. Cuckoo doing well and eats freely. Moves about the box in a clumsy way, and utters a peevish chirp, usually after being fed. Legs well developed, but apparently weak; eyes very bright. Dec. 22. Young Cuckoo died last night, much to my regret, as I was anxious to make it live through the winter."

CHRYSOCOCCYX LUCIDUS.

(SHINING CUCKOO.)

Shining Cuckoo, Lath. Gen. Syn. ii. p. 528, pl. xxiii. (1782).
Cuculus lucidus, Gm. Syst. Nat. i. p. 421 (1788, ex Lath.).
Variable Warbler, Lath. Gen. Syn. Suppl. ii. p. 250 (1801).
Sylvia versicolora, Lath. Ind. Orn. Suppl. ii. p. lvi (1801).
Chalcites lucidus, Less. Traité d'Orn. p. 153 (1831).
Cuculus nitens, Forst. Descr. Anim. p. 151 (1844).
Cuculus versicolor, Gray, Gen. of B. ii. p. 463 (1847).
Chrysococcyx lucidus, Gould, B. of Austr. iv. pl. 89 (1848).
Cuculus chalcites, Illiger, MS. in Mus. Berol., *undè*
Chrysococcyx chalcites, Licht. Nomencl. Av. p. 78 (1854).
Lamprococcyx lucidus, Cab. & Heine, Mus. Hein. Th. iv. p. 14 (1862).
Chrysococcyx plagosus, Hutton (*nec* Lath.), Trans. N.-Z. Inst. (1872), vol. v. p. 223 *.

Native names.

Warauroa, Pipiauroa, and Pipiwarauroa.

Ad. suprà metallicè viridis, æneo et cupreo nitens, supracaudalibus lateralibus latè albo semifasciatis: fronte, supercilio distincto et facie laterali albo maculatis, viridi transfasciatis: loris mentoque albidis haud viridi notatis: tectricibus alarum dorso concoloribus: remigibus brunneis, ad basin pogonii interni albidis, primariis extùs æneo nitentibus, secundariis magis conspicuè lavatis et pennis dorsalibus omninò dorso concoloribus: caudâ brunneâ, æneo-viridi nitente, fasciâ anteapicali nigricante, rectricibus tribus exterioribus ad apicem pogonii interni albo maculatis, pennâ extimâ albo conspicuè fasciatâ, penultimâ in medio vix rufescente tinctâ: pectore et subalaribus albicantibus transversim æneo-viridi fasciatis: abdomine puriùs albo, hypochondriis subcaudalibusque conspicuè æneo-viridi transfasciatis: rostro nigro: pedibus brunnescenti-nigris, plantis pedum flavicantibus: iride nigrâ.

Juv. obscurior et sordidior, minùs metallicus: tectricibus alarum brunneo marginatis: caudâ nusquam rufescente: gutture et pectore superiore fulvescenti-albis, fuscescenti-brunneo variis, vix viridi lavatis: corpore reliquo subtùs fulvescenti-albo, hypochondriis et corporis lateribus fasciis interruptis metallicè viridibus notatis: subcaudalibus maculis viridibus triquetris transnotatis.

Adult male. Upper parts bright golden green, changing to coppery purple in certain lights; frontal feathers tipped more or less with white; superciliary streak formed of irregular whitish spots; throat, sides of head, and fore part of neck white, with narrow broken bars of coppery green; breast and underparts generally white, with transverse bands of changing golden green, coppery brown in certain aspects; on the sides, flanks, and under tail-coverts these bands are very regular and conspicuous, each feather being crossed by

* "Captain Hutton says that the Chatham-Island Bronze Cuckoo is not the same as the New-Zealand one, but is *C. plagosus* of Australia, in which opinion I do not agree, after having compared a specimen from the Chatham Islands lent me by the New-Zealand Institute. The underparts show a little broader gold-green crossbands, and the second tail-feathers, instead of two well-defined rusty bands, have only indications of them; but there is no other difference, and I see no reason to separate the Chatham Island bird from the New-Zealand *C. lucidus*." (Finsch, Trans. N.-Z. Inst. vol. vii. p. 227.)

two broad, equidistant bars; the lower part of the abdomen pure white; quills dark brown, glossed with coppery brown, changing to bright golden green on the secondaries; with the exception of the three outer primaries, all the quills are yellowish white in the basal portion of the inner webs, forming a broad oblique bar on the under face of the wing; under wing-coverts and axillary plumes indistinctly barred with coppery brown; tail, when closed, bronzy green, with a broad subterminal band of purplish brown; upper tail-coverts bright golden green, the lateral ones largely marked with white on their outer webs. On spreading the tail the outermost feather on each side is found to be blackish brown, with five broad white bars on the inner web, the fifth one being terminal, and with five irregular spots of white on the basal portion of the outer web; the next feather blackish brown, slightly glossed with green, marked on the inner web with two obscure spots of rufous, darker brown towards the tip, and terminated by a round spot of white; the succeeding one similar, but without the rufous markings, and with the terminal spot on the inner web much smaller; and the median feathers coppery brown, glossed with green, and crossed by a darker subterminal bar. Irides and bill black; tarsi and toes brownish black; soles of feet yellowish. Total length 7 inches; extent of wings 11·75; wing, from flexure, 4; tail 2·75; bill, along the ridge ·5, along the edge of lower mandible ·75; tarsus ·5; longer fore toe and claw ·8, longer hind toe and claw ·65.

Young. Metallic tints of the upper parts duller; upper wing-coverts edged with brown; tail-feathers as in the adult, but with the rufous markings obsolete; throat and fore part of neck yellowish white, clouded and mottled with dusky brown, faintly glossed with green; underparts generally yellowish white, marked on the sides and flanks with fragmentary or interrupted bands of dull shining green; the under tail-coverts crossed by broad triangular spots of the same.

There is nothing more delightful, on a sultry summer's day, than to recline in some cool shade and inhale the sweet fragrance of the native woods. All is still and quiet save the hum of bees in the air and the loud drumming of the tarakihi as it clings to the bark overhead. Then there falls upon the ear the well-known cry of the Koheperoa—not the vociferous scream of the early morning, but a low sleepy cry—issuing from some lofty tree-top where the bird is resting during the heat of the day. From a neighbouring tree comes the full rich note of the Tui, uttered at short intervals like the slow tolling of a silver bell; then the low whistle of a Kaka calling to its mate to come and seek repose while the sun is at the meridian; then all is still again, and nothing is heard but the soft murmur of insects in the air and the languid cry of a solitary Fantail as it flits around with full-spread wings and tail, dancing from side to side, or the sweet trill of the Ngirungiru, full of pleasant associations. But while we are still listening, a new sound arrests the attention—a peculiar whistling cry, different from that of any other bird. This announces the arrival in our country of the Shining Cuckoo, an inhabitant of Australia, and probably New Guinea *, which appears in New Zealand (also in Norfolk Island) only as a summer migrant. Its cry is always welcomed by the colonists as the harbinger of spring; and during its short stay with us its sweet but plaintive notes may be heard in every grove throughout the long summer days. It makes its appearance, year after year, with surprising punctuality, arriving first in the extreme north, and about a fortnight later spreading all over the country. A correspondent informs me that for three successive years, at Whangarei (north of Auckland), he first heard its familiar note on the 21st September, and that on one occasion he noticed it as early as the 3rd of that month. Another correspondent, in the same locality, informs me, as the result of twelve years' careful observation, that this migrant invariably appears between the 17th and 21st of September. For a period of ten years I kept a register of its periodical arrival at Wellington, and noted its regular occurrence between the 5th and 10th of October. Mr. Potts writes to me from Canterbury that it generally arrives there on or about the 8th of October, although in one year (1865) it visited that part of the country as early as the 27th September. It usually departs about the first or second week in January: but in the far north it sometimes lingers

* *Cf.* Ramsay, Proc. Linn. Soc. N. S. W. vol. iii. p. 256.

till the end of that month. As is always the case with migratory birds, there are occasionally stragglers arriving before the appointed time or lagging behind the departing flights. For example, I have a record of their occurrence in Auckland as early as August 17th, and I have met with a solitary bird in the south as late as April.

This is undoubtedly the most tropical-looking of our birds. The glancing of the sunlight on its burnished plumage is very effective, especially, too, when the bird is seen resting on the bare stems of the quasi-tropical *Cordyline*, or feeding on the green-and-gold cicada, which is so abundant there *.

* In New-Zealand scenery there is much to remind one of a tropical country. The scattered clumps of "cabbage-trees" in the open and the nikau-palms in the deep wooded gullies have quite a tropical aspect, and the wild luxuriance of the evergreen bush brings vividly to mind the rank prodigality of a Brazilian forest. To show that this is not a mere play of fancy, I will give here a leaf from my own diary containing an account of a day at Rio de Janeiro:—

"We landed from the S.S. 'Tongariro' at 9 A.M. on the 1st April, and came on board again before midnight, having spent a very pleasant day on shore. On landing, we walked through the market-place, which was interesting, then up the principal street, through which no wheeled vehicles are permitted to pass. The passage is narrow and the balconies are overhanging, giving it the appearance of a street in Constantinople. Many of the shops are most attractive in their multifarious exhibits—feather-plumes, rare butterflies, and brilliant beetles being not the least interesting objects. At the street corner we took a tramcar, and, after one or two changes, proceeded to the railway station, passing on our way some fine public and private buildings, notably a marble palace, the property of a rich coffee-planter. Many of the gardens are very beautiful, being brilliant with tropical flowers of every hue. After a short delay at the station, we entered the railway carriage and started up the Corco Vardo line. From the commencement to the finish at the peak the trip was one of unmitigated enjoyment. The day was clear as crystal, with the sun hot and bright; and the scenery was enchanting. The railway line, which ascends spirally at a gradient of 1 in 3, is something quite unique. Looking down into the deep gullies, I was often reminded of our beautiful New-Zealand bush in the tangled richness of the vegetation. There was the same character of forest-growth, the same crowding together of the tree-tops, the same wealth of lianas, vines, and epiphytes, but all on a more luxuriant scale. In place of our *Astelia cunninghamii*, with its narrow flag leaves, the trees were laden with large clumps of some tropical species with leaves six inches in width; in place of our tiny-flowered orchids there were magnificent tropical species with gorgeous blossoms. There was along the wayside a dense undergrowth in every shade of green, but the leaves were larger and the foliage richer than in the New-Zealand woods, whilst in the places exposed to the sun beautiful flowers of brilliant hues added the charm of high colour to this sylvan picture. The ground below the forest trees was covered with vigorous young plants of many kinds; but the eye sought in vain for that ever-present charm of the New-Zealand bush, the carpet of spreading ferns and mosses. Here and there could be seen a tuft of maiden-hair or a clump of *Pteris* struggling to assert itself, but nothing to remind the observer of the glorious beds of *Lomaria*, the fields of *Asplenium bulbiferum*, and the other beautiful forms so familiar to the wanderer in our native woods. As our train moved slowly up the side of the mountain, the eye seemed never tired of gazing on this view of tropical growth, rendered the more conspicuous by clumps of banana-trees and groups of beautiful palms, lifting their tops proudly above the forest vegetation, whilst huge masses of crimson and purple flowers lightened up the rolling landscape of living green.

"Then all along the line brilliant butterflies of every size and colour fluttered in the warm sunlight; glorious morphos, with a spread of six inches and of the richest metallic blue, hovered, hawk-like, among the trees; large black-and-grey 'swallow-tails' winged their way like Swifts among the lower vegetation; crimson Danaidæ and smaller forms of different kinds—scarlet, golden-yellow, green, or spotted—rested on the leaves or hovered over the flowers, almost within reach of our delighted party, some being actually caught by the hand from the carriage windows. Then here and there a tiny humming-bird, sparkling like a ruby under the rays of the midday sun, might be seen suspended before an open flower or spinning like a moth through the air in search of its absent mate. Such were the sights of tropical loveliness through which we passed on our way to the summit of the Corco Vardo. The view from this point, which is just 2200 feet above the sea-level, baffles description. The far-reaching panorama of sea and land, the wondrous archipelago in front and the glorious amphitheatre of mountains behind—on the one hand the boundless Atlantic, on the other the towering peak of Tijuka, 3300 feet high, rising out of deep valleys filled with tropical forests; then, contracting the scope of vision to the left, the city of Rio in all its quaint oriental beauty lying before you far down in the plains, its suburbs of villas and gardens spreading away for miles and far as the eye can reach, whilst to the right, nestling as it were in an illimitable expanse of ornamental shrubbery, are the Botanical Gardens, with their double row of Imperial palms in perfect symmetry, their feathery tops reared nearly a hundred feet above the ground, presenting a picture of unparalleled beauty; then, still further contracting the scope of vision, the sides of the Corco Vardo and the deep ravine below, clothed and filled with a perfect tangle of tropical vegetation, wildly exuberant in its growth, presenting every hue of green, and enlivened with spreading floral masses of purple and white. The view which burst upon us when we took our stand on the topmost peak of the Corco Vardo was, in short, one which no human artist could depict and no words describe: a view to be gazed on once and then remembered all through life!"

During its sojourn with us it subsists almost exclusively on caterpillars, and the black leech which attacks our fruit-trees. It is therefore entitled to a place among the really useful species.

In disposition it is very gentle. On one occasion I was watching this bird from the window of my hotel, foraging in the garden below for caterpillars, while a brood of young Sparrows were doing the same. Whilst the Cuckoo rested for a moment on a slanting stick, the Cock Sparrow dropped down till it almost touched him, as if to inspect his shining coat. The object of these attentions never left his perch, but simply swerved his body and spread his outer wing, without uttering a sound. I noticed that the young Sparrows were far more active in catching caterpillars than the Cuckoo, although both birds adopted the same plan of search, darting right into the shrub-tops and bringing out their victims to batter and kill them before swallowing.

Its general attitude is that depicted in the Plate, with its tail half-spread and its wings drooping, my artist having utilized a pencil-sketch which I made of a captive bird as it rested quietly on the paper-basket in my study.

Its cry is a remarkable one, as the bird appears to be endowed with a peculiar kind of ventriloquism. It consists of eight or ten long silvery notes quickly repeated. The first of these appears to come from a considerable distance; each successive one brings the voice nearer, till it issues from the spot where the performer is actually perched, perhaps only a few yards off. It generally winds up with a confused strain of joyous notes, accompanied by a stretching and quivering of the wings, expressive, it would seem, of the highest ecstasy. The cry of the young birds is easily distinguished, being very weak and plaintive *.

I had a young bird brought to me as late as the 15th February. It appeared to be in vigorous health, with the membrane at the angles of the mouth still visible; and on being approached by any one would open its mouth in an imploring sort of way, but without making any sound.

Like the Long-tailed Cuckoo already described, this species is parasitic in its breeding-habits, and entrusts to a stranger both the hatching and the rearing of its young.

The little Grey Warbler (*Gerygone flaviventris*) is the customary victim; but exceptional cases have been recorded where the duty was entrusted to the South-Island Tomtit (*Myiomoira macrocephala*); and Captain Mair assures me that he once saw the young of this species attended and fed by a Korimako (*Anthornis melanura*). Dr. Bennett, writing of the same bird in Australia, states † that the egg of the Shining Cuckoo has been found in the nest of *Acanthiza chrysorhina*, and that he has seen a nest of this bird with five eggs, that of the Cuckoo being deposited in the centre of the group, so as to ensure its receiving the warmth imparted by the sitting bird, and thus less likely to be addled. He also narrates the following circumstance:—"A White-shafted Flycatcher (*Rhipidura albiscapa*) was shot at Ryde, near Sydney, in the act of feeding a solitary young bird in its nest, which, when examined, was found to be the chick of the Bronze Cuckoo of the colonists. . . . . It was ludicrous to observe this large and apparently well-fed bird filling up with its corpulent body the entire nest, receiving daily the sustenance intended for several young Flycatchers."

Mr. G. M. Thomson records in the 'Journal of Science' (vol. ii. p. 576) that an egg of this Cuckoo was found on November 5th in a House-Sparrow's nest which had been built in a large bramble-bush and which contained besides three legitimate eggs. He describes it as being "10 lines long, of very thin texture, and much paler than usual, being of a pale greenish white feebly marked with pale brown spots and markings" ‡.

* Captain Mair writes to me:—"Speaking from ten years' observation of this bird in the Tauranga district, I may state that it never sings after the middle of February and seldom after the beginning of that month. As late as the end of March or beginning of April, during several successive years, I have met with these birds in the Mangorewa forest between Tauranga and Rotorua, but never heard them utter a note at this season. I have seen numbers of them perched in silence on the branches of the poporo (*Solanum nigrum*), always in full feather, but absolutely songless. This I regard as a very curious fact."

† Gatherings of a Naturalist in Australasia, p. 207.

‡ Mr. Thompson states, further, that in Otago, *Gerygone flaviventris*, *Myiomoira macrocephala*, and *Zosterops cærulescens*.

As it is usual to find the Cuckoo's egg associated with those of the Grey Warbler, we may reasonably infer that the visitor simply deposits its egg for incubation without displacing the existing ones. But the young Cuckoo is generally found to be the sole tenant of the nest; and the following circumstance, related to me by the Rev. R. Taylor, sufficiently proves that the intruder ejects the rightful occupants and takes entire possession. He discovered the nest of a Grey Warbler in his garden-shrubbery containing several eggs, and among them a larger one, which he correctly assigned to the Shining Cuckoo. In due time all the eggs were hatched; but after the lapse of a day or two the young Cuckoo was the sole tenant of the nest, and the dead bodies of the others were found lying on the ground below. At length the usurper left the nest, and for many days after both of the foster-parents were incessantly on the wing, from morning till night, catering for the inordinate appetite of their charge, whose constant piping cry served only to stimulate their activity.

Since the above was written, I have had an opportunity of examining a young Cuckoo in possession, and it exhibits a droll phase of bird-life, the intruder occupying the entire cavity of the nest, with its head protruding from the opening.

I have received from Mr. W. W. Smith, of Oamaru, some interesting notes from his diary *, showing how inevitably the young Warblers, in the struggle for existence, must succumb to the more vigorous intruder.

The egg of the Shining Cuckoo is of a broad ovato-elliptical form, generally of a greenish-white or very pale olive colour, often clouded or stained with brownish grey, and measuring ·8 of an inch in length by ·5 in breadth. One taken by myself, many years ago, from the nest of a Grey Warbler, in the manuka scrub, on what is now the site of a flourishing city, was of a pale creamy colour; and another, which was laid by a captive bird in my possession, is pure white. A specimen in the Otago Museum is broadly elliptical in form, measures ·7 of an inch in length by ·5 in breadth, and is of a uniform dull olivaceous grey inclining to brown. Of two specimens in my son's collection one is rather more elliptical in form and of a uniform olivaceous brown, somewhat paler at the smaller end; the other (which came from the Chatham Islands) is pale olivaceous grey, perceptibly darker at the larger end, and very minutely granulated with brown over the entire surface.

---

are the usual foster-parents. Mr. Gould records that, in Australia, the task of incubation is often delegated to the Yellow-tailed *Acanthiza*, and adds, "I have several times taken the egg of the Cuckoo from the nest of this bird, and also the young, in which latter case the parasitical bird was the sole occupant." Mr. Potts reports (Journ. of Science, ii. p. 477) that at Ohinitahi he found an egg of this species in the nest of *Zosterops cærulescens*, together with three eggs of the dupe. He enumerates sixteen instances, between Oct. 23 and Jan. 5, of its being found in the nest of *Gerygone flaviventris*. Generally there were from one to three eggs of the dupe in the nest: in two cases (Dec. 18 and Jan. 6) the Cuckoo egg only; and in three other cases (Dec. 17, Dec. 23, and Jan. 1) the young Cuckoo only. He states further that he has in his possession an egg of this bird taken from the nest of *Gerygone albofrontata* at the Chatham Islands.

* "Oct. 7th. Found a nest of *Gerygone flaviventris*, with four small eggs and one much larger. The latter I take to be the egg of the Shining Cuckoo (*Chrysococcyx lucidus*). Left the nest, intending to return in a few days. 11th. Visited place again. The Grey Warbler flew out when I approached. Five eggs still all right. 21st. Still unhatched. 24th. Two young ones hatched; one egg lying on the ground outside the nest, containing chick quite cold and dead. 25th. Three young ones in nest; large egg unhatched. 26th. Large egg hatched—a chick of the Shining Cuckoo; very clumsy in nest, lying on top of the three young Warblers. 30th. Found one dead chick lying on the ground; two young Warblers still alive; young Cuckoo growing rapidly, being now nearly large enough to fill the nest itself; beak and legs fairly well developed. Nov. 2nd. One of the young Warblers lying dead in nest, the other alive. Young Cuckoo has now its eyes open; signs of feathers on the neck and wings, but underparts of the body perfectly bare. 5th. Visited nest again. Young Cuckoo thrust out its head to receive food when I approached. Lifted the surviving young Warbler out of the nest, and found it very feeble. 6th. Young Cuckoo lying with its head at opening of nest, having taken full possession. Its lifeless companion was lying underneath, having apparently died from starvation. 8th. Found young Cuckoo almost ready to leave its cradle. Brought both nest and bird home with me. 10th. Thriving well, being fed on small worms, grubs, flies, spiders, and very small pieces of lean meat. 15th. Has now come out of nest; eats largely three times a day, but does not care for meat; increasing rapidly in size. 20th. Nearly feathered. Placed it in a cage, but it looks sickly. 21st. Young Cuckoo died. Proved, on skinning, to be a male bird."

PLATYCERCUS NOVÆ ZEALANDIÆ.

(RED-FRONTED PARRAKEET.)

Pacific Parrot, Lath. Gen. Syn. i. p. 252 (1781).
Psittacus novæ zeelandiæ, Sparrm. Mus. Carls. pl. 28 (1787).
Psittacus pacificus, Gm. Syst. Nat. i. p. 329 (1788).
Platycercus pacificus, Vigors, Zool. Journ. i. p. 526 (1825).
Pezoporus novæ zeelandiæ, Voigt, ed. Cuv. Thierreich. p. 750 (1831).
Lathamus sparmanii, Less. Traité d'Orn. i. p. 206 (1831).
Platycercus erythrotis, Wagl. Monogr. Psitt. p. 526 (1835).
Cyanoramphus erythrotis, Bonap. Rev. et Mag. de Zool. vi. p. 153 (1854).
Cyanoramphus novæ zelandiæ, Bonap. Rev. et Mag. de Zool. vi. p. 153 (1854).
Cyanoramphus aucklandicus, Bonap. Naumannia, 1856, Suppl. p. 352.
Cyanoramphus novæ guineæ, Bonap. Naum. 1856, Suppl. p. 352.
Platycercus aucklandicus, Gray, Cat. Brit. Mus. Psitt. p. 13 (1859).
Platycercus cookii, Gray, Cat. Brit. Mus. Psitt. p. 13 (1859).
Platycercus novæ guineæ, Gray, Cat. Brit. Mus. Psitt. p. 13 (1859).
? *Cyanorhamphus saisseti*, Verr. et Des Murs, Rev. et Mag. de Zool. xii. p. 387 (1860).
Platycercus rayneri, Gray, Ibis, 1862, p. 228.
Coriphilus novæ zeelandiæ, Schlegel, Dierent. p. 77 (1864).
Euphema novæ zeelandiæ, Schl. Mus. Pays-Bas, *Psittaci*, p. 105 (1864).
Platycercus forsteri, Finsch, Papag. ii. p. 287 (1868).

Native names.

Kakariki, Kakawariki, Powhaitere, Porere, and Torete.

♂ prasinus, uropygio paullò lætiore: genis et corpore subtùs flavicanti-viridibus: pileo antico, maculâ anteoculari, alterâ supraauriculari et plumis paucis ad latera uropygii positis puniceis: occipite ad basin plumarum celatè citrino: tectricibus alarum dorso concoloribus: remigibus brunneis, alâ spuriâ lætissimè ultramarinâ: primariis extùs ad basin ultramarino, versus apicem augustè flavido marginatis: caudâ suprà lætè prasinâ, subtùs magis flavicante: subalaribus cyanescenti-viridibus: maxillâ cyanescenti-albâ, versus apicem nigricante, mandibulâ omninò nigricante: pedibus pallidè brunneis: iride rubrâ.

Adult male. General plumage bright grass-green, lighter, or rather yellowish-green, on the underparts. Forehead, crown, and streak across the eye terminating on the ear-coverts deep crimson, with a spot of the same, more or less distinct, on each side of the rump; on the nape a broad basal mark of yellowish white, observable only when the plumage is disturbed or raised. The wing-feathers are dusky black, lighter on the under surface, and crossed by an obscure yellowish band; the outer primaries and their coverts, as well as the bastard quills, bright blue on their outer webs. Irides cherry-red; upper mandible bluish white, with a black tip; lower mandible bluish black; feet pale brown. Extreme length 12 inches; wing, from flexure, 5·5; tail 6; culmen ·8; tarsus ·8; longer fore toe and claw 1·15; longer hind toe and claw 1.

Adult female. Of similar plumage to the male, although the frontal crimson cap is not so conspicuous. It is, however, somewhat smaller. Extreme length 10·25 inches; extent of wings 14; wing, from flexure, 5; tail 5; tarsus ·75; longer fore toe and claw 1.

Young. The plumage of the young bird does not differ appreciably from that of the adult.

Varieties. Like many other members of the large natural family to which it belongs, this species exhibits a strong tendency to variability of colour; and the slight differences which some of the ornithologists of Europe have recognized as sufficient specific characters are clearly of no value whatever. A specimen brought to me by a native, in the Kaipara district, many years ago, had the whole of the plumage of a brilliant scarlet-red. Another, obtained in the woods in the neighbourhood of Wellington, had the green plumage thickly studded all over with spots of red; this handsome bird was caged, and at the first moult the whole of the spots disappeared. An example of this species in the British Museum has the abdomen and under tail-coverts bright yellow mixed with green; the thigh-spots very large and bright; the rump stained, and the tail obscurely banded on the upper surface, with dull yellow.

A Southland paper thus describes a specimen which was shot in the Seaward Bush:—"One of the most beautifully plumaged native birds we have ever seen was shown us yesterday by Mr. James Morton, a taxidermist, to whom it had been handed to be stuffed. It is a variety of *Platycercus novæ zealandiæ*, and proved to be a male. Instead of the usual green hue, the feathers of the one in question are tipped and edged with green on a beautiful lemon-yellow ground—the small feathers of the wing showing a steel-blue tint at the edges, or mixed bronze and yellow. The large pinion-feathers are yellow and green, merging into bronze at the tips—the tail-feathers being similarly coloured. The beak is surmounted by a crescent-shaped patch of blood-red, and there are two others on the back."

I have in my possession a feather of rich uniform yellow with a white shaft, from the tail of a tame bird of this species, formerly in the possession of the Wellington Working Men's Club, in which all the rest of the plumage was of the normal colour. I am indebted to Mr. W. W. Smith, of Oamaru, for a curious specimen (♂) in which the back, rump, upper surface of wings, and nearly the whole of the abdomen are marked with irregular patches of pale lemon-yellow.

There are three very beautiful varieties in the Otago Museum:—

No. 1 has the entire plumage of a uniform vivid canary-yellow, except that the vertex, ear-coverts, and uropygial spots are crimson as in the ordinary bird; there are a few dashes of ultramarine blue on the tertials and some "invisible green" markings on the quills and tail-feathers, the shafts of which are white, as though the normal colours had here endeavoured to assert themselves; the bill, feet, and claws are white. The crimson markings, especially on the sides of the uropygium, are bright and conspicuous, and the bird altogether is as lovely an object as the most ardent ornithologist could desire as the type of a new species; but, alas, it is nothing but a "freak of nature" whose exact counterpart may never occur again. This specimen was obtained at Seaward Bush in October 1874, and was presented to the Museum by Mr. J. M. Broderick.

No. 2 is a beautiful instance of cyanism. The entire plumage of the cheeks, throat, and underparts is a delicate marine-blue, or isabelline, the feathers on the lower parts and sides of the body narrowly edged with dusky; supplying the place of crimson on the vertex and ear-coverts is a pale yellowish or greyish brown; the rest of the upper surface is a deeper isabelline, varied with a still darker shade of blue, and with the feathers more distinctly margined with dusky; there are no uropygial spots; the quills are marked with ultramarine as in ordinary specimens; the tail-feathers have greenish reflections, with a wash of blue down the outer vane of the lateral ones, the under surface of wings and tail being dusky brown. Bill and feet of the normal colours. This is the Parrakeet mentioned by Prof. Hutton as the "blue variety from Southland."

No. 3 is a very different looking bird, from Invercargill. The entire plumage is dirty yellow, with a varying wash of green, which is deepest on the underparts and least apparent on the quills and tail-feathers; the vertex and ear-coverts are crimson, the former having a flush of canary-yellow along its posterior edge; the shafts of the quills and tail-feathers are white, and on the primaries and tertials there is just the faintest indication of the normal colour in a delicate shade of greenish blue; the upper wing-coverts are washed on their edges with green; the crimson uropygial spots are present, and the bill and feet are the same as in ordinary specimens.

Note. The synonymy of the genus *Platycercus*, as may be seen above, has been involved in much confusion. We are indebted to Dr. Otto Finsch, of Bremen, for a complete elucidation of the subject, in his able Monograph of the Psittacidæ (Die Papageien, ii. p. 275, 1868). Examples of *P. novæ zealandiæ* vary much in size and in the depth of their colouring. The shade of the prevailing green, the brilliancy of the crimson vertex, and the extent of red colouring on the ear-coverts and of blue on the wings are alike variable.

Dr. Finsch is of opinion that *P.* (*Cyanorhamphus*) *saisetti* (Verr.) is inseparable from this species. On comparing a specimen sent by Mr. Edgar Layard from New Caledonia to the Otago Museum, I find that this bird differs from *P. novæ zealandiæ* only in having the sides of the face, throat, breast, and underparts generally greenish yellow, deepening into grass-green on the sides of the body and on the flanks. If, however, this is a constant character I accept it as specific. There is a wash of blue on the outer vanes of the tail-feathers, but this may be an accidental peculiarity. The crimson of the vertex likewise has a wash of yellow in it, to which the same remark will apply, for I have met with New-Zealand examples tinged in the same manner. The crimson uropygial spots in Layard's specimen have an admixture of yellow; and the bill is blue and black, without any of the whiteness characteristic of our bird.

I am of opinion that *P. forsteri*, admitted with some hesitation by Dr. Finsch, and founded on a single example in the British Museum, is nothing but *P. novæ zealandiæ*, with the red uropygial spots accidentally absent; and I have accordingly included it in the synoptical history of this species as given above.

There is an example in the Otago Museum with an abnormally developed bill, as shown in the accompanying woodcut. It likewise has a wash of yellow on the secondary quills.

The Red-fronted Parrakeet is very generally dispersed over the whole country, but is more plentiful in the southern portion of the North Island than in the far north, where the yellow-fronted species predominates. It frequents every part of the bush, but appears to prefer the outskirts, where the vegetation is low and shrubby, as also the wooded margins of creeks and rivers. It is often met with amongst the dense koromiko (*Veronica*) which covers the low river-flats, or among the bushes of *Leptospermum* and other scrub. It seldom ventures beyond the shelter of the woods, unless it be to visit the farmer's fields for its tithe of grain, or to reach some distant feeding-place, when it rises rather high in the air and flies rapidly, but in a somewhat zigzag course. When on the wing it utters a hurried chattering note; and when alarmed, or calling to its fellows, it emits a cry resembling the words "twenty-eight," with a slight emphasis on the last syllable. It often resorts to the tops of the highest trees, but may always be enticed downwards by imitating this note. It is gregarious, forming parties of from three to twelve or more in number, except in the breeding-season, when it is generally met with in pairs *.

Its food consists chiefly of berries and seeds; but I suspect that it also devours small insects and their larvæ; for I have observed flocks of a dozen or more on the ground, engaged apparently in a search of that kind, and it is a well-established fact that several of the Australian members of this group subsist partly on insect food. When the corn-fields are ready for the harvest, flocks of this gaily-coloured Parrakeet resort to them to feed on the ripe grain; and it is very pretty to see them, on any alarm being given, rise in the air together and settle on a fence, or on the limb of a dead tree, to wait till the danger has passed, keeping up all the time a low, pleasant chatter.

Sir William Fox, after his return from a trip through the Canterbury district in 1871, informed

* "At nesting-time the old birds often indulge in a low murmuring note to each other." (Journ. of Science, ii. p. 480.)

me that the farmers had suffered a visitation, tens of thousands of these birds having descended on their ripening crops of corn and proved almost as destructive as an army of locusts. It is difficult to account for these occasional irruptions in such numbers, in the case of a bird not otherwise plentiful. The sudden failure, or scarcity, of the ordinary food-supply in certain wooded districts is the most rational way of accounting for such unexpected visitations; but, apart from this cause, there are doubtless others directing and regulating the migratory impulse, although at present we are unable to define them. The same sort of thing is occurring, more or less, in every part of the world and in every department of the animal kingdom. Beyond laying down general principles, it is impossible to explain some of the phenomena. For example, who has been able to account satisfactorily for the sudden irruption of Pallas's Sand-Grouse (*Syrrhaptes paradoxus*) into Europe in 1863? These birds, which had scarcely ever been heard of before, came from beyond the Caspian Sea, traversing some 4000 geographical miles, spreading themselves over Europe in countless flocks like a Tartar invasion, without any apparent cause, and disappeared again just as suddenly and unaccountably as they had come. The same question may be asked of the remarkable influx of the Waxwing into England in the winter of 1849–50, an event quite unparalleled in the ornithological history of the country. To come nearer home, what naturalist was able to account, more than theoretically, for the plague of caterpillars which (up to the time of the introduction of the much-abused House-Sparrow) periodically, but at long intervals, visited our country districts, coming in countless millions, sweeping all before them, and utterly wrecking the hopes of the farmers?

This species bears confinement remarkably well, and is very docile and familiar even when taken as an adult bird. It is also very intelligent, and possesses the faculty of mimicry in a high degree.

It is quite the cottagers' friend in New Zealand. Riding or driving through the suburbs of the provincial towns—the Porirua and Karori districts for example, near Wellington—you will notice in many of the farmers' houses and roadside cottages small wooden cages of primitive construction (often merely a candle-box or whisky-case, faced with wire netting or thin wooden bars) fixed up to the front of the building or under the simple verandah. On closer inspection each of these cages will be found to contain a tame Parrakeet—the pet of the rustic home and "Pretty Poll" of the family; and I have often been quite interested at finding how attached these simple people become to their little captive.

One of these birds has been in the possession of a lady at Christchurch (Canterbury) for more than eight years. Although full-grown when first caged, it has learnt to articulate several words with great clearness. It is very tame, and displays a considerable amount of intelligence—leaves its cage every day for exercise, and returns to it immediately on the appearance of a stranger. It knows its fair owner's voice, will respond to her call, and will "shake hands" with each foot alternately in the most sedate manner. Another, in our own possession, survived confinement for more than eleven years, and appeared then in perfect health and strength, when it fell a victim to the household puss. This bird could articulate sentences of three or four words with great precision; and the loss of so intimate a family-friend was "sincerely lamented" by all our circle.

At the Foxton railway-station there used to be (and may be still) a tame Parrakeet that had learnt to say "Be quick!" and was accustomed to repeat these words with energy and clear articulation as the passengers by train crowded round the ticket-window.

In certain particular woods where, for some unaccountable reason, all other birds are scarce, this Parrakeet may always be found. One such tract lies between Cambridge and Ohinemutu, where the coach-road passes through some twelve miles of the most picturesque bush imaginable. Destitute as it generally is of bird-life, the scenery is enchanting. At intervals of a few miles there are deep wooded gorges, the eye often resting on tree-tops some three hundred feet below the spectator. The bush itself is of the usual mixed kind, with every gradation of shade in green and brown; but the dominant feature, as almost everywhere else in New Zealand, is the beautiful tree-fern, which I think I have never seen in greater beauty or abundance. In some places you come upon whole groves of *Cyathea smithii*, with its grand expanse of graceful fronds, then groups of *Dicksonia antarctica* nestling among the denser vegetation on both sides of the road. And before the eye has had time to take in the full beauty of the scene, the aspect changes and straggling clumps of *Cyathea cunninghamii* present themselves to view, with their soft and feathery fronds, some exhibiting an open coronet of slender stalks and others with their crowns depressed; then, at a fresh turn in the road, far away in the depths of the gorge, and shaded by the overhanging foliage, may be seen superb umbrella-tops of mamaku (*Cyathea medullaris*), resting on giant stems often sixty or seventy feet in height. In the more open glades of the forest the stately *Cyathea dealbata* lifts its graceful head, those in exposed positions displaying the silvery white of their under surface with every breath of wind. On nearer inspection other forms may be distinguished, there being apparently no limit to their beauty and variety. Each fern is a study in itself, and the natural grouping is such as no landscape gardener with all his artificial skill could ever produce. Some have their stems encircled with vines, ferns, and creepers, from base to summit; others have their trunks hung thickly round with the withered fronds of a former growth. Some have slender, naked stems, while others have massive pyramidal trunks. Some stand out clearly and sharply defined against the darker background, while others are almost lost in the luxuriance of their epiphytic growth. I do not mention the ever-present ground-ferns, in their infinite variety, because no New-Zealand bush could well exist without them; but I ought to notice here the most beautiful object on the road. A little more than halfway through, from the Cambridge side, our coach stopped at a point near which the crape-fern, or "Prince of Wales' feather," is known to exist. We alighted and entered the forest. At a distance of only a chain from the highroad we came upon one of the loveliest sylvan sights I have ever witnessed. This was a dense bed of *Todea superba* growing close together, each plant with beautiful deep green, velvety fronds, arranged like the feathers on a shuttlecock, each with a spread of three feet or more, and covering altogether about an acre of ground. This luxuriant bed of an elsewhere rare fern, of the richest green and of crape-like texture, closely covering the ground and protected from the sun by a thick forest canopy, presented a picture of surpassing beauty never to be forgotten.

A hole in a decaying or dead tree affords this species a natural breeding-place, the eggs being laid on the pulverized rotten wood at the bottom; for, as a rule, there is no further attempt at forming a nest *. I ought to mention, however, that in the Canterbury Museum there is a loose nest, formed of moss, and lined with fern-hair and green Parrakeet feathers, which was taken from the hollow of a tree and is assigned (I believe correctly) to this species. The months of November and December constitute the breeding-season. The eggs vary in number from three to seven; and a native stated that he once found a nest containing as many as eleven; but five is the usual number. Captain Mair informs me that a pair of these birds bred in the hollow trunk of a hinau-tree for several successive years, although robbed of their young every season, and that he has frequently observed the cock bird feeding the hen, during incubation, by regurgitating berries from his crop.

Although exhibiting a preference for hollow trees, they sometimes nest in the holes or crevices of rocks. On the Upper Wanganui the natives pointed out to me a small round cavity in the perpendicular cliff forming the bank of the river, and assured me that this was the entrance to a small chamber where a pair of Parrakeets had reared their young in security for many years. The eggs are very broadly oval, measuring 1·05 by ·85 inch; they are pure white and are very finely granulate on the surface, sometimes with minute limy excrescences near the larger end.

* Prof. Scott states that during a visit to Campbell Island he found this species of Parrakeet there "in great numbers round the shore," and that, in the absence of woods, it makes its nest under the grass-tussocks.

PLATYCERCUS AURICEPS.

(YELLOW-FRONTED PARRAKEET.)

Platycercus auriceps, Kuhl, Consp. Psittac. p. 46 (1820).
Pacific Parrot, var. C, Lath. Gen. Syn. i. p. 252 (1781).
Psittacus pacificus, var. 𝛿, Gm. Syst. Nat. i. p. 329 (1788).
Platycercus auriceps, Vigors, Zool. Journ. i. p. 531 (1825).
Platycercus novæ zelandiæ, Bourjot St.-Hilaire, Perroq. t. 37 (1837).
Euphema auriceps, Licht. Nomencl. Av. p. 72 (1854).
Cyanoramphus auriceps, Bonap. Rev. et Mag. de Zool. vi. p. 153 (1854).
Cyanoramphus malherbi, Souancé, Rev. et Mag. de Zool. ix. p. 98 (1857).
Platycercus malherbii, Gray, Cat. Brit. Mus. Psitt. p. 14 (1859).
Coriphilus auriceps, Schlegel, Dierent. p. 77 (1864).

Native names.

The same as those applied to the preceding species.

Ad. P. novæ zealandiæ similis, sed valdè minor, et vertice aureo, fronte puniceâ facilè distinguendus.

Adult male. General plumage beautiful grass-green, paler or more suffused with yellow on the underparts. A band of dark crimson connects the eyes, passing across the forehead, immediately above the nostrils; upper part of forehead and crown golden yellow; on the nape a basal spot of yellowish white, apparent only on moving the feathers; on each side of the rump a conspicuous spot of crimson; quills dusky black, crossed on their under surface with a band of pale yellow; the outer web of the bastard quills and first four primaries, with their coverts, indigo-blue, narrowly margined with yellow. Irides pale cherry-red; upper mandible bluish white at the base, black towards the tip; under mandible bluish black; feet pale brown. Extreme length 10·5 inches; wing, from flexure, 4·5; tail 5; culmen ·6; tarsus ·6; longer fore toe and claw 1; longer hind toe and claw ·9.

Varieties. Like the preceding bird, this species also exhibits abnormally coloured varieties. A young bird, brought to me from the nest, and not fully fledged, had the plumage of the body pale yellow, shaded with green on the upper parts, and the quills and tail-feathers marked with red. Another had numerous light crescentic marks on the wing-coverts. In the summer of 1863 I obtained a very beautiful variety at Manawatu. I found it in the hands of a labouring settler, who had purchased it from the natives for something less than a shilling. Finding him unwilling to part with it, I tempted him with a guinea, and secured the prize. It was a bird of the first year, and presented the following appearance:—Frontal band crimson; vertex golden yellow; space around the eyes and a band encircling the neck green; head, shoulders, and lower part of back red, the intermediate space variegated with red and green; quills dusky, obscurely banded with yellow, and margined on the outer web with blue; wing-coverts greenish yellow, barred and margined with red; tail-feathers green, obscurely barred with yellow in their apical portion; underparts green, variegated with crimson and yellow, an interrupted band of the former colour crossing the breast. Like the spotted variety of *P. novæ zealandiæ* already mentioned, within a short time it commenced to moult, and was fast assuming the common green livery of the species, when it was accidentally killed. This specimen, which still exhibits traces of its original colours, belongs now to the type collection in the Colonial Museum.

A pretty male bird obtained by Reischek near Dusky Sound, at an elevation of 2000 feet, has the entire plumage tinged with saffron-brown, which is darkest on the breast, shoulders, and upper wing-coverts; the yellow on the vertex is mixed with orpiment-orange; the blue on the bastard quills and primaries is unusually brilliant; the scapulars have a wash of yellow; and the uropygial spots are very indistinct.

I have seen several examples exhibiting marks of red on the vertex and crown; and in the Canterbury Museum there is a specimen which has the frontal band dull red instead of crimson, the crown, upper surface of wings, and the abdomen more or less marked with yellowish brown, the primaries tipped and the secondaries largely margined with paler brown.

Mr. Henry Travers obtained one on Mangare Island (at the Chathams) "with a faint tinge of yellow on the head."

A specimen obtained by Dr. Lemon at Takaka, in the South Island, and presented to the Colonial Museum, is one of the loveliest objects in the mounted collection. The whole of the plumage is of a vivid canary-yellow, which is brightest on the vertex, and is bordered by a narrow band of crimson across the forehead. The uropygial spots are large and of flaming crimson. The only indications of the normal colour are on the quills and tail-feathers. The quills are pale canary-yellow, inclining to white; the middle primaries in one wing are clouded with dark grey, and in the other wing there is a splash of green across the secondaries; in both wings the bastard quills are edged with blue; the two middle tail-feathers are stained with green, and the two succeeding on either side are green in their central portion; one of the outer laterals also is marked with green. Bill pure white; legs and feet flesh-white.

This bird, as Dr. Lemon informs me, was shot in May 1882, in Eve's Valley, Waimea, by Mr. Fabian, telegraph lineman, who had the good sense to preserve it. By the courtesy of Sir James Hector, it was brought to England, and exhibited in the New-Zealand Court at the Colonial and Indian Exhibition, 1886.

Obs. This species is very readily distinguished from all the other members of the group of *Platycerci* by its beautiful golden vertex. Individuals vary both in size and in the brilliancy of their plumage.

Some specimens exhibit the yellow vertex stained more or less with crimson. The type of *Platycercus malherbi*, in the British Museum, received from the Auckland Islands, and characterized by Souancé as "encore plus petit que l'*auriceps*," is nothing but a very small example of this species. There is an equally small one in the same collection from the Bay of Islands, New Zealand.

Professor Hutton states that two specimens brought by Mr. Henry Travers from the Chatham Islands are slightly larger than the New-Zealand bird.

The Yellow-fronted Parrakeet, although generally dispersed over the country in all suitable localities, is more plentiful than the red-fronted species in the northern parts of the North Island, and less so as we approach Cook's Strait. In the South Island, however, the two species appear to be more equally distributed.

In habits this bird closely resembles the preceding one; but it is less gregarious, being seen generally in pairs. It loves to frequent the tutu bushes (*Coriaria ruscifolia*), to regale itself on the juicy berries of this bushy shrub; and on these occasions it is easily snared by the natives, who use for that purpose a flax noose at the end of a slender rod. When feeding on the tutu-berry, the whole of its interior becomes stained of a dark purple. When the wild dock has run to seed, this pretty little Parrakeet repairs to the open fields and feasts on the ripe seeds of that noxious weed; at other seasons the berries of *Coprosma lucida*, *Fuchsia excorticata*, and other forest-shrubs afford it plentiful and agreeable nutriment.

Far up the course of the Northern Wairoa, just below Mangakahia, the banks of the river for some miles are cleared of the original forest, the land having been in years gone by occupied by Maori plantations. A new growth has covered the long-abandoned "wairengas," and, just along the margin of the stream, the soil, enriched by deposits of fine silt through the occasional overflowing of the muddy waters, supports a belt of tupakihi, intermixed with other shrubs and completely overgrown with climbing convolvulus. In no part of New Zealand have I found the Yellow-fronted

Parrakeet so numerous as in this tangled retreat, especially at the season when the tutu-berries have ripened and are hanging in drupes from every branch. I have seen a native lad enter this thicket —which is open below and matted overhead—and, armed only with a flax noose at the end of a slender rod, catch numbers of them with perfect ease, by slipping the loop over the head of the unsuspicious bird.

My son met with it in the stunted woods in the Owhaoko-Kaimanawa district, when the whole country was under snow.

At irregular periods, after intervals of from seven to ten years, this Parrakeet (in company with the preceding species) visits the settled and cultivated districts in astonishing numbers, swarming into the gardens and fields, devouring every kind of soft fruit, nibbling off the tender shoots on the orchard trees, and eating up the pulse and grain in all directions. Sir William Fox gave me a graphic account of one of these sudden irruptions in the South Island in the summer of 1870-71, when great injury was done to the crops. The last of these visitations occurred in the early part of 1886, and the one before that at the close of 1877. On each of these occasions much public interest was excited by the occurrence, and many theories (such as the devastation of the country by bush-fires) were put forward to account for this recurrent "plague of Parrakeets." Whilst the newspapers were busy with these more or less colourable theories, the birds vanished as suddenly as they had come.

There is a widespread popular belief that the movements of certain species of birds indicate approaching climatic changes, or form a sort of index to the seasons; and it would not be difficult to find and multiply apparent proofs of such a connection. But the theory, as generally accepted, is true only to a certain extent. Everyone is probably aware that birds, of all animals (except perhaps frogs), are the best natural barometers. For example, to every native colonist the vociferous cry of the Sparrow-Hawk betokens change: the altitudes at which these birds habitually fly make them susceptible to the slightest change of temperature, and to all observers of outdoor nature they announce the fact with no uncertain sound. Even our little Wood-Robin, which keeps near the ground and never leaves the seclusion of its forest home, is so ready to detect any atmospheric disturbance and to predict by its peculiar note a change of weather, that it is commonly called the "rain-bird" in many parts of the country. The presence on a calm day of the snow-white Gannet, sailing majestically over our harbours and, ever and anon, plunging headlong into the placid waters, or of a flock of playful Sea-Gulls coming inland to rest themselves in our fields and pastures, is a sure indication that a storm is brewing at sea, although there may be no actual appearance of it at the time. But, of course, it does not follow from such instances as these that any species of bird can foresee an impending change of season, or, by any ratiocinative process, prepare for it by migration. So far as I understand the facts, the case is simply this:—The failure, more or less complete, of their natural food (which in itself is often a safe indication of seasonal derangement) necessitates the migration of all birds dependent on such food-supply to other parts of the country in search of the ordinary means of subsistence. And as the migration always precedes the other evidences of climatic change, the popular notion that birds are instinctively prophetic in the matter of seasons is easily accounted for. The sudden irruption of Parrakeets in the South Island, referred to above, to such an extent as to be an actual "pest" is, it seems to me, but an illustration of this natural law of cause and effect. This pretty little Parrakeet is strictly an arboreal bird. It is an inhabitant of the woods, and, besides being well distributed, its plumage is so admirably suited to its natural surroundings by the law of assimilative colouring that, although it exists in tens or hundreds of thousands, it is rarely seen, and except to the lovers of nature and bush-craft its very existence is almost unknown to the colonists. But when, from some unknown cause, there is a failure of its everyday food-supply, the fact is proclaimed by the sudden and unexpected appearance of countless numbers of these birds in our cultivated fields, gardens, and hedgerows, all fugitives under a common calamity and becoming a nuisance by the very intensity of their hunger.

The same thing happens, although in a less pronounced manner, with the Tui and Korimako, both of which species occasionally appear in our midst, all miserably lean and in a state of absolute starvation.

What occasions this widespread failure of the natural food is generally a mystery; but that such failure is the chief factor in the migratory impulse there can be little doubt. The case of the Passenger Pigeon in the United States is strongly in point. The movements of this bird are irregular in the extreme—completely disappearing from entire districts for years together, and then returning in prodigious numbers (in flocks of hundreds of thousands), the migration being regulated entirely by the scarcity or abundance of the natural food.

In captivity it is very gentle and tractable, but it is far inferior to the larger red-fronted bird in its talking-capacity. One or two instances of its being taught to articulate words of two syllables have come to my knowledge; but, as a rule, the attempt to instruct it ends in failure.

Like its congener it nests in hollow trees, and lays from five to eight eggs, resembling those of *Platycercus novæ zealandiæ*, but smaller. Specimens in my son's collection measure ·9 of an inch in length by ·75 in breadth; others are more broadly ovoid, measuring ·85 by ·70, and are stained yellowish white, probably the result of incubation. Major Mair informs me that he watched a pair of these birds breeding in the cavity of a dead tree for three successive seasons. The first year's brood numbered five, the second eight, and the third seven.

As will be seen by the synonymy at the head of this article, there has been a considerable amount of confusion in the nomenclature of this and the preceding species, notwithstanding their strongly marked characters. I trust that the reference lists and full descriptions now given will, for the future, make it impossible to confound these forms with other members of the genus. As a brief review, however, of the types in the National Collection may help to elucidate the synonymy of the group, I will reproduce here the notes on the subject which I published in the 'Transactions of the New-Zealand Institute' (vol. xi. pp. 368, 369).

British-Museum Collection.—My examination of the types gives the following results:—*Platycercus aucklandicus* not distinguishable from *P. novæ zealandiæ*, but smaller than ordinary examples; beak decidedly smaller, being of same size as in *P. auriceps*, but lighter at base; ear-spots indistinct; frontal spot less extensive, but of same colour as in *P. novæ zealandiæ*. *P. malherbii* = *P. auriceps*, but smaller than average specimens of the latter. *P. pacificus* similar to *P. novæ zealandiæ*, but much larger, with a more robust bill. *P. erythrotis*, from Macquarie Islands, = *P. pacificus*, but with lighter plumage. *P. forsteri* = *P. novæ zealandiæ*, with the thigh-spots accidentally absent. There is another specimen marked "*Platycercus forsteri*," to which I shall refer again presently, in very different plumage. *P. cookii* = *P. pacificus*. *P. unicolor*, a much larger and very distinct species. *P. rayneri*, from Norfolk Island, is like *P. pacificus*, but larger and with a more powerful bill; the frontal spot is more extensive but lighter in colour; ear-spots small and obscure as compared with *P. novæ zealandiæ*. I think we may pretty safely conclude that *P. rayneri* is in reality *P. pacificus*, although the British-Museum specimen is both larger and lighter-coloured than ordinary specimens of the latter. *Platycercus ulietanus*, from the Society Islands, is very distinct in appearance from all those enumerated above. The so-called *P. forsteri*, before referred to, labelled as from the main island Otaheiti, appears to hold an intermediate position between *P. ulietanus* and *P. pacificus*. It has the general plumage of *P. pacificus* but of much duller tints, mixed with brown on the upper parts and clouded with a colder green on the underparts. It wants the crimson vertex; but there is a frontal patch of brownish black corresponding to the colour of *P. ulietanus*, which changes to crimson in front of the eyes; behind which, also, there is a small obscure spot of dull crimson. It has the concealed nuchal patch of yellowish white which is found in *P. pacificus*; while, on the other hand, it has the bright crimson rump which is characteristic of *P. ulietanus*. The tail has a dingy, washed-out appearance, and the colours of the plumage generally are very undecided. The bill and feet are exactly as in *P. ulietanus*, of which species this bird may be an accidental variety, or possibly a hybrid. There is likewise in this collection a specimen of our *P. novæ zealandiæ*, exhibiting much bright yellow mixed with the green on the abdomen and under tail-coverts. It likewise has the thigh-spots very large and bright; the rump stained, and the tail obscurely banded on the upper surface with dull yellow.

PLATYCERCUS ALPINUS.

(ORANGE-FRONTED PARRAKEET.)

Platycercus alpinus, Buller, Ibis, 1869, p. 39; Birds of New Zealand, 1st ed. Intr. p. xvi (1873).

Ad. P. auricipiti similis, sed minor, et fronte aurantiacâ, vertice pallidè flavo distinguendus.

Adult. Plumage bears a general resemblance to that of *Platycercus auriceps*; but the frontal band is orange and the vertex pale yellow; and there is an absence of the yellow element in the general plumage, which is of a cold pure green, much paler on the underparts; the rump-spots, moreover, are smaller and less conspicuous, being orpiment-orange instead of crimson. Extreme length 9 inches; wing, from flexure, 4·2; tail 4·5; culmen ·5; tarsus ·5; longer fore toe and claw ·85; longer hind toe and claw ·75.

Note. In treating of the members of this section in my former edition, I had recourse to Dr. Otto Finsch's elaborate Monograph on the Parrots ('Die Papageien'), a work the care and labour of which may be estimated from the fact that, of the 350 species described therein, all but 18 were examined by the author personally. Accepting the decision of so able an authority, I agreed to sink my *Platycercus alpinus* as a species, and to consider it the young state of *P. auriceps* (*vide* Birds of N. Z. 1st ed. pp. 61 & 62). The validity of the species, however, was afterwards established beyond all doubt. More than twenty specimens were brought to this country before the completion of my work; and I accordingly took the opportunity, when writing the Introduction, to rehabilitate the species (at page xvi) under the head of Platycercidæ.

This species is probably the bird mentioned by Latham as the "Buff-crowned Parrot."

THIS form differs from its near ally (*Platycercus auriceps*) both in size and in the tints of its plumage; so that we have, thus far, three species of *Platycercus* presenting a distinct gradation in size and colouring. In *P. novæ zealandiæ* the frontal spot, ear-coverts, and rump-spots are deep crimson, while the general plumage is dark green. In the smaller species (*P. auriceps*) the frontal band is crimson and the vertex golden, while the general plumage is a warm yellowish green. In *P. alpinus*, which is smaller, again, than the last-named species, the frontal band is orange, and the vertex pale yellow, while there is a further modification of the body-plumage as described above. On comparing the bills of the two species the difference is very manifest, that of *P. alpinus* being fully one third less than that of *P. auriceps*. A fourth species has yet to be mentioned, in which a size intermediate between *P. auriceps* and *P. alpinus* is combined with the well-defined plumage of *Platycercus novæ zealandiæ*.

The present bird was originally described by me, under the above name, from specimens obtained in the forests of the Southern Alps, at an elevation of from 2000 to 2500 feet. In its native haunts it may be found frequenting the alpine scrub, in pairs or in small parties, and is very tame and fearless. It is by no means uncommon in the wooded hills surrounding Nelson.

Mr. Reischek met with this little Parrakeet in the scrub on the summit of Mount Alexander (above Lake Brunner); and he met with the species again on the Hen, where he shot two, and on the Little Barrier, where he observed another pair on the highest peak and killed the male. It does not exist on the opposite mainland, nor indeed, so far as I am aware, in any part of the North Island.

At Nelson I saw many caged birds of this species, and one in particular was remarkable for the clear manner in which it articulated the words "pretty Dick," repeating them all day long in the most untiring way.

PLATYCERCUS ROWLEYI.

(ROWLEY'S PARRAKEET.)

Platycercus rowleyi, Buller, Trans. N.-Z. Instit. vol. vii. p. 220 (1874).

Ad. P. novæ-zealandiæ similis, sed conspicuè minor.

Adult male. Similar in plumage to *P. novæ zealandiæ*, but considerably smaller. Total length 10 inches; wing, from flexure, 4·75; tail 5; bill, along the ridge ·55; tarsus ·45; longer fore toe and claw 1; longer hind toe and claw ·9.

Female. Slightly smaller than the male, but differing in no other respect.

Young. A specimen from Dusky Sound has the frontal spot of crimson mixed with green, and a line of undeveloped feathers in silvery shields along the base of the upper mandible; the aural bar of crimson very small and indistinct; the abdomen pale yellowish green; the bill greyish white, tinged with blue on the sides of the upper and base of the under mandible. The culmen measured along the curve only ·45 of an inch.

Obs. There is an appreciable difference in size between this bird and the type of Bonaparte's *P. aucklandicus*.

WHEN I was in England superintending the publication of the first edition of this work, the late Mr. Dawson Rowley of Brighton sent me for examination the skin of a red-fronted Parrakeet received from the South Island, and remarkable on account of its small size. On comparing the specimen with the type of Bonaparte's *Platycercus aucklandicus* in the British Museum, I came to the conclusion that although Mr. Rowley's specimen was somewhat less in size, both were referable to *P. novæ zealandiæ*, being only exceptionally small examples of that species. On my return, however, to the colony, my attention was directed to a very large series of Parrakeet skins collected by the late Mr. F. R. Fuller in the provincial district of Canterbury; and, after making due allowance for the great individual variation which some members of this group exhibit, I found it impossible to resist the conclusion that there does exist another species, having similar plumage to *P. novæ zealandiæ*, but so much smaller in size as to be even less than ordinary examples of the Yellow-fronted Parrakeet (*P. auriceps*). Mr. Fuller, who had skinned some hundreds of Parrakeets for the Canterbury Museum, assured me that the bones of this smaller red-fronted bird could be readily distinguished from those of *P. novæ zealandiæ*, being weaker and more slender, and more like the bones of *P. alpinus*. He likewise informed me that all his specimens of this small form had come from Canterbury North; and it seemed to me a significant fact that although *P. novæ zealandiæ* is a very common species in the North Island, none of the very small examples have been recorded there.

We have thus a regular gradation in the following sequence: *Platycercus novæ zealandiæ* (red-fronted), *P. auriceps* (yellow-fronted), *P. rowleyi* (red-fronted), and *P. alpinus* (orange-fronted).

In selecting a specific name to distinguish this diminutive form, I thought I might appropriately dedicate it to Mr. Dawson Rowley, who first called my attention to its existence, and whose interest in New-Zealand ornithology found expression in a charming little museum of rarities, numbering among its treasures the unique specimen of the Moa's egg from the Kaikoura sepulchre *.

Reischek met with this small form on the Hen, but on none of the other islands in the Hauraki Gulf, although *P. novæ zealandiæ* was abundant everywhere.

* There is an excellent figure of this species in Rowley's 'Ornithological Miscellany,' vol. ii. facing p. 115.

PLATYCERCUS UNICOLOR.

(ANTIPODES-ISLAND PARRAKEET.)

Platycercus unicolor, Vigors, Proc. Zool. Soc. 1831, p. 24.
Platycercus viridis unicolor, Bourj. St.-Hilaire, Perr. t. 34 (fig. fide Lear), 1837.
Cyanoramphus unicolor, Bonap. Rev. et Mag. Zool. 1854, p. 153; id. Naumannia, 1856.
Platycercus unicolor, Gray, Gen. of B. ii. no. 19 (1845); id. List of Psitt. 1859, p. 14; id. Ibis, 1862, p. 229.
Platycercus unicolor, Finsch, Die Papag. 1868, p. 289.
Platycercus unicolor, Buller, Trans. N.-Z. Instit. vol. vi. p. 121 (1873).
Platycercus fairchildii, Hector (*in litt.* 1886).

♂ *ad.* omninò prasinus, vertice capitisque lateribus lætioribus: dorso et corpore subtùs flavido lavatis: alâ spuriâ et primariis exterioribus extùs cyanescentibus: caudâ sordidè viridi, subtùs flavicanti-brunneâ: rostro nigro, versus basin albido: pedibus brunnescentibus: iride flavicanti-rubrâ.

♀ mari simillima, sed valdè minor et pallidior: maxillâ cinerascenti-albo, versus apicem nigricante, mandibulâ omninò cinerascenti-albâ.

Adult male. General plumage grass-green, brighter on the crown, sides of the head, face, and ear-coverts; back, rump, and all the under surface strongly tinged with yellow; primaries bright green on their outer vanes; the margins of the outermost primaries, as well as their coverts, and the whole of the bastard quills, indigo-blue; tail-feathers dull green, olivaceous or yellowish brown on their under surface. Bill black, greyish white towards the base of lower mandible; legs and feet dull brown. Total length 13·25 inches; wing, from flexure, 6; tail 6·25; culmen 1·25; tarsus ·9; longer fore toe and claw 1·4; longer hind toe and claw 1·25.

Female. Of smaller size and paler plumage than the male. Bill greyish white, the upper mandible brownish black in its apical portion, and with a clouded bluish spot in front of each nostril. Wing 5·75 inches; culmen 1; tarsus ·8.

Obs. My description of the male is taken from the type specimen in the British Museum; that of the female from the specimen referred to below.

Note. One of the specimens collected by Captain Fairchild (as stated below) was sent to the Canterbury Museum; and of this Prof. Hutton has sent me the following note:—"It answers very well to your description of the bird in Trans. N.-Z. Inst. vol. vi. p. 122, except that in the bill it is the basal half of the *upper* mandible that is greyish white, and not the lower. The measurements are as follows, but taken from the skin after it had been mounted:—Extreme length 14·25 inches; wing, from flexure, 6·25; tail 6; culmen 1; tarsus ·9; longer fore toe and claw 1·18; longer hind toe and claw ·95. The foot seems smaller than in the British-Museum specimen; but I measured with a pair of compasses from point of claw, while you may have measured round the curve."

Sir James Hector sends me the following account of two specimens received at the Colonial Museum:—"General plumage yellowish green, lighter on the underparts; forehead and cheeks with minute feathers of intense verditer-green; first three quills and coverts dull blue on outer web, rest dusky black; tail-feathers green above, dusky below; under surface ash-colour and very downy; upper mandible pale blue, with black margin and top; lower black; mandible not grooved. Legs and feet black; irides yellowish red."

On my first visit, in company with the late Mr. G. R. Gray, to the fine collection of Parrakeets in the galleries of the British Museum, a mounted specimen standing on the same shelf with *Platycercus novæ zealandiæ* and *P. auriceps* immediately arrested my attention. My companion informed me that this was the type of *Platycercus unicolor* (Vigors), and that it was supposed to have come from New Zealand. On further inquiry I found that the bird had come to the Museum from the Zoological Society's Gardens, where it had lived for some time, that its origin was quite unknown, and that the specimen was unique.

Mr. Gray had included the species in his "List of the Birds of New Zealand" (*l. c.*); but in the absence of any positive evidence as to the habitat I felt bound to omit it from my former edition.

The home of *Platycercus unicolor* has at last been discovered. Captain Fairchild, of the Government steamboat 'Hinemoa,' on a visit to Antipodes Island in March 1886, found the bird comparatively common there and brought several specimens back with him to New Zealand. One of these was forwarded to me by Sir J. Hector; and this has enabled me to add the description of the female to that of the hitherto unique specimen of the male bird in the British-Museum collection.

Although this type specimen (which has been in the Museum for upwards of fifty years) had no ascertained habitat it was always supposed to have come from New Zealand, and Mr. G. R. Gray included it in his list of our avifauna, published in 'The Ibis' (1862).

Captain Fairchild, who is an excellent observer, reports that on Antipodes Island he found it inhabiting a plateau 1320 feet above the sea. It was very tame and easily caught. He never saw it take wing, which he attributes as much to the boisterous winds that sweep over this exposed island as to its naturally feeble powers of flight. It habitually walks and climbs among the tussock-grass, reminding one of the habits of the Australian Ground-Parrakeet (*Pezoporus formosus*).

Besides collecting several good specimens, Captain Fairchild brought with him to Wellington a live one. Sir James Hector sends me the following account of this interesting bird, for which he had proposed the name of *Platycercus fairchildii*:—"It is a *ground* Parrakeet, *i. e.* a Parrakeet that resembles a Kakapo. It is twice the bulk of *P. novæ zealandiæ*, flies feebly, does not care to perch, climbs with its beak and feet, and walks in the same waddle-and-intoed fashion as the Kakapo."

So far as external characters go there is absolutely nothing by which to separate this bird from *Platycercus*. An investigation of its skeleton (of which the Colonial Museum has fortunately secured a specimen) may perhaps bring to light some new character showing its relation to a different group. But my own view at present is that the apparent inability to use its wings for purposes of flight is just another of those remarkable cases where the muscles have in some degree atrophied through long-continued disuse. Even in the case of *Pezoporus* from Australia, neither Mr. Sharpe nor I can find anything, apart from the different style of coloration, by which to distinguish the genus.

Sir George Grey tells me that forty years ago the natives assured him of the existence of a strange Parrot on Cuvier Island, and described the sexes as differing from each other. Excepting only Mair Island, Cuvier is the most seaward point in the Hauraki Gulf. It is a mountainous island of a few thousand acres, rising abruptly from the ocean and clothed to the very summit with dense vegetation. It is difficult of approach, but there are several practicable landing-places in fine weather. Tamihana Te Rauparaha and other natives of the present generation declared to Sir George that they had in their youth visited the island and actually seen these Parrots. He suggested to me that they might be descendants of some stragglers from the South-Sea Islands; but if such birds do really exist there, it seems far more likely that they are the last survivors of a species that has become extinct on the mainland, for, as before remarked, expiring forms linger longest on sea-girt islands remote from the coast, where the struggle for existence is less severe.

NESTOR MERIDIONALIS.

(KAKA PARROT.)

Southern Brown Parrot, Lath. Gen. Syn. i. p. 264 (1781).
Psittacus meridionalis, Gm. Syst. Nat. i. p. 333 (1788).
Psittacus nestor, Lath. Ind. Orn. i. p. 110 (1790).
Psittacus australis, Shaw, Mus. Lever. p. 87 (1792).
Nestor novæ zealandiæ, Less. Tr. d'Orn. p. 191 (1831).
Centrourus australis, Sw. Classif. of B. ii. p. 303 (1837).
Nestor meridionalis, Gray, in Dieff. Trav. ii., App. p. 193 (1843).
Psittacus hypopolius, Forst. Descr. Anim. p. 72 (1844).
Nestor australis, Gray, Gen. of B. ii. p. 426 (1845).
Nestor hypopolius, Bonap. Rev. et Mag. de Zool. 1854, p. 155.
Nestor occidentalis, Buller, Ibis, 1869, p. 40; Hutton, Cat. of N. Z. Birds, p. 20 (1871); Buller, Birds of New Zealand, 1st ed. p. 50 (1873).

Native names.

Kaka; varieties distinguished as Kaka-kura, Kaka-kereru, Kaka-pipiwarauroa, Kaka-reko, and Kaka-korako.

Ad. pileo albicanti-cinereo, plumis nuchæ brunneo marginatis: torque collari aurantiaco et coccineo mixtâ: facie laterali fusco-brunneâ, regione auriculari aurantiacâ et genis anticis sordidè coccineo notatis: dorso superiore olivascenti-brunneo, interdum olivaceo-viridi nitente, plumis omnibus nigro marginatis: uropygio et supracaudalibus sordidè coccineis, plumis lætiore coccineo fasciatis et nigro terminatis: tectricibus alarum pallidè brunneis, nigro marginatis: remigibus pallidè brunneis, pogonio interno dilutè coccineo transfasciatis: caudâ pallidè brunneâ, suprà vix distinctè olivaceo vel rubro tinctâ, sed subtùs hôc colore lavatâ et ad basin coccineo irregulariter fasciatâ: pectore toto cinereo-fusco, plumis nigro terminatis: abdomine toto cum hypochondriis et subcaudalibus pallidè brunneis, plumis omnibus coccineo et ad apicem nigro transfasciatis; subalaribus et axillaribus coccineis, plus minusve aurantiaco tinctis, et minimis brunneo transfasciatis: rostro cyanescenti-cinereo, mandibulâ versus basin fulvescenti-brunneâ: pedibus cyanescenti-cinereis, plantis pedum flavicanti-brunneis: iride saturatè brunneâ.

Juv. torque nuchali indistinctiore: alâ subtùs fusco transfasciatâ.

Adult. General plumage olivaceous brown, each feather margined with darker brown, flushed on the lower parts of the body with dark red, the plumage of the upper parts sometimes with a metallic green tinge; crown and sides of the head grey, margined with dusky brown; ear-coverts orpiment-orange, margined with brown; feathers projecting over the lower mandible dark vinous red, with black hair-like filaments; on the nape the feathers are dingy red, margined with yellow and black, and forming a broad collar with blending edges; feathers of the lower part of the back, rump, sides, abdomen, upper and lower tail-coverts, in their outer portion, dark blood-red, of varying shades, and more or less tinged with yellow in different examples; on the underparts these feathers are narrowly margined with black, on the upper they are banded alternately with black and a lighter shade of red, quills light olivaceous brown, toothed on the inner web with pale yellowish red, and the secondaries washed, on their inner surface, with pale red; lining of the wings, as well as the axillaries, brilliantly coloured with scarlet and yellow, varying in shade in almost every specimen, and differing in their markings according to age. In the fully mature bird all these soft feathers, excepting

KAKA PARROT AND VARIETY "KAKA KURA".
NESTOR MERIDIONALIS.

the longer ones underlying the primaries, are of a bright scarlet, variegated more or less with yellow, especially towards the outer edge of the wing, where the ground-colour changes to olivaceous; in some specimens the yellow tint predominates, while in rare instances the whole of this plumage is of a uniform bright canary-yellow, the axillaries alone being tinged with scarlet, and the toothed markings on the quills almost white, or only tinged with orange. The long axillary plumes are always bright scarlet, barred with olivaceous brown, and sometimes tipped with yellow; tail-feathers light olivaceous brown on their upper surface, with a broad transverse band of dark brown near the tips, obscure vinous red on the under surface, with toothed markings of brighter red on their inner webs, and with the subterminal band very distinct. Irides dark brown; eyelid dull yellow; bill dark bluish grey, the lower mandible sometimes yellowish brown towards the base; legs bluish grey; soles of feet yellowish brown. Total length 18·5 inches; extent of wings 32; wing, from flexure, 11; tail 7; bill, along the ridge 2·25, along the edge of lower mandible 1; tarsus 1·25; longer fore toe and claw 2·5; longer hind toe and claw 2·25.

Young. In the younger birds the scarlet lining on the under surface of the wings is marked by numerous transverse bars of dusky brown; and towards the carpal edges the feathers are olivaceous brown, barred and margined with orpiment-orange; the long soft feathers underlying the secondaries are dusky grey, with faint bars of scarlet. In some examples the nuchal collar is very indistinct, being simply indicated by a tinge of yellow, while in others it is fully as conspicuous as in the adult.

Nestling. The newly hatched nestling is covered with soft white down, thinly distributed, and very short on the underparts; abdomen entirely bare; bill whitish grey, the upper mandible armed near the tip with a white horny point; cere pale flesh-colour; rictal membrane greatly developed and of a pale yellow colour; legs dull cinereous. The bill and feet seem disproportionately large, giving the nestling a very ungainly appearance. The fledgling (Feb. 5) has the membrane at the angle of the mouth and the rim encircling the eyes yellow.

Obs. In this species of *Nestor* the cere is very prominent, and towards the head generally has an abraded appearance, as if the feathers had been rubbed off. The two mandibles are connected at the base by a tough elastic membrane, capable of much expansion, the mandibles being more than an inch apart when fully extended. The tongue, which, like the beak, is bluish grey, is hard and smooth on the under surface, having the appearance of a human finger-nail much produced, along the terminal edge of which there is a fine brush-like development. The upper surface of the tongue is soft, rounded on the edges, with a broad central groove. In adult birds the denuded shaft of the tail-feathers is produced to a fine point a quarter of an inch or more beyond the web. Freshly killed birds have a peculiar woody odour, which is sometimes very strong. During the season that the rata is in bloom the long feathers of the cheeks and the light parts of the lower mandible, as well as the bare membrane at its base, are stained a rich orange-colour by contact with the juice of these flowers, which evidently contain strong colouring-matter.

Apart from the strongly marked varieties to be presently noticed, individual specimens exhibit a considerable amount of variation in the details of their colouring. The nuchal collar varies not only in extent, but in colour, from pale orpiment-orange to a dark wine-red margined with yellow; and there is much difference in the colour of the ear-coverts and of the filamentous feathers overlapping the under mandible. Examples also vary in size, a small one in my possession measuring only 16·5 inches in length; wing, from flexure, 10; tail 6.

Varieties. The members of the genus *Nestor* show a great tendency to individual variation, examples even of *Nestor productus* (which is confined in its range to a single rocky island) presenting such differences of plumage as almost to induce a belief in the existence of more than one species. But this variability of character is developed to the highest degree in *Nestor meridionalis*. Although it may be necessary, or convenient, to recognize a larger and a smaller race, the former confined to the South Island, and the latter having a wider dispersion, I have come to the conclusion that the following are merely aberrant varieties of the typical form, and, although sometimes recurrent in different localities, are not entitled to recognition as distinct species.

Var. a. *Nestor superbus*, Buller, Essay on New-Zealand Ornithology, p. 11.

This is one of the most beautiful of the many varieties to be noticed. Owing to the discovery, at the same time and in the same locality, of several examples, all in the same brilliant plumage, I felt no hesitation in characterizing the species as new, under the above designation. Several connecting forms, however, have since been found, and I now feel bound to sink *N. superbus* as a species. The following description of this supposed species appeared in my 'Essay' (*l. c.*) :—"Crown, hind neck, breast, scapulars, and upper wing-coverts canary-yellow of different shades, and tinged with scarlet; upper surface of wings whitish yellow, the primaries inclining to pale ash; upper surface of tail, when closed, pale ashy yellow, the sides being bright canary-yellow with a scarlet tinge; sides, abdomen, lower tail-coverts, axillaries, lining of wings, lower part of back, and upper tail-coverts bright scarlet, varied on the underparts, and minutely edged on the upper tail-coverts with canary-yellow; cheeks, throat, ear-coverts, and a broad nuchal collar paler scarlet, largely mixed on the ear-coverts and collar with bright yellow. The under wing-coverts are beautifully marked with alternate bands of scarlet and yellow; the primaries, on their under surface, are ashy, marked on their inner vane with triangular spots of scarlet and yellow; under surface of tail-feathers pale scarlet for two thirds of their extent, and banded on their inner vane with brighter, ashy beyond, and yellowish towards the tip. Bill and legs dark bluish grey."

There are two specimens (said to be ♂ and ♀) in the Canterbury Museum. They differ slightly in the details of their colouring. In one the nuchal collar of scarlet and yellow is much broader and brighter than in the other, while the crown of the head is paler, being of a dull yellowish white. The lower part of the back is equally brilliant in both; and the peculiar ashy white, which is characteristic of albinism, is very strongly apparent in the primaries and tail-feathers, although tinged on the latter with yellow. One has the bill considerably larger and stronger than the other, while in both the tail-feathers have denuded tips, or, more properly, the shaft is produced half an inch beyond the webs.

An example in my collection, obtained on Banks Peninsula (Canterbury), corresponds exactly with the supposed male above described.

There is another specimen (obtained in the Tararua ranges) in the possession of Wi Parata at Waikanae. It is well mounted in a glass case, and exhibited with other novelties in his elegant *Whare-puni*. The general plumage is white, with a wash of canary-yellow, shading into crimson on the cheeks and feathers overlapping the lower mandible; a narrow nuchal collar of crimson and golden yellow intermixed; the feathers of the breast and the small wing-coverts tipped with bright yellow; the whole of the abdomen, flanks, and under tail-coverts bright crimson, and the under surface of tail-feathers flushed with the same. Bill white; legs and feet grey.

Var. δ. *Nestor esslingii*, Souancé, Rev. et Mag. de Zool. 1856, p. 223.

M. de Souancé, the original describer of the supposed species, says :—"Le *Nestor* dont nous allons donner la description est, sans contredit, l'oiseau le plus remarquable de la collection Masséna. Intermédiaire entre le *N. hypopolius* et le *N. productus*, ce magnifique Perroquet réunit, dans son plumage, des détails caractéristiques de ces deux espèces. Coloration générale semblable à celle du *N. hypopolius*."

Mr. Gould, in the Supplement to his 'Handbook to the Birds of Australia,' says of it :—"A single specimen only of this magnificent Parrot has come under my notice; and this example is perhaps the only one that has yet been sent to Europe. It formerly formed part of the collection of the Prince D'Essling, of Paris, but now graces the National Museum of Great Britain. It is in a most perfect state of preservation, and is, without exception, one of the finest species, not only of its genus, but of the great family of Parrots. The native country of this species is supposed to be New Zealand; but I, as well as M. de Souancé, have failed to learn any thing definite on this point. In size it even exceeds the great Kaka (*Nestor hypopolius*), which it resembles in the form of its beak, while in its general colouring it closely assimilates to *Nestor productus*."

Dr. Finsch, on the other hand, states, in his Monograph, that *Nestor esslingii*, De Souancé (of which the type is in the British Museum), is in size and general colour the same as *Nestor meridionalis*, but has the breast ash-grey, with brown terminal margins and a broad yellowish-white transverse band straight across the belly. He adds that he was not able to make such an examination of it as he wished, owing to its being in an hermetically closed glass case, but quotes Souancé to the effect that the red marks on the inner vane of the quills and tail-feathers are precisely as in *Nestor meridionalis*; whereas Mr. Gould distinctly says that while the tail-feathers in *N. meridionalis* and *N. productus* are strongly toothed on the under surface with red, "in *Nestor esslingii* no such marks occur, the toothing on the inner webs of the primaries is not so clear and well-defined, and the light-coloured interspaces are more freckled with brown."

Referring to these several accounts, I expressed the following opinion, in the 'Transactions of the New-Zealand Institute' (vol. iii. 1870, p. 51) :—"Assuming Dr. Finsch's description to be strictly correct—that it most nearly resembles *Nestor meridionalis*, from which it is only distinguishable by the broad yellowish-white band across the underparts of the body—and considering the extreme tendency in that species to variability of colour, I should be inclined to regard the British-Museum bird as an accidental variety of the common Kaka. Among the numerous abnormally coloured examples which I have seen, from time to time, varying from an almost pure albino to a rich variegated scarlet, I remember one which, although like the common bird in its general plumage, had a broad longitudinal band of yellowish white on the abdomen. The specific identity of this specimen with *Nestor meridionalis* was unmistakable."

It only remains for me to add that the examination which I have since made of the type specimen in the British Museum has entirely verified this conclusion. It may be mentioned that this bird furnished Mr. Gould with a subject for a beautiful picture in the Supplement to his 'Birds of Australia.'

My son saw one at Owhaoko with a white tail, the rest of the plumage being dingy brown. He endeavoured in vain to shoot it.

Var. γ. *Nestor montanus*, Haast.

This is a larger race than the common Kaka, and is generally much brighter in colour. It appears to be confined to the South Island, whence all the examples that have come under my notice have been obtained. No doubt some naturalists will be disposed to regard this larger race as a distinct bird; and for a considerable time my own inclinations were in that direction; but, looking to the extreme tendency to variation in this species, and to the difficulty of drawing a clear line between the larger and smaller races, in consequence of the occasional intermediate or connecting forms, I feel that I am taking a safe course, concurrently with Dr. Finsch, in refusing, for the present at least, to separate these birds *.

* While adhering to the view expressed above, I think it only right to quote the following opinions as to its claims to take rank as a distinct species:—

Sir Julius von Haast in forwarding me a specimen wrote:—

"I send you another skin of our Alpine Parrot. Even judging from its habits alone, it is quite distinct from the common Kaka. It is never found in the Fagus forest, whilst the other never goes above it into the sub-alpine vegetation. Near the glacier sources of the Waimakariri, where I was in the latter part of March, I saw them frequently in the alpine meadows—4000 to 5000 feet high—feeding on the large red berries of *Coprosma pumila* and *nivalis*, two dwarf plants lying close to the ground. We found these berries in the gullets of those we opened. They evidently had their nests with young ones among the crags of the nearly perpendicular rocky walls (about 6000 feet above the sea), and I repeatedly observed them flying backwards and forwards as if feeding their young. After the first day's shooting they got exceedingly shy, and could not be approached within gun-shot."

Sir James Hector informs me that it was to this bird (and not to the so-called *Nestor occidentalis* as previously quoted) that he intended the following note to refer:—

"I never met with it in the forests of the low lands. It is more active in its habits and more hawk-like in its flight than the common *Nestor*. It often sweeps suddenly to the ground; and its cry differs from that of the common Kaka in being more shrill and wild."

Mr. Fuller (Taxidermist to the Canterbury Museum) also stated, as the result of very careful observation, that "the manner of flight is quite different from that of the common Kaka, for they soar after the manner of the Kea (*Nestor notabilis*)."

Mr. Reischek, to whom I am indebted for some fine specimens, of all ages, obtained at Dusky Sound, is strongly of opinion that this is a distinct species. He says (Trans. N.Z. Inst. vol. xvii. p. 194):—

"This bird represents *Nestor meridionalis* in the sounds, but it is not very plentiful. I have found them alone or in pairs or with their young, from two to four. They breed in hollow trees. The nest consists of a deepening lined with wood-dust and feathers out of the parent birds. They lay their eggs from the beginning of March till April. Male and female hatch and rear the young birds together; in August the young are full-grown. This bird is not so gregarious as its ally *meridionalis*, also different in plumage and construction of the skeleton [?] and habits. The cry and whistle is shriller; the male is fiery red under the wings, the female golden yellow and a little smaller. These birds are very bold. On the 13th April, 1884, I found in a hollow tree a female with one egg and three young birds, which she pluckily defended by biting and scratching. At the cry of the female the male came swooping several times past my head. This species is the finest of the three existing species of *Nestor*."

Among the specimens received from Mr. Reischek is a nestling covered with grey down; but it differs in no respect from that of the common Kaka, except perhaps that the downy covering is a shade darker. An egg which he submitted to me differs, however, slightly from that of *Nestor meridionalis*; it is creamy white, the surface covered with extremely fine pores, making it almost granulate, of a regular ovoid form, and measuring 1·5 inch in length by 1·25 in breadth.

There are some beautiful examples of this larger form in the Canterbury Museum. One of these has the crown silvery grey; the sides of the head and neck washed with sea-green; the ear-coverts glossy golden yellow; the feathers overlapping the lower mandible, and the whole of the throat and fore neck, rich vinous red with paler centres; the nuchal collar very broad, and composed of various shades of scarlet and yellow beautifully blended; the breast and sides varied with crimson and yellowish olive, blending on each feather, and across the former an indistinct pectoral band of yellowish grey; the rump, flanks, abdomen, upper and lower tail-coverts as in ordinary specimens, but brighter in colour. In another example the small wing-coverts are pale orange-red, terminally margined with black; while in a third the abdomen has a conspicuous, irregular patch of canary-yellow. An unusually fine specimen forwarded to me by Sir Julius von Haast for examination had the forehead of a rufous-orange colour; but this proved to be entirely the result of flower-stains, as I had no difficulty in demonstrating. This bird measured 20 inches in length, wing from flexure 12, tail 7·5, culmen 2·75, tarsus 1·5. The plumage of the upper parts was faded and snow-beaten, the ends of the primaries and tail-feathers being much worn and jagged. Crown and sides of the head grey tinged with dull metallic green; ear-coverts bright golden yellow with darker edges; breast and sides olivaceous brown, with a reddish hue; feathers composing the nuchal collar dull red, with golden tips; those covering the shoulders marked in the centre with a large irregular spot of red, and stained with golden yellow; rump and upper tail-coverts dull arterial red, each feather with a narrow terminal margin of black; under surface as in ordinary specimens, but more largely suffused with yellow.

In another example of the southern bird (in my own collection, which contains a good series) the crown and hind part of the head are light grey edged with darker grey; the feathers composing the nuchal collar are rich orange-red, narrowly barred with yellow and black; ear-coverts bright orpiment-orange, changing into deep vinous red on the cheeks; the feathers overlapping the lower mandible edged with black; the fore neck, breast, shoulders, and upper wing-coverts olivaceous brown margined with darker brown, and having, more or less, a green metallic lustre; sides, abdomen, rump, and upper tail-coverts dark red, banded with bright arterial red and dusky brown; under tail-coverts dull red, tipped with brighter red, olivaceous brown at the base; quills olivaceous brown, lighter on the outer web, largely toothed on the inner one with pale orange-red; lining of wings and axillary plumes bright scarlet tipped with yellow, and banded, more or less distinctly, with brown; tail-feathers olivaceous brown, darker in their apical portion, washed on their under surface with dull vinous red, and toothed with pale scarlet. Bill uniform bluish grey; tarsi and toes dark bluish grey.

In another specimen the general colours are altogether duller; but there is more of the metallic lustre on the wings; the arterial-red bands on the rump and abdomen are wanting, the plumage of these parts being dark red edged with dusky brown or black; the lining of the wings is less brilliant; the toothed markings are paler on the quills, and far less distinct on the tail-feathers.

A beautiful specimen in Mr. Reischek's collection (♂) has the light feathers of the crown tipped with yellow, the feathers of the nape deeply margined with oil-green, the nuchal collar broad and very richly coloured, the whole of the chin, fore neck, and breast flushed with crimson; abdomen, sides of the body, and under tail-coverts flaming crimson with transverse bands of a lighter colour; small wing-coverts metallic green, flushed in their apical portion with crimson and terminally margined with a narrow band of black; rump and upper tail-coverts same as abdomen and crissum, but darker.

Var. δ.

The following brilliantly coloured variety of *N. meridionalis* was obtained more than twenty years ago in the Wanganui district, and is now in the author's collection, in the Colonial Museum, at Wellington. General plumage bright scarlet-red, deepest on the lower part of back, sides, and abdomen, and variegated with orpiment-yellow on the nape, sides of the neck, and breast. Crown greenish yellow, with a metallic gloss, each feather centred with brown; feathers overlapping the under mandible, and a broad patch on the throat, dark reddish brown, as in ordinary examples. The feathers of the breast are stained in the centre with dull ashy brown, and, as well as those of the upper parts, are narrowly bordered with black. Primaries dark olivaceous brown, largely marked in their basal portion with yellowish white; secondaries and their coverts pale scarlet, variegated with yellow, olivaceous brown in their apical portion; all the quills on their under surface pale orange in their basal portion, but without the toothed markings; lining of wings vivid scarlet, varied with yellow. Tail-feathers pale scarlet with a broad terminal band of olivaceous brown; under tail-coverts darker scarlet. On the bright upper surface of the tail-feathers there are obsolete bars, and on the under surface there is a broad olivaceous margin; but the "toothed" character peculiar to the species is entirely wanting. Bill bluish grey; feet dark grey, paler on the soles; claws black.

A specimen in the possession of Mr. W. Luxford, at Wellington, has the prevailing colour a bright scarlet; but on the back and wings each feather has a narrow terminal band of blackish brown; head and throat rusty brown; breast darker rust-colour, each feather broadly margined with yellow. Primaries canary-yellow on the outer web for one third of their length, then brown; upper wing-coverts brown margined with scarlet. About two thirds of the tail pale scarlet; there are then a few interrupted bands of brown, and the terminal portion is of that colour. This bird was shot in the hills near the town of Wellington in the early days of the colony, and before the requirements of the settlers had led to the destruction of the surrounding woods.

Under this section may be placed a gorgeous example obtained in the Hawke's Bay district, and sent by Mr. J. Baker to the recent Colonial and Indian Exhibition at South Kensington. It is somewhat similar to my Wanganui bird, but is more brilliant. Feathers of the vertex and crown orpiment-orange centred and narrowly margined with brown; throat, cheeks, and many of the upper wing-coverts much as in ordinary specimens, being olivaceous brown, the normally coloured feathers irregularly marked and margined with pale scarlet; shoulders and interscapulars olive-brown washed with crimson, banded with golden yellow, and narrowly margined with brown; on the head and neck the brown centres almost disappear, whilst the feathers composing the nuchal collar are entirely scarlet, with broad golden-yellow margins; the whole of the rump, abdomen, and lower sides of the body, with the upper and under tail-coverts, fiery scarlet, very narrowly and sparsely tipped with dusky black; the breast is a mixture of dark brown, scarlet, and orpiment-orange, the latter predominating; primaries and tertials dark olivaceous brown, the outer vanes pale canary-yellow towards the base; secondaries pale scarlet for two thirds of their length, then olivaceous brown; tail-feathers with a similar extent of pale scarlet, then blackish brown with olivaceous tips; but the colouring gets paler on the lateral feathers, fading to canary-yellow on the outermost vane and presenting only a tinge of scarlet on the succeeding one. Bill and feet as in ordinary examples.

The tail-feathers in the Canterbury Museum found near Cass river (mentioned in Trans. N.-Z. Instit. vol. iv. p. 148) are exactly similar to those here described.

Var. ε.

The following is the description of a very light-coloured variety obtained by the natives near the burning mountain of Tongariro, and presented to me by Mr. H. W. Woon, R.M.:—

General plumage pale canary-yellow; the crown tinged with grey; ear-coverts bright orange-yellow; feathers of the throat, hind part of the neck, and some of the upper wing-coverts margined with the same; feathers on the lower part of the cheek, and those overlapping the lower mandible, yellowish red, with paler shafts; sides, abdomen, rump, upper and lower tail-coverts vivid scarlet, the feathers of the underparts narrowly margined with yellow; lining of wings bright yellow tinged with scarlet; axillary plumes, and the soft feathers underlying the secondaries, bright scarlet, tipped with yellow; quills pale canary-yellow on their upper surface, ashy on their under surface, with broad toothed markings of pale red, obsolete on the outer remiges, and diminishing on the secondaries; tail-feathers ashy yellow, with brighter margins, tinged with orange in the centre and along the tips, changing on their under surface to orange-yellow, in their basal portion with narrow toothed markings of scarlet. Bill white horn-colour. Irides dark brown. Tarsi and toes pale brown or flesh-coloured; claws white horn-colour.

The late Rev. R. Taylor, who resided more than twenty years on the Wanganui river, and who published many interesting notes on the natural history of the country, informed me that he had seen several examples of this beautiful variety from the same locality as the one noticed above. The natives assured him that they always pair together, nesting in the crevices of the rocks.

Var. ζ.

I am indebted to Sir Julius von Haast for a specimen showing a very decided tendency to albinism, although still exhibiting the bright scarlet facings which adorn the others. In this bird the crown is greyish white, with pale yellow margins; the nape dull crimson, with yellowish tips, forming a broad nuchal collar; ear-coverts bright orpiment-orange stained with red; feathers overlapping the lower mandible, and those covering the throat, pale vinous red; fore neck and upper part of breast smoky grey, washed with red, and each feather tipped with dull yellow; back and upper surface of wings smoky yellow tinged with gamboge; lining of wings and axillary plumes bright scarlet-red; quills dark yellowish grey, obscurely toothed, and washed at the base with pale scarlet; sides, flanks, and abdomen scarlet-red, tipped more or less with dusky and yellow; tail-feathers yellowish brown, with paler edges, washed on the under surface with scarlet, marked with dusky freckles, but not toothed; upper and lower tail-coverts bright gamboge, crossed near the tip by a band of bright red. Bill very narrow and fine; yellowish grey in colour, bluish at the tip. Tarsi and toes dark grey; claws bluish horn-colour.

Under this head may be placed the creamy-white Kaka with scarlet rump and abdomen, and a narrow nuchal collar of canary-yellow, which was shot in the Makereru ranges near Waipawa, and sent by Mr. J. Baker to the Colonial and Indian Exhibition of 1886.

Var. η.

A specimen obtained by Mr. Henry Travers in the Provincial district of Marlborough is remarkably small, as compared with ordinary examples from the same locality, and is differently coloured.

Crown of the head hoary grey; fringed behind the eyes and on the occiput with pale sea-green; ear-coverts golden yellow tinged with red; mantle, scapulars, and wing-coverts dull olivaceous green, margined with black; nuchal collar dull vinous red, with lighter tips; neck above dark olivaceous brown; cheeks, throat, front and sides of the neck dark brown, strongly tinged with red; breast, sides, abdomen, and under tail-coverts of different shades of arterial red shaded with brown; lower part of the back, rump, upper tail-coverts, and thighs dark arterial red banded with lighter red, and tipped with black; lining of wings and axillary plumes beautiful scarlet, transversely barred with dusky black. Quills and tail-feathers olivaceous brown, with paler edges, toothed on their inner webs with pale orange-red.

Var. θ. "Kaka-kereru" of the natives.

The following description is taken from a specimen in my collection, which was obtained in the vicinity of Wellington, in 1856:—

Upper parts generally tinged with oil-green, and each feather narrowly margined with black; crown light grey, with darker shades, varied with deep sea-green over the eyes and on the hinder part of the head; nape sea-green, mixed with brown and yellow; nuchal collar, which is nearly two inches broad, dark crimson, each feather faintly margined with yellow and black. Upper wing-coverts and upper portion of the tail-feathers tinged with olivaceous. The ear-coverts are orpiment-orange varying in shade; while the cheeks and throat are dark vinous red, each feather having a bright centre; feathers of the neck and breast dark brown, with a marginal tinge of crimson; rump, upper and lower tail-coverts, thighs, and abdomen deep crimson, with lighter crescentic bands and narrow terminal margins of black. This bird was shot with a flock of twelve others (all bagged), and was the only one presenting this character of plumage.

In another example, obtained at Otaki in September 1862, all the tints of the plumage are very rich, and the red of the underparts extends to the breast, each feather having two bright crescentic bands of arterial red and a terminal margin of dusky black; the ear-coverts are gallstone-yellow, and the nuchal collar, which is much extended, is of the same colour intermixed with red; the secondaries and lesser wing-coverts are pale metallic green, narrowly edged with black; and the whole of the dark upper plumage is tinged with the same colour.

Var. ι.

In June, 1870, I received from Manawatu a very beautiful specimen of the variety known among the natives as "Kaka-pipiwarauroa." The whole of the plumage was most handsomely variegated, each feather having a brownish-black centre, and the margins broadly edged with orange-red and yellow. These bright markings were most conspicuous on the nape and upper surface of the wings. The sides of the face and the ear-coverts were of a bright golden yellow, changing to red on the long feathers overlapping the lower mandible; the sides, thighs, and lower part of the abdomen arterial red, with lighter bands; the lining of the wings brilliant scarlet, banded with yellow and black. The natives had this beautiful bird in their possession for many months; and the delighted settler who wrote apprising me of it described it as "a bird with all the colours of the rainbow." I ultimately induced the owner to part with it, giving him in return a block of the much-prized greenstone, weighing more than 20 lb. I designed this *rara avis* for the Zoological Society of London, and shipped it accordingly with every care; but it appeared to suffer from the extreme cold, and, unfortunately, perished before it was out of sight of the New-Zealand coast.

Var. κ.

In the Otago Museum there is a remarkable specimen, obtained in the south, in August 1874, and presented by Mr. J. Coulan. This bird (which is a male) has the plumage of the upper parts smoky yellowish brown, and, except on the crown, each feather has a dusky margin; the feathers of the crown, wings, and tail pale yellowish brown, the latter flushed with scarlet, and the outer ones edged with yellow; underparts generally of a darker hue, shaded with brown and flushed with crimson; the sides of the face, nuchal collar, rump, upper and lower tail-coverts, abdomen, flanks, and inner lining of wings all very highly coloured, the crimson feathers forming the collar being prettily rayed with orpiment-orange; bill and claws white horn-colour.

Var. λ.

A fine bird received from Catlin river (likewise preserved in the Otago Museum) has the hind part of the crown and the whole of the nape and hind neck rich canary-yellow of varying shades, the normal nuchal collar only appearing at the outer edge of this gorgeous hood. The ear-coverts are bright orpiment-orange; and the filamentous feathers overlapping the mandibles are crimson with light shafts; so also are the chin-feathers, under which there is a band of rich canary-yellow suffused with crimson, spreading over the throat and connecting the two sides of the head. On the breast and underparts of the body there are numerous canary-yellow feathers interspersed irregularly with the ordinary plumage. The upper surface is in the plumage of the "Kaka-kereru" (var. θ), being highly flushed or burnished with metallic green.

Var. μ. *Nestor occidentalis*, Buller, Birds of N. Z. 1st ed. p. 50.

To the above numerous varieties I feel bound now to add the form which, with some hesitation, I kept distinct under the above name in my former edition. As stated in the text, my reason for then rejecting the supposition of its being a mere aberrant variety of the common species was the account of its habits and peculiar cry furnished by Sir James Hector, who found it "frequenting the precipitous wooded cliffs in the neighbourhood of George Sound and thence along the coast to Milford Sound." As however, during the last fifteen years, no further examples have been obtained, and no additional evidence to support its recognition as a species, it will be safer to treat this as another instance of congenital variation. For its more exact definition I will quote here my original description when proposing to differentiate the species:—

"Upper surface dark olivaceous brown, tinged with yellow on the wing-coverts, each feather margined with dusky black; feathers of the nape dull red, margined with yellow and black, and forming a narrow nuchal collar; rump, tail-coverts, and abdomen dark arterial red, the feathers of the latter banded with a brighter tint; ear-coverts pale orpiment-orange; feathers projecting over the lower mandible tinged with red; throat, neck, and breast dark olivaceous brown; lining of wings and axillary plumes bright scarlet, obscurely barred with black, and tipped with golden yellow; quills and tail-feathers russet-brown, the former toothed with yellow on the inner web; bill and feet dark olivaceous grey. Length 16·5 inches; wing, from flexure, 10·5; tail 6; tarsus 1; longer fore toe and claw 2·25; longer hind toe and claw 2·1; bill, following curvature 2·25, along edge of lower mandible 1·5.

"Apart from the difference of plumage this species is appreciably smaller than the common one, while the bill is more slender and has the upper mandible produced to a finer point. The two specimens obtained by Dr. Hector on the west coast of the South Island differ very slightly in the details of their colouring, and there is scarcely any perceptible difference in their size."

Note. To illustrate the brilliancy and beauty of some of these accidental forms, I have given a portrait of the brighter of the two specimens sent by Mr. Baker to the Colonial and Indian Exhibition, both of which are now in my collection.

General Remarks. To MM. Blanchard and Pelzeln belong the credit of having first determined the true affinities of the genus *Nestor*. It bears a close relation to the Australian Lories, and the New-Guinea form known as Pecquet's Parrot (*Dasyptilus pecqueti*) appears to exhibit the transitional or connecting link between these two well-marked groups.

In habits and structure the members of the genus *Nestor* are true flower-suckers, the tongue being furnished at its extremity with a fine brush-like development for that special purpose. The common Kaka of New Zealand is the type of the genus.

Modern systematists, as a rule, have placed it in the subfamily *Trichoglossinæ*; but I accept Prof. Garrod's view that its proper station is among the typical Parrots *. Its decidedly aberrant characters, however, cannot be denied; and I have thought it the safest course to place the genus in a separate family under the name of Nestoridæ.

* Proc. Zool. Soc. 1872, pp. 787–789.

Sprightly in its actions, eminently social, and more noisy than any other inhabitant of the woods, the Kaka holds a prominent place among our native birds. Being semi-nocturnal in its habits, it generally remains quiet and concealed during the heat of the day. If, however, the sportsman should happen to find a stray one, and to wound instead of killing it, its cries of distress will immediately rouse the whole fraternity from their slumbers, and all the Kakas within hearing will come to the rescue, and make the forest echo with their discordant screams. Unless, however, disturbed by some exciting cause of this sort, they remain in close cover till the approach of the cooler hours: then they come forth with noisy clamour, and may be seen, far above the tree-tops, winging their way to some favourite feeding-place; or they may be observed climbing up the rough vine-clad boles of the trees, freely using their powerful mandibles, and assuming every variety of attitude, or diligently tearing open the dead roots of the close epiphytic vegetation in their eager search for insects and their larvæ. In the spring and summer, when the woods are full of wild blossom and berry, these birds have a prodigality of food, and may be seen alternately filling their crops with a variety of juicy berries, or sucking nectar from the crimson flowers of the rata (*Metrosideros robusta*—a flowering branch of which is depicted in the Plate) by means of their brush-fringed tongues.

With the earliest streaks of dawn, and while the underwoods are still wrapped in darkness, the wild cry of this bird breaks upon the ear with a strange effect. It is the sound that wakes the weary traveller encamped in the bush; and the announcement of his ever active Maori attendant, "Kua tangi te Kaka," is an intimation that it is time to be astir. But although habitually recluse during the day, it is not always so. During gloomy weather it is often very active; and, sometimes, even in the bright sunshine a score of them may be seen together, flying and circling about, high above the trees, uttering their loud screams and apparently bent on convivial amusement. When the shades of evening bring a deeper gloom into the depths of the forest, and all sounds are hushed, save the low hoot of the waking Morepork, or the occasional *cheep-cheep* of the startled Robin, the Kaka becomes more animated. It may then be heard calling to its fellows in a harsh rasping note, something like the syllables "*t-chrut, t-chrut,*" or indulging in a clear musical whistle with a short refrain.

It is strictly arboreal in its habits, and subsists to a large extent on insects and their larvæ, so that it is probably one of our most useful species. Where they exist in large numbers, they must act very beneficially on the timber-forests; for in the domain of nature important results are often produced by apparently trivial agencies. Like all the honey-eaters, while supplying their own wants, they do good service with their brush-tongues, by fertilizing the blossoms of various trees, and thus assisting in their propagation; while, on the other hand, the diligent search they prosecute for insects and grubs, and the countless numbers daily consumed by each individual, must materially affect the economy of the native woods.

I am aware that in some parts of the country there is a prejudice against the Kaka on account of its alleged injury to forest trees by barking them; but this animus is quite undeserved *. On the

* Against this unmerited charge the Kaka is well defended by Mr. Potts, who writes:—"Although so often accused of injuring trees by stripping down the bark, from careful observation we do not believe a flourishing tree is ever damaged by its beak. It is the apparently vigorous, but really unsound tree that is attacked, already doomed by the presence of countless multitudes of insects, of many varieties, of which it is at once the food and refuge, either in their perfect or larval state. In the persevering and laborious pursuit of this favourite food, the Kaka, doubtless, lends his assistance in hastening the fall of decaying trees; the loosened strips of bark dissevered admit to the exposed wood rain and moisture collected from dews and mists, to be dried by evaporation by the heat of the sun, by the desiccating winds, only to become saturated again. Under this alternation the insidious fungi take root, decay rapidly sets in, the close-grained timber gives place to a soft spongy texture, branches drop off, and gradually the once noble-looking tree succumbs to its fate: but its gradual decay and fall, the work of years, has proved beneficial to the surrounding plants: the dropping of the branches admits light and air to the aspiring saplings, assists in checking the undue spread of lichens and epiphytes; and when the old stem falls, tottering down from its very rottenness, its place is supplied by vigorous successors."

contrary, the bird does good service in this respect, for its vigilant eye is the first to detect the insidious attacks of the small longicorn beetles and other borers, and it is to search out and devour these enemies to forest conservation that it takes the trouble to break the bark at all. I have seen a vigorous limb cut to a depth of two inches or more by the Kaka's beak, for that purpose only, the cutting or rasping being effected by means of the under mandible, the hook of the upper giving the necessary leverage. Such a wound in a healthy tree rapidly heals over and thus the limb is saved through the intervention of the bird at the right moment.

This is one of our highly characteristic forms and is met with, more or less, in every part of the country. Far away in the depths of the forest—where the trees are clad with rich mosses, cryptogams, and lycopods to their very tops—where, as if to hide the mouldering decay of nature, huge masses of green vines and creeping plants cover the aged trunks and bind the bush together—where the sunlight, struggling through the leafy tops, discloses here and there a feathery tassel of *Asplenium flaccidum* hanging from the branches or a clump of the scarlet-flowered mistletoe—there the Kaka is at home and may be studied to advantage. So long as he does not know he is watched, he may be seen twisting and turning among the sprays, hopping Cockatoo-fashion along a branch, then climbing higher with graceful agility; resting for a moment to whistle for his mate and, when she has joined him, expressing his pleasure in a sharp chuckling note, like the striking together of two quartz pebbles; then, as if suspecting some treachery below, he suddenly takes wing with loud cries of *ka-ka* and glides smoothly through the leafy maze, closely followed by his spouse. On a near view the brilliant plumage under the wings is very conspicuous when the bird is flying; but when the bird is climbing or hopping, in the manner habitual to it, the wings are kept closed. Then on the outskirts of the forest you meet with him again—more generally in the early morning—hunting diligently for his insect food or regaling himself on ripe berries of various kinds in the thick underwood; and towards evening three or more of them may be seen in company, flying high above the forest level; then alighting on the withered, naked top of some lofty kahikatea or kauri tree—always perching on the highest points—resting a few moments, and taking wing again till they are fairly out of sight. In the early watches of the night, too, especially during the breeding-season, and just before the break of dawn, its peculiar cry betrays its wakeful restlessness.

In the dark *Fagus*-forests, both north and south, it shares the domain with the stealthy Woodhen, descending often to the ground to hunt for grubs and insects among the moss-covered roots and decaying wood. In the low-lying woods, where the climbing kiekie (*Freycinetia banksii*) attaches its rooted stems to the larger trunks and, spreading upwards its tufted coils, wraps the whole tree in a flowing mantle of brilliant green, there too at flowering-time the Kaka will be found, feasting on the sugary bracteæ and fleshy-white spadices of this remarkable plant. He fills his crop with this delicious food, and then betakes himself to some leafy shade to avoid the heat of the noonday sun. In more open places, on the outskirts of the bush, where huge clumps of *Astelia* fasten themselves to the dead or withering branches, the Kaka may sometimes be seen eagerly tearing open the matted roots of this parasite, in quest of the worms and beetles which find an abode there, attracted by the moisture.

In the South Island, during certain seasons, it frequents the open land, alternately perching on the rough blocks of trachyte and feeding among the grass and other stunted vegetation. I remember on one occasion, some years ago, counting upwards of twenty at a time on the Port hills which divide Lyttelton from Christchurch.

On its feeding-habits Captain Mair writes to me:—"In June 1875 I was at Tuhua in the upper Wanganui. I found the Kakas there so fat that they could not fly. I actually caught fifteen of them on the ground, as they were unable to take wing."

Mr. Buchanan informs me that he has seen the Kaka stripping off the bark from a green tree

(*Panax colensoi*), and sucking up with its tongue the gummy matter underneath, in the same manner that it extracts the honey from the flowers of the *Phormium tenax*; and Mr. Potts has observed it luxuriating on the viscid nectar which fills the blossoms of this tree in spring time, till sated at last it cleanses its beak against a neighbouring bough, and then, with grateful clatter, glides off to join its fellows.

It is said also to feed on the sweet honey-like substance which exudes copiously from the bark of the *Fagus* when it is attacked by the fatal grub.

When migrating from one part of the country to another, the Kakas travel in parties of three or more, and generally at a considerable height, their flight being slow and measured and their course a direct one. They occasionally alight, as if for the purpose of resting, and in a few minutes resume their laboured flight again. On these occasions the bleached and bare limbs of a dry tree are always selected, when one of the requisite elevation is within reach, as affording most fully that which they appear to delight in, an unobstructed prospect.

A curious circumstance in the natural history of the Kaka was mentioned by me, on the authority of an eye-witness, in a communication to the Wellington Philosophical Society *. At a certain season of the year, when this bird is excessively fat, large numbers of them are found washed ashore in Golden Bay, or on the Spit which runs out from it. They are generally dead, but if not, are so exhausted as to be unable to take wing. The apparent explanation is that the Kakas in their migration across Cook Strait, which is widest at this part, are unable to maintain the long flight, owing to their fat and heavy condition, and fall into the sea. The set of the current being towards Cape Farewell, the bodies of the perishing birds are swept in that direction and finally cast ashore.

It is surprising how seldom one meets with dying birds in their natural or wild state. Like Macgillivray's wounded Gull, seeking some quiet retreat in order to "pass the time of its anguish in forgetfulness of the outer world," birds in general, and indeed all wild animals, have the faculty of hiding themselves away when the time of their dissolution approaches. During the many months I have spent in the New-Zealand woods I never but once picked up a bird that had died from natural causes, and this was a little Riroriro at the base of a kauri tree, as mentioned on page 45. On one occasion, however, at Omahu, about the end of July, a native brought in a Kaka which he had caught by the hand at the roadside. It seemed sickly, drooping its wings and uttering its "*kete-kete*" when touched. My friend, Renata Kawepo, put it on a parrot-perch as a *mokai*, but it died that night.

On the ground it generally moves by a succession of hops, after the manner of the Corvidæ, and not with the awkward waddling gait peculiar to most Parrots. In the trees, where it is more at home, it is perpetually on the move, often walking deliberately along a branch, and then climbing to another by a dexterous use of both beak and feet, or silently winging its way to a station in a neighbouring tree. Its alarm-cry resembles that of the Sulphur-crested Cockatoo of Australia. During the pairing-season the two sexes are always together, and when on the wing keep side by side, both calling as they go. In the neighbourhood of their nests they have a low call-note, like *ki-i-to, ki-i-to*, and a very soft whistling cry.

Possessing excellent powers of mimicry, and useful to the natives as a decoy-bird, the Kaka is much sought after, and almost every native village has its "mokai." Like most Parrots, it is a long-lived bird; and one which had been in the possession of the Upper Wanganui tribes for nearly twenty years presented the curious feature of its overgrown mandibles completely crossing each other. This was no doubt attributable to the fact of its having been constantly fed with soft food, thereby depriving the bill of the wear and tear incident to a state of nature. It is not so easy, however, to

* Trans. N.-Z. Instit. 1878, vol. xi. p. 369.

account for this abnormal growth in a wild specimen, obtained by Mr. Lambert at Akaroa, which presents the same feature, and in an exaggerated degree, both mandibles being quite deformed.

It would seem that in this species there is a natural tendency to a deformity of growth in this respect. This will be manifest from the drawings of two remarkable examples which I gave in the 'Transactions of the New-Zealand Institute' for 1876 (vol. ix. page 340).

One of these represents a specimen in the Canterbury Museum; and the other a case of natural deformity in the British Museum, which was brought under my notice by Dr. Günther.

The tame Kaka is very susceptible to kindness, and forms strong attachments. It soon learns to distinguish its keeper's voice, and will respond to his call. It often, however, proves a mischievous pet, especially if it gets access to the orchard, where I have known it, in a single day, nip off thousands of blossoms from a promising pear-tree. I have seen it treat a favourite vine in a similar manner and apparently from a sheer love of mischief.

If it be allowed the freedom of the house, it will destroy the furniture in the most wanton manner with its powerful beak and proclaim itself a nuisance in a variety of ways *.

When the korari-flower (*Phormium tenax*) is in season, the Kakas repair in flocks to the flax-fields to feast on the flower-honey; and on these occasions numbers of them are speared by the natives as an article of food. In the woods also at certain periods they are captured in abundance by means of an ingenious snare called a "tutu" worked by a decoy-bird.

I have seen it climbing among the crimson flower-stalks of the tree-honeysuckle (*Knightia excelsa*), gathering the honey most carefully with its long brush-fringed tongue. At another season it feeds on the pollen of the kowai (*Sophora grandiflora*), when the feathers of its head become stained with the yellow juice.

The tame village Kaka is not the useless pet that Parrots generally are. It may amuse the young people by its wonderful articulation of Maori words and phrases, and by its whistling powers, but it has far more substantial attractions for the owner. It is a source of profit and subsistence to him; and as it requires the experience of several seasons to give it a proficiency as a decoy-bird, it acquires a specific value according to its age and training. I have known a native refuse an offer of £10 for a well-trained "mokai," although an aged bird and in a very ragged condition of plumage.

* Mr. W. T. L. Travers writes:—"The habits of the Kaka are in many respects remarkable. In its absolutely wild state it is fearless and inquisitive. I have often, whilst resting on the banks of a stream which falls into the lake [Guyon], and runs through forest frequented by these birds, seen several of them gravely take post upon some trees close to me, eyeing me with the utmost apparent curiosity, and chattering to themselves as if discussing the character and intentions of the intruder. After the lapse of a few minutes they have darted away, uttering loud cries, as if proclaiming to the rest of the forest the presence of a stranger, who was either to be avoided or not, as the case might be. During the winter season the wild birds often unhesitatingly enter the house for food, making themselves thoroughly at home, and even roosting on the cross-beams in the kitchen on specially inclement nights. Two of these in particular soon learnt how to open the door of the dairy, which they were fond of getting into, in order to regale themselves on cream and butter, both of which they appeared to like excessively. I have had several of these wild birds billing on the eaves of the house in the evening, waiting to be fed, and coming readily to receive from the hand pieces of bread spread thickly with butter, and strewed with sugar. But they rarely eat any of the bread itself, dropping it as soon as they had cleared off the butter and sugar. If one bird happened to have finished his portion before the others, he unhesitatingly helped himself to a share of some neighbour's goods, which was always yielded without the slightest demur. They are fond of raw flesh, and I have seen them hovering in front of a sheep's pluck hung on a tree, precisely as a Humming-bird hovers in front of a flower, eating fragments which they tore off, giving the preference to the lungs. When anxious to get into the house, they take post on the window-sills and beat at the window with their beaks until admitted. They are very mischievous, however, invariably cutting off all the buttons from any article of clothing which may happen to be left within their reach. I regret to say, indeed, that in some instances their familiarity degenerated into such gross impudence, that my manager was obliged to kill them, in order to prevent their constant mischief." (Trans. N.-Z. Inst. 1871, vol. iv. pp. 209, 210.)

These pets are never caged, but are secured to a perch by means of a "poria" made of bone, in the form shown in the accompanying woodcut, the bird's foot being squeezed through the ring, so as to make the latter encircle the tarsus, and a thong of plaited flax-fibre, of convenient length, being then attached to the outer process and tied to the perch*.

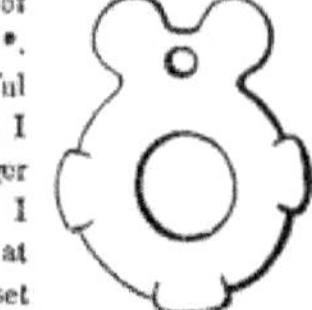

As will be seen by the full descriptive notes given above, very beautiful varieties of the Kaka are met with. I have never seen a pure albino; but I am assured by the natives that they are occasionally found, and Major Messenger of Taranaki has the skin of one which he has kindly promised to send me. I am informed that a bird very nearly approaching that condition was shot at Whanawhana (in the county of Marsden) in the summer of 1863. The value set on these rare varieties by the natives may be inferred from the following circumstance:—A "kaka-korako" was seen by a party of Rangitane in the Upper Manawatu, and followed through the woods as far as the Oroua river, every effort being made to take it alive. The Oroua people (of another tribe) then took up the chase, and followed the bird to the foot of the Ruahine range; and although carrying guns, to their credit they allowed it to escape rather than shoot it, in the remote hope that it might hereafter reappear in their district. Nor were they disappointed. Two seasons later the bird came back to the Oroua woods, and was taken alive by a native trapper. It was forwarded to Wellington by Mr. Alexander MacDonald, and, after passing through several hands, was ultimately sent to Europe. Finally it came into the possession of the late Mr. Dawson Rowley of Brighton, and formed the subject of the beautiful plate which faces page 26 of his 'Ornithological Miscellany,' part i.

From some unaccountable cause the Kaka has always been a comparatively scarce bird in the forests north of Auckland, although there is no lack of its ordinary food supply. In some other districts it is less common than it formerly was; but it still exists, in very considerable numbers, in various parts of the country. In the months of December and January when the rata is in flower, thousands of these birds are trapped by the natives, in the manner already indicated, and which I will presently describe more fully. Partly owing to this cause and partly to the extension of settlement in some districts, where, in former years, they were excessively abundant, their cry is now seldom or never heard; but in the wooded parts of the interior they are as plentiful as ever. Certain wooded ranges are noted as Kaka-preserves, and are very jealously protected by the native tribes owning them, who annually resort to them for the purpose of trapping these birds as an article of food. Nor is this its only practical value. Some half-dozen of the pillows in my house are filled with the feathers of the Kaka; and they are so delightfully soft and elastic that it is a positive luxury to sleep on them. These feathers were obtained at Raukawa, in the Upper Manawatu, some twenty years ago, when Kakas were far more plentiful in that part of the country than they are now. With the march of settlement, roads have been made, townships have sprung up, and a railway-line is being laid down within a mile or two of Raukawa, thus altering the whole face of the country. At the time to which I refer this place could be only reached by a canoe journey of some eighty or ninety miles from Foxton, or by a rude bush-track—one of the Maori war-paths of former times. The Manawatu gorge, lying just above, has now become a highway of busy traffic; the telegraph-wire already connects it with the commercial centres, and so, indirectly, with every

* The Rev. W. Colenso, F.R.S., remarks:—"The poor prisoners had not the common chance allowed them of biting and tearing their perch, or any wood (and this from mere thoughtlessness and carelessness, or long-continued custom, on the part of their Maori owners), for they were invariably kept fastened by a bone ring or carved circlet around one leg, and thus tied securely, but loosely, with a strong, short cord to a slender, polished, cylindrical hard-wood spear, up and down which, for the space of 2 or 3 feet, the poor bird ran and danced and flapped his wings, always without water, and frequently in the hot burning sun, without any shade."

part of the civilized world; steamboats are on the river, and the forest solitudes have exchanged the Kaka's scream for the whir of the saw-mill and the sound of the woodman's axe. This is, of course, the inevitable result of progress all over the country. What the condition of our avifauna will be in another twenty years it is not very hard to predict, especially when we reflect that, in addition to this legitimate pressure from without, the Government is threatening to bring about the extinction of all the terrestrial native species by the wholesale introduction of stoats and ferrets.

The Kaka is particularly abundant in the Urewera country, and during the short season the rata is in bloom the whole Maori population, old and young, are out Kaka-hunting. An expert bird-catcher will sometimes bag as many as 300 in the course of a day; and at Ruatahuna and Mangapohatu alone it is said that from 10,000 to 12,000 of these birds are killed during a good rata season, which occurs about every three years *.

There are several modes adopted for catching the Kaka, but the commonest and most successful is by means of a trained *mokai* or tame decoy, the wild birds being attracted to artificial perches, skilfully arranged around the concealed trapper, who has simply to pull a string and the screaming Kaka is secured by the leg, as many as three or four being often taken at the same moment. At the close of each day the dead birds are buried, and when a sufficient number have been collected they are unearthed, stripped of their feathers, fried in their own fat, and potted in calabashes for winter use, or for presents to neighbouring tribes. The perches used for Kaka-trapping are often elaborately carved and illuminated with *paua* shell.

It may be mentioned that the birds manifest extreme fastidiousness in the matter of these perches (or *tutu-kaka* as they are called), alighting very readily on some, and avoiding others in the most careful manner.

They commence breeding in the early part of November; and at Christmas the young birds are old enough to be taken from the nest, although, if unmolested, they probably do not leave it before the second week of January, or even later. The place usually selected for depositing their eggs is the deep hollow of a tree the heart of which is completely decayed. There is very little attempt at forming a nest, the eggs being placed on the dry pulverized wood which these cavities usually contain.

Mr. James Edwards of Kihi-kihi assures me that on more than one occasion, when taking wild honey from old Kaka holes, he has found the skeletons of the young birds underneath the honey-comb, showing that the bees had dispossessed the birds and appropriated the nest as a natural hive. I have heard similar accounts from Maoris in various parts of the country, and have no reason to doubt the fact.

The eggs are generally four in number (sometimes, according to the natives, six), broadly ovoid, measuring 1·6 inch in length by 1·2 in breadth, with a slightly glossy surface, and pure white in colour till they become soiled and stained in the process of incubation. Much care, and even some degree of fastidiousness, is displayed in the choice of a suitable tree; and once decided on, it is often resorted to by the same birds for many seasons in succession. On this account the natives set a high value on their "rua Kaka." The mere robbing of the nest, if accomplished with caution, does not cause the birds to abandon it; but the natives consider it of importance not to *breathe* into the cavity or to touch any part of it with their hands, for fear of "polluting" the nesting-place and causing its desertion. A nest which I discovered in the Otairi range, on my journey to Taupo, on the 23rd December, contained two young ones, apparently about ten days old. In a large maire tree with a decayed heart, about three feet from the ground there was a long narrow opening (measuring 2 feet in length by only 14 inches in width) leading into an inner chamber more than a foot in diameter. The walls of this chamber were smooth; and on the floor there was a deep layer of decayed wood,

* "Rarangi tahi" is the Maori proverb in allusion to this periodic recurrence of the Kaka season.

mixed with fragments of dry rata-bark, evidently collected by the birds and brought into the cavity. The natives state that two females, attended by one male bird, sometimes breed in the same cavity, their nests being placed side by side. The fact that during the breeding-season three birds are frequently seen in company, appears to give some colour of truth to this statement.

Mr. Enys informs me that, on Sir Charles Clifford's station at Stonyhurst, he found two nests of the Kaka, one of them situated in the crevice of a rock in a low mountain-gully, and the other in a deep cavity under the roots of a tree. This was on the 24th of December; and both nests contained young birds.

An egg of this species received from Reischek is much soiled on the surface, being more or less of a dark brown colour, as if stained by contact with decayed wood or some other colouring-matter. It was taken (late in December) from a hollow pukatea tree in the central part of the Little Barrier, just below the high pinnacle of rocks so distinctly visible at sea. The nest contained four eggs, all of which, as my correspondent assures me, were stained in a similar manner.

An egg from Dusky Sound yields somewhat larger measurements than those given above, is creamy white in colour, and marked with extremely fine points, making the surface almost granulate; but these may be regarded as mere individual variations. It was taken on the 13th April, from a tree-hole, in which a soft nest had been formed by means of pulverized rotten wood and feathers, apparently plucked from the bird's own breast. Besides this egg there were three young birds in the nest, and the mother, who remained in possession, defended her offspring in a very plucky manner.

A nest of four eggs in my son's collection presents this difference, that all of them are slightly larger than ordinary examples, measuring 1·7 inch in length by 1·25 in breadth. They were evidently freshly laid when taken, the surface being beautifully white.

Besides *Nestor meridionalis*, there is another very distinct species (*N. notabilis*) inhabiting New Zealand, a full account of which will follow next in order. But, in addition to these, there recently existed on Phillip Island a closely allied form (*Nestor productus*) which is now extinct, although many specimens of it are preserved in public and private museums in this country. Another (*N. norfolcensis*), formerly inhabiting Norfolk Island, although recognized by Dr. Finsch as a distinct species*, was probably only a local variety of this highly variable form. These are the only known representatives of the genus *Nestor*†.

* *N. producto* similis; at pileo et cervice viridibus; dorso ac tectricibus alarum olivaceo-viridibus, horum maculis apicalibus nigris triangulis; genis flavis; pectore abdomineque superiore ochraceis, unicoloribus; rostro insigniter longo, introrsum curvato.

† The following "Note on the Tongue of the Psittacine genus *Nestor*" was communicated to the Zoological Society by the late Prof. Garrod, F.R.S.—"On the death of a specimen of *Nestor hypopolius* in the Society's Gardens, a short time ago, Mr. Sclater kindly directed my attention to the peculiarity of its tongue, and referred me to Dr. Finsch's work on the Parrots, where *Nestor* is placed among the *Trichoglossinæ*, though the author states that he is not possessed of any very precise information on the subject.

"Mr. Gould, in his 'Birds of Australia' (vol. v. pl. vi.), partly describes the tongue of this bird, and shows that it is not that of a Lory; but he has omitted to notice its chief peculiarity.

"Dr. Buller, in the recently published 1st part of his 'Birds of New Zealand' (1st edit.), has also described the tongue quite correctly, though not much in detail—but nevertheless places *Nestor* close to the Lories, mentioning that this affinity was first shown by MM. Blanchard and Pelzeln. As, however, the tongue of *Nestor* does not in reality resemble that of the *Trichoglossi* at all, it may be of interest to describe it more fully.

"As far as I have had opportunity of observing, in all Parrots the fleshy tongue ends anteriorly by a dilated portion, supported on a narrower neck. This tip is much like the end of a human finger, as mentioned by most observers: and its function is similar also; for it is employed by the bird as a third prehensile organ in connexion with the upper and lower beak, any solid substance being held by the tongue and upper beak, while the mandible is freed to give another bite. Continuing the simile of the finger, the tip is directed forwards with the nail-like portion downwards, the part corresponding to the free edge of the nail appearing along the lower margin of the anterior rounded surface. This unguis, or nail-like portion, appears to me further to resemble a nail in that its anterior edge is not quite regular and is free, while the posterior margin is continuous with the neighbouring epithelium, which is almost enough to show that it grows forwards, and is worn down, as is a nail, by constant contact with foreign substances. In the typical Parrots this unguis of the tongue is broader than long, horny in texture, semicylindroid, with its lateral margins extending up the sides of the organ and encroaching on the borders of the superior surface for a short distance, not imbedded at the sides as is a nail. Its anterior border is nearly straight.

"In the *Trichoglossi* this horny plate is also present, and is similarly constructed; but on the superior surface of the tongue, between the lateral edges of the unguis, in the part which in others is covered by a smooth longitudinally plicated epithelium, there is an arrangement of retroverted papillæ forming a spinous covering; and their mechanism is such that when the tongue is protruded beyond the mouth to grasp any object, the papillæ stand upright or are even directed somewhat forward.

"In *Nestor* there are no papillæ of this description, but the tongue is here, as Dr. Buller says, 'soft, rounded on the edges, with a broad central groove,' and it is as smooth as in other Parrots. Therefore the Kaka Parrot cannot in this point be said to approach the *Trichoglossi* (badly so called).

"The peculiarity of the tongue of *Nestor* consists in the fact that the anterior edge of the unguis, always free (though for a very short distance) and jagged, as mentioned above, in the other birds of the class, is here prolonged forwards, beyond the tip of the tongue, for about $\frac{1}{5}$ inch as a delicate fringe of hairs, with a crescentic contour. This fringe seems to result from the breaking up into fibres of the forward-growing plate, which is always marked by longitudinal striations, clearest anteriorly, the result of unequal density and translucency of the tissue composing it, though on making a cross section I was not able to find any of the longitudinal papillary ridges which are present in the human nail and which the striation led me to expect. The unguis is also longer than broad, and very narrow considering the size of the bird, as is also the whole tongue, though the length is greater than in others of the class. In the living bird the mouth is moist, as in the Lories, and not, as in the Cockatoos and others, dry and scaly.

"From these considerations, and a comparison of the tongues of *Stringops*, *Nestor*, and *Trichoglossus*, it is evident that the structure of this organ would lead to the placing of *Nestor* among the typical Parrots, though an aberrant one, and not with the *Trichoglossinæ*; and other points in its anatomy favour this conclusion." (P. Z. S. 1872, pp. 767–769.)

Nestlings of *Nestor notabilis*. (See page 174.)

NESTOR NOTABILIS.

(KEA PARROT.)

Nestor notabilis, Gould, P. Z. S. 1856, p. 94.

Native names.

Kea and Keha; "Mountain-Parrot" of the colonists.

♂ superne omninò olivaceo-viridis: plumarum omnium scapis et marginibus nigricantibus: pileo paullò dilutiore, vix canescente: facie laterali magis brunnescente: dorso postico, uropygio et supracaudalibus sordidè cruentatis versus apicem angustè flavicantibus et nigricante marginatis, his imis olivaceo-flavicantibus: scapularibus et tectricibus alarum dorso concoloribus, his majoribus extùs vix cyanescente lavatis, remigibus nigricantibus, alâ spuriâ et primariis versus basin cyanescentibus, secundariis olivaceo-viridibus dorso concoloribus: remigibus subtùs pogonio interno versus basin citrino transfasciatis: caudâ olivaceo-viridi, supra sordidè cyanescente lavatâ et fasciâ nigrâ anteapicali transnotatâ, rectricibus subtùs flavicante tinctis, et pogonio interno citrino vix aurantiaco dentatis: corpore toto subtùs olivaceo-viridi, plumis nigricante marginatis, abdomine dilutè aurantiaco lavato: subalaribus et plumis axillaribus laetè scarlatinis, minimis flavicantibus, angustè nigricante terminatis: rostro cinerascenti-brunneo, mandibulâ ad basin laetè flavicante: pedibus flavicanti-olivaceis.

♀ mari simillima, sed sordidior et plumis nigricante latiùs marginatis.

Adult male. General plumage dull olive-green, brighter on the upper parts, with a rich gloss; each feather broadly tipped and narrowly margined with dusky black, with shaft-lines of the same colour, except on the head, where there is merely a darker shaft-line; ear-coverts and cheeks olivaceous brown, with darker margins; feathers on the sides strongly tinged with orange-red; primaries dusky brown, the outer webs light metallic blue in their basal portion, largely toothed on the inner web with bright lemon-yellow; secondaries greenish blue, changing to olive on their outer webs, dusky brown on their inner, and toothed with orange-yellow; lining of the wings and axillary plumes vivid scarlet, with narrow dusky tips; inner coverts towards the flexure washed with lemon-yellow; rump and upper tail-coverts bright arterial-red mixed with olive, and prettily vandyked at the tips with dusky black, this colour being richest on the middle tail-coverts and changing on the lateral ones to bright olive shaded with red and tipped with brown; tail-feathers olive-green on their upper surface, with a fine metallic gloss, paler at the tips, inclining to blue on the outer feathers, the whole crossed near the extremity by a broad band of blackish brown; the under surface pale olive-green, with the subterminal band less distinct, and broadly toothed on their inner webs with bright lemon-yellow; under tail-coverts dull olive-green, tipped with brown. Irides black; bill greyish brown; lower mandible rich wax-yellow in its basal portion; eyelid and cere dull yellow; feet yellowish olive, with paler soles. Total length 19·5 inches; wing, from flexure, 13·5; tail 7·5; bill, along the ridge 1·75, along the edge of lower mandible 1; tarsus 1·5; longer fore toe and claw 2·25; longer hind toe and claw 2.

Female. Similar to the male, but having the tints of the plumage generally duller, and the dusky margins of the feathers broader.

Obs. The bill is very variable, measuring in one of my specimens 2·25 inches along the ridge, and 1·5 along the edge of lower mandible. In some examples the lower mandible, instead of being yellow at the base, is dark brown, like the upper one, with only a faint line of lighter brown down the centre. This is probably characteristic of the young bird.

I examined a very brightly coloured specimen in the Otago Museum, the markings being unusually distinct. On the upper parts each feather has a narrow subterminal crescent of dull yellow, bordering the black and imparting a very pretty effect. The nuchal collar is heavily margined with brownish black, giving it the appearance of a collaret of looped lace-work. The feathers covering the rump and the short upper tail-coverts are dull crimson shading into green, then bordered by bright crimson and terminally margined, in a deeply notched manner, with black; on the tail-coverts there are generally two bands of bright crimson, and the larger coverts are uniform olive-green with black margins. These margins are very conspicuous on the back and mantle; but the blue on the outer webs of the primaries is less vivid than in many other specimens I have seen. Bill dark grey, without any tinge of yellow; the sides of the lower mandible paler grey. (Presented by Mr. Spence, Aug. 1877.)

Varieties. As with the other members of the genus *Nestor*, individuals vary much in the brilliancy of their tints. In July 1883, Mr. J. H. Berryman sent me the following description of a specimen procured by a friend of his in the interior of Otago:—"Bright canary-yellow, with a few red feathers interspersed throughout the plumage; vivid red on the rump and upper tail-coverts, as well as under the wings. Such a gorgeous bird has never been seen in the district before."

Remarks. Apart from the difference in plumage, this species differs from *Nestor meridionalis* in having more pointed wings; it likewise has a longer, slighter, smoother, and less curved bill, without any notch. The subjoined woodcuts will best illustrate the divergence of character in this respect.

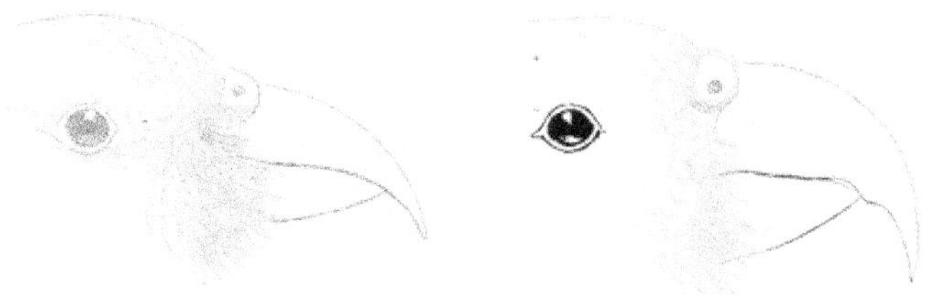

Nestor notabilis. *Nestor meridionalis.*

The first recorded examples of this interesting bird were obtained in 1856 by Mr. Walter Mantell, one of the early explorers of New Zealand, to whom we are indebted for many valuable discoveries in natural history, and who is now one of the patrons of science in his adopted country. Two specimens, from the Murihiku district, in the South Island, were forwarded by that gentleman to Mr. Gould, who thereupon characterized the species in the Proceedings of the Zoological Society, and figured it in the Supplement to his 'Birds of Australia.' Nothing more was heard of the Kea till the year 1859, when Dr. Haast received a fine example which had been caught on Mr. Tripp's station, near Mount Cook, and forwarded it, preserved in spirit, to Professor Owen. In the winter of the following year I first made the acquaintance of the species on a station near the Rangitata Gorge, where a live one, which had been snared by a shepherd and partially tamed, was frequenting the premises. Of late years, however, owing to the spread of colonization and to the development of a new character in the bird itself, to be presently mentioned, we have become better acquainted with this remarkable Parrot *.

It is essentially a mountain species, inhabiting the rugged slopes of the Southern Alps, and

* *Cf.* remarks on the skeletons of *Nestor notabilis* and *Stringops habroptilus*, with illustrative plates, by L. v. Lorenz, SB. Ak. Wiss. Wien, Bd. lxxxiv. Abth. 1, pp. 624–633, pls. i.–iii.

descending to the plains only during severe winters when its customary haunts are covered with snow and its means of subsistence have consequently failed *.

That distinguished explorer and geologist, the late Professor von Hochstetter, in describing the physical features of New Zealand, gives the following graphic account of the grand scenery in the South Island:—"High, precipitous, craggy mountain ranges, intersected by narrow longitudinal valleys, run parallel to each other from Foveaux Strait to Cook's Strait. They are connected by transverse ridges and intersected by the deep transverse valleys of the various rivers. In the centre of this range are seen, towering up in majestic grandeur, the peaks of Mount Cook, Tasman, and the adjacent mountain giants, glistening with perpetual snow and ice, to a height of 13,000 feet above the level of the sea, almost as high as Mont Blanc. Splendid glacier streams, lovely mountain lakes, magnificent cataracts, mountain passes, and gloomy ravines with roaring mountain streams rushing through them—such are the charms of a wild and uninhabited Alpine region but seldom trodden by human foot!" But this furnishes only a passing glimpse of our noble southern Alps, with their lofty peaks, capped with perpetual snow, flanked with glaciers of almost measureless depth, and presenting some of the finest mountain scenery in the world. In the deep valleys which divide these upheaved zones of stratified rocks of different ages luxuriant forests flourish, and on the high mountain-slopes there is the characteristic low vegetation, becoming more and more stunted as it approaches the line of perpetual snow. Such are the haunts of the Kea! I have seen it soaring or flying—often in parties of three or more—from one peak to another, high above the wooded valley; but it is more generally to be met with on the open mountain side, flying from rock to rock, or hopping along the ground amongst the stunted alpine vegetation, in quest of its natural food.

Sir James Hector found it everywhere in the snow-mountains of Otago during his topographical survey of that region in 1861–62. As a rule these birds were so tame there that he had no difficulty in knocking them over with a stone, or other missile, when he wanted to replenish his larder.

For many years the Kea ranked amongst our rarest species, and it is not very long ago that a specimen fetched £25 in the London market. But all this is changed, and, although still of very rare occurrence in the northern parts of the South Island, and quite unknown in the North Island, it has become, as will presently appear, an absolute pest in the middle and southern districts.

At the heads of all the principal rivers in the provincial district of Canterbury it is comparatively common; and especially near the sources of the Rangitata river.

The late Sir Julius von Haast sent me, from time to time, some beautiful specimens obtained in these localities, in the course of his geological and other scientific work. He informed me that in disposition it is most inquisitive, prying into and examining everything that comes in its way. On one occasion he left a large bundle of valuable alpine plants, which he had collected with much labour, lying exposed on the summit of a lofty mountain-crag. During his temporary absence a Kea

* "The rigour of a hard winter, when the whole face of the alpine country is changed so as to be scarcely recognizable under a deep canopy of snow, is not without its influence on the habits of this hardy bird. It is then driven from its stronghold in the rocky gully, and compelled to seek its food at a far less elevation, as its food-supply has passed away gradually at the approach of winter, or lies buried beyond its reach. The honey-bearing flowers have faded and fallen long before; the season that succeeded, with its lavish yield of berries, and drupes that gaily decked the close-growing *Coprosmas*, the trailing *Pimeleas*, or the sharp-leaved *Leucopogon*, has succumbed to the stern rule of winter. Nor has this change of season affected the flora of the Alps alone; the insect world, in a thousand forms, which enlivened every mountain-gully with the chirp and busy hum of life, now lies entranced in its mummy state, as inanimate as the torpid lizard that takes its winter sleep, sheltered beneath some well-pressed stone. Under the effects of such a change, that cuts off the supply of food, the Kea gradually descends the gullies, where a certain amount of shelter has encouraged the growth of the kowhai that yields its supply of hard, bitter seeds, the beautiful *Pittosporums* with their small hard seeds packed in clusters, and the black-berried *Aristotelia*; these and numerous other shrubs or trees, such as the pitch-pine and totara, furnish the means of life to the Parrot. It is during the continuance of this season that we have had the best opportunities of becoming somewhat familiar with it."—*Out in the Open*.

came down, and, with a supreme disregard for botanical science, tumbled the collection of specimens over into the ravine below, and quite beyond recovery. Mr. Potts also records an instance of this bird's extreme inquisitiveness. On one occasion a shepherd's hut was shut up, and left for a day or two, the man being required elsewhere. On his return he was surprised to hear something moving about within the hut; and on entering, he found that the noise proceeded from a Kea, which had gained access by the chimney. On a closer survey, the worthy shepherd discovered that his visitor had been exercising its powerful mandibles to some effect on his slender stock of goods and chattels. Blankets, bedding, and clothes were grievously rent and torn, pannikins and plates scattered about; and everything that could be broken was apparently broken very carefully, even the window-frame having been attacked with great diligence.

When hunting for food in its wild mountain home, it may be seen perched for a few moments on a jutting rock, then descending to the ground to hunt for grubs and insects, or to gather the ripening seeds from certain alpine plants, it disappears for a time and then mounts to the summit of another rock, just as I have seen the Common Raven doing in the higher parts of the Bernese Alps.

On the level ground their mode of locomotion is similar to that of the Kaka, consisting of a hopping rather than a walking movement. Like that bird also, they are semi-nocturnal, exhibiting much activity after dusk and in the early dawn.

The cry of the Kea, as generally heard in the early morning, has been aptly compared to the mewing of a cat; but it likewise utters a whistle, a chuckle, and a suppressed scream, scarcely distinguishable from the notes of its noisy congener.

But the most interesting feature in the history of this bird is the extraordinary manner in which, under the changed conditions of the country, it has developed a carnivorous habit—manifesting it, in the first instance, by a fondness for fresh sheep-skins and other station offal, and then, as its education progressed, attacking the living sheep for the purpose of tearing out and devouring the kidney-fat, and inflicting injuries that generally prove fatal *. This habit, confined at first to only a few of the more enterprising birds, soon became general, and it is a common thing now for whole parties of them to combine in this novel hunt after live mutton! So destructive, indeed, have they become on some of the sheep-runs that the aid of Parliament has been invoked to abate the nuisance by offering a subsidy to Kea-hunters †.

* The first announcement of this strange development of character in the Kea was made in the 'Otago Daily Times' newspaper, in the following terms:—"For the last three years the sheep belonging to a settler, Mr. Henry Campbell, in the Wanaka district (Otago), appeared afflicted with what was thought to be a new kind of disease; neighbours and shepherds were equally unable to account for it, having never seen anything of the kind before. The first appearance of this supposed disease is a patch of raw flesh on the loin of the sheep, about the size of a man's hand; from this, matter continually runs down the side, taking the wool completely off the part it touches; and in many cases death is the result. At last a shepherd noticed one of the Mountain-Parrots sticking to a sheep, and picking at a sore, and the animal seemed unable to get rid of its tormentor. The runholder gave directions to keep watch on the Parrots when mustering on the high ground; the result has been that, during the present season, when mustering high up on the ranges near the snow-line, they saw several of the birds surrounding a sheep, which was freshly bleeding from a small wound on the loin; on other sheep were noticed places where the Kea had begun to attack them, small pieces of wool having been picked out. . . . . The birds come in flocks, single out a sheep at random, and each, alighting on its back in turn, tears out the wool and makes the sheep bleed, till the animal runs away from the rest of the sheep. The birds then pursue it, continue attacking it, and force it to run about till it becomes stupid and exhausted. If, in that state, it throws itself down, and lies as much as possible on its back to keep the birds from picking the part attacked, they then pick a fresh hole in its side; and the sheep, when so set upon, in some instances dies . . . . Where the birds so attack the sheep, the elevation of the country is from 4000 to 5000 feet above the sea-level; and they only do so there in winter-time. On a station owned by Mr. Campbell, about thirty miles distant from the other, and at the same altitude, in the same district, and where the birds are plentiful, they do not attack the sheep in that way."

† The following statement appeared in one of the leading newspapers of the colony:—"In one instance a foal was attacked in this manner, and would have died had it not been rescued; in another, 200 out of 500 choice sheep were destroyed by these

On the surgical operation performed on the living sheep by the Kea, an interesting paper was read before the Pathological Society of London, in November 1879, by the distinguished surgeon, Mr. John Wood, F.R.S. He exhibited the colon of a sheep in which the operation known as colotomy had been performed by this Parrot, of which likewise he produced a skin, both specimens having been sent to him for that purpose by Dr. De Latour of Otago. Mr. Wood was informed by his correspondent that when the sheep are assembled, wounds resulting from the Kea's "vivisection" are often found on them, and not unfrequently the victims present an artificial anus—a fistulous opening into the intestine—in the right loin.

"The specimen exhibited was from a sheep that had been so attacked. It consisted of the lumbar vertebræ and the colon, showing the artificial anus between the iliac crest and the last rib on the right side—just in the place, that is, where modern surgeons perform the operation known to them as Amussat's; below the wound the intestine was contracted, while it was enlarged and hypertrophied above. The sheep was much wasted. The *modus operandi* was described as follows:—The birds, which are very bold and nearly as large as Rooks, single out the strongest sheep in the flock; one bird, settling on the sacrum, tears off the wool with its beak, and eats into the flesh till the sheep falls from exhaustion and loss of blood. Sometimes the wound penetrates to the colon, when, if the animal recovers, this artificial anus is formed; it may be on the left, but is more frequently on the right side *. It has been suggested that the bird aims at the colon in search of its vegetable contents; but the Kea's carnivorous appetite has been too frequently noticed to necessitate any such hypothesis. This strange phase of development through which the Kea has gone since the European colonization of New Zealand, and the consequent introduction of sheep to islands in which indigenous mammals are almost unknown, by which it has come to prefer an animal to a vegetable diet, was first described in 1871 by Mr. T. H. Potts ('Nature,' vol. iv. p. 489); but it was reserved for Dr. De Latour to discover the interesting result which Mr. Wood has just introduced to English naturalists."—*Zoologist*, Feb. 1880.

Before the full development of the raptorial habit described above, the penchant for raw flesh exhibited by this Parrot in its wild state was very remarkable. Those that frequented the sheep-stations soon manifested a distaste for all other food and lived almost exclusively on flesh. They took possession of sheep's heads that were thrown out from the slaughter-shed, and picked them perfectly clean, leaving nothing but the bones. An eye-witness thus described this operation:—"Perching itself on the sheep's head, or other offal, the bird proceeds to tear off the skin and flesh, devouring it piecemeal, after the manner of a Hawk, or at other times holding the object down with one foot, and with the other grasping the portion it was eating, after the ordinary fashion of Parrots."

At this period of its history the plan usually adopted on the stations for alluring this bird was to expose a fresh sheep-skin on the roof of a hut; and whilst engaged in tearing up the bait it was easily approached and shot.

---

birds, which are the more difficult to shoot from their nocturnal habits. Two or three runs in wild districts have been abandoned in consequence of the ravages of these harpies. This is a remarkable instance of change of habits, under altered conditions, for, of course, it is only within a few years that sheep have been introduced into the part of the country the Kea inhabits, and there was formerly no indigenous animal for it to prey on. In the summer the Kea lives on honey and berries. It is in the winter, when these fail, that it descends from the mountains and harries the flocks." Another newspaper, by way of comfort, adds:—"The Keas have found rivals in Seagulls, which are now to be seen in the Lake Country, Otago, driving away the Keas from the carcases of sheep these birds are devouring."

* *Cf.* an interesting article on *Nestor notabilis*, 'Zoologist,' 1881, pp. 290-301; also the figure facing p. 164 of 'Out in the Open.' This illustration represents a scene from the alpine country when under snow; a well conditioned merino is attacked by a Kea, and the animal in its terror, rushing hither and thither, has broken away from a small mob of sheep and is undergoing the first experience of torture from the beak of the Parrot.

In connection with the flesh-eating propensity of *Nestor notabilis*, I may mention a very remarkable case that occurred within my own experience, in which a whole fraternity of caged Parrots took to "cannibalism," if I may so term the killing and devouring of one another, without necessity, and in defiance of their natural habits and instincts. I had the following Parrots associated together in one compartment of my aviary at Wanganui, viz. two King Lories (*Aprosmictus scapulatus*), a pair of Rosellas (*Platycercus eximius*), a pair of Blue Mountain-Parrots (*Platycercus pennantii*), and a Grass-Parrot (*Platycercus semitorquatus*), all of them species indigenous to Australia. For nearly two years they lived together on terms of perfect amity and friendship, feeding from the same seed-troughs, often playing and coying with each other, and forming a constant source of attraction by their noisy clamour and the glittering of their rich plumage in the sunlight. One species alone (the last-named) was moody and shy, generally retiring to the highest perch under the domed roof, and disputing its possession with the rest. At length one of the pretty Rosellas met with an accident, which, in part, disabled it in the wing. The attention that it received from its partner was quite touching to witness. The maimed bird being unable to reach its perch, and therefore compelled to roost at night on the lower framework of the aviary, its mate forsook its sheltered perch under the dome and took up its position beside it; and during the day it was constantly chattering to it in a low confidential sort of manner. But this mishap led to a series of disasters that proved fatal, in the end, to the whole company. The Grass-Parrot, still retaining his sulky demeanour, began to persecute the disabled Rosella, and ultimately killed and partly devoured it. There was abundance of grain and other food in the troughs; but the Blue Mountain-Parrots followed suit, and whetted their appetites on the defunct Rosella. Attributing this *contretemps* to the weakly condition of the victim, I simply removed the mutilated body, and left the murderer in the aviary. On the following morning, however, I found, to my dismay, that he had killed and partly eaten one of the beautiful Blue Mountain-Parrots, and was murderously pursuing the surviving Rosella. I at once removed the author of all this mischief, and hoped to see harmony restored in the family; but the spirit of evil had been fairly roused, and I next found that the surviving "Blue Mountain" had killed the male King Lory and was devouring his body. Then I witnessed another touch of nature; for the mate of the last-named bird fretted and moped, refused her food, and died of a broken heart. Finally, the bereaved Rosella, as if to seek revenge for the murder of his sickly mate, made open war on the surviving female Blue Mountain-Parrot, and succeeded in killing her. I found this valiant little Parrakeet standing on the body of its vanquished enemy, and whistling in the most excited manner. And thus, within the limits of a single week, a group of Parrots that had lived together so long on the most satisfactory terms had, during a contagious passion for killing and devouring, come to utter grief, and only a solitary male Rosella remained! This bird shortly afterwards gained its liberty; and thus terminated my first and last experiment with Australian Parrots. But it must be remembered that this was an abnormal development of character under domestication, or at any rate under the artificial restraints of confinement. The difference in the case of the Kea is that, in its wild and natural state, it readily feeds on raw meat, and seems to prefer that to its proper vegetable diet.

When the Kea first began to frequent the outstations and sheep-yards, it was very unsuspicious and tame. Mr. Potts, Jun., snared a number of them by means of a simple flax noose at the end of a long rod. He describes them as exhibiting great boldness and confidence, clambering about the roof of his hut, and allowing a very close approach, for they had not yet learnt to regard man as their natural enemy. When caught (he tells us) they remained quite still, without any of the noisy fluttering which usually accompanies the capture of birds, even when managed with adroitness. One of the birds caught by him was placed on the floor under an inverted American bucket, the places for the handle not permitting the rim of the bucket to touch the ground. The Kea, taking advantage of this, wedged its beak into the space, and using its head as a lever, it moved the bucket, raising it sufficiently to effect its escape *.

The 'New Zealand Herald' of Sept. 12, 1880, contained the following announcement, which shows how rapidly the Kea nuisance had spread through the southern part of the country:—

"Mr. D. A. Cameron, one of the oldest runholders in Lake Country, Otago, is throwing up his run at the Nokomai through the Keas, which, if not more numerous, are, according to report, becoming greater adepts at the destruction of sheep. Formerly the birds used to annoy and worry, but now they kill outright. There is not a run which includes mountainous country, but is more or less plagued with the infliction, and on one spur alone on one mountain range in the Wakatipu, a runholder lost no less than 1000 sheep during last year."

From the KeKenzie country Mr. W. W. Smith reports in 1883:—"The estimated number of sheep annually destroyed by these birds is fifteen thousand. Formerly they attacked only the weak and dying sheep, caught in the snow-drift; but now the strongest and weakest suffer alike, both in summer and winter."

The war which is now being waged against this Parrot must, in the end, bring about its extermination. On some of the sheep-runs a bonus of three shillings a head is paid to the men for all they kill. Mr. Rolleston informed me that on his own little run at the Ashburton he had paid in one season for as many as 800; and I noticed, as far back as April 1884, a newspaper report that at the previous meeting of the Lake County Council no less than 2000 Keas' beaks were paid for.

In March 1884, Mr. R. Bouchier, the Sheep Inspector at Queenstown, reported that on a station on Lake Wanaka a mob of hogget-sheep were attacked by Keas, and in one night no less than 200 of them killed. Most of the birds, however, were afterwards destroyed by the shepherds, whose zeal in this work was stimulated by the bonus. The Inspector reported further that at the subsequent shearing hardly a sheep was marked, while the death-rate had been reduced by nearly one half. In the meantime the beaks of 1574 birds had been delivered at his office, for payment of the reward.

It is the fashion for cabinet ornithologists to declaim against the destruction of this "interesting form." But there is a good deal to be said on the other side. In some parts of the country the Kea nuisance has reached such a pitch that the runholders have been fairly driven off their country. In places where a few years ago only occasional birds were seen they now appear in hundreds, attracted of course by the sheep †. They are most numerous in winter, when, as already explained, they are driven down from their natural home in the mountains by the severity of the climate; and so bold do they become in their depredations, that, as I have been assured by credible eye-witnesses, they will actually attack a mob of sheep whilst being driven to the yards!

As a rule they confine their attentions to the latter animal; but there is at least one well-

* At a meeting of the Wellington Philosophical Society, a paper was read by Mr. Alexander McKay, who related a number of personal observations on the Kea, which went to prove that this bird possesses a high degree of intelligence. The author expressed his own conviction, as the result of careful observation, that the Keas had the power of communicating ideas among themselves. He related an anecdote within his own experience in support of this view. He stated that on one occasion a number of Keas, after a consultation, delegated one bird twice in succession to untie the knot in a string which fastened one of their number to a pick-handle. This statement," the report continues, "evoked some discussion. Mr. W. M. Maskell expressed great astonishment that even an intelligent bird like the Kea should know how to untie a knot at first sight; but Dr. Hector, Mr. W. T. L. Travers, and other gentlemen who were present related instances of still more surprising sagacity on the part of native birds."

† There occurs the following singular confusion of two well-known New-Zealand species (the Kakapo and the Kea) in Mr. A. R. Wallace's 'Australasia,' at page 561:—"Another remarkable bird is the Owl-Parrot (*Stringops habroptilus*) of a greenish colour, and with a circle of feathers round the eyes, as in the Owl. It is nocturnal in its habits, lives in holes in the ground under the tree-roots or rocks, and it climbs about the bushes after berries, or digs for fern-roots. It has fully developed wings, but hardly ever flies, and has lately exhibited a singular taste for flesh, picking holes in the backs of sheep and lambs" (!).

authenticated case of their coming in force and attacking a valuable mare, which they seriously injured in the loins.

On the habits of this species Mr. John George Shrimpton, of Southbrook, Canterbury, has sent me the following very interesting notes:—

"While residing at the Wanaka Lake, I received a letter from my brother Walter (of Matapiro) to the effect that you would like a specimen of the Kea or Mountain-Parrot, and any notes of their habits which I might be able to afford you. My time there was so short after receipt of his letter that, although many Keas were killed, I only succeeded in getting one fair skin, which I forwarded to you by mail a few days ago, and trust it has reached you safely. By this mail I forward a water-colour sketch of some young ones drawn from nature by Mr. Huddleston. In the rocky cavern, high up on the mountain, whence these were obtained, were several broods of young ones of various ages and sizes.

"I believe the Kea does not come farther north than the Rakaia River, Canterbury, and is strictly confined to the central range and its spurs as a rule, but may occasionally and will probably be more seen on those hills adjacent to the main range which attain an elevation of five thousand feet and upwards. There is no doubt that, in spite of the war waged against them, they are increasing very rapidly, probably owing to the plentiful supply of food in the shape of mutton which they can get, and to which they help themselves most liberally. Fifteen years ago, when I first knew the Lake country, it was a rare thing to see these birds on the hills even in their chosen home among the snow; but now you meet them in flocks of fifty even, and so bold have they become that they will attack sheep under the shepherd's immediate care. Not that they were ever very wild; on the contrary, I think they are the tamest birds in New Zealand; and it is their insatiable curiosity that has probably led them to find out the taste of mutton. At first, they contented themselves with tearing up tents, blankets, and sheep-skins, the usual impedimenta of a musterer's camp. They have now so improved upon that, that nothing less than the primest mutton will suit their fastidious tastes. Though so tame that you can often knock them down with a stick, and apparently so inoffensive, a single Kea will swoop down on the strongest fat wether or hogget, fix himself firmly on its back, generally facing the sheep's tail, and commence digging his daily meal. Sometimes the sheep runs till exhausted, sometimes contents itself by trying to dislodge its adversary by a series of contortions only, but the Kea troubles himself very little about either; he hangs on till the sheep gives in. He then digs away, carefully avoiding the backbone, till he reaches the kidney fat. This is his choicest relish. His cries soon attract others, and between them the poor sheep is soon fitted for a museum. Sometimes a sheep gets away from a timid or perhaps less experienced workman; but he carries with him an indelible scar. On some stations about 5 per cent. of the whole flock are mustered in at shearing-time more or less marked in this manner, and the death-rate is almost incredible. I have no hesitation in saying that, on the runs bordering the Wanaka and Hawea Lakes, the loss from Keas alone is nothing short of from fifteen to twenty thousand sheep annually, and these the primest of the flocks. Although Keas are seen openly enough in the daytime, there is no doubt they work their mischief mostly at night, a bright moonlight one being preferred. A severe winter, with sheep snowed in, is their great opportunity; and this they avail themselves of to the uttermost. Although, like other Parrots, they are given to anything in the shape of fun or mischief (and, on one occasion they killed a young Kaka, tethered), I have never known them to *seriously* attack any animal other than a sheep. But as a moiety of them have advanced so far in the course of the last eight or ten years, it is impossible to say to what lengths they may aspire in the future.

"I cannot state for certainty that there are no Keas north of the limits I have here assigned as their habitat. I can only say that I have travelled over a considerable portion of that country without either seeing or hearing of them. But as to their habits and destructiveness in the neighbourhood of the great lakes south, I can speak from a long and painful experience."

I have reproduced, on a smaller scale, in the woodcut given on page 165, the spirited drawing received from Mr. Shrimpton, exhibiting a pair of ungainly nestlings in their alpine nursery.

There is a fine living specimen of the Kea in the "Parrot-house" at the Zoological Society's Gardens, which appears to thrive in spite of the unnatural semi-tropical heat to which it is subjected *.

This bird was received from Dr. De Latour, who sent the following interesting account of it to 'The Field' prior to its departure from the colony:—

"A shepherd in bringing down a mob of sheep was annoyed by one of these Keas attacking the sheep while he was driving them down the mountains; being angry, he threw a stone at it and knocked it over. He succeeded in capturing it alive; he did not kill it, and in return the Kea made great havoc with his clothes. However, after cutting its wings and tying its legs together, he brought it down to his camp. There the shepherd broke his own leg, and came under my care, and the Kea came down shortly after. He was in an ordinary cage made of wood and small iron wire. He was only a day and a half coming down eighty-four miles, but in that time the cage was all but destroyed, the wires bent, some broken in two, as though cut with pliers, and the woodwork was reduced to tinder, and it was just a piece of luck that he did not escape. I had a strong cage of galvanized iron and stout wire built for him, and he has now been with me for two years. The cage is a big one, about 3 feet high, 2 feet across, and 18 inches deep, so that he has lots of room to move about in. He is rather expensive to keep, as he generally gets a mutton chop every day; he does not like cooked meat, and will only take it if very hungry; he will not touch beef if he can get mutton, but is not averse to pork. Some say the Keas only want the fat, but this bird takes lean and fat impartially; indeed I find the fat parts often left on the bone, but never any of the flesh. I have tried him with canary and hemp seed, but he does not seem to care for it, only scattering it about as though for mischief, and they are very mischievous. I am told that when they get into an empty hut—and there are many of these huts used only on occasions when the shepherds are out mustering and away from home for some days—if any blankets, tin pots, sacks, &c. are left, the Keas tear the blankets and sacks to pieces, and bend the tin pots until they are useless.

"My Kea does not care much for vegetable food; give him a lettuce or cabbage and he only tears it up and throws it away; he is, however, fond of the seeds of the sowthistle. I see that you say in your article that a specimen was received by the Zoological Society in 1872, which only survived a few days. It has struck me that my bird having been in captivity for two years, and being now tame, and we will suppose reconciled to his lot, would be in a favourable condition to bear the

* The advent of this Parrot was thus chronicled in the London press:—

"There is now in the 'Zoo' a very remarkable bird, the *Nestor notabilis*, or Mountain Kea, of New Zealand. It is a parrot of strong frame and powerful bill and claws, which were used like those of all Parrots for obtaining a vegetable diet, until the colonists introduced sheep and pigs. As soon as this was done the Kea seems to have abandoned vegetable food, and to have taken entirely to flesh-eating. He attacks sick or dying or disabled sheep, and with his powerful cutting beak opens a passage through the back, and eats the intestines. Even healthy animals are sometimes assailed by the *Nestor notabilis*, and there are sheep-runs in New Zealand where considerable losses have been incurred through these strangely degenerated birds. The specimen in the Zoological Gardens gave as much trouble to capture as an Eagle, tearing the clothes of the shepherd who knocked it down while pouncing on a lamb, and lacerating his hands. The Kea scorns cooked meat, biscuits, fruit, or seeds, and likes raw mutton better than any food. He will tear the skin and flesh from a sheep's head after the furious fashion of a Vulture, leaving nothing but the bare skull. He at one time holds the morsels in his lifted claw, after the style of Parrots, and at another grips them under his feet while rending with his beak like a Hawk. This is a very curious example of change of habit, for there is every reason to believe that before sheep and pigs were introduced into New Zealand the Kea was as frugivorous in its meals as most, if not all, other Parrots. He will now eat pork and beef, as well as mutton, and has become, in fact, utterly and hopelessly carnivorous. It is to be feared, after this example, that temptation is often fatal to birds and beasts as well as men. Had it not been for Captain Cook and the English sheep flocks, the *Nestor notabilis* would have lived and died innocent of crime; but now its bloodstained carcase is suspended outside many a sheepfold near Otago."

voyage. I shall be very pleased to present him to the Society—that indeed has been my intention always—but I imagined that they would have specimens.

"My purpose in writing has been mainly to acquaint you with the habits of the bird in captivity, and somewhat of what I have learnt of its habits in the wild state; and also to ask you for hints as to sending the bird home should the Zoological Society care to have him.

"I have just now another specimen of a sheep attacked by these birds; it is of even greater pathological interest than the other one which I sent home, for in this case the opening is into the rumen or large stomach; the sheep survived for a long time. There are also several other living sheep that have been injured waiting for favourable opportunities to be sent down. I want one with an opening into the rumen, so as to be able to watch the process of digestion. I think it would be very interesting."

Although, as already shown, very easily captured, it is difficult to detain the bird against its will. My brother, during his residence in the back Mackenzie country, obtained, at various times, no less than eight live specimens for me; but in every instance they managed to escape, either by eating their way out of the wooden cage, or in some other, unaccountable manner, before reaching their destination *. If taken young, however, they are readily tamed and become very tractable pets. Dr. Finsch, during his travels in New Zealand, was accompanied by one which was daily allowed to leave its cage, and could be handled with impunity. I never heard whether Dr. Finsch sent it to Europe, as he then proposed doing, or whether it remained to share the vicissitudes of his consular life in the South Pacific Islands.

On being removed from its cage and fondled with the hand it crouched down and ruffled up its feathers after the manner of an Owl. I noticed that whilst in the cage it had a habit of dancing up and down in true *Nestor*-fashion. It seemed very prying and inquisitive, trying the quality of anything within its reach by means of its well-curved beak.

A live one in the possession of Mr. J. Baker, at Waipawa, became perfectly tame and was allowed the freedom of the establishment.

The inference I ventured in my former edition, that, judging from its general economy, the Kea nests in the crevices and crannies of the rocks in its wild alpine haunts, has since been verified, many nests of this Parrot having from time to time been met with, and always in such situations †.

An egg in my son's collection, being one of two found in a Kea's nest "under a high cliff at Forest Creek," is of similar form and appearance to that of *Nestor meridionalis*, but is appreciably larger, measuring 1·75 inch in length by 1·3 in breadth; it is pure white, with a slightly glossy surface.

* "A Kea has been seen by his gratified captor to eat his way out of a wooden cage almost as quickly as it had been coaxed to enter it. Two which had been tamed by a neighbouring friend were permitted to wander at large. They regularly returned to his house for their meals and then rambled away again, scrambling and clambering amongst the trees and outbuildings. Any kind of food appeared to suit their accommodating appetite, but a piece of raw meat was evidently the bonne bouche."—*Out in the Open.*

† The following account is given ('Zoologist,' 1883, p. 276) of an egg obtained by Mr. H. Campbell:—"The specimen, with three others, was taken from a nesting-place, in an almost inaccessible fastness of rocks, high up the mountains near Lake Wanaka. An egg was broken in getting out; two of those remaining have also come to grief. Placed among a series of eggs of the Kaka (*N. meridionalis*) it can be picked out at once; it is larger, rougher, the surface being granulated, dotted over irregularly with small pits, a very few slight chalky incrustations towards the smaller end. The shell is very stout and thick, exceeding in that respect any examples that I have seen of the eggs of the Kaka. It is broadly ovoid, measuring one inch seven lines in length; in width it is one inch three lines."

STRINGOPS HABROPTILUS.

(OWL PARROT.)

Strigops habroptilus, Gray, P. Z. S. 1847, p. 62.
Strigopsis habroptilus, Bonap. Consp. Gen. Av. i. p. 8 (1850).
Stringops habroptilus, Van der Hoeven, Handb. Zool. ii. p. 466 (1856).
Stringopsis habroptilus, Schl. Mus. Pays-Bas, *Psittaci*, p. 107 (1864).
Stringops habroptilus, Finsch, Papag. i. p. 246 (1867).

Varieties.

Strigops greyii, Gray, Ibis, 1862, p. 230.
Stringops greyi, Finsch, Papag. i. p. 253 (1867).

Native names.

Kakapo, Tarapo, and Tarepo; "Ground-Parrot" of the colonists.

Ad. viridis: plumis pilei dorsique medialiter pallidè flavidis, irregulariter nigricanti-brunneo transfasciatis et transvermiculatis: uropygii plumis latiùs viridescentibus: loris plumisque rictum obtegentibus pallidè fulvescenti-brunneis, medialiter albicantibus: regione auriculari brunneâ, rachidibus plumarum fulvescentibus: facie laterali brunneâ, plumis medialiter latè flavicantibus: remigibus nigricanti-brunneis, primariis extùs et intùs flavicante maculatis, secundariis irregulariter flavido fasciatis variis et extùs olivascenti-viridi lavatis: caudâ olivascenti-brunneâ, ubique nigricante fasciolatâ: subtùs magis flavicans, viridi lavatus, abdomine puriùs flavicante: pectoris plumis paullò nigricante variis, hypochondriis magis conspicuè fasciatis: subalaribus olivascenti-flavis, obscurè brunneo fasciatis: subalaribus flavicantibus, minoribus nigro variis: rostro flavicanti-albido, ad basin saturatiore: pedibus flavicanti-brunneis, unguibus saturatioribus.

Adult. General colour of the upper surface dark sap-green, brighter on the wings and lower part of back, and largely varied with dark brown and yellow; on their under surface the feathers of these parts are light verditer-green towards the tip, with a fine metallic lustre; on the crown and nape the centre of each feather is blackish brown, with a narrow shaft-line of dirty yellow and a broad terminal band of dull green; on the back, rump, and upper surface of the wings, each feather is silvery brown at the base, pale lemon-yellow beyond, changing to sap-green on the sides and towards the tip, and crossed by numerous broken bars and vermiculations of dark brown; on the anterior portion of the back these bars are regular and distinct, but on the other parts they are interrupted by a broad shaft-line of lemon-yellow. These details of colouring, however, can only be observed when the plumage is disturbed, the general effect on the surface being as already described. The feathers at the base of the upper mandible, lores, sides of face, and feathers projecting over the lower mandible dull yellowish brown, with darker filaments; ear-coverts darker brown, mixed with yellow; fore neck, breast, and sides of the body yellowish sap-green, varied with pale yellow and brown, the distribution of colouring on each individual feather being the same as on the upper parts, but with more yellow down the shaft; lower part of abdomen, thighs, and under tail-coverts light greenish yellow, the longer coverts obscurely barred with light brown; lining of wings pale lemon-yellow, blotched and streaked with dark brown; primaries dark brown, largely toothed on their outer webs with dull lemon-yellow, and on their inner with paler; secondaries and their coverts dull greenish yellow, rayed and freckled with dark brown on the outer webs; dusky brown on the inner webs, with broken transverse markings of lemon-yellow; tail-feathers yellowish brown, with arrow-shaped markings along the shaft, and largely

freckled and mottled with blackish brown. Irides black; bill yellowish white, darker at the base and along the fluting of the lower mandible; tarsi and toes yellowish brown; claws darker. Extreme length 26 inches; wing, from flexure, 12; tail 10; bill, along the ridge (from base of cere) 2, along the edge of lower mandible 1; tarsus 1·75; longer fore toe and claw 3; longer hind toe and claw 2·5.

Obs. The sexes are alike in plumage. Individuals vary a good deal both in the brilliancy of their tints and in the details of their colouring. The ground-colour of the upper parts varies from a dull sap-green to a bright grass-green, and in some examples the whole of the plumage of the underparts is strongly suffused with lemon-yellow. The barred character of the individual feather is more defined in some specimens, while in others the light markings on the quills and tail-feathers are softened to a pale yellow. Individual birds also differ perceptibly in size, owing probably to conditions of age and sex.

Captain Preece, R.M., has in his possession the skin of a Kakapo obtained at Hikurangi, in the North Island. Its plumage is in no respect different from that of the southern bird. Length 25 inches; wing, from flexure, 11; tail 8; tarsus 2; longer fore toe and claw 2·5.

Varieties. Of this species there is a beautifully marked variety in Mr. James Brogden's collection of New-Zealand birds, at Porthcawl. The whole of the plumage is largely suffused with yellow, especially on the underparts, where each feather has a broad irregular central spot of pale yellow, edged with dusky brown; towards the tips the feathers are greenish yellow. The upper parts are bright green, prettily rayed with black, and varied more or less obscurely with yellow, the feathers of the nape and sides of the neck having spear-head points of bright yellow near the tips. The tail is conspicuously marked at regular intervals with vandyked bars of clear lemon-yellow, getting darker towards the tips; these yellow markings are edged with black, and the interspaces are yellowish brown, more or less freckled and marbled with black. The primaries and secondaries are similarly marked on their outer webs, but the yellow is not quite so clear.

A specimen in my collection has the cheeks of a bright reddish brown, this colour fading away on the edges. There is a somewhat similar example in the Otago Museum, with the crown, sides of the face, chin, and upper part of the throat dingy reddish brown. I suspect that this coloration results from some vegetable stain, inasmuch as in this specimen I observe that the ridge and sides of the upper mandible and the fluted grooves in the lower are similarly stained.

In Mr. Silver's fine collection of New-Zealand Birds at Letcomb Manor there is an abnormally small specimen, the measurements being:—Total length 20 inches; wing, from flexure, 10·5; tail 7·5; bill, along the ridge 1·5, along the edge of lower mandible ·75. The plumage as in the ordinary bird, except that, on the left cheek, there is a patch of yellow about an inch in extent, completely covering the ear-coverts and extending downwards.

I have examined the type specimen of Mr. G. R. Gray's *Strigops greyii* in the British Museum and have come to the conclusion that it is simply an accidental variety, although a very singular one, of the true *S. habroptilus*. The specimen is in very bad condition, the quills being much worn and abraded, and the tail worn down to a mere stump; indeed the whole of the plumage is dingy and soiled, apparently the result of long confinement. The feathers of the upper parts, instead of being sap-green at the ends, are of a dull greenish blue, changing in certain lights to a purplish blue. There is, moreover, somewhat less of the terminal colour; and as the barred markings on the basal portion of the feathers are fulvous white instead of yellow, the back has a more variegated appearance. The entire plumage of the underparts is a pale yellowish fulvous, mottled, except on the abdomen, with brown. The cheeks and feathers overlapping the lower mandible are the same as in ordinary examples, but without any yellow tinge. On the sides and flanks the feathers are slightly tinged with blue, but of a duller tint than on the upper parts; thighs deeply stained with yellow. The newest of the tail-feathers (*i. e.* the stumpy portion that remains) is rayed in the same manner as in ordinary examples, but without the yellow element, showing a decided tendency to albinism. In the wing-feathers, in which also the yellow colour is absent, the bars appear at first sight more regular and distinct; but on closer examination it will be found that in both wings the broad inner secondaries and the scapulars have been torn out (doubtless due to the bird's captivity), and the barred effect is therefore more conspicuous. Although, among the numerous examples that have come under my notice, I have never seen one in any degree approaching this condition, yet I have detected in some a tendency in the feathers of the back to assume a bluish margin, and in all specimens these feathers have a bright metallic lustre on their under surface. There is no means of determining the exact length of the wing, as the long primaries, on both sides, have been broken off; but the specimen does not appear to differ in size from ordinary small examples of *S. habroptilus*. But what tends more than any thing else to convince me that the so-called *S. greyii* is merely an abnormal or accidental variety of the species under review is the fact that some of the small coverts on both wings, and the feathers of the crown, have assumed the normal sap-green colour, thus betraying a strong tendency to reversion. In the absence of any other examples in a similar condition of plumage, this fact appears to me of itself fatal to the recognition of the species. At the same time, I should add that the difference in colour was so manifest and striking, that Mr. G. R. Gray was perfectly justified in characterizing it provisionally as a distinct species, although (as appears from his Catalogue of Psittacidæ, 1859) he was himself of opinion that it might ultimately prove a mere variety. Even Dr. Finsch, who is scrupulously careful in all his identifications, states (in his valuable Monograph of Parrots) that, after a careful examination of the type specimen, he felt bound to admit *S. greyii* as a good species. It only remains for me to say that I regret that my convictions compel me to sink a name designed by the describer as a compliment to Sir George Grey, who has always taken so zealous a part in the furtherance of ornithological science.

In Reischek's collection there is a specimen with a single canary-yellow feather among the scapulars; and another has a bluish glint on the feathers of the upper parts, somewhat like that described above.

I examined a remarkable variety from Dusky Bay, this example having been obtained (as I was assured) at a considerable elevation. This bird had the crown of the head uniform dark green, the cheeks dull greenish brown, the markings on the upper surface generally very obscure, and the plumage of the underparts dull greenish yellow, with faint marbled markings of a paler colour, presenting a very soft appearance, whilst the flanks were prettily marked with numerous narrow bars of brown; the bill was pale yellow, the sides of the under mandible inclining to brown.

Mr. J. D. Enys sends me the following note:—"Mr. G. Müller, the Chief Surveyor of Westland, has a Kakapo with the entire plumage yellow. It came from Jackson's Bay. Have you heard of it?"

Mr. Reischek, who spent six months in the West Coast Sounds, brought back with him some very beautiful specimens, differing from the common Kakapo in having the entire upper surface rayed with narrow transverse, more or less wavy, bars of brownish black, and the markings on the wing-feathers very regular and distinct, being of a pale lemon-yellow. Of this bird he writes:—"The Alpine Kakapo—so called by me, as I have never found this beautiful bird anywhere except on high mountains—is considerably larger and much brighter than the ordinary Kakapo. The young ones are much duller in plumage than their parents. These alpine birds are rare, but I was fortunate in securing about a dozen of them. Amongst them was a specimen of a beautiful varied plumage: on the top of the head very light green; back, wing-coverts, and tail yellowish green with crimson spots; round the bill crimson; throat, breast, and abdomen yellow with crimson spots; bill light yellow; legs silver-grey; eyes dark brown."

Several of these fine specimens are now in my collection, and although I fully appreciate the difference in the plumage of the upper surface, yet, with my knowledge of the extreme variability of this form, I am unable, however willing, to recognize a new species. As to individual size, that counts for very little, for I have in my collection even larger specimens in the ordinary plumage. Again, one of the alpine birds received from Mr. Reischek, in which the colours are particularly brilliant, has little thread-like tufts of down adhering to the tips of the secondaries; it is obviously a very young bird, and does not conform to Mr. Reischek's description as quoted above.

On the accompanying Plate my artist has represented this Alpine form, in the distant figure, just emerging from its burrow.

Young. The young Kakapo assumes the adult plumage from the nest, although the colours are duller than in the mature bird and with a less admixture of yellow; the ear-coverts are darker and the facial disk less conspicuous. The bill, instead of being horn-coloured, is of a delicate bluish-grey colour.

Nestling. In the Otago Museum there is a Kakapo chick apparently just extruded from the shell. It is extremely small for such a bird, and is covered with thick fluffy down of a creamy-white colour; bill and feet white. It was obtained at Dusky Bay, in April 1877, by Mr. Docherty, who presented it to the Museum. I have seen more advanced nestlings covered with greyish down. (See woodcut on page 191.)

General remarks. In the peculiar form which constitutes the unique member of the genus *Stringops* the bill is broad and powerful; the upper mandible has a peculiar rasp-like character within, while the lower mandible is deeply fluted on its outer surface, with a worn, notched process near the extremity. The plumage is soft but compact; the wings apparently well developed, but useless for purposes of flight, with the quills much curved or bent; the tail long and slightly decurved, the feathers composing it acuminate and sometimes with the tips abraded; the projecting feathers on the cheeks loose, with disunited filaments and shafts much produced; the legs strong and well formed; the tarsi covered with elevated rounded scales; the toes similarly protected in their basal portion, scutellate towards the end; the claws strong, well-arched, sharp on their inner edge, and with fine points.

This is one of the very remarkable forms peculiar to New Zealand, and has been appropriately termed an Owl Parrot. Dr. Sclater refers to it as "one of the most wonderful, perhaps, of all living birds." As its name *Stringops* indicates, its face bears a superficial likeness to that of an Owl. In all the essential characteristics of structure it is a true Parrot; but in the possession of a facial disk (in which respect it differs from all other known Parrots), in the soft texture of its plumage, and especially in its decidedly nocturnal habits, it betrays a striking resemblance to the Owl tribe. Its toes, as in all other members of the order, are zygodactyle; but, as pointed out by Mr. Wood in an interesting article communicated to the 'Student' (1870, p. 492), the foot of an Owl, when the bird is perched, considerably resembles that of a Parrot, as the outer toe is then placed backwards with the hind one, so that the bird's feet may be said to be temporarily zygodactyle, whereas those of the Parrot are permanently so.

Although it may, perhaps, be morphologically incorrect to say that this form supplies a quasi-connecting link between the Owls and the Parrots, there can be no doubt that the Kakapo, in some of its external characters as well as in its mode of life (as Mr. A. R. Wallace has well expressed it), "imitates the Owl" in a very remarkable manner.

Although exclusively a vegetable-eater, its habit of hiding during the day in holes of trees and dark burrows exhibits a further point of resemblance to the nocturnal birds of prey. As these latter are in reality night Hawks, so is this bird, what the native name indeed implies, a night Parrot; and the analogy thus presented harmonizes with the idea expressed above.

The feathers surrounding the eyes and filling the lores differ from those on the other parts of the body not only in being of a lighter colour, but also in form and structure, being narrow and penicillate, with the shaft considerably produced. Those overlapping the base of the lower mandible are more stiff and elongated.

All who have studied the bird in its natural state agree on this point, that the wings, although sufficiently large and strong, are perfectly useless for purposes of flight, and that the bird merely spreads them to break the force of its fall in descending from a higher point to a lower when suddenly surprised; in some instances (as one of the writers quoted below informs us) even this use of them is neglected, the bird falling to the ground like a stone.

We are naturally led to ask how it is that a bird possessing large and well-formed wings should be found utterly incapable of flight. On removing the skin from the body it is seen that the muscles by means of which the movements of these anterior limbs are regulated are fairly well developed, but are largely overlaid with fat. The bird is known to be a ground-feeder, with a voracious appetite, and to subsist chiefly on vegetable mosses, which, possessing but little nutriment, require to be eaten in large quantities; and the late Sir J. von Haast informed me that he had sometimes seen them with their crops so distended and heavy that the birds were scarcely able to move.

These mosses cover the ground and the roots or trunks of prostrate trees, requiring to be sought for on foot; and the bird's habit of feeding at night, in a country where there are no indigenous predatory quadrupeds, would render flight a superfluous exertion, and a faculty of no special advantage in the struggle for existence. Thus it may be reasonably inferred that *disuse*, under the usual operation of the laws of nature, has, in process of time, produced the modification of structure which distinguishes this form from all other known Parrots and thereby occasioned this disability of wing.

The sternum, which in all other birds of its class has so prominent a keel, is so completely altered that it presents almost a flat surface, although the symmetry of the skeleton does not appear to have suffered in any other respect.

Prof. W. K. Parker says:—"Like all those who glory in 'high degree,' the Parrots have a poor relation or two to abate their pride. The Owl-billed Parrot (*Stringops habroptilus*) of New Zealand is as lowly as 'the younger son of a younger brother.' If birds were to be classified by the sternum only, then the *Stringops* should be put near the *Apteryx* and the Tinamou attached to the train of the Peacock."

The late Prof. Garrod has pointed out that the Parrots, as an order, are peculiar for the variation that occurs in their carotids, which show four different arrangements, and that *Stringops* is one of those forms in which the two carotids run normally *.

Conformably also with the doctrine of natural selection, we have here another striking instance of the law of assimilative colouring, which obtains more or less in every department of the animal kingdom. Nature has compensated this bird for its helplessness when compelled to leave its hiding-place in the daytime, by endowing it with a mottled plumage so exactly harmonizing with that of the green mosses among which it feeds, that it is almost impossible to distinguish it.

Although the existence of a large ground-Parrot was known to the early colonists of New Zealand from the reports of the natives, who set a high value on the feathers for purposes of decoration, it was not till the year 1845 that a skin of this bird reached Europe; and this was purchased by the Trustees of the British Museum for the sum of £24.

* The same distinguished anatomist, in one of his earlier papers on the muscles of Birds, pointed out that the ambiens may be present normally, or it may be differentiated in the thigh, but fail to cross the knee, being lost in the fascia over it, or it may be absent; and he stated that in *Stringops habroptilus* it is present but does not cross the knee. In a subsequent paper "On the Anatomy of the Parrots" (P. Z. S. 1874, pp. 586–598) he says:—"I have twice had the opportunity of dissecting *Stringops habroptilus*. As a Parrot it is not so strikingly peculiar as many seem to think. Its wings are useless, and the carina sterni is correspondingly reduced, it is true; but as points of classificational importance, I regard these as insignificant. The points of special anatomical interest which it does possess, however, are particularly instructive. The proximal ends of the incomplete furcula are well developed, so much so that it might at first sight seem that the symphysial ends are only lost in correlation with the excessive reduction of the powers of flight; though this is probably not the case, because the allied similarly modified genera *Euphema* &c. do not keep to the ground. Further, in the Society's specimen above mentioned, though the ambiens muscle did not cross the knee, yet its fleshy belly was well differentiated on both sides, its thin tendon being lost over the capsule of the joint. In the College of Surgeons' specimen, however, this muscle was entirely absent in the only knee which was in a fit state for dissection, the other being much shot. It is only in the genus *Œdicnemus* that I have elsewhere found a similar partial loss of the ambiens. The partial development of this muscle in this particular instance shows that the tendency to lose it is not of great antiquity: and it is to be noted that there is no other Parrot with normal carotids in which any trace of an ambiens is to be found. These considerations suggest, what may perhaps be the case, as is suggested by the peculiarities of their geographical distribution, that *Agapornis* may be the representative among the normal-carotid Parrots of the Platycercine branch from the Ariinæ, whilst the Stringopinæ proper (including *Geopsittacus*, *Melopsittacus*, and *Euphema*) are more direct continuations of the main stem, *Stringops* itself being the nearest living representative of the common ancestor of the whole suborder." And in a postscript (dated Dec. 8, 1874) he adds:—"On the 25th of last month, from the death of one of the specimens of *Stringops habroptilus*, recently purchased by the Society, I have had an opportunity of dissecting a third individual of the species. In it the ambiens muscle is complete, of fair size, at the same time that it crosses the knee as in *Psittacus*. This makes me feel more convinced that the arrangement indicated by me is the correct one, and that the main stem has given rise to three instead of two branches—the Stringopinæ being the nearest representatives of the ancestral form, some of its members (*Geopsittacus*, *Melopsittacus*, *Euphema*, and *Cyanorhamphus*) having quite recently lost, whilst *Stringops* itself is just now on the point of losing the ambiens muscle. It is, however, quite possible, if external resemblances and geographical distribution are left out of consideration, that *Stringops* must stand as the sole representative of the Stringopinæ, thus conforming with generally received ideas."

According to native tradition, the Kakapo was formerly abundant all over the North * and South Islands; but at the present day its range is confined to circumscribed limits, which are becoming narrower every year. In the North Island it is rarely heard of; but it still exists in the Kai-Manawa ranges, and, as I have been assured by the chief Herekiekie, it is still occasionally met with in various parts of the Taupo district †.

Until within the last few years the Kakapo abounded in the Urewera country, and the natives were accustomed to hunt them at night with dogs and torches. The Maori proverb, "Ka paru a putaihinu," relates to the former abundance of this bird. The natives say that the Kakapo is gregarious, and that when, in the olden time, numbers of them congregated at night their noise could be heard to a considerable distance. Hence the application of the above proverb, which is used to denote the rumbling of distant thunder ‡.

The first published account of this singular bird is that given by Dr. Lyall, R.N., in a paper read before the Zoological Society of London, on the 24th of February, 1852, and which I have transcribed from the 'Proceedings' of that year:—"Although the Kakapo is said to be still found occasionally on some parts of the high mountains in the interior of the North Island of New Zealand, the only place where we met with it during our circumnavigation and exploration of the coasts of the islands in H.M.S. 'Acheron,' was at the S.W. end of the Middle Island. There, in the deep sounds which intersect that part of the island, it is still found in considerable numbers, inhabiting the dry spurs of hills or flats near the banks of rivers where the trees are high and the forest comparatively free from fern or underwood. The first place where it was obtained was on a hill nearly 4000 feet above the level of the sea. It was also found living in communities, on flats near the mouths of rivers close to the sea. In these places its tracks were to be seen, resembling footpaths made by man, and leading us at first to imagine that there must be natives in the neighbourhood. These tracks are about a foot wide, regularly pressed down to the edges, which are two or three inches deep amongst the moss, and cross each other usually at right angles.

"The Kakapo lives in holes under the roots of trees, and is also occasionally found under shelving rocks. The roots of many New-Zealand trees growing partly above ground, holes are common under them; but where the Kakapo is found, many of the holes appeared to have been enlarged, although no earth was ever found thrown out near them. There were frequently two openings to these holes; and occasionally, though rarely, the trees over them were hollow for some distance up. The only occasion on which the Kakapo was seen to fly was when it got up one of these hollow trees and was driven to an exit higher up. The flight was very short, the wings being scarcely moved; and the bird alighted on a tree at a lower level than the place from whence it had come, but soon got higher up by climbing, using its tail to assist it. Except when driven from its holes, the Kakapo is never seen during the day; and it was only by the assistance of dogs that we were enabled to find it. Before dogs became common, and when the bird was plentiful in inhabited parts of the islands, the natives were in the habit of catching it at night, using torches to confuse it. It offers a formidable resistance to a dog, and sometimes inflicts severe wounds with its powerful

* Te Heuheu's father, Ngatoroirangi, a renowned Maori naturalist of former times, was a successful Kakapo-hunter. He was (so the natives relate) accustomed to lie in ambush near the beaten tracks of these birds, and capture them, in the early dawn, on their way to their hiding-places. This good old chief is said to have attempted the introduction of the Snapper into the Taupo Lake. He planted the island of Mokoia, in the Rotorua Lake, with *tataru*, and left behind him other evidences that he was a "scientific man" far in advance of his time.

† Through the kindness of Mr. White, R.M., I obtained a native-prepared skin of the Kakapo from Taupo, for comparison with examples from the South Island. It was a very small specimen, measuring only 21 inches in length and 8·5 in the wing; but I was able to satisfy myself of the real identity of the species in both islands.

‡ *Cf.* Note on *Stringops habroptilus* and its skeleton by E. Deslongchamps, Ann. Mus. H. N. Caen, i. pp. 40–50; also skeleton as figured by A. B. Meyer, Abbild. Vögelsk.

claws and beak. At a very recent period it was common all over the west coast of the Middle Island; but there is now a race of wild dogs said to have overrun all the northern part of this shore, and to have almost extirpated the Kakapos wherever they have reached. Their range is said to be at present confined by a river or some such physical obstruction; and it is to be feared that, if they once succeed in gaining the stronghold of the Kakapo (the S.W. end of the island), the bird may soon become extinct. During the latter half of February and the first half of March, whilst we were amongst the haunts of these birds, we found young ones in many of the holes—frequently only one, never more than two, in the same hole. In one case where there were two young ones, I found also an addled egg. There was usually, but not always, an old bird in the same hole with the young ones. They build no nest, but simply scrape a slight hollow amongst the dry dust formed of decayed wood. The young were of different ages, some being nearly fully fledged, and others covered only with down. The egg is white and about the size of a Pigeon's.

"The cry of the Kakapo is a hoarse croak, varied occasionally by a discordant shriek when irritated or hungry. The Maoris say that during winter they assemble together in large numbers in caves, and that at the times of meeting, and again before dispersing to their summer haunts, the noise they make is perfectly deafening. A good many young ones were brought on board the ship alive. Most of them died a few days afterwards, probably from want of sufficient care; some died after being kept a month or two; and the legs of others became deformed after they had been a few weeks in captivity. The cause of the deformity was supposed to be the want of proper food, and too close confinement. They were fed chiefly on soaked bread, oatmeal and water, and boiled potatoes. When let loose in a garden they would eat lettuces, cabbages, and grass, and would taste almost every green leaf that they came across. One which I brought within six hundred miles of England (when it was accidentally killed), whilst at Sydney ate eagerly of the leaves of a *Banksia* and several species of *Eucalyptus*, as well as grass, appearing to prefer them all to its usual diet of bread and water. It was also very fond of nuts and almonds, and during the latter part of the homeward voyage lived almost entirely on Brazilian ground-nuts. On several occasions the bird took sullen fits, during which it would eat nothing for two or three days at a time, screaming and defending itself with its beak when any one attempted to touch it. It was at all times of an uncertain temper, sometimes biting severely when such a thing was least expected. It appeared to be always in the best humour when first taken out of its box in the morning, hooking on eagerly with its upper mandible to the finger held down to lift it out. As soon as it was placed on the deck it would attack the first object which attracted its attention—sometimes the leg of my trousers, sometimes a slipper or a boot. Of the latter it was particularly fond: it would nestle down upon it, flapping its wings and showing every symptom of pleasure. It would then get up, rub against it with its sides, and roll upon it on its back, striking out with its feet whilst in this position. One of these birds, sent on shore by Capt. Stokes to the care of Major Murray, of the 65th Regiment, at Wellington, was allowed to run about his garden, where it was fond of the society of the children, following them like a dog wherever they went.

"Nearly all the adult Kakapos which I skinned were exceedingly fat, having on the breast a thick layer of oily fat or blubber which it was very difficult to separate from the skin. Their stomachs contained a pale green, sometimes almost white, homogeneous mass, without any trace of fibre in it. There can be little doubt but that their food consists partly of roots (their beaks are usually more or less covered with indurated mud), and partly of the leaves and tender shoots of various plants. At one place where the birds were numerous we observed that the young shoots of a leguminous shrub growing by the banks of a river were all nipped off; and this was said by our pilot, who had frequented these places for many years in a whaling-vessel, to be the work of the Kakapo. Their flesh is white and is generally esteemed good eating."

Sir George Grey, two years later, sent the following interesting account of the Kakapo to Mr. Gould, who gave it a place in the Appendix to his 'Birds of Australia':—

"During the day it remains hid in holes under the roots of trees or rocks, or, very rarely, perched on the boughs of trees with a very dense thick foliage. At these times it appears stupid from its profound sleep, and if disturbed or taken from its hole immediately runs and tries to hide itself again, delighting, if practicable, to cover itself in a heap of soft dry grass; about sunset it becomes lively, animated, and playful, issues forth from its retreat, and feeds on grass, weeds, vegetables, fruit, seeds, and roots. When eating grass it rather grazes than feeds, nibbling the grass in the manner of a rabbit or wombat. It sometimes climbs trees, but generally remains upon the ground, and only uses its short wings for the purpose of aiding its progress when running, balancing itself when on a tree or in making a short descent, half-jump, half-flight, from a higher to a lower bough. When feeding, if pleased with its food, it makes a continued grunting noise. It is a greedy bird and choice in its food, showing an evident relish for any thing of which it is fond. It cries repeatedly during the night, with a noise not very unlike that of the Kaka (*Nestor meridionalis*), but not so loud.

"The Kakapo is a very clever and intelligent bird—in fact, singularly so; contracts a strong affection for those who are kind to it; shows its attachment by climbing about and rubbing itself against its friend, and is eminently a social and playful bird: indeed, were it not for its dirty habits, it would make a far better pet than any other bird with which I am acquainted; for its manner of showing its attachment by playfulness and fondling is more like that of a dog than a bird.

"It builds in holes under trees and rocks, and lays two or three white eggs, about the size of a pullet's, in the month of February; and the young birds are found in March. At present (1854) the bird is known to exist only in the Middle Island of New Zealand, on the west coast, between Chalky Harbour and Jackson's Bay, and in the Northern Island about the sources of the Wanganui and in part of the Taupo countries. It was, within the recollection of the old people, abundant in every part of New Zealand; and they say it has been exterminated by the cats introduced by the Europeans, which are now found wild and in great numbers in every part of the country. They say also that the large rat introduced from Europe has done its part in the work of destruction.

"The natives assert that, when the breeding-season is over, the Kakapo lives in societies of five or six in the same hole; and they say it is a provident bird, and lays up in the fine season a store of fern-root for the bad weather. I have had five or six of the birds in captivity, but never succeeded in keeping them alive for more than eighteen months or two years. The last I had I sent home as a present to the Zoological Society; but it died off Cape Horn."

Mr. G. S. Sale (now Professor of Classics in the New-Zealand University) succeeded, in 1870, in bringing, for the first time, to England a live specimen of the Kakapo. This bird was deposited for a short period in the gardens of the Zoological Society, and excited much interest *. An excellent portrait of it appeared in the 'Field' newspaper of October 15, 1870, accompanied by a short article on the subject, in which the readers of that journal were informed that "unfortunately for the gratification of the curiosity of visitors, the Kakapo in the gardens obstinately persists in indulging in its nocturnal habits. During the day it remains concealed; and it is only at night, when the visitors have departed, that the singular movements and habits of this animal can be studied with advantage." This notice called forth a letter † from Mr. Sale, the owner of the bird, in which further interesting particulars of its history are recorded. After explaining that the bird had been in his possession for several months before he deposited it in the gardens, and that he had carefully observed its habits, Mr. Sale continues:—"Sir G. Grey exactly hit the chief characteristics of the Kakapo when he

* The Council of the Society offered a sum of £50 for this bird, but were unable to come to terms with the owner.

† 'Field' newspaper, November 12, 1870.

spoke of its affectionate and playful disposition. During the whole time that the bird has been in my possession it has never shown the slightest sign of ill-temper, but has invariably been good-humoured and eager to receive any attention. Its playfulness is remarkable. It will run from a corner of the room, seize my hand with claws and beak, and tumble over and over with it exactly like a kitten, and then rush back to be invited to a fresh attack. Its play becomes sometimes a little severe; but the slightest check makes it more gentle. It has also, apparently, a strong sense of humour. I have sometimes amused myself by placing a dog or cat close to its cage; and it has danced backwards and forwards with outstretched wings, evidently with the intention of shamming anger, and has testified its glee at the success of the manœuvre by the most absurd and grotesque attitudes. One trick especially it has, which it almost invariably uses when pleased: and that is to march about with its head twisted round, and its beak in the air—wishing, I suppose, to see how things look the wrong way up; or, perhaps, it wishes to fancy itself in New Zealand again. The highest compliment it can pay you is to nestle down on your hand, ruffle out its feathers, and lower its wings, flapping them alternately, and shaking its head from side to side; when it does this it is in a superlative state of enjoyment. I do not think it is quite correct to say that it has dirty habits; certainly it is not worse in this respect than an ordinary Parrot.

"I am surprised to find that during the time it was in the Zoological Gardens it very rarely showed itself in the daytime. My experience has been the reverse of this. It has generally been lively enough during the greater part of the day, though not quite so violent and noisy as at night. I had this bird at Saltburn, in Yorkshire, during the summer; and any of your readers who were at that place in the month of August, will remember seeing this bird at the bazaar held in aid of the district church, on which occasion its playfulness never flagged during the whole day. This may partly have been due to excitement at seeing so many strange faces; but it also, no doubt, felt the excellence of the cause (recollect, Sir G. Grey testifies to its cleverness and intelligence), and exerted itself accordingly to help the Church-building Fund."

In another account of the habits of the particular bird in his possession, Mr. Sale remarks:—"I observe that it rarely makes any noise by day; but about dusk it usually begins to screech, its object being apparently to attract attention; for if let loose from its cage and allowed to have its usual play, it ceases to make any noise. It also makes a grunting noise when eating, especially if pleased; and I have myself attracted it to me by imitating the same sound. It also screeches sometimes when handled, not apparently from anger, but more from timidity." In a note he adds:—"The sound of the bird is not a shrill scream, but a muffled screech, more like a mingled grunt and screech."

Sir James Hector found the Kakapo very numerous on the west coast of the Otago Province during his exploration of that country in 1861–62; and his collection of birds in the Otago Museum contains many beautiful specimens of it. He succeeded in bringing some live ones to Dunedin; but although they had become perfectly tame, they did not long survive their confinement. Having had good opportunities of studying this bird in its native haunts, the following additional particulars from his pen will be read with interest:—

"The name of Owl Parrot is very appropriate, from the aspect of its head and face, as the bill is short and almost buried among feathers and long bristly hairs like the whiskers of a cat. These whiskers, no doubt, are used in the same manner, as delicate feelers for distinguishing objects in the dark, as the Kakapo is strictly nocturnal in its habits—never stirring from the holes and burrows in which it rests during the day until nightfall. They then emerge from the woods to the sides of the rivers; and, as they feed, their harsh screams can be heard at intervals until they return at daybreak to the depths of the forest. Notwithstanding the shortness of their legs and large size of their feet, they run at a good pace, with a waddling duck-like gait; and though they climb with great facility, and rapidly take to trees when disturbed or pursued, they never make any attempt to fly. They are found on the mountains at all elevations; but their favourite haunts are either on the flats by the sides of the rivers, or at 3000 to 4000 feet elevation, where the forest is very scrubby and dense and merges into open ground, and where the spurs that lead to the precipitous and rocky ridges are covered with coarse grass. In their nocturnal rambles on the mountain-tops—which the Kakapos seem at some seasons to indulge in—they appear to keep in line along the spurs and ridges, as they beat down broad tracks which it would be quite excusable to mistake for the well-frequented paths leading to some encampment in the woods. They seem strictly herbivorous, their food being principally grass and the slender juicy twigs of shrubs, such as the New-Zealand broom (*Carmichælia*), which they chew up into a ball without detaching it from the plant—satisfying themselves with the juice which they extract. Their haunts are therefore easily recognized by the little woolly balls of chewed fibre which dangle from the branches of the shrubs, or strew the ground where they have been feasting on the succulent grasses. It is stated by the Maoris that in winter they assemble in large numbers, as if for business; for after confabulating together for some time with great uproar, they march off in bands in different directions. However, they are not gregarious at all seasons of the year, but are generally found in families of two or three together. They breed in February, having two eggs at a time, which they lay in the holes they scrape for dormitories under the roots of decayed trees and fallen rocks.

"The Kakapo can only be successfully hunted with dogs. The best time for hunting these birds is in the early morning, as soon as it is sufficiently light to permit of the sportsman passing rapidly through the bush, as at that time the scent is still fresh of the birds that were abroad during the night. The Maori dogs enjoy the sport very keenly, and follow it largely on their own account—so much so that, when the Maoris encamp in a locality where these nocturnal birds abound, the dogs grow fat and sleek, and the birds are soon exterminated. The Kakapo is esteemed a great delicacy by the natives; but its flesh has a strong and slightly stringent flavour."

Probably no New-Zealand explorer enjoyed more favourable opportunities for investigating the natural history of the Kakapo than the late Sir Julius von Haast, whose observations on the subject were embodied in a paper, full of scientific interest, read before the Canterbury Philosophical Society on the 4th June, 1863.

A German version of this paper was contributed by the author to the 'Verhandlungen' of the Zoological and Botanical Association of Vienna, of October 10, 1863. A translation appeared in 'The Ibis' of the following year (pp. 340–346); and, curiously enough, a retranslation was published in the 'Journal für Ornithologie' for 1864 (pp. 458–464). But the paper as originally written has never been published; and as the author favoured me at the time with a private copy of it, I have much pleasure in finding room for the following copious extracts:—

"So little is known of this solitary inhabitant of our primeval forests, that the following short narrative of observations which I was fortunate enough to make during my recent west-coast journey may interest you. Although I was travelling almost continuously for several years in the interior of these islands, it was only during my last journey that I was enabled to study its natural history. I was well acquainted with its call, and had often observed its tracks in the sands of the river-beds and in the fresh fallen snow, but I had not actually seen it. The principal reason for this was, that formerly I had no dog with me; and consequently it would only be by the greatest accident that this bird, not at all rare in those untrodden regions, could be obtained.

"The true habitat of the Kakapo is the mossy open *Fagus*-forest, near mountain-streams, with occasional grassy plots: but it also lives both on the hill-sides, amongst enormous blocks of rock, mostly overgrown with roots of trees and a deep covering of moss, and on wooded flats along the banks of the larger rivers, liable to be inundated by heavy rainfalls or by the sudden melting of the snow . . . . . It is a striking fact that, with the exception only of the valley of the river Makarora, forming Lake Wanaka, I never found the Kakapo on the eastern side of the Alps, although extensive *Fagus*-forests exist there also. It appears to have crossed the main chain at the low wooded pass which leads from the source of the Haast to that of the Makarora, and reached the mouth of this river at Lake Wanaka, where probably the absence of forest put a stop to its further advance. It is very abundant in the valley of the last-mentioned river, and is found even in the Makarora bush, notwithstanding that numerous sawyers are at work there. When camped on the borders of that forest, we continually heard its call near our tents; but none of the sawyers had any idea of the existence of such a large bird in their neighbourhood, although the irregular shrill call had sometimes attracted their attention. It also occurs in the valley of the Wilkin, but is less numerous there, which may be accounted for by the existence of wild dogs in this locality. We may therefore safely assume that from the junction of this river with the Makarora the Kakapo ascended toward the sources of the former. In the valley of the Hunter, only divided by a mountain-range of great altitude but with some low saddles, no sign of it was to be observed, although large *Fagus*-forests would appear to offer a propitious abode. . . . . . This bird has hitherto been pronounced to be of true nocturnal habits; but I think, from observations I was able to make, that this opinion ought to be somewhat modified. It is true that generally an hour after sunset, the dense foliage of the forest giving additional darkness to the country, its call began to be heard all around us. It then commenced to rove about, and, attracted by the glare of our camp-fire, frequently came close to our tent, when the heedless bird was immediately caught by our dog. But as we met with it on two occasions in the daytime, occupied in feeding, and as I observed that it knew and understood perfectly well the danger which approached, we may safely assume that it has, at least in this respect, some relation to diurnal birds. In order to show why I come to this conclusion, I will particularize the two occurrences I have mentioned, especially as they appear to bear directly upon some other important points in the structure of this bird. When returning from the west coast, we observed, in the afternoon (the sky being clouded), a Kakapo sitting on the prostrate trunk of a tree in the open forest. When about ten yards from it, the bird observed us, and disappeared instantly in its hole, from which, with the aid of the dog, we afterwards took it. It is clear that in this case the bird was not overtaken by the coming day, when far away from its abode, but that it left its retreat voluntarily during daylight. The second instance I shall mention is more striking, and shows that the Kakapo feeds also during the day. It was towards evening, but still broad daylight, when we passed along the hill-side near a deep rocky gorge, and saw a large Kakapo sitting on a low fuchsia tree, about ten feet from the ground, feeding on the berries. When close to it, the bird saw us, and instantly dropped down, as if shot, and disappeared amongst the huge fragments of rocks strewed along the hill-side. But the most remarkable circumstance was, that the frightened bird did not open its wings to break its fall, but dropped as if it did not possess any wings at all . . . . . . In order to see whether they would fly, or even flutter, when pursued by an enemy, I placed on the ground a full-grown specimen, which had been caught by the dog without being hurt. It was on a large shingle-bed; so that the bird had ample room for running or rising on the wing, if for this purpose it wanted space. I was not a little astonished to observe that it only started running towards the nearest point of the forest, where a dark shadow was apparent—and quicker than I had expected, considering the position of its toes and its clumsy figure, resembling closely a Gallinaceous bird in its movements. As I was standing sideways to it, I thought that it kept its wings closed upon its body, so little were they opened; but my companion, who was equally anxious to see how our prisoner would try to escape, and who stood a little behind it, observed that it opened its wings slightly, but without flapping them in any degree, using them apparently more for keeping its balance than for accelerating its movements. This would almost lead to the conclusion that the Kakapo does not travel far, especially as I have already shown that its whole structure is ill adapted for running. But having myself frequently followed the tracks, and found them to extend a great distance over the sandy reaches along the river, such a conclusion as that suggested above would be erroneous. It must be exceedingly fond of water, because in many localities its tracks were observed for half a mile over shingle and sand to the banks of the river; and I am unable to explain the curious fact, unless the object be to mix river-water with the enormous mass of pulpy vegetable matter which is to be found in its crop. With the exception of two specimens, the crops of which were filled with the large berries of a small-leaved *Coriaria*, by which their flesh was flavoured, all the birds examined by me had their crops widely distended by a mass of finely comminuted vegetable mosses, weighing many ounces . . . . . . I carefully examined the subterranean abode of this bird. From the account given by the natives, I thought that it would be found living in well-excavated holes, resembling in their construction those of the fox or badger, that the entrance would be so small as to enable only the inhabitant to enter, and thus to exclude larger animals from persecuting it. This, however, is not the case, because, with one exception, all the specimens obtained were either in fissures amongst rocks, or in cavities formed by huge blocks, tumbled one over another, and overgrown with moss, or in holes formed by the roots of decayed trees. The cavities in the rocks were generally sufficiently large to allow of my dog (a good-sized retriever) freely entering them. The openings to the other holes being smaller, it was sometimes necessary to cut away a few roots at the entrance. Inside, the cavity was invariably of very large size, because we could plainly hear the dog advancing several yards before commencing his scuffle with the occupant; and on returning, with the bird in his mouth, he always emerged head foremost, thus proving that the chamber was large enough to enable him to turn himself round. Before he had become accustomed to the work, the dog was often punished severely by the bird's powerful beak and claws; but he ultimately became quite an expert, always seizing his prey by the head and crushing the skull. He appeared to take a delight in searching for these birds, and was never tired of providing for us in this manner . . . . . The holes or abodes of the Kakapo were not only on the mountain-sides, but also on the flats near the river-banks, which are liable to be overflowed. There can be no doubt that, when a sudden inundation takes place, the bird can save itself upon a bush or neighbouring tree. I do not think, however, that it can climb the boles of standing trees, because it never resorted to them during the night or when persecuted by the dog—except in one single case, when the bird ascended a leaning tree close to our camp, and remained till the dog had given up the attempt to obtain it. But, notwithstanding that almost all the abodes that came under examination were natural cavities, I met with one hole which seemed to have been regularly mined. On the northern bank of the river Haast, just below the junction of the river Clarke, a large flat occurs, formed by deposits of sand, over which a thin layer of vegetable mould is spread, and on which a luxuriant vegetation has sprung up. The river, in washing against these deposits, has in some places formed nearly perpendicular banks, about six to eight feet high. At one spot, about two feet below the surface, several rounded holes were observed; and the dog tried in vain to enter them. After carefully scenting the ground, he began to scratch the surface with his paws, and soon succeeded in widening the entrance sufficiently to admit his body; and he immediately afterwards emerged with the bird in his mouth. There is no doubt, in my own mind, that this hole, at least, had been excavated; and the burrowing-faculty of the bird may be considered so far established. On a flat, in the valley of the Makarora, the dog brought one from the interior of a hollow drift-tree, which was lying amongst sedges and grasses in an old river-channel. There was never more than one individual in the hole, although very often within twenty or thirty yards of it another specimen would be scented out by the dog, the two being generally of opposite sexes. At night-time, in visiting our camp-fire, they generally came in pairs, the two being successively caught by my dog, a single or sometimes a repeated angry growl from the bird informing us that he had hold of it. These circumstances lead me to conclude that during the day each inhabits separately its hole, and that only after dark do they meet for feeding and for social intercourse."

In his Nelson report *, the same naturalist informs us that "in former years the Maruia Plains were a celebrated hunting-ground of the Maoris for this bird. They generally went there on fine moonlight nights, when the berries of the tutu (*Coriaria sarmentosa*), a favourite food of the *Stringops*, were ripe, and ran them down partly with dogs, or even killed them with long sticks upon the tutu bushes. Another mode of capture was, when they had found their holes, to introduce a long stick, to which they had fastened several strong flax snares. Feeling the bird with the end of it, they twisted the stick until some part of the bird was caught in the snares, and thus drew it out. The cry of the Kakapo, heard during the night, very much resembles the gobble of a Turkey."

The following notes were contributed to 'The Ibis' (1875, pp. 390, 391) by the Baron A. von Hügel:—

"One thing I can boast of already is having been in the midst of the Kakapos: but I did not accomplish this without some trouble; for the *Stringops*, unfortunately, is driven yearly further and further up country by the settlers, and now it is only met with in the most lonely mountain-districts. But I hardly think that any trouble and labour would be too great to see the bird as I saw it, at home, and, what is even better, procure a fine series of specimens. My trip was undertaken from Invercargill, and consisted of forty miles by rail, twenty-four in a coach, and some fifty more on horseback, with finally a ten-mile row up and across Lake Te-Anau. This brought me into the midst of the Parrots. The whole ground in the bush, which is covered with thick moss, is honeycombed with their burrows—which emit a strong scent, a sort of greasy essence of Parrot-bouquet. The entrance to each—as in fact is the whole ground—is strewn with their excrement, so as almost to make one believe that a flock of sheep had been grazing there. I had an old Scotch shepherd and his dog with me, and they both proved very useful. The latter caught the birds very cleverly by the back, and invariably brought them already killed to us with their feathers in perfect order; but some we lost through his killing them in the bush instead of on the open tract of bracken where we were posted, and then feeding on them quietly before we could make out his whereabouts. The note of the *Stringops* is very peculiar, quite unlike that of a bird. I think it is when feeding that they indulge in a series of the most perfect porcine squeals and grunts. It is really as like a young pig as any thing can be. Then their other note, which I think answers more to a call or warning, is a very loud aspirated scream, with a sort of guttural sound mixed in with it, almost impossible to describe. Then, when pursued and caught by the dog, it emits a low harsh sort of croak; but some were perfectly silent to the last . . . . . The food I found to consist of the bracken (*Pteris aquilina*), both frond-tips and roots, but chiefly the former. I examined six; and all were crammed with it; but what surprised me much was to find parts of two moderate-sized lizards in the gizzard of an old male. I think this is quite a new fact in the *Stringops* life-history" †.

Mr. Reischek says (Trans. N.-Z. Inst. vol. xvii. pp. 195–197):—"In April 1884 I found under the root of a red birch, in a burrow, two young Kakapos, covered with white down. During the same month I found several other young birds of this species. So late in the season as the 12th May, Mr. Docherty found a Kakapo's nest containing a female sitting upon an egg, with a chick just

* *Loc. cit.* p. 7.

† In this communication the Baron mentioned that although he was unable then to give a complete life-history of the Kakapo, his observations did not altogether agree with those recorded by me on the authority of Sir J. von Haast, who was the first to study the bird in its native haunts. As, however, he did not either then or subsequently point out in what respects the account appeared to him faulty, I have, since my arrival in England, written to Baron von Hügel asking him for the desired information, and he has kindly sent me the following:—"The notes I promised, about the habits of the Kakapo, to 'The Ibis' (which fact until you reminded me of it I had completely forgotten) were never sent. I do not remember now the point in which I thought Haast in error."

hatched. He kindly pointed out the nest, which I measured. The burrow had an entrance from both sides and two compartments. Both entrances led to the first compartment, the second and deeper chamber being connected with the first by a small burrow of about a foot. The nest was in the outer compartment, and was guarded by very strong rocks, rendering it difficult to open it. The distance from the entrances to the nest was two feet and three feet respectively. The first chamber was twenty-four inches by eighteen inches, and twelve inches high; the inner compartment was fourteen inches by twelve inches, and only six inches high. The nest was formed by a deepening, lined with wood-dust, ground by the bird as fine as sawdust, and feathers, which the female had evidently plucked from her own breast which was quite bare. From my observations I am of opinion that the male bird takes no part in the hatching or the rearing of the chicks, as in all cases the female was the sole attendant from first to last. I did not see a male near a breeding burrow nor did I, in any single instance, find two grown-up birds in one burrow, though I have seen them in pairs on their nocturnal rambles. Whenever two males meet they fight, the death of the weaker sometimes resulting. The female is much the smaller (probably about three-fourths the weight) and duller in plumage. These bush Kakapos are very common in various parts of the Sounds district. . . . I was particularly anxious to observe the manner in which the Kakapos make their tracks; I therefore hid myself on several occasions in proximity to one of the tracks, and in such a position that I could see every bird as it passed along. It was very amusing to watch these creatures—generally one at a time—coming along the track feeding, and giving a passing peck at any root or twig that might be in the way. Thus the tracks are always kept clean; in fact they very much resemble the native tracks, with the exception that they are rather narrower, being from eight to fourteen inches wide. The Kakapos generally select the tops of spurs for the formation of their tracks. I was curious to know how the birds would manage when their tracks should be covered with snow. Opportunities were afforded of satisfying my curiosity. I found that they travelled on the surface of the frozen snow, and that their tracks were soon plainly visible, though not more than an inch below the level of the surrounding snow. In many places the scrub, which consists of silver pine, akeake, and other alpine vegetation, is so dense that the snow cannot penetrate it. The Kakapos take advantage of this to make their habitations under the snow-covered scrub, where it is both dry and warm.

"The Kakapos leave their burrows after sunset and return before daylight. If they cannot reach their own home during the darkness, they will shelter in any burrow which may be unoccupied, as they travel long distances. They consume large quantities of food, which consists of grass, grass-seed, and other alpine vegetation. In July they are in splendid condition, those found having as much as two inches of fat upon them. I was much surprised and interested to find in the intestines of the old birds parasites from six inches to two feet long. These parasites are flat, about a quarter of an inch wide, milky white, and jointed very closely. I have found three of these parasites knotted together and many single ones tied in three or four knots.

"In the spring, when the sun begins to shed its warmth, the Kakapos emerge from their burrows, and select some favourable spots in the sunshine, where they crouch down and remain the whole day. In September I selected a suitable day for observing this peculiarity. The snow had disappeared from all the sunny places. I found three birds in different places, sitting upon low silver-pine scrub. They took no notice of my approach until I had them safely in my hand, when they endeavoured to release themselves by biting and scratching."

On two occasions I have myself kept a live Kakapo, for the purpose of studying its habits. My first bird was somewhat vicious and would not allow itself to be handled. It had a great penchant for raw potato, of which it could stow away a surprising quantity. It was an adult bird when caught and did not long survive the complete change in its condition of life. My other Kakapo was brought to me as a young bird, being readily distinguishable as such by its dark cheeks, with little or no admixture of yellow, and its delicate pale grey bill. This proved a far more satisfactory pet, and soon became attached to the surroundings of its new home. On our first interview, it gave me a severe bite in the hand, fairly cutting out a piece of flesh; but at the end of a day or two, on being taken from its cage, it would spring upon my arm, mount to the shoulder, and nibble my beard in a playful way. It would partake freely of almost any kind of food, eating as readily of fat mutton as of a green apple or potato; but it seemed most at home when nibbling grass and other succulent herbage. It uttered at times a low grunt, and when excited a peculiar growling sound. It was sometimes allowed the freedom of the garden; and on these occasions I have been much struck with its wonderful assimilation of colouring to the surrounding vegetation, it being quite impossible at a little distance to distinguish it, the singular distribution of green and brown markings in the plumage being very deceptive.

To illustrate the extreme vitality of this Parrot, I may mention that on one occasion the Kakapo's cage was left for a whole day in an outhouse where some painting operations had been carried on. The fumes of red lead and probably the absorption of the poison by the water which the bird had been drinking produced their natural effect; and at nightfall the Kakapo was found at the bottom of his cage, lying on his side quite helpless and to all appearance *in articulo mortis*. Restoration to fresh air, aided by a small quantity of spirit poured into his crop, brought the bird out of this state of asphyxia; and, although he continued very weak and tottery for twenty-four hours, he ultimately regained his full vigour and sprightliness.

On its general conduct I find the following entry in my note-book:—It is decidedly nocturnal in its habits, making for any dark corner or shaded recess immediately on being liberated from its cage. It walks in a measured deliberate way, and when hurried expedites its movements by flapping its wings. Sometimes it utters a scream not unlike that of the Kaka when excited or alarmed. It partakes freely of every kind of vegetable food: it nibbles grass, rolling up and detaching a blade at a time in a very deliberate manner; it devours with avidity lettuce, ripe tomatoes, apples, and raw potatoes; it sucks up the contents of ripe grapes with great relish; and it is at all times ready to make a substantial meal off fat mutton or soaked bread; so that, in point of fact, the bird is omnivorous.

It loves to move about among the herbage in our shrubbery, exploring with its bill and nibbling off the leaves, but never attempting to climb. In the evening it becomes more active in its movements, perambulating its cage when confined, and showing every inclination to be abroad. One night it succeeded in effecting its escape by twisting some of the wire bars, and after foraging about to its heart's content it voluntarily returned before daylight to its prison-house, squeezing its body through the aperture it had made.

Its distinguishing characteristic, however, is its playfulness. When not permitted to climb one's arm and "make-believe" at biting, it thrusts its head into the little tin drinking-vessel, visor-like, and struts about its cage, with every appearance of delight.

On examining my captive Kakapo at night, by the aid of a candle, I was much struck with the resemblance of its general contour to that of the Laughing Owl. It had the same habit of standing almost bolt upright, with the feathers of the head raised and the brows arched, as if in an attitude of contemplation. I mention this as among the many superficial characters justifying the appellation of Owl Parrot.

Mr. Kirk, the well-known botanist, informs me that this bird, as observed by him in captivity, evinces a great partiality for the male flowers of *Pinus pinaster*.

A specimen, sent to England by Mr. Murdoch, the Inspector of the Bank of New Zealand, lived for a considerable time in the Zoological Society's Gardens, but in the same retired way as its predecessor, closely concealing itself in its box by day, exhibiting itself to the public only under coercion of the keeper, and then manifesting the utmost impatience to regain its dark retreat.

A life-sized drawing of this species was given in Gray and Mitchell's 'Genera of Birds' (1842), admirably coloured, but placed in an attitude quite foreign to the habits of the bird. Mr. Gould gave a portrait of it in the Supplement to his 'Birds of Australia,' executed in his usual masterly style; and other figures, of less note, have appeared at various times. The coloured drawing of this bird in the 'Student' for 1870, as well as the woodcut in the 'Field,' although in other respects excellent pictures, possess a fault in common—namely, in having the tail broad and fan-like, instead of being compressed, narrow, and inclined inwards. This, as I have been informed, was owing to the damaged condition of the tail in the particular bird from which both of these figures were taken.

The egg of the Kakapo, of which there is a figure (from the pencil of Mr. Wolf) in the 'Proceedings' of the Zoological Society for 1852, is broadly ovoido-conical in form, and of pure whiteness till discoloured in the process of incubation. A specimen in the Canterbury Museum, much stained and slightly damaged, measures 2 inches in length by 1·4 in its greatest breadth: the surface of the shell is smooth, but without any gloss or polish; and on close inspection it is found to be finely granulate. Another in the Otago Museum is of almost exactly similar size, measuring 2 inches in length by 1·45 in breadth: this specimen is somewhat discoloured, probably by contact with the bird's feet; the shell is minutely granulate, having a slightly rough surface to a sensitive touch. Another in my son's collection is appreciably smaller, measuring 1·85 inch in length by 1·35 in breadth, and, originally of a greenish-white colour, is stained and discoloured, though somewhat unequally, to a pale yellowish brown.

A specimen from Preservation Inlet is rather larger than the last mentioned one, measuring 2 inches in length by 1·4 in breadth. It is yellowish white and somewhat soiled, the surface being without any gloss, and slightly granulate, or marked with extremely fine points.

Kakapo chick, just hatched: natural size. (See page 178.)

SPILOGLAUX NOVÆ ZEALANDIÆ.

(NEW-ZEALAND OWL, OR MOREPORK.)

New-Zealand Owl, Lath. Gen. Syn. i. p. 149 (1781).
Strix novæ seelandiæ, Gm. Syst. Nat. i. p. 296 (1788, ex Lath.).
Strix fulva, Lath. Ind. Orn. i. p. 65 (1790).
Noctua zelandica, Quoy & Gaim. Voy. de l'Astrol. Zool. i. p. 168, t. 2. fig. 1 (1830).
Athene novæ seelandiæ, Gray, Voy. Ereb. & Terror, p. 2 (1844).
Athene novæ zealandiæ, Gray, Cat. Brit. Mus. Accipitr. p. 52 (1844).
Noctua venatica, Peale, U. S. Expl. Exp. p. 75 (1848).
Spiloglaux novæ seelandiæ, Kaup, Isis, 1848, p. 768.
Ieraglaux novæ zealandiæ, Kaup, Tr. Zool. Soc. iv. p. 218 (1852).

Native names.

Ruru, Koukou, and Peho; "Morepork" of the colonists.

Ad. suprà chocolatinus, scapularibus maculis fulvis plus minusve celatis notatis: loris, genis anticis et supercilio distincto fulvescentibus: regione auriculari chocolatinâ: tectricibus alarum medianis et majoribus extùs fulvo vel albo maculatis: remigibus brunneis, extùs albo maculatis, et saturatè brunneo transfasciatis: caudâ suprà brunneâ, subtùs pallidiore, fasciis distinctis saturatè brunneis transnotatâ: collo laterali et corpore subtùs toto lætè fulvis, medialiter latè brunneo striatis: abdomine imo, hypochondriis et subcaudalibus pulchrè albo marmoratis: cruribus et tarsorum plumis lætè ferrugineis: rostro nigro, culmine albicante: pedibus flavis, digitis setis nigricantibus indutis: iride aureo-flavâ.

Adult male. Crown of the head and all the upper parts dark umber-brown, obscurely spotted on the scapulars and wing-coverts with fulvous white; lores and region of the bill white, with black produced filaments; forehead, fore neck, and upper part of the breast light fulvous, mixed with brown; underparts generally fulvous, with triangular spots of dark brown disposed in rows and blending; under tail-coverts fulvous barred with white; quills and tail-feathers dark brown obscurely banded, the former touched on the outer webs with fulvous white; feathers covering the tarsi fulvous. Irides golden yellow; toes yellow, with dark hairs; bill black, white on the ridge. Length 12·5 inches; extent of wings 25, wing, from flexure, 8; tail 5·75; bill, along the ridge 1, along the edge of lower mandible ·75; tarsus 1·5; middle toe and claw 1·25.

Female. The female is slightly smaller, and the markings of the plumage are less distinct than in the male.

Nestling. A nestling obtained at Westland (and apparently a fortnight old) is covered with thick, fluffy down, of a sooty-brown colour, with loose white filaments; inclined to tawny on the underparts, and whiter on the sides of the head and neck; bill dark brown, with a whitish ridge; legs and feet yellow. They assume the full plumage before quitting the nest.

Fledgling. In my collection there are two specimens of different ages:—

No. 1 has the forehead, chin, and sides of the face destitute of feathers; the crown of the head and all the upper surface sooty brown, or almost black, without any light markings; the plumage extremely soft and fluffy, with the downy white filaments still adhering to it, and more abundantly on the head, neck, and rump; underparts sooty brown, mixed with fulvous; on the thighs thick fluffy plumage of a dull tawny

colour; the tarsi thickly covered to the toes with white down, having the appearance of stockings; quills and their coverts just developing, the rounded white spots on the latter being very conspicuous; bill greenish black; toes yellow; claws dark brown.

No. 2, which is apparently ten days or perhaps a fortnight older, is in a condition to leave the nest: plumage as in the adult, but duller, and mixed with dark-coloured down on the breast; head well-feathered, but with less white about the chin and facial disk; feathers very fluffy and with downy filaments still adhering on some parts of the body; white spots on wings more regular than in the adult, forming two parallel diagonal series, following the order of the coverts; bill dark brown.

Varieties. Examples from different localities present slight but uniform differences of plumage. Specimens from the Nelson district are, on comparison with those from the north side of Cook's Strait, invariably found to be more largely marked with white around the eyes and on the feathers surrounding the bill. As we proceed further south the variation is still more apparent, the whole plumage partaking of a lighter character. There is also considerable variation in size; and a specimen received by me from Mr. W. T. L. Travers is not only unusually small in all its proportions, but has the whole of the plumage deeply stained with ferruginous. A beautiful albino was shot at Te Whauwhau (Whangarei) in the winter of 1871.

In Mr. J. C. Firth's fine collection of New-Zealand birds, at Mount Eden, Auckland, there is a beautiful specimen (obtained at Coromandel) in partial albino plumage. The whole of the body is marked with white, presenting a mottled appearance, and particularly so on the underparts, where the white is softly blended with the normal tawny colour, producing a very pretty effect; each wing has two white primaries, but the tail-feathers are as in ordinary examples.

EVERY New-Zealand colonist is familiar with this little Owl, under the name of "Morepork"*. It is strictly a nocturnal species, retiring by day to the dark recesses of the forest, or hiding in the crevices of the rocks, and coming abroad soon after dusk to hunt for rats, mice, and the various kinds of moths and beetles that fly at night. It is common in all parts of the country, although not so numerous now as it formerly was; and the familiar cry from which it derives its popular name may often be heard in the more retired parts of our principal towns, as well as in the farmer's country home or in the rustic Maori "kainga": I have even known several instances of its voluntarily taking up its abode in a settler's house or, more frequently, in the barn, and remaining there a considerable time.

When discovered in its hiding-place during the day, it is found sitting upright, with the head drawn in, the eyes half closed, and the feathers of the body raised, making the bird appear much larger than it really is. It will then allow a person to approach within a few yards of it, and, if disturbed, will fly off noiselessly for a short distance and attempt to secrete itself. It will often

* "This bird gave rise to rather an amusing incident in the Hutt Valley during the time of the fighting with Mamaku and Rangihaeata, and when, in anticipation of a morning attack, a strong piquet was turned out regularly about an hour before daylight. On one occasion the men had been standing silently under arms for some time, and shivering in the cold morning air, when they were startled by a solemn request for 'more pork.' The officer in command of the piquet, who had only very recently arrived in the country, ordered no talking in the ranks, which was immediately replied to by another demand, distinctly enunciated, for 'more pork.' So malapropos a remark produced a titter along the ranks, which roused the irate officer to the necessity of having his commands obeyed, and he accordingly threatened to put the next person under arrest who dared make any allusion to the obnoxious beast. As if in defiance of the threat, and in contempt of the constituted authorities, 'more pork' was distinctly demanded in two places at once, and was succeeded by an irresistible giggle from one end of the line to the other. There was no putting up with such a breach of discipline as this, and the officer, in a fury of indignation, went along the line in search of the mutinous offender, when suddenly a small chorus of 'more pork' was heard on all sides, and it was explained who the real culprits were.

"At the attack in the Bay of Islands by Heke and Kawiti, the native parties, in moving to their positions about the block-houses and town before daybreak, communicated their whereabouts to one another by imitating the cry of this bird, which the sentries had been so accustomed to hear of a morning that it did not attract their notice." (Captain Power's 'Sketches in New Zealand,' 1849.)

remain many days, or even weeks, in the same piece of bush. In the volcanic hills or extinct craters that surround the city of Auckland there are numerous small caves, formed by large cracks or fissures in the ancient lava-streams, the entrance to them being generally indicated by a clump of stunted trees growing up among loose blocks of scoria. These gloomy recesses are a favourite resort of the Morepork in the daytime.

On the approach of night its whole nature is changed: the half-closed orbits open to their full extent, the pupils expand till the yellow irides are reduced to a narrow external margin, and the lustrous orbs glow with animation, while all the movements of the bird are full of life and activity. It then sallies forth from its hiding-place and explores in all localities, preferring, however, the outskirts of the forest, where nocturnal insects abound, and the bush-clearings in the neighbourhood of farms, or the ruins of Maori villages, these places being generally infested with rats and mice, on which it chiefly subsists. Like other birds of prey, it afterwards regurgitates the hair and other indigestible parts of these animals in hard pellets. That the Morepork also preys on small birds there can be no reasonable doubt, although it has been frequently called in question. Captain Mair has seen one, at sunset, seated on the branch of a tutu bush (*Coriaria ruscifolia*) with a live Korimako in its claws, and in the act of killing it; and a native once told me that he had seen one of these Owls killing and devouring a Parrakeet. Mr. Drew, of Wanganui, informs me that the stomach of one which he skinned contained the entire body of a House-Sparrow. Captain Robinson, of Manawatu, further attests the fact; for on one occasion, when walking in his garden after sunset, he saw a Morepork emerge from a blue-gum and spring upon a Kingfisher, firmly grappling it in its claws. The bird uttered a cry of pain or terror; and on my informant advancing towards the spot, the Owl released its victim and flew off, but immediately afterwards made a second attack, securing the Kingfisher firmly in its grasp, and only relaxing its hold at the moment of being seized.

Mr. J. T. Stewart informs me that, in his own garden at Foxton, he has witnessed two instances of the Owl attacking and vanquishing the Kingfisher, this happening on both occasions towards evening.

I have been informed by Sir George Grey that, of nearly a hundred Diamond-Sparrows which he liberated on the island of Kawau, very few survived the ravages of this little Owl, and that some other importations suffered in like manner. Sir Edward Stafford, who had for many years interested himself in the introduction and acclimatization of useful birds, has also given evidence against the Morepork on this charge; for he has assured me that on one occasion, having turned out a large number of insectivorous birds in his grounds at Wellington, an unusual number of Owls sought harbour there, and preyed on the little immigrants till scarcely a single one remained. For a considerable time, however, it was doubted whether the Morepork was destructive to acclimatized birds; and a lengthy controversy on the subject appeared in the Auckland newspapers. The careful observations of Mr. Brighton, the Curator of the local Acclimatization Society, at length placed the matter beyond all discussion. Frequently he had to forego his night's rest in order to watch the aviaries, and during a period of only a few months he shot no less than fourteen of these birds. Some of these were surprised in the act of attacking the aviaries, and all of them in the immediate vicinity. He repeatedly found the dead and lacerated bodies of Sky-Larks and Chaffinches lying on the wooden ledge just inside the eave of the wire-roofing; and the abundance of Morepork-feathers found entangled in the netting afforded a clue to the perpetrator of these murderous attacks. From the appearance of the feathers, and the mutilated condition of the dead birds, it was evident that the Morepork had tried hard, but unsuccessfully, to pull them through the wire netting in the roof. The following account, by the Curator, renders this perfectly intelligible:—

"The aviary is constructed in the usual manner, on the model of a bird-cage, of wire netting over a wooden framework, with a sloping roof, also of wire netting. Attached to the framework comprising the wall-plates, on either side, there are wooden ledges, resembling shelves, on which the Larks rest at night, while the Chaffinches roost upon twigs planted within the aviary, and reaching within a few inches of the wire netting of which the roof is composed. During moonlight nights the Moreporks have been seen to fly upon the roof of the aviary, and after making, as it were, a reconnaissance of the defences, to pounce repeatedly against the wire, causing a loud vibration, and startling the feathered inmates. These, in their fright, fly towards the light, dashing themselves against the wire netting, until the Morepork, by hopping about on the roof, succeeds in fastening upon one of them, and, of course, making short work of him."

In addition to the above evidence, sufficient of itself before any common jury to convict the culprit, I may mention that on one occasion in Christchurch I saw a Morepork, towards the cool of the evening, enter the verandah of the house in which I was staying and boldly attack a Canary whose cage was suspended there, vainly endeavouring to clutch it as it fluttered against the wires. I heard of another instance in which the depredator actually succeeded in tearing off a limb of the occupant in its efforts to pull it through the bars.

There has, in consequence, been a crusade against the Morepork in many parts of the country. But whether this wholesale destruction of an indigenous species, on account of these predatory habits, is wise, or even prudent, may be seriously questioned. The Morepork, as we have already shown, not only preys on rats and mice, but is also a good insectivorous bird, with a voracious appetite. Its habit of feeding largely on the nocturnal lepidoptera is of itself an inestimable benefit to the agriculturist, as it tends to check the spread of the caterpillar, whose ravages are becoming more severely felt every year. It is a dangerous thing to disturb the balance of nature by violent means; and, in a new country especially, we must be careful that in removing one evil we are not opening the door to an immeasurable greater one. For my own part, I consider the killing of a single Owl a positive injury to the farming industries of the country, and scarcely compensated for by the introduction of a score of soft-billed insectivores in its place.

I have sometimes found this species, at night, among the rocks along the sea-margin, from which it may be inferred that crabs and other small crustacea contribute to its support. In the stomachs of some I have found remains of the large wood-beetle (*Prionoplus reticularis*); and those of others I have found crammed with moths of all sizes, or with nocturnal coleoptera. I examined some castings of the Morepork in the Canterbury Museum. They are hard pellets, of an oval form, and of the size of a Sparrow's egg, composed chiefly of the hard elytra and heads of various coleopterous insects, among which I noticed particularly the shining covering of the mata (*Feronia antarctica*), a handsome ground-beetle which is found on the Canterbury plains, but does not occur in the North Island.

I have noticed that individual birds are very local in their disposition, often fixing on a particular roost or hiding-place by day, to which they will regularly resort for weeks or perhaps months together, the ground immediately below the perch becoming at length quite foul with their accumulated droppings.

Judge Munro informs me that some years ago on opening a bird of this species he found in its stomach a specimen of the weta-punga, or tree-cricket (*Deinacrida heteracantha*), with a body as large as a magnum-bonum plum; and the stomach of another which I obtained in the Rimutaka Ranges, in the month of March, was filled with broken remains of the small weta (*D. thoracica*).

The flight of the bird is light, rapid, and so noiseless that, I verily believe, it could surprise and capture a mouse at the very entrance to its burrow. On examining the feathers of the wing, it will be found that they are furnished with a soft or downy margin, and are specially adapted for this manner of flight. From an examination of the orifice of the ear we are led to infer that the power of hearing in this Owl is very acute. It is therefore the more surprising that, on two occasions after dark, I have succeeded in seizing this species with the hand, when perched on the eaves of a verandah, over which its tail projected. It is comparatively easy to capture it on the wing by a dexterous use of a strong insect-net. When caught, it manifests its anger by a repeated clicking of the mandibles, while it dexterously uses its beak and talons in its appeals for liberation. The ordinary call of this Owl at night consists of two notes uttered with vigour, and having a fanciful resemblance to the words "more pork," from which it derives its popular name. These notes are repeated at regular intervals of from eight to ten seconds, as I have ascertained by timing the performance with my watch. Sometimes the bird breaks off at the end of half an hour, probably to go in quest of food; at other times he keeps up this hooting for a couple of hours or more continuously, especially on clear nights. Besides this cry it has a peculiar call of *ke-e-e-o ke-e-e-o*, repeated several times; and when disturbed or excited a scream, which is not unlike the alarm-cry of the Australian Rosehill Parrakeet (*Platycercus eximius*), but louder and more shrill. At dusk also, before leaving its retreat, it utters a low croaking note, quickly repeated, which is responded to by the other Owls within hearing. This note resembles the syllables *kou-kou*, uttered from the chest; and among the northern tribes the bird is usually called by a name resembling that cry. It is, however, more generally known as the "Ruru," and in some districts as the "Peho" *.

At night two rival males may be heard answering each other from neighbouring woods, or, as Longfellow expresses it,

"Talking in their native language,
Talking, scolding at each other."

Although habitually nocturnal I have occasionally seen it abroad in the daytime, but only during very dull weather. On the occasion of my last visit to Auckland, about 5 o'clock one afternoon, I observed a Morepork, in broad daylight, sail across the public highway, in the very midst of the busy traffic, and take refuge in some trees in the old College grounds, in a spot where (although it no longer forms part of the school enclosure) thirty years before I had played cricket and football with the friends of my youth. A few evenings later I heard another screaming among the chimney-tops in Shortland Crescent, in the very heart of the city—facts showing conclusively, I think, that this species has not been much affected by the spread of civilization in its native country.

Although naturally very fierce, I have known at least one instance of its becoming quite tame in confinement and taking food from the hand of its keeper.

It nidificates, as a rule, in hollow trees; but in the Mackenzie country, where there is little or no timber, nests have been found under the shelter of loose boulders. The young leave the nest about the beginning of January, and may be heard during every night of that month uttering a peculiar, sibilant, snoring sound, sometimes sufficiently sharp to resemble the stridulous song of the native cricket. But the breeding is occasionally delayed to a much later period of the year; for I have heard young Owls in the woods at Palmerston North on the 6th March, and on one occasion, at the North Shore (Auckland), I both heard and saw a young bird so late as the 11th of April. On the other hand, there are sometimes very early broods; for the downy nestling, of which a figure is given on the opposite page (and which is now in my collection), was taken from the cavity of a tree near Dunedin in the month of November.

Mr. J. D. Enys writes to me that he met with a nest of the Morepork at the Ohunga river, containing three eggs; and I have a similar report from Mr. W. Fraser, who found a nest in a hollow puriri (*Vitex littoralis*), containing three young birds. The Owls continued to breed there for three successive seasons. Captain Mair found a nest of this species in the hollow of a dry hinau tree

* According to Maori legend, this bird was one of the first winged inhabitants of New Zealand:—"He kopara te manu nana i noho tuatahi te puhi o te rakau; he ruru to te po; no muri nga manu nunui i noho ai ki te motu, te kaka me te kereru, me nga manu katoa."

(*Elæocarpus dentata*), containing two very young birds, which were "covered with soft white down, plumbeous beneath." In a clump of wood on the banks of the Wairoa river I found a nest, also containing two fully fledged young ones. I sent my native lad, Hemi Tapapa, up the tree to capture them; and while he was so engaged the parent birds came forth from their hiding-place, and darted at his face with a low growling note, making him yell with fear *. The Maoris share in the almost universal feeling of superstition regarding the Owl. Hemi's conscience was troubled; and as the shades of night were closing in upon us with the call of "more pork!" in every direction, he handed me the captives and hurried away from the scene of his exploit, evidently sharing, in some degree, the horrors of that luckless wight, immortalized by Mr. Stevenson in his 'Birds of Norfolk,' who, having killed the church Owl as it flitted past him, ran shrieking home and confessed his awful crime—"I've been and shot a Cherubim!"

There are two eggs of this species in my son's collection. One of these is almost spherical, the other is slightly ovoid, measuring 1·5 inch in length by 1·2 in breadth; they are perfectly white, with a very slight gloss on the surface.

* "Once the writer had an unusual adventure with one of these birds. It was early evening in the summer-time. The Owl was sitting on a gate. Anxious to watch and study its motions we sat down close by it; soon it left its perch, making a sudden swoop at the intruder. This manœuvre it continued to repeat time after time, most perseveringly, and with great gravity and deliberation. Only once was a blow felt; after each attack the bird resumed its perch on the gate. After a while the writer rose and walked up a dark ferny gully at some distance, when the Owl followed and again attacked him. This is the only instance we have met with in which this species has shown any symptoms of boldly resenting an intrusion on its privacy."—*Out in the Open.*

Young of New-Zealand Owl.

SCELOGLAUX ALBIFACIES*.

(LAUGHING-OWL.)

Athene albifacies, Gray, Voy. Ereb. & Terror, p. 2 (1844).
Sceloglaux albifacies, Kaup, Isis, 1848, p. 768.
Ieraglaux albifacies, Kaup, Tr. Zool. Soc. iv. p. 219 (1852).
Athene ejulans, Potts, Trans. New-Zeal. Inst. vol. iii. p. 63 (1870).

Native names.

Whekau, Ruru-whekau, and Kakaha; "Laughing-Jackass" of the colonists.

Ad. suprà lætè fulvescens, plumis omnibus medialiter latè nigro striatis: uropygio lætiùs fulvo: scapularibus et dorso postico brunnescentioribus, latè albido marmoratis: tectricibus alarum magis ferrugineo tinctis, fulvo marmoratis: remigibus brunneis, extùs ferrugineo lavatis et fulvo maculatis: caudâ brunneâ, fasciis fulvis conspicuè transnotatâ: fronte, superciliis, gulâ cum collo laterali griseo-albidis, angustè nigro striatis: regione oculari et auriculari brunnescentibus: corpore reliquo subtùs lætè aurantiaco-fulvo, plumis medialiter brunneo striatis: tarso plumulis albidis induto: rostro nigro, versus apicem corneo: pedibus carneo-brunneis, setis fulvescentibus ornatis, unguibus nigricantibus: iride rufescenti-brunneâ.

Adult. Forehead, throat, ear-coverts, and sides of the head greyish white, with black shafts and hair-like filaments; sides of the neck white, each feather having a narrow central streak of black; upper parts dark brown, the feathers of the crown and nape broadly margined with yellowish brown towards the tip; those of the lower part of the back streaked, spotted, and barred with fulvous and white; lower part of the fore neck and the whole of the breast dark brown, each feather narrowly margined with bright fulvous or yellowish brown; on the abdomen, sides of the body, and under tail-coverts the latter colour predominates, the centre of each feather being dark brown; the soft ventral feathers and the short plumage covering the thighs and tarsi light fulvous, without any dark markings; primaries dark brown, marked on the outer web with equidistant angular spots of white, and on the inner web with obsolete bands; secondaries dark brown, with broad transverse bands of white, and clouded in the centre; scapulars dark brown, handsomely variegated with ocellated spots of white. The feathers forming the mantle are all differently marked, some having two broad approximate lateral bars of white, others a double series of spots on each web, while others again have a narrow lateral bar of white on one side of the shaft, and broad angular spots on the other; a few of them are transversely barred and margined with a narrow terminal crescent; upper wing-coverts dark brown, with numerous oval spots of fulvous white more or less distinct; tail-feathers dark brown, with five equidistant transverse bands and a terminal margin of fulvous white. Irides dark reddish brown; toes fleshy brown, and covered with coarse yellow hairs; bill black, horn-colour towards the tip; claws black. Extreme length 19 inches; wing, from flexure, 11; tail 6·5; bill, along the curvature to anterior edge of cere, 2·75; cere ·25; middle toe and claw 1·6; hind toe and claw ·75.

Obs. The above description is taken from one of the specimens in the Colonial Museum. In the British-Museum example, figured in my former edition, there is less of the spotted character on the upper surface, and the plumage is stained with ferruginous. The accompanying drawing is from a fine specimen, in my own collection, obtained near Timaru in 1874.

* Inadvertently named *Sceloglaux novæ zealandiæ* on the accompanying Plate.

Nestling. When freshly hatched the young bird is sparsely covered with coarse yellowish-white down, the abdomen being bare.

Varieties. Examples differ from each other in the minute details of their colouring. The two specimens in the Canterbury Museum have less white about the face; the soft feathers forming the facial disk are tawny white, with black shaft-lines and hair-like filaments; and along the exterior edge of the disk there is a narrow crescent of pure white, each feather marked with a narrow brownish streak down the centre. In one of these examples the lengthened spots or fusiform markings on the upper surface are less distinct, while in the other they are wholly wanting; but in the latter the fulvous-white bars on the primaries are very conspicuous, and add much to the beauty of the plumage. In this specimen the feathers of the upper surface are blackish brown, with a broad tawny margin, those forming the mantle, scapulars, and upper wing-coverts having, on each web, a broad oblique ear of fulvous white. A specimen more lately received at the Canterbury Museum, and forwarded to Europe, and another in my own collection are sufficiently white about the face to justify the specific name bestowed by Mr. G. R. Gray. In ordinary examples, however, this is quite a subordinate feature. One of those figured in Mr. Dawson Rowley's 'Ornithological Miscellany' has an entirely white face; the other exhibits a strong wash of rufous. The North-Island bird (in the Colonial Museum) is several shades darker than those from the South Island, the whole of the plumage being deeply stained with ferruginous. The feathers at the base of the upper mandible, and those immediately above the eyes, are white, with black shaft-lines; but the facial disk is washed with fulvous. There is an entire absence of the white markings on the upper surface; underparts rich tawny fulvous, with a dark brown stripe down the centre of each feather; tail dark brown, crossed by five broad V-shaped bands of tawny fulvous.

A specimen obtained from the Albury Rocks is inclined to albinism, there being a number of white feathers on the head, shoulders, and mantle, giving the bird a very pretty appearance.

This bird was originally described by Mr. G. R. Gray, in the 'Voyage of the Erebus and Terror,' under the name of *Athene albifacies*; and Dr. Kaup afterwards made it the type of his genus *Sceloglaux*, of which it still remains the sole representative. Mr. Gould, in treating of this singular form, has already pointed out that its prominent bill, swollen nostrils, and small head are characters as much Accipitrine as Strigine, and that its short and feeble wings indicate that its powers of flight are limited, while its lengthened tarsi and shortened toes would appear to have been given to afford it a compensating increase of progression over the ground; and it does, at first sight, appear strange that a bird specially formed by nature for preying on small quadrupeds should exist in a country which does not possess any. It must be remembered, however, that when the Laughing-Owl was more plentiful than it now is, New Zealand was inhabited or, rather, overrun by a species of frugivorous rat, which is now almost, if not quite, extinct. The *kiore maori*, which has been exterminated and replaced by the introduced Norway rat (*Mus decumanus*), formerly abounded to such an extent in the wooded parts of the country that it constituted the principal animal food of the Maori tribes of that period. It was a ground-feeder, subsisting almost entirely on the fallen mast of the tawa, hinau, towai, and other forest-trees; and it would therefore fall an easy prey to the *Sceloglaux*. The fact that the extinction of the native rat has been followed by the almost total disappearance of this singular bird appears to warrant the conclusion that the one constituted the principal support of the other*. Be that as it may, the Laughing-Owl, as it has been termed, in allusion to its cry, is at the

* On this point Mr. Smith writes to the 'Journal of Science,' vol. ii. pp. 86, 87:—

"The suggestion of Dr. Buller that the kiore maori (native rat), before its extermination, may have constituted the principal food of this Owl, is an important one; and my researches among the rocks at Albury, and experiments with the living birds in captivity, are greatly in support of this. In several of the crevices where I explored them, I found an ancient conglomerate of exuviæ ranging from three to twelve inches thick. From the under surface, and through the mass to nearly the upper surface, this conglomerate is thickly studded with Owl's castings, composed entirely of light brown hair (which is unquestionably that of the kiore maori) and small bones. The castings more recently deposited among the rocks are composed of elytra and legs of beetles."

present day one of our rarest species. There are three specimens in the British Museum, and one in the fine collection of raptorial birds formed by Mr. J. H. Gurney, and presented by him to the Norwich Museum. The Colonial Museum at Wellington and the Canterbury Museum * contain two specimens each; and there is a fifth in the local Museum at Dunedin. There are three fine specimens in the late Mr. Dawson Rowley's private museum at Brighton, and a still finer series in my own collection. All these examples, but one, were obtained in the South Island—the exceptional one having come from Wairarapa, in the provincial district of Wellington.

My first acquaintance with this Owl in the live state was made in the Acclimatization Society's Gardens at Christchurch. Unfortunately this Owl, which had lived in the Gardens for upwards of two years, was stone-blind, and its large eyes had a dead, glassy appearance; but I saw quite enough to satisfy me that, in its natural state, it is strictly a ground-feeder. Its appearance was very full and rounded, the feathers of the head and neck being puffed out to a considerable extent. Although it had the freedom of a commodious shed, I observed that it remained constantly on the ground, standing high on its feet, the strong, feathered tarsi being vrey conspicuous. It manifested much impatience or, rather, restlessness, striding with rapidity along the ground, or sometimes moving by a succession of hops, and generally in a rotatory manner, which may have been due to its blindness. The keeper informed me that this bird was a very poor eater, refusing fresh meat, and taking nothing but newly killed birds and live mice. A young mouse, quite paralyzed with fear, was crouching near the ground awaiting its fate, but the Owl took no heed of it; and in another part of its shed there was lying the half-devoured body of a hen Pheasant. I remarked of this bird that the feathered tarsi were much broader and stronger than they appear to be in the dried specimens. It walks quickly and with long strides, the body being held very erect; and when its speed is increased, the wings are raised with a quivering motion. During the whole time of its confinement, the keeper had never heard it utter a sound, except once, when it startled him with its loud mocking cry.

It should be mentioned that this bird, which was obtained near the source of the Cass River, in the county of Westland, was much darker in plumage than the specimens in the Canterbury Museum, and more nearly resembled the North-Island example mentioned above. As the colours underwent no change during its long confinement, it is sufficiently clear that the dark plumage is not a condition of immaturity.

The late Sir J. von Haast believed latterly that the large Owl captured by his dog amongst the rocky precipices in a creek near the Lindis Pass, and noticed by me, on his authority, under the provisional name of *Strix haasti* †, was in reality a bird of the present species. Professor Hutton wrote informing me that this was the Owl referred to in the following passage, in his account of the Birds of the Little Barrier Island ‡:—"Another bird also lives on the island, apparently in the cliffs, and comes out

* Of the examples in the Canterbury Museum, one was procured from the Kakahu Bush, near Arowhenua; and the other, killed at the Levels Station, near Timaru, was presented to the Museum by Mr. Donald McLean. Mr. Potts writes:—"In May 1857, while living in a tent on the Upper Ashburton, we were constantly disturbed at night by their doleful yells amongst the rocky mountain-gullies. When disturbed on the ground, it bursts forth its weird-like cry immediately after taking wing. Its robust form, thickly clothed with soft feathers, is admirably adapted for encountering the severities of climate to which it must be frequently exposed whilst securing its wild hunting-grounds. Far less arboreal than its smaller congener, it roams over the bleakest tracts of country in many districts where bush of any extent is rarely to be met with, finding shelter among the numerous crevices in the rocks of rugged mountain-gullies. Being strictly nocturnal in its habits in pursuit of its prey, it must brave the icy blast of the alpine snowstorm at the lowest temperature. The severity of the climate in these elevated regions would scarcely be credited by those who have only known the mildness of the coast-line. As may be inferred, the real home of this hardy raptorial bird is amongst the fastnesses of the Southern Alps, from whence it makes casual excursions by the numerous river-beds to the lower-lying grounds, these occasional visits extending as far as the plains. Although well known from its cry, not many specimens have been obtained."

† Essay on New-Zealand Ornithology, 1867.

‡ Transactions of the New-Zealand Institute, vol. i. p. 162 (1868).

only in the evenings. Its cry is a peculiar kind of laugh in a descending scale, and is very ridiculous to hear. I saw it twice by the light of the fire." But he afterwards found reason to modify this opinion (Ibis, 1874, p. 35). My own belief is that there has been some misconception on this point, and that the "series of dismal shrieks, frequently repeated, waking the tired sleeper with almost a shudder," as described by one writer on the subject, are not due to this Owl, but probably to a nocturnal species of Petrel (*Procellaria affinis*); for during the very long periods that captive birds were kept by Mr. Smith and myself, although habitually noisy, they were never guilty of this "convulsive shout of insanity." That they do, however, when on the wing produce a sound not unlike laughter, is beyond question; and when several of them are hunting together they seem to laugh in unison. This is specially noticeable on very dark nights.

Mr. Enys informs me that it has been seen at the Bealey Police Station (in the Southern Alps), and that it sometimes utters a note "something like that of the Morepork, but just as if he had his mouth full."

Mr. W. W. Smith, formerly residing on the Albury estate near Timaru, and now settled at the Ashburton, has sent me from time to time very interesting notes on this rare Owl. He has not only been exceptionally fortunate in getting specimens, but he has likewise been successful in his endeavours to make them breed in captivity. The following extracts from one of his earliest communications on the subject (already published by me in the Trans. N.-Z. Instit. vol. xvi. pp. 308–311) will show what a good observer Mr. Smith is, and how keen his love of natural history. I have received many letters from him since, all replete with interesting facts, chiefly relating to this species; and I am also indebted to him for several fine specimens of the bird, together with the eggs and a newly-hatched chick:—

"February 8, 1882. In compliance with your request I have much pleasure in writing a short account of my experience in trying to breed the Laughing-Owl. The drawing of the bird made a great impression on me when I saw it for the first time in your 'Birds of New Zealand,' and since then I had been searching for over five years, trying to procure a specimen; but I was never successful until April of last year, when I succeeded in finding a very handsome one. In June I found another pair; and again in September I found two more. They have been a great source of pleasure and instruction to me. I found the birds in fissures of the limestone rocks at this place (Albury), but they are certainly very difficult to find. I first discovered that they were about the rocks by finding several fresh pellets, and being anxious to secure a specimen, I procured long wires and felt in the crevices, but with no good results. I, however, discovered a plan which proved successful. I collected a quantity of dry tussock grass and burned it in the crevices, filling them with smoke. After trying a few crannies, I found the hiding-place of one, and, after starting the grass, I soon heard him sniffing. I withdrew the burning grass, and when the smoke had partly cleared away, he walked quietly out, and I secured him. I obtained four birds by this means. I explained in a former letter how very tame they became in a short while after being captured. I also mentioned their call, which varies considerably during the year. When I captured the second pair (male and female) their call for a long time, in waking up in the evening, was, as formerly stated, precisely the same as two men 'cooeying' to each other from a distance. The voice of the male is much harsher and stronger than that of the female, and he is also a much larger and stronger bird. During the period of hatching he is very attentive in supplying his mate with food, as no sooner had the food been put into the large apartment of their house, than he would regularly carry every morsel into the dark recess; when feeding her she would utter a low peevish twitter and rise off her eggs. I may here correct a mistake which I made in writing to you on a former occasion. I stated that 'the male sits by day, the female by night.' I only saw the male twice on the eggs, and it was at this time I wrote the letter; but I certainly was mistaken, as the female performs most of the duty of hatching. I also ascertained the difference of the sexes by separating them at night until the second egg was laid. The females are much shier and more timid than the males, as they hide themselves on hearing the least noise. After sitting nine days on her first egg, the female forsook them, and all efforts to induce her to sit again were unavailing. She laid two more eggs a month afterwards, and had sat seven days, when, I regret to say, I had to leave home for medical treatment at Timaru. When I returned, eight days afterwards, she was still sitting and continued to sit until the 17th November, when she left the eggs without bringing out the young. The eggs must have been allowed to get cold when eight or nine days sat-on, as when I tried to blow them I found they contained embryo chicks. I am glad, however, that I succeeded in getting the eggs; another season I may succeed in getting young birds. I supplied them with many different articles of food, such as beetles, lizards, mice, rats, rabbits, and mutton, of all of which they partook freely; but they have the greatest preference for young or half-grown rats. They are a little slow and clumsy in capturing living prey, but their want of proper exercise and freedom may account for this; it may be otherwise in their wild state. After what I have pointed out, there can be no doubt that the *Sceloglaux* inhabits the dry warm crevices of rocks. All the birds I captured I found in such places, generally five or six yards from the entrance, perfectly dry, and where no wet could possibly enter. One thing surprised me much—the very narrowness of the entrance to their cranny. In some instances the birds must have forced themselves in. I noticed, however, that the crevices widened as they extended into the rock. The bottoms are covered with soft sand crumbled down from the sides, and affording comfortable resting places.

"Regarding the nidification of this bird, I am no longer surprised that so little is known, and likewise of its natural habits, considering that it conceals itself in such inaccessible places, and where few would think of searching for it. As a rule they could lay their eggs and hatch their young unseen and unmolested.

"The breeding-season may be said to extend over September and October. I found the bird mentioned in my last letter sitting on an egg on the 25th September; but it must have been laid about the beginning of the month, as it contained the chick I sent you. I discovered the bird by reaching a long stick with a lighted taper into the crevice. My captives laid on the 23rd, 27th, and 29th September, and again on the 20th and 22nd October *. The birds were very restless and noisy for a fortnight before nesting. They began to moult in December, and are not yet (Feb. 8) in full plumage. When casting their feathers they have a very curious appearance, as they become almost naked. At this stage two of my birds were stung to death, a month ago, by a swarm of bees passing through the fine wire netting and taking up their quarters on the roof of their dark recess. I was very sorry to lose them, as I cannot now send you a living pair. I have one very fine male I will send you in April. I am going to Lyttelton at that time, and I will forward it by the first steamer bound for Wellington. I will likewise send you another Owl's egg, but hardly such a fine specimen as either of the two I sent before. I intend to search the rocks carefully for more birds, and, if I succeed in finding more, I will not fail to send you a pair. You may, however, rely on getting a second specimen from me. I should mention that I have collected a quantity of pellets at different times, composed of the hair of rats and mice and the elytra of beetles. Three large species of the latter swarm among the débris beneath the main rock, and certainly constitute part of the bird's food."

From Mr. Smith's further notes I have extracted the following account, merely modifying it for convenience of narrative:—

* "There is an error in the account given by Mr. Potts in his article 'On Oology' published in 'Nature' in regard to this species. He describes an egg in the possession of Master C. Richardson as having been 'laid early in January.' As I procured the specimen I may state that it was laid on the 4th of October. The writer of the above-mentioned article was evidently misinformed.—W. W. S."

"I first heard the Laughing-Owl on a very dark, damp night; and I frequently afterwards found its castings before I was able to discover the bird. After repeated searches I was at length fortunate enough to capture a very handsome one. He had secreted himself in a deep fissure in the rocks, from which I dislodged him by burning some tussock grass at the entrance—in fact I smoked him out. I think I was never so pleased at capturing any bird. I brought it home and put it in a comfortable cage, where its demeanour was very quiet. It was in beautiful plumage, with the facial disk grey, shading off to white on the outer edges. I remarked that the eyes were conspicuously large, and the iris bright hazel. From the blunted condition of its claws it was evidently a fully matured bird, and to all appearances a male. During the first night of his incarceration he remained perfectly quiet, and refused to take any food. On the following night he moved restlessly about his cage, and once in the evening uttered a loud hailing call, as if wishing to communicate with an absent mate. By this time hunger had overcome his scruples, and before morning he had devoured two live mice which I placed in his cage, besides several pieces of mutton. After a few days' confinement he appeared to become more reconciled to the restraint, ceased to run about when approached with food, and indulged in a loud calling note when waking up in the evening. On one occasion I placed four live mice in the cage, and cautiously watched the result. After intently looking at the mice for a time, the Owl seized one of them, and, after bruising its head, tore it from the body, and swallowed it, and then devoured the other parts, tearing them to pieces before swallowing. After a pause of a few minutes he repeated the same operation on another mouse; but, although quick in despatching its prey, it is not so active as the nimble little Morepork. The latter species, instead of tearing a mouse to pieces, will reduce its head to a soft state and then swallow the animal whole. I tried my captive with some large lizards, which he immediately began to consume. I then offered him some beetles. After a long pause he commenced to eat them, with a quick snap of the bill. It was interesting to observe the rapidity with which he caught and swallowed them in succession, the elytra flying from the bird's mandibles like sparks from a blacksmith's forge. Eleven days afterwards I picked up in the cage a hard pellet composed of mouse-hair and the wings and legs of beetles: the rejectum of this savoury feast. On a subsequent occasion I gave the Owl three live mice: he treated two of them in the manner described above, and swallowed the other whole. I tried my bird with a live rat, but he failed to kill it after many attempts. I then despatched the rat and cut it up into small pieces, which the Owl readily devoured. At the end of a fortnight he had become quite tame, would watch all my movements very attentively, and with every appearance of confidence.

"On the 19th April I was lucky enough to capture two more birds. These were together, in one fissure of the rock, and were undoubtedly male and female. I had considerable trouble in dislodging them from their hiding-place. When I caught the female bird (the smaller of the two) she uttered a peevish twitter, and bit my hands severely. I placed them in a roomy cage, with a good supply of beetles and lizards. On the following morning I found that they had consumed all the food, and that they had already settled down to their new quarters in a spirit of contentment. I then gave them some pieces of mutton, two live mice, and a lizard, all of which they disposed of during the night.

"I placed all three Owls together, and although for a few days they appeared to agree very well, they afterwards commenced fighting; so I removed my first captive to a separate house, and left the pair together. The latter seemed perfectly happy in each other's company, and on waking up every evening, both of them joined in the peculiar hailing-call already mentioned.

"On the 26th July I made a fresh excursion among the rocks, in the evening, in the hope of seeing this Owl in its native haunts, but without success. Later in the night I heard the laughing call from several birds simultaneously. They evidently fly a considerable distance from the rocks, as

I heard them several miles down the river. A few days later I heard one 'laugh' while passing on the wing close over my Owl-house, possibly attracted thither by the call-note of my captives. They appear to fly very high, and to laugh every few minutes, particularly on dark and drizzly nights.

"On the evening of August 23rd, when I went as usual to attend to my captives, I noticed that one of them did not come out to be fed, and on looking into their dark recess I found the female sitting on an egg. On the next evening when I dropped the food into the cage the larger bird was alone; and picking up a piece of the meat, he walked into the dark recess with it, uttering all the while a low, hoarse, croaking sound. I gently looked in and saw that as he approached the sitting female she rose from the nest with a very peevish twitter, and taking the meat from his bill dropped it at her side. This operation was repeated over and over again, till all the pieces of meat were strewed around the nest. On the 27th I found a second egg in the nest, scarcely equal in dimensions to the first one laid, and more oblong in form *. I particularly observed that during the breeding-time both birds were habitually silent, scarcely ever uttering a sound of any kind, except when the male was feeding his mate in the manner described. This touch of nature was very pleasant to witness, and the gentleness and caution he displayed at this time were remarkable. At the slightest noise the female would utter her peevish scream, and would sometimes rise from the eggs. When all was quiet again, she would settle down, and the male bird would then retire to the outer house, and would remain there, apparently keeping watch over his mate."

On the 22nd September Mr. Smith found a nest containing one egg in a deep natural fissure near Rocky Peninsula. The parent bird was in the nest, and he left it in the hope of getting a chick, as the Owl was incubating. He continued, at frequent intervals, to visit the nest till the 17th of the following month, when, for some unaccountable reason, the bird had abandoned it.

A fine male bird received from him lived for a considerable time in my aviary, and afforded me much interest. During the spring months it was accustomed to make a peculiar barking noise all through the night, just like the yelping of a young dog. At times the cry changed, resembling that of a Turkey calling in the peculiar key that denotes it is about to roost. It was a melancholy cry, and is perhaps aptly described as that of a "disconsolate Owl seeking a mate." But it ceased altogether at the end of December.

Subsequently two more examples (male and female) were received from Mr. Smith, and were placed in the aviary with the previous occupant, who manifested his pleasure, but not in a very demonstrative way, at seeing old faces once more.

They seemed in perfect health, and partook readily of all the food offered to them. I contemplated, with some degree of certainty, being able to forward them to the Zoological Society of London, but my hopes were destroyed. Through an unfortunate accident to the temporary cage in which these birds were being removed to my new residence on Wellington Terrace, they both escaped one

* The following further observations on this case are from Mr. Smith's diary:—

"Sept. 28. Bird sitting closely on the eggs. To-night I found the female in the outer house, and the male in the recess, standing over the eggs. I retired for an hour, and on my return I found, to my surprise, that the male was sitting on the nest. 29th. To-night I found in the nest a third egg, which I removed. Again the male bird relieved the female in the task of sitting. 30th. The female resumed her duty on the nest, and the male bird carried every piece of meat into the dark recess, his mate responding with a weak call, and, taking the meat from him, dropped it again on the stones close to the nest, but did not leave the eggs. Oct. 7. To-night both birds were in the outhouse, and on looking into the recess I found the two eggs forsaken and perfectly cold. I attribute this to the intense inclemency of the weather on the previous night. Oct. 8. I confined the female to the dark recess in the hope of inducing her to sit, but to no purpose. Oct. 9. Eggs still cold, and I accordingly removed them. Oct. 20. On feeding the birds to-night I observed the male acting precisely as he did a month ago, and on opening the lid of the recess I found that the female had laid another egg. Oct. 21. Female sitting closely, and male carried every morsel of food to her. Oct. 22. Examined nest, and to my delight found a second egg. Oct. 23. Owl still sitting on eggs, but becoming exceedingly timid. Oct. 28. Nest abandoned, and on removing the chilled eggs, found that they contained well-developed embryos."

stormy night, and were never seen again. Active search was made in the vicinity, day and night, for several weeks, but without any satisfactory result. Many persons declared having heard them, from time to time, on the neighbouring hills, and guided by these reports the fugitives were traced through Sir James Prendergast's grounds to the Episcopalian Cemetery, where the scent was hopelessly lost, although the old sexton solemnly averred he had heard "most all kinds o' noises among them graves"!

Owing to my absence from home when the last-mentioned pair arrived, I never had an opportunity of studying them; but my son has furnished me with the following interesting note:—

"The three birds agreed very well together from the first; but after the first few days I noticed that our old bird was scarcely considerate enough to the lady, 'wolfing' all the meat and leaving her to take her chance. So I separated them, placing the new couple in the adjoining compartment, with only wire netting between. It was interesting to see them come out of their boxes towards dusk, which appears to be their favourite feeding-time, and take up their station on their respective rocks. On a piece of meat being thrown to one of them, it will stoop down and gaze very reflectively at it for a minute or more, and then march off to its perch to devour it. I have noticed that they frequently make a whistling noise, and sometimes a note very much like a Turkey chuckling. Another sound they produce is exactly like the mewing of a cat. Solemn as they are, they seem to be inquisitive birds. If you make a whispering noise, all three of them will turn round and gaze steadfastly at you, remaining as motionless as a statue, until the whispering has ceased, when they immediately relax. During the day they remain concealed in the boxes, but they appear to keep up a constant low chatter with each other. Altogether they are very amusing birds in an aviary."

There are two specimens of the egg in my son's collection. One of these is almost spherical, measuring 1·70 inch in length by 1·55 in breadth; the other is broadly oval, measuring 1·9 by 1·5. They are perfectly white, and the spherical one has some minute granular papillæ on its surface. I have examined several other specimens, and the former seems to be the more typical one.

The two forms of *Strigidæ* described above are the only ones inhabiting New Zealand of which we have, as yet, any positive knowledge *. But the natives are acquainted with another species, which they describe as being very diminutive in size, and strictly arboreal in its habits. This is, no doubt, the bird indicated by Mr. Ellman as *Strix parvissima* † ('Zoologist,' 1861). Mr. J. D. Enys informed me that he once captured an Owl "standing only five inches high," and that it was perfectly tame and gentle. Mr. Potts records, on hearsay evidence, several instances of the occurrence in the provincial district of Canterbury of an Owl "about the size of a Kingfisher"; and the accounts which he has received appear to confirm one another in all material points, the gentleness of this Owl when captured being in singular contrast to the habitual fierceness of *Spiloglaux novæ zealandiæ*.

In the British Museum Catalogue (Birds, vol. ii. p. 43) Mr. Sharpe refers *Strix parvissima*, Ellman, to *Scops novæ zealandiæ*, Bonaparte; but I can find no evidence that the unique specimen of the latter in the Leiden Museum ever came from New Zealand, the only authority for this being a label in Temminck's handwriting, "Nouvelle Zélande," but without locality.

* Dr. Finsch says:—"Mr. Sharpe includes *Strix delicatula*, Gould, in the avifauna of New Zealand ('Erebus and Terror,' 2nd edition, p. 23) on account of my statement (Journ. für Ornith. 1867, p. 318). But I long ago stated (Journ. für Ornith. 1870, p. 245) that I had made a mistake on this point."

† "Amongst the desiderata of our public collections, a very small Owl (*Athene parvissima*) has for some time held a place. Many doubt its existence, few have seen it, fewer still have formed any note or observation concerning it. From the information that has been gleaned about this rare bird, it would appear that one of its habitats used to be the woods about the Rangitata river. One was captured with the hand on the bank of a creek, at no great distance from Mount Peel forest."—*Out in the Open*.

CIRCUS GOULDI.

(GOULD'S HARRIER.)

Circus assimilis, Gray, Voy. Ereb. and Terror, Birds, p. 2 (1844, nec J. & S.).
Circus gouldi, Bonap. Consp. Gen. Av. i. p. 34 (1850).
Falco harpe, Haast, Layard, Taylor (nec Forst.), 1859–1861.
Falco aurioculus, Ellman, Zoologist, 1861, p. 7464.
Circus approximans, Gray, Hand-l. of B. i. p. 36 (1869).

Native names.

Kahu and Manutahae; in some districts Kahu-maiepa and Kahu-komokomo; also Kahu-korako and Kahu-pango*, to distinguish the very old and the young birds.

Ad. suprà brunneus, sub certâ luce cupreo nitens, dorsi plumis plus minusve fulvo lavatis et terminatis: pilei plumis medialiter et longitudinaliter nigris, ferrugineo marginatis: nuchâ cum collo postico et laterali clariùs fulvescentioribus: regione oculari nigrâ: facie laterali brunneâ, plumis medialiter nigris: radio faciali saturatè brunneo, ferrugineo tincto et fulvescenti mixto: dorso postico brunneo, plumis latè fulvo terminatis: uropygio imo et supracaudalibus albis, his fasciâ fulvâ anteapicali transnotatis: tectricibus alarum dorso concoloribus, minimis fulvo et albo lavatis: alâ spuriâ cinereo lavatâ: remigibus brunneis, ad apicem saturatioribus, extùs argenteo-cinereo lavatis, saturatè brunneo transfasciatis: caudâ cinereâ, rectricibus exterioribus ferrugineis, plus minusve albicantibus, pennis centralibus distinctè, exterioribus irregulariter brunneo transfasciatis, omnibus ad apicem albis: caudâ subtùs albicante, fasciis brunneis interruptis notatâ: subtùs lactescenti-albus, paullò fulvescens: gulâ brunneâ, plumis medialiter nigris: pectore toto distinctè brunneo longitudinaliter striato: cruribus paullò ferrugineo tinctis, suprà angustè ferrugineo striatis: subalaribus albis, maculis ferrugineis et brunneis notatis: cerâ et pedibus flavis: rostro et unguiis nigris: iride latè flavâ.

♀ mari paullò major et ferè pallidior: scapularibus rufescenti-albo terminatis.

Juv. chocolatinus, cupreo nitens, pileo vix nigricantiore: nuchâ albicanti-fulvo notatâ: subtùs ferrugineo tinctus: caudâ subtùs albicante, suprà chocolatinâ, ferrugineo marmoratâ: remigibus subtùs ad basin lactescentibus, plus minusve brunneo marmoratis: cerâ et pedibus flavis: iride saturatè brunneâ.

Adult male. Upper parts dark brown, the feathers of the head and neck broadly margined with reddish fulvous, the wing-coverts and scapulars terminally edged with pale rufous brown; quills black, with the outer web silvery grey, obscurely banded; tail, when closed, light silvery brown, with interrupted transverse bars and a subterminal band of dark brown; the lateral tail-feathers washed with rufous: the bars more conspicuous when the tail is spread; upper tail-coverts pure white, barred near the tip with rufous brown; superadjacent feathers tipped with rufous. Underparts generally pale fulvous, with a broad dash of rufous brown down the centre of each feather, these markings being thickest on the breast and sides; tibial plumes paler fulvous, with the central streak much reduced; the axillary plumes, which are remarkably long, pale rufous, barred

* Mr. Gurney has sent me the following note:—"The circumstance which you mention (page 11 of 1st edition) of *Circus gouldi* being called by the natives 'kahu-pango' strikes me as very curious, as *C. macrosceles* bears the name of 'papango' in Madagascar, and *C. maillardi* in Réunion (*vide* Ibis, 1863, p. 338 and note). The fact of the Réunion Harrier being called 'papango' was also mentioned to me by a resident there."

with darker rufous; under surface of wings and tail light fawn-colour varied with grey. A narrow white fringe, varied with brown, encircles the throat, terminating behind the ear-coverts; bill bluish black; cere dull greenish yellow, brighter on the ridge; legs and feet bright lemon-yellow; claws black; irides bright yellow. Length 22·5 inches; extent of wings 52·5; wing, from flexure, 17; tail 10; tarsus 4; middle toe and claw 2·5; hind toe and claw 1·75; bill, along the ridge 1·5, along the edge of lower mandible 1·5.

Adult female. Slightly larger than the male, but differing very little in plumage. The tints generally are lighter, the edges of the scapulars are rufous white instead of brown; and the wings are varied with rufous and white, especially towards the flexure. Length 23 inches; extent of wings 54.

Young. In the young bird the whole of the plumage is chocolate-brown, darker on the upper parts, and edged with paler brown; hind part of the neck varied with white, and tinged with rufous; upper tail-coverts rufous brown, with paler tips and fulvous at the base, sometimes white barred with rufous brown. Cere and legs yellow; irides dark brown.

Nestling. Covered with very thick or woolly down of a buffy white or pale yellowish cream-colour, darker on the upper surface. Bill and legs yellow. The feathers appear first on the shoulders, wings, and tail; these are blackish brown, the tail-feathers with rufous tips.

Fledgling. Has the plumage of the underparts much suffused with brown, the primary wing-coverts and the scapulars with a filamentous fringe of rufous; a similar fringe on the secondaries but paler; tail-feathers and their upper coverts largely and somewhat irregularly marked at the tips with rufous; feathers of the nape edged with darker rufous.

A well-feathered fledgling in my collection, with rectrices more than four inches long, has still some fulvous-white down adhering to the crop, flanks, and upper edges of wings. Claws well developed and very sharp.

Progress towards maturity. Upper parts dark brown with a purple gloss; the tail with five rather obscure bars of black, about half an inch apart, and darkest towards the tip; upper tail-coverts delicate fawn-colour, with the centre of each feather brown, shaded off on the sides. The wing-coverts have a coppery hue, and the longer ones, together with the scapulars, are narrowly tipped with rufous white. Underparts bright chocolate-brown, tinged with rufous, especially on the neck and abdomen; tibial plumes rufous brown. Cere and legs yellow; beak and claws black; irides bright yellowish brown.

Obs. It must be noted that individuals differ, more or less, in the details of their colouring during their progress towards maturity *. With extreme age, the fulvous of the lower parts changes to white, and the brown markings become much narrower, being almost obsolete on the tibial plumes. The silvery grey on the quills and tail-feathers increases, while the rufous colouring diminishes, and the lining of the wings becomes pure white, with narrow shaft-lines of dark brown. There is a beautiful albino specimen in the Nelson Museum.

The present species is spread over a wide geographical area; for not only is it found in all parts of our own country, but it also occurs in Australia and Tasmania, and extends eastward to the Fiji Islands. Mr. J. H. Gurney has already drawn attention (Ibis, 1870, p. 536) to the fact that our Harrier is exactly the same species as that figured by Mr. Gould in the 'Birds of Australia' under the name of *Circus assimilis*. The true *Circus assimilis* of Jardine and Selby (Ill. Orn. ii. pl. 51) has proved, however, to be only the young of *Circus jardinii*, also figured in the 'Birds of Australia' (pl. 27); and therefore the New-Zealand Harrier bears the name of *Circus gouldi*, Bonap. (*l. c.*) †.

* My eldest son, writing to me from Horowhenua on the 6th of May, 1881, says:—"I shot a beautiful Harrier yesterday, winging it when very nearly out of range. The plumage is handsomely mottled, and on the upper surface of the wings there is a steel-blue lustre; the breast yellowish white; lower part of body and tibials nearly pure white. Instead of the unpleasant odour peculiar to these carrion-feeders, it has a 'woody' smell like that of the Kaka."

† Dr. Finsch writes:—"A comparison of specimens in the Leiden Museum from Australia, New Zealand, Fiji, and New Caledonia has fully convinced me of their identity. The specimen from New Caledonia (*C. wolfi*, Gurney) does not show a

It is a very common bird in New Zealand, being met with on the fern-covered hills, in the plains, among the marshes of the low country, and even along the open seabeach, where it feeds on carrion. It is seldom, however, found in the dense bush, although I once surprised one there in the act of picking a large Wood-Pigeon *.

Like all the other members of the genus, it hunts on the wing, performing wide circles at a low elevation from the ground, and sailing over meadows, fern-land, or marshes in quest of lizards, mice, and other small game. Its flight is slow but vigorous and well sustained. The small size and specific gravity of its body, as compared with the great development of wings and tail and corresponding muscles, enable it to continue these wanderings for a whole day without any apparent fatigue. When sailing, as it often does, at a high elevation, the wings are inclined upwards so as to form a broad obtuse angle (with the tail half spread), and there is no perceptible motion in them, except when the bird alters its course. A pair may often be seen sailing thus in company, mounting higher with each gyration, and emitting a peevish whistle as they cross each other's course. On these occasions I have sometimes seen the birds close in upon and attack each other, the upper one making the first swoop, and the lower one instantly turning on its back, with upstretched talons, to receive him, and, after thus parrying the attack, wheeling upwards and becoming in turn the assailant. Whether it be the angry meeting of rival males, or the amorous gambols of raptorial lovers, I have never been able to determine; but this aerial encounter, whether in earnest or in play, has a very pretty effect. A correspondent informs me that he once observed five of these birds engaged together in this manner, at the commencement of the breeding-season, and that it was one of the prettiest sights of the kind he had ever witnessed.

It is worthy of remark that the birds of the first year are apparently incapable of the peculiar sailing flight which I have described, their locomotion being effected entirely by slowly repeated flappings of the wings. This circumstance, taken in conjunction with the dark colour of the young bird (appearing perfectly black at a little distance), has led to the common belief that there are two distinct species.

When gorged with food, the Harrier takes up its station on a rising knoll, a projecting stump, or the naked limb of a detached tree standing in the open, when it assumes an erect posture, with the head drawn closely in and the wings folded (as represented in the accompanying Plate), and remains perfectly motionless for a considerable time. When thus reposing, it is possible to get within gun-range of a "Kahu-korako," or very old bird; but at other times it is extremely difficult to obtain a shot. Hawks are known to be long-lived; and they appear to gain more experience of the world as they grow older. The dark-plumaged Harrier falls an easy prey to the gunner; it may be winged as it sails above him at an easy elevation, or it may be approached quickly and surprised when it descends to the ground to capture and devour a mouse or lizard. But the wary old "White Hawk" carries with him the experience of many dangers, and is not so easily

---

single character by which it can be specifically distinguished. As the true *C. assimilis*, Jard. & Selby, is undoubtedly the same as *C. jardinii*, Gould (which, therefore, must bear the former appellation), the New Zealand Harrier must stand as *approximans*, Peale." But Mr. J. H. Gurney, who is a recognized authority in regard to the Accipitres, has arrived at a different conclusion; and even were the matter entirely free from doubt, I should hesitate before disturbing a name so generally understood and accepted as that of *Circus gouldi*.

* I am indebted to Mr. J. A. Wilson for the following interesting information:—In March 1884 there was a violent eruption from the crater of White Island in the Bay of Plenty. For some weeks there was a continuous discharge of volcanic débris from the pit of the crater, with the usual accompaniments; and the heat thus evolved had the effect of driving out the rats which abound there in prodigious numbers (a small black rat, supposed by some to be the true *Kiore maori*). This exodus, strange to say, was the signal for the appearance on the island of the Harrier, which came over in large numbers from the mainland, as many as seventy having been counted on the wing together in one spot alone.

taken. I have followed one for the greater part of a day before I have succeeded in shooting it. These old birds, notwithstanding the extreme abundance of the species, are comparatively rare, and they are called Kahu-korako by the natives, in allusion to their hoary plumage. Birds in ordinary adult plumage are also somewhat shy: but on horseback I have often approached near enough to detect the colour of the cere and legs.

Besides devouring carrion of all kinds, the Harrier subsists on rats, mice, lizards, feeble or wounded birds, and even grubs and spiders. One, which I had confined in an outhouse, subsisted for several days entirely on spiders, for which he made a systematic search among the cobwebs that covered the walls. At the close of each day I found him with a matted circlet of spiders' webs surrounding the base of his bill. On my offering him the body of a Wood-Robin (*Miro australis*) he struck his talons into it, and, holding it firmly down, plucked off the feathers with his beak with remarkable rapidity, and then, tearing it to pieces, devoured it—the whole proceeding occupying only a few minutes. Captain Mair, who kept several of these birds in confinement for a considerable time, fed them frequently with freshwater fish, which they devoured with great avidity; and he assures me that he has observed them, in the wild state, capturing mullets in a shallow fish-pond.

The Harrier secures his prey by grappling it in his talons, sometimes bearing it off with him, but more generally remaining on the spot to devour it. On newly ploughed land he may occasionally be seen regaling himself on grubs and earthworms. It may be noticed that on these occasions, instead of walking, he moves by a succession of hops, the toes being turned inwards, in order, as it would appear, to protect the fine points of his grappling-instruments.

When the winter rains have inundated the low-lying flats and filled the lagoons, these places become the favourite resort of Wild Duck, Teal, Pukeko, and numerous other waterfowl; but this Hawk also puts in his appearance with the new comers, and is a perpetual terror to them. I have frequently seen one attack a full-grown Pukeko (*Porphyrio melanotus*), attempting to grapple it in its talons—its long tarsi and legs being stretched downwards to their full extent, accompanied by much noiseless fluttering of the wings. The Pukeko, anticipating the attack, springs upwards with open mouth and outstretched neck, and generally succeeds in warding off its assailant till it reaches cover and hides in the sedge. Audubon, in his 'Birds of America,' states that he has seen the *Circus cyaneus* attack the Marsh-Hen (*Rallus crepitans*) in the same manner. Young birds, and those wounded by the sportsmen, suffer most. On one occasion I fired at and disabled a large Pukeko, which at once took refuge in some rushes on the edge of the lagoon; but before I could get round to the spot, one of these Hawks had killed, plucked, and partly devoured it.

Once I saw a Harrier boldly attack a party of seven Pukekos. The birds crowded together, as if for mutual protection, on a dry clump in the midst of the swamp, and eventually succeeded in warding off their assailant.

But although, under press of extreme hunger, it will thus attack live birds, it is in reality a very cowardly representative of its tribe; for I have seen one chased by a pair of Australian Magpies (*Gymnorhina tibicen*) whose nest was in danger and driven ingloriously off the field, the pursuers assaulting it in a most determined manner and from opposite directions. An observant friend assured me that on one occasion he witnessed an attack made by four or five of these Magpies, acting in concert, and that the Harrier was not only vanquished but actually killed by them.

In the spring months it may be seen skimming low along the edges of the lagoons in pursuit of young Ducks, ever and anon swooping down among a swimming brood, but not always with success, the young birds instinctively diving under water on the approach of their natural enemy.

I have known the Harrier, when urged by excessive hunger, visit the poultry-yard and snatch up a chicken in its talons; and I have occasionally seen it attack both the wild and the domestic duck; and Mr. Gould, in writing of this species in Australia, declares that it is addicted to the stealing of eggs. On the other hand, I have seen it assailed by the Common Sea-Gull (*Larus dominicanus*) on approaching the nest of this bird, and put to an ignominious flight.

It is worth recording that the Harrier will sometimes pursue on the wing. Riding along the road near the Whenuakura river, on one occasion, I observed a Kahu pursuing a small bird (apparently a Ground-Lark) high in the air. The pursuit was continued for a considerable time, the Hawk making frequent swoops and the small bird eluding its grasp by suddenly altering its course and thus gaining on its pursuer. When nearly out of sight the Hawk was joined by another, both in pursuit of the same bird, from which circumstance I concluded that the raptor was foraging for hungry ones at home. This might account for the eagerness of the pursuit, and for a mode of chase which I had never observed before during a very long acquaintance with this species.

Mr. Hamilton, of Petane, states that he has on two occasions surprised the Harrier in the act of devouring an eel in the bed of a shallow creek.

When travelling through the Waikato district in July 1883, I observed one of these birds hawking in the rain. Although a heavy shower was falling the Harrier continued to hover without any apparent inconvenience, only occasionally shaking the raindrops off its tail.

It is said to be very destructive on the sheep-runs during the lambing-season; and I have been assured by eye-witnesses that three or four of them will sometimes detach a lamb from the flock, and then, assailing it from different points, tear out the animal's eyes and ultimately kill it. I am of opinion, however, that these attacks are confined to the weakly or sickly lambs of the flock, and occur only in times of great famine. Be that as it may, the practice of poisoning Hawks in the lambing-season has now become very general; and I have known upwards of a hundred of them destroyed in this manner, during that season, in a single locality. It is accomplished by rubbing a small quantity of strychnine into the body of a dead lamb or piece of offal, and leaving it exposed on the run. The poison takes immediate effect, and often eight or ten birds are thus destroyed in the course of an hour. As stated in my former edition, on one station alone in Canterbury upwards of a thousand Hawks per annum were destroyed in this manner during the preceding two or three years, and, as an almost necessary corollary of this, rats became excessively abundant on this particular sheep-run. I have always been of opinion that the wholesale killing of Hawks in a country like this is a questionable policy, from a utilitarian point of view, as it tends to alter the balance of nature, and to interfere with the general conditions of animal life, already too much disturbed by the operations of Acclimatization Societies. The rapacious birds have an important part to perform in the economy of nature: and species like the present, which are partly insectivorous, are too valuable to the practical agriculturist to be destroyed with impunity, although they may occasionally attack a sickly lamb in the flock, or swoop on an inviting young turkey. The damage to a flock where these Hawks abound is, no doubt, greatly overrated. It is true, however, that this species does sometimes hunt in packs, for I have counted as many as twenty of them at one time hovering over a small mob of sheep detached from the main flock; and three of them have been seen to attack a full-grown turkey, and, acting in concert, to overpower and kill their quarry.

The natives take this species by means of flax snares, arranged in such a manner that the bird, in attempting to grapple the bait, gets its legs entangled in a running noose, which its efforts to escape only serve to tighten. I have frequently taken it alive by means of a steel trap, with muffled edges, baited with a dead rat or chicken. When shot at, and wounded in the wing, it attempts to escape by a succession of leaps along the ground, and, on being overtaken, defends itself vigorously with beak and claws, its beautiful golden eyes sparkling with passion. In captivity it is at first fierce, throwing itself backwards when approached, and striking forwards with its long talons; but it soon becomes reconciled to the situation, and permits itself to be stroked with the hand. The late Captain

Buck, 14th Regiment, informed me that, while stationed at Napier, one that he had winged became so tame that, on recovering health and liberty, it was accustomed to return every evening to his garden and roost in the arbour.

The peculiar whistling note already alluded to is only heard when two or more of these birds are in company. The young has a cry resembling the hoarse note of our Stilt-Plover. Professor Hutton informs me that the cry of this Hawk is very similar to that of the Govinda Kite of India, which he has frequently heard in that country.

I have observed that in very old birds of this species the feathers of the upper parts present a faded and ragged appearance, from which it may be inferred that the moulting-power becomes impaired as age advances. A specimen that came under my examination, in the flesh, presented the following singular condition, for which I was quite unable to account, although probably the result of disease. A space on the breast and the whole surface of the sides were entirely denuded of feathers, these parts being covered by a thick growth of white down; on the back also there was simply a narrow strip of feathers down the line of the spine. The head of this bird was greatly infested with parasitic ticks.

There is a very beautiful albino variety in the Nelson Museum, presented by Mr. Goodall, of Riwaka, where the bird was obtained. The whole of the plumage is of a very delicate white ash-colour, the underparts having a rosy-purple tinge. The primaries are ashy grey; and both these and the tail-feathers present, on the under surface, obsolete bands, as though they had been washed out. The shafts of all the feathers on the upper parts are dark grey, presenting the appearance of finely pencilled lines. The bill, as also a superciliary line of hairs and those covering the lores, black; cere, tarsi, and toes yellow. The taxidermist to whom this handsome specimen was entrusted, with a full appreciation of its value, charged the modest sum of eight guineas for stuffing it, and had to be compelled to give it up by process of law.

During a visit to the lake district, in the autumn of 1877, I saw another, apparently very like the last-mentioned bird, hovering over the fern ridges that close in the intensely blue waters of Tikitapu. As he swooped down upon a rat or lizard in the fern his underparts appeared to be perfectly white, and the upper surface of the body and wings ashy. Major Mair informs me that, in 1885, he observed a similar one at Lake Rotoiti.

This species prefers a swamp for its breeding-place, and generally builds its nest on the ground, though sometimes in a tussock. It often repairs to the same place for several successive seasons, the old nest forming a foundation for the new one, which is usually constructed of the dry blades of *Arundo conspicua* and the flower-stalks of the Spaniard-grass rudely placed together and overlaid with dry grass*. The breeding-months are October and November; but as late as Christmas Day (1863) I saw, in Matene Te Whiwhi's house at Otaki, a very young one that had been taken from a nest (containing two) about three weeks previously. It was about the size of a half-grown

* Mr. C. H. Robson, of Cape Campbell, has sent me the following interesting note:—"In the spring of 1873, I observed a very large female Hawk of a brighter colour than usual, with very distinct markings, and presenting quite a yellow appearance as compared with the ordinary Hawk. She rose, the first time I saw her, out of a piece of swampy ground near the beach, and, on a subsequent occasion, finding her in the same place, I hunted about and found her nest in a tussock, with two white eggs in it. Being anxious to secure the young birds, I did not handle the eggs, but visited the nest every week, each time coming quite close to the bird. In due time one of the eggs hatched out a little yellow-white chick, but a few days later, to my great regret, it was taken, I presume, by a rat. On flying off the nest the Hawk was joined by the male bird, not nearly so large as herself, and always too high in the air for me to observe his plumage."

"In November 1884 in one of the large swamps in the Hind district, on the Canterbury Plains, a nest of this Harrier, built on a large tuft of coarse growing rushes (*Juncus*) was knocked over by a 'mob' of cattle. The nest being set up again and the eggs put back the Hawk returned and resumed incubation. The nest contained five eggs; another nest in the Horoatu district also contained five eggs."—*Zoologist*, 1885, p. 424.

gosling, and was covered with thick cottony down of a dirty white colour inclining to buff, with feathers beginning to show themselves on the back, wings, and tail; cere and legs yellow. It opened its mouth for food on being approached, and when provoked would strike forward or upward with its well-armed feet. It made one aware of its presence by its rather fetid odour, as well as its occasional cry, which was like a half-suppressed whistle.

When there are two young birds in a nest there is often a remarkable disparity in their size. They are always very savage when molested, throwing themselves on their back and striking vigorously with their talons at the hand of the intruder.

A nest found by a Wanganui settler contained, in addition to two full-grown young birds, the remains of 11 Pheasants, 5 rats, 3 Quail, and a Weka.

The eggs are from two to four in number, but generally three, ovoido-conical in form, with a smooth or finely granulate surface, perfectly white, till stained by the bird's feet during incubation, and measuring 1·9 inch in length by 1·5; my largest example measures 2 by 1·6. At first sight they appear to be disproportionately small for the size of the bird; but they are not so in reality, for the body of this Hawk, when stripped of the feathers, is almost ridiculously small. After being blown, if held up against the light, the interior of the shell presents a surface of a beautiful clear green.

Before passing on to the next group, I may mention that in a case of mounted Raptores which I had the pleasure of presenting, some years ago, to the Colonial Museum there is a fine specimen of the White-bellied Sea-Eagle (*Ichthyaëtus leucogaster*), which I received from the late Mr. Gould as having been obtained in New Zealand. This species has been observed along the whole southern coast of Australia, from Moreton Bay on the east to Swan River on the west, including Tasmania and all the small islands in Bass Strait; and as it is a powerful flier there is no physical reason why it should not occur sometimes as a straggler on the New-Zealand coast. Mr. Gould had satisfied himself that this specimen was obtained there, although unable to ascertain the precise locality. In corroboration of its presumed occurrence, I may mention that an officer of the 14th Regiment, who was a good sportsman and a tolerable naturalist, assured me that he had actually seen and fired upon a "Sea-Eagle" on the rocks near the entrance to the Wellington harbour.

Two other species of Accipitres, the *Falco subniger* (a rare bird inhabiting South Australia) and the *Milvus isurus*, or Australian Kite, have had New Zealand assigned as their habitat, on the authority of Mr. J. H. Gurney, who, in a letter to 'The Ibis' (1870, p. 536), offers the following explanation:—"My authority for quoting New Zealand as a habitat for the former (*Falco subniger*) was the veteran ornithologist, M. Jules P. Verreaux, who informed me that a New-Zealand specimen had passed through his hands. With regard to the latter (*Milvus isurus*), the Norwich Museum possesses a specimen, which I obtained from Mr. A. D. Bartlett, who assured me, at the time, that he had received it from New Zealand, and had satisfied himself that it had been killed in that country. Probably both these species, if not indigenous to New Zealand, may occasionally occur there as accidental visitors from the Australian continent." In support of Mr. Gurney's surmise, I may state that the account sent to me, many years ago, by Sir Julius von Haast, of a Hawk observed by him in the Southern Alps, although unfortunately not secured, seems to accord with that given by Captain Sturt of the Australian *Falco subniger*.

QUAIL HAWK (ADULT AND YOUNG)
HIERACIDEA NOVÆ-ZEALANDIÆ.

HARPA NOVÆ ZEALANDIÆ.

(QUAIL-HAWK.)

New-Zealand Falcon, Lath. Gen. Syn. i. p. 57 (1781).
Falco novæ seelandiæ, Gm. Syst. Nat. i. p. 268 (1788, ex Lath.).
Falco australis, Hombr. et Jacq. Ann. Sci. Nat. 1841, p. 312.
Hypotriorchis novæ zealandiæ, Gray, Gen. of B. i. p. 20 (1844).
Falco harpe, Forst. Descr. Anim. p. 68 (1844).
Hieracidea novæ zealandiæ, Kaup, Isis, 1847, p. 80.
Harpe novæ-zealandiæ, Bonap. Comptes Rendus, xli. p. 652 (1855).
Ieracidea novæ zealandiæ, Gray, Hand-l. of B. i. p. 22 (1869).
Hieracidea novæ zealandiæ, Buller, Birds of New Zealand (1st ed.), p. 1 (1873).
Harpa novæ zealandiæ, Sharpe, Cat. Brit. Mus. Birds, vol. i. p. 372 (1874).

Native names.

Karearea, Kaiaia, Kaeaea, Kakarapiti, Karewarewa, and Tawaka.

♂ suprà nigricanti-brunneus, pileo unicolore saturatiore: dorso fasciis irregularibus fulvescentibus transnotato: remigibus nigricanti-brunneis, pogonio interno albo transfasciato: secundariis extùs fasciis angustis albidis notatis: caudâ nigricanti-brunneâ, albido angustè et interruptè transfasciatâ: facie laterali nigricante, supercilio indistincto et genis imis rufescentibus: gutture fulvescenti-albo, scapis plumarum nigro indicatis: corpore reliquo subtùs lætiùs fulvescente, pectoris plumis saturatè brunneo medialiter striatis et fulvo plus minusve distinctè ocellatis: hypochondriis imis cum cruribus et subcaudalibus lætissimè castaneis: subalaribus fulvescentibus, castaneo tinctis, his et axillaribus fulvescenti-albo ocellatis: rostro cyanescenti-nigro, ad basin mandibulæ corneo: cerâ pallidè flavâ: pedibus flavis: iride sordidè flavâ.

♀ mari similis, sed paullò major.

♂ *juv.* suprà fuliginoso-brunneus, pileo magis cinerascente: caudâ minùs distinctè transfasciatâ: gutture fulvescenti-albo, angustè brunneo striato: subtùs fuliginoso-brunneus, pectore paullò nigricante et hypochondriis cruribusque vix castaneo tinctis: pectore medio albido obscurè maculato: abdomine imo crissoque fulvescentibus: hypochondriis distinctè fulvo ocellatis: cerâ et plagâ oculari cyanescenti-albis: pedibus plumbeis: unguiis nigricantibus.

Pull. lanugine plumbeâ indutus.

Adult male. Crown of the head and nape glossy black; upper surface generally brownish black, barred on the scapulars and tail-coverts with rufous, and narrowly on the wing-coverts with rufous grey; a line over each eye, and sides of the neck, varied with rufous; facial streak and ear-coverts black; throat fulvous white, with narrow black shaft-lines, broadening out towards the breast; fore part of the neck and breast fulvous varied with rufous, and having the centre of each feather brown; sides of the body dark brown varied with rufous, and with large rounded spots of fulvous white; abdomen and vent rich fulvous; under tail-coverts and tibial plumes rufous brown, with narrow black shaft-lines; quills and secondaries obscurely marked on their outer webs with grey; tail with eight narrow interrupted bars of greyish white, and slightly tipped with rufous; under surface of quills and tail-feathers dusky, the former largely toothed and the latter barred with white. Bill bluish black; base of lower mandible horn-colour; cere pale yellow; legs brighter yellow; claws black; irides brownish yellow, becoming purer yellow with advancing maturity. Extreme length 19 inches; extent of wings 31; wing, from flexure, 11·25; tail 8·25; culmen 1·2; tarsus 2·5; middle toe and claw 2·75; hind toe and claw 1·75.

Adult female. The plumage is similar to that of the male, excepting, perhaps, that the spotted markings on the sides are more distinct; but there is a slight difference in the size. Extreme length 19·5; wing, from flexure, 11·5; tail 8·5; tarsus 2·75.

Young. Crown of the head and upper parts generally brownish black, glossed with grey in certain lights; line over each eye reddish fulvous; throat fulvous white, with a central line of brown on each feather; sides of the neck, breast, lining of wings, and underparts generally dark brown varied with fulvous; sides marked with rounded spots of fulvous white, very obscure in some specimens; tibial plumes reddish brown; lower part of abdomen, vent, and inner side of thighs fulvous; under surface of quills and tail-feathers dusky, with numerous transverse bars of white. Cere and bare space around the eyes bluish white; irides black; legs dark grey, with black claws.

Nestling. Covered with plumbeous-grey down.

Obs. The above measurements were taken from a pair of birds of this species formerly in the Christchurch Acclimatization Gardens, and now preserved in the Canterbury Museum, the sex in both cases having been carefully ascertained by dissection. The figure of the adult female is from a fine specimen obtained in the South Island, and now in my collection. Examples vary in the details of their colouring. In some the light spots on the sides are far more conspicuous and the tibial plumes are of a brighter rufous than in others. As a rule, the white bars on the tail-feathers, although interrupted in the middle, are conterminous on each side of the shaft. In a specimen, however, obtained by Mr. Travers in the South Island the bars are alternate on each web, as was also the case with another, shown to me at Ohinitahi; but this character is quite exceptional.

The synonymy given above will serve as a tolerably complete guide to the scientific and literary history of the present species; but much confusion has arisen at various periods with regard to the nomenclature employed, and a few words in further explanation of the subject appear to be necessary.

In Mr. G. R. Gray's 'List of the Birds of New Zealand,' published as an Appendix to Dieffenbach's 'Travels' (1843), that naturalist recognizes only two species of Accipitres, which he calls respectively *Falco harpe*, Forst., and *Falco brunneus*, Gould, thereby intending, of course, to indicate the existence of two distinct species of true Falcons in New Zealand; but in this list there is no mention whatever of the Harrier (*Circus gouldi*), a common and well-known bird in our country. In adding the native names an unfortunate mistake occurred; for *Falco harpe* was stated to be the bird known to the inhabitants as "Kahu" and "Kahu-papango," whereas these are in reality the native appellations for the Harrier, which, as already stated, had been omitted from the list. This will, no doubt, account for the mention of Gould's Harrier, in the earlier writings of Layard, Haast, and Taylor, under the erroneous title of *Falco harpe*. Mr. Gray himself afterwards, in his 'Birds of New Zealand' (Voy. Ereb. and Terror), partially rectified this error by introducing the *Circus* in its proper place; but the misapplication of the native names was continued. In that work Mr. Gray substituted the prior title of *Falco novæ zealandiæ*, Gmel., for *F. harpe*, Forst., with *F. australis* (Homb. et Jacq.) correctly added as a synonym; he likewise reduced Gould's *F. brunneus* to the rank of a synonym; but in a subsequent list (Ibis, 1862, p. 214) he recognized it again as a distinct species, and equivalent to *F. ferox* of Peale (U. S. Explor. Exped. 1848), referring both forms to Kaup's genus *Hieracidea*. Unfortunately Mr. Gould's description of *H. brunnea* was founded on an immature bird, in a condition of plumage exactly corresponding with the young of *H. novæ zealandiæ*. This circumstance, together with the great difference in size between the male and female, led me, among others, to the conclusion that the two birds were referable to one and the same species *. Dr. Otto Finsch (Journal für Ornithologie, 1867, p. 317) expressed his belief that *H. brunnea* was the female of *H. novæ zealandiæ*—a decision based (as he has since informed me) on Forster's account of the bird; but in a subsequent paper (*op. cit.* 1870), referring to my observations on the subject, he adopts the view of its being the young of that species, quoting, at the same time, Dr. Haast's opinion to the contrary. In Captain Hutton's 'Catalogue' † only one species is admitted, the author remarking that it is very variable in size, and that "a large male can be distinguished from a small female by its more slender legs, which are 0·6 of an inch in circumference in the male, and 0·88 of an inch in the female." On the other hand, several excellent local observers have always held that they could distinguish a larger and a smaller species, the former differing in some of its habits from the common Bush-Hawk, and frequenting the open country in preference to the woods. Mr. Gurney also called attention to the subject in a letter to 'The Ibis' (1870, p. 535), in which he gave the dimensions of various examples that had come under his notice. Of these, the small specimen of *H. brunnea*, in the Norwich Museum, marked ♀ (measuring 14·5 inches in total length, wing 9·25), is, no doubt, as Mr. Gurney suggests, incorrectly labelled; for I have never met with so small an example of that sex; and it must be confessed that conclusions based on a mere examination of skins, in the absence of a positive determination of the sex, are very unsatisfactory.

It will be seen, on reference to the measurements I shall give in treating of the smaller species, that the sexes differ very much in size, the female, as is always the case with members of this family, being the larger bird. The fact that a male of the present species (of which the sex was carefully ascertained by Dr. Haast) was actually larger than the female of *H. brunnea* appeared to me sufficient of itself to warrant a specific separation. Having, however, brought with me to England good examples of both forms for illustration in my former edition, I compared them with the fine series of specimens in the British Museum (about twenty in number) and with Forster's original drawings, and came to the conclusion that there were in reality two distinct species, closely resembling each other in plumage in both the young and adult states, but differing appreciably in size. In this examination I was kindly assisted by Mr. J. H. Gurney, an ornithologist who, as is well known, has made Birds of Prey his special study; and as he entirely concurred in the conclusion arrived at, I felt that I could publish it with some degree of confidence.

Mr. Sharpe afterwards pointed out (Ibis, 1873, p. 327) that the name of *Falco brunneus* of Gould had been preoccupied by Bechstein, who thus called the common Kestrel of Europe, and that consequently our small bird, if allowed to be distinct from *H. novæ zealandiæ*, must bear another title. He considers that this should be *Hieracidea australis* (Homb. & Jacq.); but it seems to me that this is only a synonym of the older species and that the right name to fall back upon for the former is *Falco ferox* of Peale. In his official catalogue of the Accipitres in the British Museum, under the generic name of *Harpa*, he not only gives *H. australis* the precedence, but commits (as I venture to think) the further error of making it a "subspecies," or constant variety, of *H. novæ zealandiæ*. The two birds are either specifically distinct or they belong to one and the same species.

Professor Hutton contributed to 'The Ibis' for October 1879 a table of measurements for the purpose of showing that there existed only one species; but in my reply to that paper (Ibis, 1881, p. 458) I pointed out that his argument was quite inconclusive, inasmuch as "his ♀ specimen B gives a wing-measurement only ·25 of an inch longer than that assigned by me to the female of the smaller species."

* Vide Trans. N.-Z. Instit. vol. i. p. 106 (1868).

† 'Catalogue of the Birds of New Zealand,' by F. W. Hutton, Geol. Survey of N. Z. (1871).

Since that time the question has received much attention at the hands of local ornithologists; and although there may be still some difference of opinion as to the propriety of keeping the birds distinct, nearly all the subsequent evidence is in support of my contention.

Apart from the manifest difference in size already mentioned, the Quail-Hawk may be distinguished from the smaller species by the colour of the irides, which become yellow in the fully adult bird, whereas in *Harpa ferox* they are dark brown.

This larger form is seldom if ever met with in the North Island, where the other is comparatively plentiful. The only specimen ever obtained by me there was shot in the Kaipara district, more than five-and-twenty years ago, and this is preserved in my old type-collection in the Colonial Museum. It is met with in suitable localities all over the South Island.

Its food consists of birds, rats, mice, lizards, and the larger kinds of insects. It often takes its prey on the wing, swooping down on its terrified quarry with the rapidity of an arrow. It never feeds on carrion or offal.

I have been informed by a credible eye-witness that on one occasion a Quail-Hawk swooped down upon a man who was carrying a dead Pigeon, and, striking the bird forcibly out of his hands, retired to its station in a puriri tree to wait the course of events. It unfortunately fell a victim to its intrepidity, as it was instantly shot.

The late Sir J. von Haast, who always believed in the existence of two species, stated that their habits differ in the manner of taking their prey; and his collector, the late Mr. Fuller, assured me that he had invariably found the large birds paired together in the plains, and the small ones in the bush.

Mr. Reischek, who has been collecting for eight years in every part of the country, declares that all the examples obtained by him in the North Island were undoubtedly referable to the smaller form. He has collected both species in the South Island, where he invariably found the Quail-Hawk on the plains and lower ranges of hills, and the Bush-Hawk near the summits of the wooded ranges. Even on the Hen Island (in the Hauraki Gulf) he found the latter species frequenting only the tops of the hills. Having studied the birds in their native haunts and shot and compared scores of specimens in every condition of plumage, he unhesitatingly affirms that the two forms are specifically distinct.

Mr. Smith, whose full notes on the subject were communicated by me to the Wellington Philosophical Society *, writes that having procured upwards of thirty specimens and worked out the subject for himself he is "decidedly in favour of the existence of two species." He states that he had nestlings of both, and that those of *H. ferox* never attained to the size of *H. novæ zealandiæ*, although he kept them four months longer. In disposition, too, they differed, being fiercer and more untamable than the larger form.

Mr. Potts, who also recognizes two species, makes the following pertinent remarks:—

"If the cabinet ornithologist will not permit the fauna to possess two species, *Falco ferox* = *F. brunnea* must be the young state of *Falco novæ zealandiæ*. In this case we must try to believe that the greatest boldness and audacity in attacking, the greatest activity and swiftness of wing in pursuing, is exhibited by the Quail-Hawk before it has reached the adult state; neither may we have regard to the difference of size which specimens of either sex very often present.

"In November 1868 two sets of young Falcons were found on Lake Coleridge by Mr. Oakden's shepherd; they were taken from the nesting-place and presented by Mr. Oakden to the Canterbury Acclimatization Society. He stated to the writer that the birds from one nest were readily distinguishable from those of the other nest even from the first. In size there was a marked difference,

* Trans. N.-Z. Instit. vol. xvi. pp. 318–322.

perhaps of about one-third, this contrast of size being maintained up to the time when some of the birds were shipped for export to England. The writer has seen numbers of both species, and has a series of many specimens that have been collected in the course of some years. In life, besides the marked difference in size and in robustness of frame, the Sparrow-Hawk (*F. ferox*) looks flatter about the head and carries the wings more prominently forward, this carriage giving the bird a less rounded appearance than is observable in the larger species. The smaller Falcon is more savage and resolute, and swifter in flight than its congener *.

"The Quail-Hawk exhibits great perseverance in pursuit of its prey, and almost unequalled audacity. I have known it pursue and strike down a large Spanish hen in a stockyard, not relinquishing its hold till killed with the blow of a stick. I have also known it pursue its prey into the inner room of a small cottage. When Quail-shooting, years ago, I have been on different occasions attended by this dauntless fowler, and have shot an individual in the act of pouncing on the flying Quail. I have seen a female of this species bear off a Tui trussed in her talons, and carry it some distance without a rest, the male bird apparently keeping watch and ward, soaring within easy distance. I remember also seeing a Quail escape the rapid pursuit of one of these Hawks by dropping like a stone, at the very instant that I expected to see it trussed up in the talons of its pursuer, so close was the chase before the Quail adopted its last resource for escape."

On the breeding-habits of this species, the same observer has communicated the following particulars:—"At present it is in the 'back country' only that we can hope to find its breeding-place, which is usually in a ledge of rock commanding a prospect over some extent of country. Such an outlook gives an advantage of no little value, of which the Falcon is not slow to avail itself, should such a bird as a Tui or Pigeon appear in sight. Several of the breeding-places which we have had opportunities of examining have presented, in a remarkable degree, very similar conditions as regards situation. Amongst bold rocks, on the mountain-side, somewhat sheltered by a projecting or overhanging mass, appears to be its favourite site for rearing its young. The eggs very closely resemble those of *Falco peregrinus* of Europe in colour, size, and shape, are usually three in number, and are deposited on any decayed vegetable matter that wind or rain may have collected on the rocky ledge; for the efforts of this bird in the way of nest-building are of the feeblest description." He gives October, November, and December as the breeding-months; and states that above the upper gorge of the Ashburton or Haketere River he discovered a nesting-place on the bare soil, sheltered by a large isolated rock. It contained two young Hawks covered with grey down; and the old birds were very bold in defence of their offspring.

From my brother, in Canterbury, I received a very handsome pair of eggs belonging to this species. Although taken from the same nest, they differ somewhat from each other, both in size and in the details of their colouring. One of them measures 2 inches in its longer axis, by 1·4 in diameter; is elliptical in form; mottled and blotched with dark brown on a lighter ground, and encircled at the larger end with a broad zone of very rich brown, varied with blotches of a paler or

* "We once had the gratification of witnessing a most interesting trial of powers between a Sparrow-Hawk and the Brown Parrot (*Nestor meridionalis*). It was near the shore of that most romantic sheet of water Lake Mapourika. Standing just within the trees that fringe its margin, we heard the alarm-cry of the Kaka, and swiftly there came in sight, crossing a corner of the open space above the placid waters, two birds in active contest, the Parrot labouring heavily, wheeling and clumsily gliding aside, as its fierce pursuer drove at it with its talons. Then the rapid shifting of colours—now one saw the olive-brown of the Kaka's back, then the blood-red markings of its soft under-plumage, almost hidden the next instant with the dark brown, blackish pinions of the Falcon. Borne downwards with the momentum of a last stroke, the Hawk occupied some time in regaining 'the air,' whilst the terror-stricken Kaka hastened at its topmost speed towards the friendly cover of the wood. Once more its persevering enemy darted towards it with almost incredible swiftness, but the persecuted bird seemed to tumble amongst the trees that ensured its safety, quite regardless of appearances, so that it reached an asylum."—*Out in the Open.*

reddish tint. The other is more broadly elliptical, measuring in its axis 1·9; diameter 1·45. It wants the well-defined dark zone of the former, the whole surface being more or less mottled and blotched with reddish brown on a paler ground.

The fine series of eggs of this species in the Canterbury Museum exhibit considerable individual variation. Two specimens, taken from the same nest, are more ovoido-conical than ordinary examples, having an appreciably smaller end. One of these is of a rich reddish brown towards the larger end, with darker blotches, and towards the other end pale brown, profusely sprinkled and mottled with dark reddish brown. The other is somewhat similar, but more blotched with dark brown in its median circumference, and with the ground-tint towards the smaller end reduced to a whitish cream-colour. In two other examples (also from one nest) the whole surface is reddish brown, stained, mottled, and blotched with darker brown; but one of them has the brown of a richer tint, and the mottled character more distinct.

Among the more recent additions to this collection there is a singular specimen of the egg of this species. It is very ovoido-elliptical in form, measuring 2·25 inches by 1·4, of a warm sepia-brown, prettily freckled and spotted, more thickly so in the middle, and confluent in a large patch at the larger end, with reddish brown varied with darker brown.

A very handsome specimen in my son's collection (obtained at Oamaru) is broadly ovoido-conical, measuring 1·9 inch in length by 1·3 in breadth; it is of a rich cream-colour, thickly spotted, speckled, and freckled over the entire surface with dull reddish and chocolate-brown, these markings becoming entirely confluent at the larger end, which is entirely reddish brown smudged and daubed all over with chocolate-brown.

On the subject of the systematic position of this form, Dr. Finsch published the following remarks in the 'Journal für Ornithologie' for March 1872, which I have translated from the German:—"*Falco novæ zealandiæ* must be ranged among the Tree-Falcons, and follows next in order to *Falco femoralis*, having, like the latter, a long tail, which is only half covered by the wings. . . . Third primary longest; second shorter and somewhat longer than fourth; first and fifth equal. Tarsi covered in front with ten hexagonal scutes in double rows. Middle toe very long, being with the claw nearly as long as the leg; lateral toes equal, the points of their claws scarcely reaching to the base of the middle-toe claw. A subgeneric distinction appears justifiable."

Mr. Sharpe, who contributed to 'The Ibis' (1873, p. 327) some critical notes on the subject, says:—"The New-Zealand *Hieracideæ* are rather abnormal members of the Falconine series; for it is rare to find a bird which, when young, is uniform above, and becomes barred when it is old; nor do they here closely coincide with their Australian congeners, excepting as regards their uniformly cloudy breasts when young." He afterwards (Cat. Birds Brit. Mus. vol. i.) adopted Bonaparte's genus for our bird, merely altering the termination, for classical accuracy, and making it *Harpa*.

In a communication to the Wellington Philosophical Society, in September 1878 *, I took exception to the proposed generic separation of our bird from that inhabiting Australia; but I have lately gone into the question with Mr. Sharpe himself and have come to the conclusion that the distinction he makes is a reasonable one. I have accordingly adopted *Harpa* in lieu of *Hieracidea*, although my Plate of the species, which had already been worked off, bears the latter name, being that by which the bird has been hitherto known in the Colony.

Mr. Gurney in his 'Diurnal Birds of Prey' (p. 95) says, in reference to the smaller species:—"Mr. Sharpe applies to this Falcon the specific name of '*australis*' proposed by MM. Hombron and

* Trans. N.-Z. Inst. vol. xi. pp. 306, 307.

Jacquinot in the 'Annales des Sciences Naturelles,' 2nd series, vol. xvi. p. 47; but, according to the letterpress of the 'Voyage au Pôle Sud,' Zool. vol. iii. p. 47, this name was given to the species inhabiting the Auckland Islands as well as New Zealand, which is *H. novæ zealandiæ*. I therefore agree with Dr. Buller in considering '*Falco australis*' a synonym of the larger species." He also questions Mr. Sharpe's right to sink the specific name of *brunnea*, for he argues that "its having been proposed for a species of the genus *Tinnunculus* does not render its employment illegitimate when it is applied to a bird belonging to another and distinct genus." As will be seen, however, I have followed Mr. Sharpe in this respect, so as to avoid all possible confusion of names in the future.

Mr. Gurney, after a careful study of the series of specimens in the Norwich Museum, wrote to me saying, "I am sure you are right about the distinctness of the two New-Zealand *Hieracideæ*"; but Professor Hutton, who still adheres to the contrary opinion, says in one of his last letters:—"I examine and measure carefully every specimen of *H. novæ zealandiæ* that comes in. So far as my present measurements go they indicate one species only."

Before passing on, however, to my account of *Harpa ferox*, I will give here the results of a comparison of two carefully selected birds which I exhibited at a meeting of the Philosophical Institute of Canterbury, as recorded in the 'Transactions of the New-Zealand Institute':—

"Among Hawks generally—and the genus *Hieracidea* is no exception to the rule—the female is both larger and more handsomely marked than the male. Such being the case, let us for our present argument compare an adult female of *Hieracidea novæ zealandiæ* with an adult female of *H. ferox*. This will afford us the fairest mode of determining their relative size, and the best means of ascertaining any differences in the plumage of the two species.

"For this purpose I shall lay before the meeting two specimens selected from the type collection in the Canterbury Museum. The larger of these birds was obtained at Castle Hill, and the other on the Bealey—well known localities within this province—and both individuals proved on dissection to be females. The following is a comparative statement of their measurements:—

	H. novæ zealandiæ. inches.	*H. ferox.* inches.
Extreme length	19·5	16
Wing from flexure	12	10·5
Tail	8·5	7·75
Culmen (from cere to tip)	1	·8
Tarsus	2·5	2·2
Middle toe and claw	2·8	2·25
Hind toe and claw	1·85	1·35

"It will be seen from this that *Hieracidea novæ zealandiæ* is a considerably larger bird than *H. ferox*. It has a proportionately powerful bill, while its legs and feet are decidedly more robust. In the colours and markings of the plumage there is a general similarity between them; but on a close comparison of the two examples exhibited it will be seen that *H. novæ zealandiæ* has the bars on the upper surface far more distinct and numerous, besides being of a brighter rufous, the tail-coverts are more conspicuously marked, the bars on the tail are broader and whiter, and there is a larger amount of white on the throat, breast, and abdomen. In the present example of *H. ferox* the breast is much darker than in the other bird, the middle portion of each feather being occupied by a broad lanceolate mark of blackish brown, and there is less of the buff and rufous stains which impart so warm an effect to the breast of *H. novæ zealandiæ*. There are other minute points of difference, but these may be mere individual peculiarities. Enough has, however, been pointed out to show that the two species may be readily distinguished from each other; and this is the only point at issue.

"Of course the whole value of this evidence depends on the accuracy of the 'sexing' in each case. I think this, however, is placed beyond all doubt, for the larger bird was determined by Mr. J. D. Enys, who obtained it, while the smaller one was received at the Museum in the flesh, and was dissected by the taxidermist for the express purpose of ascertaining the sex. Mr. Fuller assures me that he was most careful in his examination, and that the specimen exhibited is to an absolute certainty a female."

HARPA FEROX.

(BUSH-HAWK.)

Falco brunneus *, Gould, P. Z. S. 1837, p. 139.
Falco ferox, Peale, U. S. Expl. Exp. p. 67 (1848).
Hieracidea brunnea, Gray, Ibis, 1862, p. 215.
Harpe brunneus, Gray, Hand-l. of B. i. p. 22 (1869).
Hieracidea brunnea, Buller, Birds of New Zealand, 1st ed. p. 6 (1873).
Harpa australis, Sharpe, Cat. Birds Brit. Mus. vol. i. p. 373 (1874).

Native names.

The same as those applied to the preceding species, but sometimes distinguished as Karewa rewa-tara. "Sparrow-Hawk" of the colonists.

♂ similis *H. novæ zealandiæ*, sed valdè minor: suprà magis cinereus: caudæ fasciis angustioribus et obscurioribus: subtùs pallidior, distinctiùs striatus et maculatus.

♀ maris staturam conspicuè superans.

Juv. a specie præcedente haud distinguendus, sed subtùs obscurior.

Adult male. Upper parts generally greyish black, darkest on the head and nape; shoulders, scapulars, and small wing-coverts narrowly barred with greyish white, the back and upper tail-coverts with small crescentic bands of rufous; throat yellowish white; ciliary bristles, ear-coverts, and the facial streak black; a line over each eye, and the sides of the neck, reddish brown, varied with fulvous and black; breast and sides fulvous, varied with reddish brown, and largely marked with black. On the breast each feather has a central dash of black; and on the sides these markings assume a triangular form, giving a spotted character to the surface of the plumage. The wing-feathers are marked, on their outer web, by narrow transverse bands of greyish white; and the tail-feathers, which are black with a purplish reflection, have a series of seven narrow white bars disunited at the shaft, and are tipped with rufous brown; axillars dark rufous brown, with a series of round white spots on each web; abdomen and vent pale fulvous; tibial plumes rufous, with black shaft-lines. Bill black, white at the base of lower mandible; irides very dark brown; cere, lores, and eyelids bright lemon-yellow, slightly tinged on the cere with green; legs and feet paler yellow and more tinged with green; claws black. Extreme length 16 inches; extent of wings 26·5; wing, from flexure, 9; tail 6·5; tarsus 2·25; middle toe and claw 2·3; hind toe and claw 1·3; bill, along the ridge ·85, along the edge of lower mandible 1.

Adult female. Differs from the male in its somewhat larger size and in the darker and richer colouring of its plumage; but in other respects the sexes are alike. Extreme length 16·75 inches; wing, from flexure, 10; tail 7·7; tarsus 2·4.

Young. The young of this species bears a general resemblance in its plumage to that of the preceding bird; but on a close comparison it will be observed that the brown of the underparts is darker, while the spotted markings on the sides are rather more conspicuous. The tibials, moreover, are of a brighter rufous, and are crossed with numerous arrow-shaped marks of brown.

Nestling. Covered with bluish-grey down; bill black; tarsi and toes leaden grey.

* Preoccupied by Bechstein, as mentioned on page 215.

Var. Individuals exhibit the usual variation in the details of their markings. A young example from the Bay of Islands, which I had the opportunity of examining, was peculiar in being largely marked with pale fawn-colour on the throat, breast, and abdomen, the lower part of the body being entirely of that colour.

A beautiful adult male specimen, from the Seventy-mile Bush, which came into my possession alive, differed slightly in its dimensions from that described above. Total length 15·5 inches; extent of wings 27·25; wing, from flexure, 9·5; tail 6·5. Another, from Wainuiomata, measured 16 inches in length; 28·5 in extent; tail 7.

Obs. This species closely resembles *Harpa novæ zealandiæ*, but is decidedly smaller, and has more slender legs and claws; otherwise it would perhaps be impossible to distinguish the two birds.

ALTHOUGH not so common as it formerly was, the Bush-Hawk is more frequently met with than its congener. The high wooded lands of the interior appear to constitute its favourite haunts; and on the southern mountain-ranges of the North Island, as well as in the subalpine woods of the Canterbury provincial district, I have found it comparatively abundant. The skin of a Hawk from Macquarie Island, sent to me by Mr. Bourne of the Otago Museum, proved on examination to be identical with this species.

It is a spirited little hunter, and subsists by the chase, its food consisting principally of mice and small birds. During the breeding-season it is more than usually bold and fearless, assailing with fury all intruders upon its nest or young. Some remarkable instances of its courage are mentioned by the late Sir J. von Haast in his interesting 'Journal of Explorations in the Nelson Province'*.

"One day," says this traveller, "walking along near the margin of the forest in Camp Valley, my hat was suddenly knocked off my head, and at the same time I heard a shrill cry. On looking up, I found it was one of these courageous little Sparrow-Hawks that had attacked me, and which, after sitting for a moment or two on a branch, again pounced on me; and, although I had a long compass-stick in my hand, with which I tried to knock it down, it repeated its attack several times.

We met with another instance of the courage of these birds in the Matakitaki Plains. A White Crane, of large size, standing in the water, was attacked by three of them at once; and they made frequent and well-concerted charges upon him from different quarters. It was admirable to behold the Kotuku (White Crane) with his head laid back, darting his pointed beak at his foes with the swiftness of an arrow, while they, with the utmost agility, avoided the spear of their strong adversary, whom at last they were fain to leave unmolested. Another day, in the same neighbourhood, a Cormorant (*Graculus varius*) passing near a tree on which two of these Sparrow-Hawks were sitting, was pounced upon by them and put to hasty flight with a shrill cry of terror, followed closely by his small but fierce foes; and all three were soon out of sight."

The ordinary flight of this Hawk is direct and rapid; but it may sometimes be seen soaring high in the air, with the wings almost motionless and the tail spread into a broad fan. On the wing it often utters a prolonged petulant scream. This is the signal for a general outcry among the small birds within hearing; and the Tui and Korimako will often rise in large flights and follow him into the air. But the little Hawk, heeding not their menaces, pursues his course, and the excitement among the feathered fraternity gradually subsides till all is quiet again. The appearance of an Owl in the daytime produces a similar commotion among the small birds of the forest; and I have often been guided to the hiding-place of the unfortunate "Morepork" by the clamour of the persecuting mob.

Besides the prolonged shrill note which is generally uttered on the wing, this species has also a low peevish cry, exactly like the squealing of a young pig, which is peculiar, I believe, to the breeding-season.

* Report of a Topographical and Geological Exploration of the Western Districts of the Nelson Province, New Zealand, undertaken by the Provincial Government. Nelson: 1861.

It is well known, as already stated, that birds are good natural barometers. The height to which they rise in the air renders them susceptible to the slightest change in the temperature of the atmosphere; and they are thus warned of approaching changes in the weather. Thus the continuous screaming of the Bush-Hawk is understood by the natives to be a sure indication of change; and they have a common saying "Ka tangi te Karearea &c." (If the Karearea screams in fine weather, 'twill soon rain; if in rainy weather, 'tis about to clear). Wilson, the American ornithologist, in treating of the Fish-Hawk (*Pandion haliaëtus*), states that when these birds are seen sailing high in air, with loud vociferations, "it is universally believed to prognosticate a change of weather, often a thunder-storm in a few hours. . . . . On the faith of the certainty of these signs, the experienced coaster wisely prepares for the expected storm, and is rarely mistaken." I have met with some remarkable instances of this unerring instinct in the species under consideration, and this, at times, when the glass gave no indication of a coming change.

The Bush-Hawk is generally met with on the outskirts of the woods or among the dead timber of native "wairengas," these localities being favourable for mice, on which it largely subsists. I once observed a young male of this species playing in the air with mice, after the manner of a cat; and the sight was as pretty as it was novel. When I first observed the bird, he was perched on the naked limb of a tree, apparently engaged in examining his quarry. Then mounting in the air with a mouse in each of his talons, and expanding his wings and tail to their full extent, he dropped first one mouse and then the other, and instantly darted after them, catching them in his talons before they reached the ground, then mounting high in the air again to renew the feat. Ultimately losing one of the mice, he discontinued his play, and, returning to the tree, killed and devoured the remaining one.

Formerly this spirited little Hawk was very common in the Hutt Valley and in the wooded suburbs of Wellington; now it is rarely, if ever, seen there. The last instance I know of was in April 1883, when a Sparrow-Hawk, after sailing inquisitively over the city and hovering for a time above the Colonial Museum—uttering all the time its shrill cry, as if in defiance of taxidermists and naturalists in general—eventually settled in the blue-gums in my garden, where it remained for half an hour; and then, after another rapid survey of the town, disappeared over the hills in the direction of Makara. A few years more, and the clarion cry of this fierce little hunter will be a thing of the past! Its appearance on this occasion was quite unusual, for my gardener, who is an old Wellington settler, declared he had not seen or heard the bird for more than ten years before.

I may mention that this species, unlike the generality of Hawks (so far as I am aware), may be attracted by an imitation of its cry. Riding along alone one fine autumn evening through the country at the northern end of Lake Taupo, on my way to Ohinemutu, I saw what appeared to be a Bush-Hawk come out of the woods at some distance and descend into an old or deserted Maori garden. By way of experiment I imitated the clamorous cry of this bird when on the wing; and in a few minutes the Hawk (a fine young male) came sailing up to me and performed several circuits in the air immediately overhead, and then took up his station on the dry limb of a tree close by the road, where he remained till I was out of sight.

The natives state that this little Hawk usually builds its nest in a bunch of puwharawhara, often at a great elevation from the ground, forming it rudely of loose materials; that it lays generally two, but sometimes three eggs; and that the young birds remain on the tree for several days after quitting the nest. The puwharawhara (*Astelia cunninghamii*) is a parasitical plant, with short, thickly set flag-leaves, radiating upwards from a clump of roots by which it adheres firmly to the parent tree. These plants, which often attain a circumference of many feet, are very common on the forks and naked branches of aged or withered trees on the outskirts of the forest, a single tree sometimes supporting twenty or more of them. A better situation for a Hawk's nest than the centre of one of these plants could hardly be selected, combining as it does the requisites of warmth, security, and shelter; and the Bush-Hawk seems to be instinctively aware of this. Some years ago I was informed that a pair of these birds had bred for several successive seasons in a nest placed as described, and situated in the high fork of a dead kahikatea tree near the Horowhenua Lake. Having waited for the breeding-season, I offered the natives a half-sovereign each for the eggs; but, although excellent climbers, they failed in all their attempts to reach the nest. They afterwards observed the Hawks carrying mice, lizards, and small birds to their young; and the latter, on quitting the nest, were shot and destroyed. When I last visited the spot the old kahikatea was still standing, and the bunch of withered *Astelia*, which had cradled several successive broods, was still clinging to the tree; but the persecuted Hawks had quitted their exposed eyrie for some more secure retreat.

In the summer, however, of 1867, during a visit to Taupo, I was fortunate enough to find the nest of this species. We had fixed our bivouac for the night on the banks of the Waitangi Creek, only a few miles from the base of the grand snow-capped Ruapehu. Our native companion soon detected the old Hawks carrying prey to their young, and on the following morning he discovered the nest *. It was situated on the ground, under cover of a block of trachyte, which cropped out of the side of the hill. There had been no attempt to form a proper nest; but the ground was covered with the feathers of birds (almost entirely those of the Ground-Lark) on which the young Hawks had been fed. The latter were three in number, of different sizes, the largest being apparently three weeks old, and the smallest scarcely a fortnight. They were extremely savage, striking vigorously with their sharp talons and uttering a peculiar scream. While we were engaged in securing them in a basket the old birds were flying to and fro, occasionally dashing up to within a few feet of us, and then off again at a sharp angle, alighting at intervals, for a few moments only, on the rugged points of rock above us, but never uttering a sound. They were in perfect plumage; and when they occasionally poised their bodies overhead, with outspread wings and tail, they presented a very beautiful appearance. During our journey of forty miles through the bush, the gun supplied the young Hawks with a sufficiency of food; but they were very voracious, two large Pigeons per diem being scarcely enough to appease their joint appetites. Fifty miles more by canoe, and about forty on horseback, brought the captives to their destination, when they were placed in a compartment of the aviary. They continued to be very vicious, punishing each other severely with their claws. The youngest one was an object of constant persecution, and ultimately succumbed to a broken back. A small tame Sea-Gull that had unwittingly wandered into the aviary, through an open doorway, was instantly pounced on, although the young Hawks, in their unfledged condition, could only move by hopping along the ground. In about three weeks these birds (which proved to be male and female) had fully assumed the dark plumage; and for about two months after they were very clamorous, especially during wet or gloomy weather. By degrees they became less noisy, till at length they were perfectly silent and moody, never uttering a sound for weeks together, with the exception of a peculiar squeal when they were fighting. A more quarrelsome couple never existed. The female, being the larger and stronger bird, generally came off best, leaving the male severely punished about the head. At the end of six months the climax was reached by her actually killing and devouring her mate. I found the aviary strewn with feathers, and the skeleton of the poor victim picked clean! The surviving bird underwent a partial moult in the month of September following, and the plumage began to assume a spotted character. The legs also became slightly tinged with yellow. By the beginning of March in the following year she had acquired the full adult plumage, except that the throat and spots on the sides were not so light as in more mature examples. The legs had changed to a pale greenish yellow, and the irides from lustrous black to a dark brown colour—the cere retaining its pale blue tint, but with indications of a change to yellow. After two months' absence I

* Captain Mair, writing to me in February 1880, says:—"The Sparrow-Hawks still build at the cliff on the Waitangi Stream where we obtained our young birds in 1867."

again saw the bird, and noticed that the lores were becoming tinged with yellow, while the colour of the legs had deepened. Unfortunately, at this stage she was found dead on the floor of the aviary; and on dissection I found in the cavity of the back an amazing number of parasitical worms, many of them measuring from six to eight inches in length. A wild specimen, which I afterwards examined, was similarly afflicted.

The result of my observations is, that the Bush-Hawk attains the mature livery during the second year, the plumage being liable to some slight variations as the bird gets older. The irides had undergone very little perceptible change at the time of the bird's death, but the eyes were large and somewhat sparkling.

This bird, a stranger to liberty from the very nest, had become quite attached to its aviary. It never attempted to escape when the door was accidentally left open; and on one occasion when it did get out it remained perched on the dome of its house, and voluntarily re-entered it. It partook readily of all kinds of meat, cooked or raw, although preferring the latter. Beef, pork, or mutton were alike acceptable; but a preference was always shown for birds. On a live bird being offered to it, the Hawk would eye its quarry intently for a short time and then make a sudden swoop upon it, seizing with the talons of one or both feet, according to the size and strength of the object. It would then proceed cautiously to destroy life by crushing the head of its victim in its powerful beak, only relaxing its hold when life was quite extinct. While thus employed, its eyes were full of animation, and its whole body quivered with excitement.

A pair of these birds bred for two successive seasons on a rocky crag at Niho-o-te-kiore. They guarded their nest with great vigilance, fiercely attacking all intruders; and on both occasions brought up their brood in safety.

The description of the male is taken from a fine specimen shot in the Karori Hills, near Wellington, in 1869, and of which I sent, at the time, a descriptive notice to the Linnean Society. Its much smaller size led me to suppose that it was distinct from *Harpa novæ zealandiæ*; and it was not then known that Mr. Gould's *H. brunnea* was founded on an immature example. That such was really the case is sufficiently proved by the account given in the foregoing pages, and previously recorded in the Transactions of the New-Zealand Institute (1868, vol. i. p. 106).

The eggs resemble those of *H. novæ zealandiæ*, but are somewhat smaller and lighter in colour. There are three examples in the Canterbury Museum, differing in the details of their colouring: but they may be defined as yellowish brown, stained and mottled with reddish brown, and having a rather soiled appearance. In one of them the blotched character is more apparent at the smaller end; in another it is equally dispersed, while in the third the dark brown markings present a smudgy character over the whole surface. They measure 1·9 inch in length by 1·45 in breadth.

In the same fine collection there is a beautiful specimen of the Bush-Hawk's egg from the Chatham Islands. It is of a rich or warm reddish brown, freckled and slightly smudged with darker brown, presenting a close resemblance to the Merlin's egg, broadly ovoido-conical in form, and measuring 1·95 inch by 1·5 inch. There is another egg of the same species, from Paringa River (South Westland), differing very perceptibly in being of a dull cream-colour, freckled and stained all over with brown. It is of the same size as the Chatham-Islands specimen, but is slightly more oval in form.

A specimen brought by Mr. Reischek from Martin's Bay, in the South Island, measures 1·8 inch in length by 1·5 in breadth, being of a regular ovoid form; the whole surface is pale reddish brown, blackish brown, and cream-colour mixed together in an irregular way, being decidedly darker at the larger end, and the light markings at the smaller end having the appearance of abrasions or scratches on the surface. The nest from which it was taken was placed in the leafy crown of a high forest tree.

COTURNIX NOVÆ ZEALANDIÆ.

(NEW-ZEALAND QUAIL.)

Coturnix novæ zealandiæ, Quoy et Gaim. Voy. de l'Astr. (Zool.) i. p. 242, pl. 24. fig. 1 (1830).

Native names.—Koreke and Kokoreke.

♂ *ad.* supra rufescenti-brunneus: dorsi plumis medialiter fulvo striatis, utrinque nigro marginatis, plumis quibusdam nigro irregulariter maculatis aut vermiculatis: pileo saturatiùs brunneo, supercilio et lineâ verticali fulvescentibus: collo postico et laterali fulvescente: facie laterali et gutture toto castaneis, genis et regione auriculari paullò nigricante variis: tectricibus alarum minimis et medianis dorso concoloribus, his magis fulvescentioribus: remigibus nigricantibus, secundariis angustè fulvo vermiculatis: rectricibus nigris, fulvo transfasciatis, scapis etiam rufescenti-fulvis: subtùs albicans, pectore superiore et abdomine imo fulvescentibus: pectoris plumis nigro marmoratis, fasciâ latâ nigrâ transfasciatis, abdominis plumis fasciis sagittiformibus nigris notatis: hypochondriis rufescenti-fulvis nigro transversim irregulariter fasciatis, et conspicuè medialiter albo striatis: crisso et subcaudalibus nigro notatis et fasciatis: subalaribus albidis, angustè brunneo marginatis, margine alari brunneo vario: rostro nigro, versùs apicem dilutiore: pedibus pallidè carneis: iride pallidè brunneâ.

♀ *ad.* mari similis, sed paullò major, ubique dilutior: facie castaneâ et pectore nigro absentibus: facie laterali gutturéque fulvescenti-albis, illâ brunneo maculatâ: corpore reliquo subtùs rufescente, abdomine medio albicante, plumis omnibus nigro marginatis, pectoris plumis et hypochondriis medialiter albo lineatis.

♂ *juv.* similis feminæ adultæ, sed facie laterali et gutture pallidè rufescentibus: corporis subtùs plumis latiùs nigro marginatis.

♀ *juv.* similis feminæ adultæ, sed corporis subtùs plumis magis distinctè nigro marginatis.

Adult male. Crown of the head and nape dark brown edged with paler, a series of feathers down the centre and on the sides marked in the middle with yellowish white; shoulders, mantle, and all the upper surface rufous brown, beautifully varied with black, and marked with numerous lanceolate stripes of white. On closer examination it will be found that this effect is produced by each feather having a broad lanceolate mark of white down the shaft, bordered on each side with black, dark brown on the webs, fancifully rayed, or banded transversely, and largely tipped with rufous brown. Lores, line over the eyes, sides of head, and throat rufous, with a lunar mark from the ear-coverts on each side, and an anterior edging or border of black; lower part of the neck mottled or obscurely spotted with black and white, the former preponderating; examined separately, however, each feather is black crossed by irregular bands and largely tipped with white; sides and long plumage overlapping the thighs rufous brown, each feather margined and marked down the centre with white, and handsomely streaked and barred on the webs with brownish black; abdomen fulvous white, the under tail-coverts barred with black; primaries and outer secondaries dark brown, the latter rayed on their outer webs with zigzag lines of paler brown; inner secondaries and all the wing-coverts, as well as the tail-feathers, greyish brown, varied with pale rufous, each feather with a narrow shaft-line of white. Irides light hazel; bill black, paler at the tip; tarsi and toes pale flesh-brown. Total length 8·5 inches; extent of wings 14; wing, from flexure, 4·25; tail 1·5; bill, along the ridge ·5, along the edge of lower mandible ·6; tarsus 1; middle toe and claw 1·25.

Adult female. In the female there is no rufous colour on the face or throat; the upper surface is light ferruginous brown mixed with fulvous, and handsomely varied with black; the lanceolate stripes are yellowish white, changing to fulvous on the longer secondaries and on the lower part of the back; the throat, fore neck, sides, and flanks ferruginous brown, and the breast fulvous white, all more or less varied with black; on the neck and breast each feather is marked near the tip with a broad crescent, and on the webs with irregular spots of brownish black; the feathers covering the sides, and the long feathers overlapping the thighs, have a broad stripe of white down the shaft and are streaked and marbled on both webs with black; the abdomen is white, the sides fulvous, and the under tail-coverts dark fulvous varied with black. The female is, moreover, slightly larger than the male in all its proportions.

Young male. In the young male the prevailing colour of the upper surface more nearly approaches that of the adult female. The rufous colouring on the cheeks and throat is very pale, and the lunate marks are less distinct than in the adult. The plumage of the underparts is largely washed with fulvous, and the dark crescents are broader and more conspicuous.

Young female. The only perceptible difference in the markings of the young female is that the dark crescents on the under surface are better defined and less blotched than in the adult bird. In my old collection (now in the Colonial Museum) there are two young females from the same nest, in one of which the prevailing tint of the plumage resembles that of the adult female, while in the other it approaches very near to that of the adult male.

Very young state. Crown of the head light fulvous varied with dark brown; ear-spots black; back and upper surface of wings yellowish brown, with dull black markings, each feather with a lanceolate stripe of fulvous white down the centre; throat and fore neck buffy white; breast and underparts pale buff, each feather marked near the tip with two converging elongate spots of a dull black colour. Bill, tarsi, and toes pale brown.

Obs. A beautiful male specimen obtained many years ago at Whangarei, in the North Island, and presented to me by Major Mair, differs from all my South-Island examples in having the whole of the plumage darker, the breast being almost entirely brownish black, relieved only by a few touches of fulvous white; the rufous colour on the face and throat is brighter, the lanceolate markings on the upper surface are very distinct, and the abdomen is fulvous.

This handsome species—the only indigenous representative in New Zealand of the order Gallinæ—was "on the verge of extinction" when I published my former edition. It is probably now extinct, for no specimen has been heard of for at least twelve years. In the early days of the colony it was excessively abundant in all the open country, and especially on the grass-covered downs of the South Island. The first settlers, who carried with them from the old country their traditional love of sport, enjoyed some excellent Quail-shooting for several years; and it is matter of local history that Sir D. Monro and Major Richmond, in 1848, shot as many as forty-three brace in the course of a single day within a few miles of what is now the city of Nelson; while a Canterbury writer has recorded that "in the early days, on the plains near Selwyn, a bag of twenty brace of Quail was not looked upon as extraordinary sport for a day's shooting." But, partly owing to the introduction of dogs, cats, and rats, and partly to the prevalence of the so-called "bush-fires" or burning of the runs (a necessary incident of sheep-farming in a new country), the Quail rapidly disappeared, and if not so already, it will ere long be numbered among the many extinct forms of animal life in New Zealand. Its place, however, has been more than adequately supplied by several introduced species, all of which appear to thrive well and multiply in their new home. Among these we may enumerate the following as being now permanently established in the country, viz. the common English Pheasant (*Phasianus colchicus*), the Chinese Pheasant (*P. torquatus*), the Partridge (*Perdix cinerea*), the Californian Quail (*Ortyx californicus*), and the Australian Quail (*Coturnix pectoralis*)*. The last-mentioned bird closely

* To these may now be added the Swamp-Quail (*Synoicus australis*), which has rapidly spread itself over the North Island, being plentiful even in the Taupo country. Three specimens of this bird (obtained at Tauranga) were sent to me by the Hon. Dr. Pollen, the then Premier, and I afterwards handed them over to the Colonial Museum with the following note:—"Two of

resembles the subject of this notice both in appearance and habits; and it will be curious to observe whether it will succeed in resisting for any length of time those physical conditions which have proved so fatal to the indigenous species.

According to the Maoris, even in the North Island it was formerly very abundant, certain grassy plains, like the Murimotu, in the Taupo district, being noted for them. Even at the present day, in the investigation of title in the Native Lands Court, the older generation of Maoris, when giving their evidence, often refer to the Quail preserves of former times in support of the tribal title *.

Sir Edward Stafford related to me the following circumstance in illustration of the suddenness with which the Quail disappeared from localities where it had once been plentiful:—On one occasion about the year 1848, accompanied by two other sportsmen, he went out to his own estate, about thirty miles from Nelson, for a day's Quail-shooting; and in the course of a few hours the party bagged 29½ brace. In the hope of preserving the game, he prohibited any shooting over this ground during the following year; but in the ensuing season, when he naturally looked for some good sport, there was not a single Quail to be found!

Sir Frederick Weld (the present Governor of the Straits Settlements), about the same period, tried a similar experiment on his property at Stonyhurst, but with no better success. Finding the Quails very abundant in a particular locality, and being anxious to preserve them, he protected a suitable cover of about 2000 acres, never allowing the sheep upon it, nor permitting fires to overrun it. When this protection was first extended, there were almost incredible numbers of Quails on the land; but in less than a year they had all disappeared. In 1851 Dr. Shortland found it very numerous on the open downs of Waikouaiti; and as late as 1861, as we learn from Haast's 'Journal of Exploration in the Nelson Province,' it was "still very abundant on the grassy plains of the interior, rising close to the feet of the traveller at almost every step."

A specimen was shot by Major Mair at Whangarei in 1860; Sir James Hector reports the taking of a pair at Mangawhai in 1866; Captain Mair saw one at Maketu in 1867; and the Hon. J. C. Richmond met with some in the Taranaki district in the months of November and December 1869. These are, I believe, the last recorded instances of its occurrence in the North Island. In the more retired portions of the South Island it was occasionally to be found down to 1875; but it had before that entirely disappeared from the settled country on the eastern side of the Alps.

In the autumn of 1860 I met with a bevy of nine on a dry grassy ridge in the midst of some shallow swamps about two miles from Kaiapoi (in the provincial district of Canterbury); and having with me a good pointer, I fortunately succeeded in bagging the whole of them. They afforded capital shooting, rising quickly and, after a low rapid flight of fifty yards or more in a direct line, dropping suddenly into the grass again. The stomachs of those I opened contained green blades of grass and a few bruised seeds, as well as some small fragments of quartz. The bevy consisted of an adult male and female, with seven birds of the first year; and as we may infer from the circumstances under which they were found that they comprised a single family, we have some evidence that this species is not less prolific than the other members of the extensive tribe to which it belongs.

Mr. Potts, writing of the bird before it had become rare, says:—"They often give utterance to a low purring sound that one might suppose to proceed from an insect rather than from a bird. The

these are in the normal plumage of the ♂ and ♀; the other is a remarkable instance of melanism. The entire plumage is a brownish slate-colour, paler on the underparts; on the crown and nape there are obsolete shaft-lines, and the whole of the upper surface is obscurely varied and mottled with blackish brown, washed with chestnut-brown on the wings. It is slightly smaller than the other specimens and proved on dissection to be a male." (Trans. N.-Z. Inst. vol. xiv. p. 534.)

* Extract from Aperahama Te Kume's evidence in the Tokorua case, at Cambridge, June 1880:—"Wahineiti died at Tauranga. He said to Ngatiki, 'Don't take me to Maungatautari but to Tokoroa, that the maku of my land may grow over me, and that my body may drink the dews of Tokoroa.' This occasioned the name of Horohau; hence, too, the proverb 'Nga wi o Tokoroa.' These plains were famous for the abundance of Quail."

call is indulged in most frequently during moist or wet weather; it sounds something like 'twit, twit, twit, twee-twit,' repeated several times in quick succession. In very stormy gusty weather these birds appear dull and silent, secreting themselves among thick tussocks. When flushed, they do not rise perpendicularly, but still very straight for a few feet from the ground. In confinement they are fond of picking about amongst sand, and thrive well on soaked bread, grain of various kinds, and the larvæ of insects. The male is not an attentive mate at feeding-time; and where several are kept in the same enclosure, constant little bickerings take place without actual hostilities being indulged in. The eggs require twenty-one days' incubation; and the chicks are most active directly they emerge from the shell. They grow very rapidly; and at about four months old the young cannot very readily be distinguished from adult birds, either by contrast of size or plumage."

It may be interesting to mention, as showing the value attaching to extinct or rapidly expiring forms, that a skin of this bird (and that, too, a female) sent from the Canterbury Museum to Italy fetched as much as £75. My own collection contains an adult male and female (from the North and South Islands respectively), a young male of the first year, and another in the "very young state" described above. The last-named bird was one of a clutch of four, and I am indebted for this, among other rare specimens, to my lamented friend Sir Julius von Haast, the announcement of whose death in New Zealand reached me whilst these pages were passing through the press.

There is a specimen of the egg of this species (probably the only one in Europe) in Professor Newton's fine collection at Cambridge; and there are five examples in the Canterbury Museum which exhibit a slight variation in form and a considerable difference in colour. Two of them (presumably from the same nest) are of a regular oval form and of equal size, measuring 1·3 inch in length by 1 in breadth; these are of a pale yellowish-brown or buff colour, thickly marked with umber, the dark colour often preponderating and having the appearance of daubs or smudges on the outer surface of the shell. Two others (also exactly alike) are of a slightly larger size and of a thicker or broader form; these are of a dull cream-colour, sprinkled and minutely dotted all over with blackish brown. In one of them the spots are confluent at the larger end, forming a greyish-brown patch nearly half an inch in diameter; and in both the more conspicuous spots have a light or faded centre. The fifth egg is smaller and more rounded than any of the rest; it is of a yellowish-white colour, covered all over, but more thickly at the ends, with small smudgy spots of umber; and it has likewise a more glossy appearance than the others. On comparing the eggs of this species with those of *Coturnix pectoralis*, of Australia, there is a manifest difference, those of the latter bird being, as a rule, creamy white, with very obscure surface-spots.

After the above article had been sent to press, I received from the Colony the welcome intelligence that the last refuge of this well-nigh extinct species had lately been discovered. During the recent expedition of the Government steamboat 'Stella' to the Kermadec Islands, for the purpose of annexing them to New Zealand, Captain Fairchild, on his return voyage, landed on the Three Kings, a group of small islands situated about 32 miles W.N.W. of Cape Maria Van Diemen, the largest of them being only 1¾ miles long by ¾ of a mile in width, and rising about 900 feet above the sea-level. There is no "bush" on this island, but the surface is covered with stunted *Leptospermum*, fern, flax, and sedges, with here and there a grassy flat. Notwithstanding the scantiness of the vegetation, no less than five plants were discovered entirely new to the New-Zealand flora, and these have since been described and named by Mr. Cheeseman, F.L.S., not the least interesting one being *Pittosporum fairchildi*. But, from an ornithologist's point of view, the most important discovery made was the existence there of several bevies of New-Zealand Quail, which were comparatively tame and fearless; and the explorers being fortunately without firearms they were left unharmed.

It is to be earnestly hoped that prompt steps will be taken by the Government to save and perpetuate this last remnant of an expiring race!

CARPOPHAGA NOVÆ ZEALANDIÆ.

(NEW-ZEALAND PIGEON.)

New-Zealand Pigeon, Lath. Gen. Syn. ii. pt. 2, p. 640 (1783).
Columba novæ seelandiæ, Gm. Syst. Nat. i. p. 773 (1788).
Columba zealandica, Lath. Ind. Orn. ii. p. 603 (1790).
Columba spadicea, Less. Voy. Coq. i. p. 710 (1826).
Columba spadicea leucophæa, Hombr. & Jacq. Ann. Sci. Nat. xvi. p. 319 (1841).
Carpophaga novæ seelandiæ, Gray, in Dieff. Trav. ii., App. p. 194 (1843).
Columba argetræa, Forst. Descr. Anim. p. 80 (1844).
Hemiphaga novæ-zealandiæ, Bonap. C. R. xxxix. p. 1077 (1854).

Native names.—Kuku, Kukupa, and Kereru.

Ad. dorso æneo-ferrugineo: pileo antico lætè metallicè viridi, posticè cum nuchâ et colli lateribus magis æneo nitentibus, his cyanescente tinctis: dorso postico et uropygio nitidè viridibus cyanescente lavatis, supracaudalibus olivascenti-viridibus æneo lavatis: tectricibus alarum minoribus et majoribus dorso proximis æneo-ferrugineis dorso concoloribus, majoribus et medianis exterioribus nitidè viridibus: remigibus nigris suprà cyanescenti-viridi nitentibus, secundariis æneo lavatis: caudâ nigrâ suprà saturatè viridi lavatâ, subtùs nigricante, pennis omnibus versùs apicem cinerascentibus: facie laterali cum gutture toto et pectore superiore lætissimè metallicè viridibus: corpore reliquo subtùs purè albo: subcaudalibus cinerascentibus: subalaribus cinereis: rostro coccineo, versùs apicem flavicante: pedibus coccineis: iride coccineâ, annulo ophthalmico pallidè rubro.

Adult male. Head, neck, and fore part of breast shining gold-green, changing according to the angle of view; nape, shoulders, and upper surface of wings, as far as the carpal joint, coppery purple, with bright metallic reflections where this colour blends with the green of the surrounding parts; back and rump greyish green, with dull metallic reflections; quills and their coverts bronzy green, with the inner webs dusky, the secondaries tinged with coppery purple; an obscure band of grey (more conspicuous in the young bird) crossing the outer webs of the primaries, being widest on the fifth and sixth quills; tail-feathers black, with blue reflections on their edges, and terminally margined with brown; under surface of tail-feathers silvery grey towards the base, especially on the outer ones, blackish in their apical portion, with lighter tips; their upper coverts dull shining green; underparts from the breast downwards pure white, the lower tail-coverts tinged with yellow; lining of wings delicate ash-grey. The line of demarcation between the lustrous green and the white is well defined, crossing the breast with an easy curve and terminating immediately above the insertion of the wings, so that when the bird is at rest a narrow margin of white appears over the bend of each wing. Irides and feet carmine-red; soles yellow and covered with small flattened papillæ; claws black; bill carmine-red in its basal half, changing to yellow towards the tip; eyelids pale red, with a reticulate margin, imparting to the brilliantly coloured eyes a very soft expression. Total length 21 inches; extent of wing 32; wing, from flexure, 10·75; tail 8·5; bill, along the ridge ·75, along the edge of lower mandible 1·4; middle toe and claw 2·25; the lateral toes equal, being ·75 shorter; hind toe 1·4.

Female. Hardly distinguishable from the male, but with the metallic tints of the plumage somewhat duller.

Young. The bronzy plumage of the neck and breast has much less iridescence than in the adult, but the hind neck and smaller wing-coverts are of a rich metallic purple shot with blue and changing in different lights;

the throat is greyish, each feather with a terminal margin of fulvous; the white of the underparts is washed with cream which deepens to fulvous, or sometimes pale rufous, on the flanks and under tail-coverts; the inner lining of the wings is uniform dark grey, and along the carpal flexure there are a few touches of fulvous. The size is appreciably less than that of the fully-grown bird; the bill and irides are of a less decided or lighter colour; the feet instead of being carmine are of a bright coral-red, and the soles pale brown instead of yellow.

Fledgling. A specimen in my collection has the chin and upper part of throat greyish brown, the feathers minutely tipped with whitish grey; the white plumage of the underparts washed with cream-yellow; the under tail-coverts stained with pale rufous; nape and hind neck shaded with coppery and vinous brown; lining of wings clear ash-grey.

Nestling. A very young chick, which I examined as a dried specimen, was covered sparingly with yellowish-white down, looking very much like flax tow, and perfectly bare on the abdomen.

Obs. Before arriving at full maturity the plumage is subject to slight variations. It is not unusual to find the under tail-coverts pale rufous and the white plumage of the underparts clouded or marked with grey.

Varieties. There is a lovely albino in the Colonial Museum, from the Wairarapa, the entire plumage being of a pure milk-white, the small wing-coverts alone presenting a slight tinge of yellowish brown; bill and feet carmine-red. Partial albinos, or light-coloured varieties, are occasionally met with. A specimen presented to me by the late Mr. Edward Hardcastle, R.M., has the head, neck, fore part of the breast, and all the upper parts pale yellowish brown, more or less glossed with purple; the wing-coverts and scapulars stained towards the tips with coppery brown; the quills and tail-feathers uniform pale yellowish brown, tinged with vinous, the tips of the latter paler. In another specimen, shot at Maungakaramea, near Whangarei, and for which I was indebted to the late Mr. Henry Mair, the neck, shoulders, back, upper tail-coverts, scapulars, and wing-coverts present scattered feathers of pure white, imparting to the plumage of the upper parts a spotted appearance. Both of these specimens are now in the Colonial Museum. A third example, in the possession of Mr. William Luxford, of Wellington, has the head, neck, shoulders, and upper wing-coverts coppery brown, and the rest of the upper parts pale grey; the primary quills tinged with brown at the tips; the underparts of the body white. Another in the Colonial Museum has the head, neck, breast, and upper parts generally pale vinous brown, without any gloss, and becoming darker on the inner webs of the quills and tail-feathers; the shoulders and smaller wing-coverts dark velvet-brown, fading off on the outer feathers, this dark patch upon the lighter plumage forming a very conspicuous feature; bill and feet almost white. Another in my own collection, presented by Mr. W. Marshall, who obtained it in the Upper Rangitikei district, is very similar, but there is a larger admixture of brown in the general plumage, and the velvet-brown extends over the entire mantle but is relieved by a light feather here and there; the quills on their inner webs and all the tail-feathers except the middle one are rufous brown with pale tips. In each wing two of the quills are entirely dark.

There is a very curious example in the Auckland Museum (marked ♀) and obtained from the Waikato in June 1884:—The head, neck, breast, upper surface of wings and tail pale vinous brown, relieved by touches of creamy white; the hind neck shaded with darker brown with a very faint gloss; shoulders, mantle, and smaller wing-coverts bright coppery brown, shaded with ashy brown, the central part of each feather being of that colour; this darker colour prevails on the scapulars, which are entirely dark brown with paler tips, delicately glossed with purple, and with whitish shaft-lines; primaries and tail-feathers dark vinous brown on their inner vanes, and paler brown shading off into creamy white on their outer; the larger wing-coverts pale vinous brown with whitish margins. Underparts pure white. Irides, bill, and feet as in the normal bird. The distinguishing feature is the bright coppery brown mantle, which is very conspicuous. The outer tail-feathers are much abraded and worn.

Mr. Cheeseman showed me a very finely coloured specimen shot by himself in the neighbourhood of Auckland. In this bird all the colours were highly iridescent, even the tail-feathers having a fine edging of metallic blue *.

* Mr. T. W. Kirk gives the following description of another remarkable variety shot by Mr. Greville in the Seventy-mile

Differing again from all the foregoing is a partial albino obtained at Ngunguru and sent to me by Capt. Mair. In this bird the shoulders, back, rump, and upper tail-coverts have a rich appearance, the white predominating. Some of the wing-feathers and their coverts are wholly white, with bronzed edges and clouded with grey, while others again present the normal coloration. The distribution of colours, however, is quite irregular, the white largely predominating in the right wing.

In the Natural History Museum of the Jardin des Plantes there is a curious variety, from the collection of MM. Hombron and Jacquinot, marked "Akaroa, ♀." In place of the bronzy green the general plumage is dark cinnamon-brown, shaded with vinous brown on the smaller wing-coverts; underparts white with a slight wash of cinnamon which is darker on the under tail-coverts. The wing-feathers are uniform cinnamon-brown; so are the tail-feathers, but these are darker in their central portion and have whitish tips. Bill and feet yellow. This is, I believe, the type of *Carpophaga spadicea leucophæa* of those naturalists. This was suggested by me in my former edition (p. 158), but I had not at that time examined and identified the bird, as I have since done.

Remarks. The head is small, the neck of moderate length, and the body full, with a prominent and rounded breast; the primaries graduate upwards to the third and fourth, which are generally of equal length; the fifth is slightly shorter, and the rest are rapidly diminished; the secondaries are broad and rounded; the tail-feathers large and even, forming together an ample fan when the tail is expanded. The plumage is thick and compact, and each feather is furnished with a dense undergrowth of downy plumules of extreme fineness, which branch laterally from both sides of the shaft. This peculiarity is most fully developed in the long plumage of the back, where only the tips of the feathers assume the surface character. By this wise provision of nature, the bird is perfectly clothed in a thick undercovering of soft down, and much warmth imparted to the body. The tarsus is completely concealed. On moving the lowest feathers, however, two broad scutella are exposed; on the middle toe there are 11 scutella, on the outer toe 10, on the inner toe 7, and on the hind toe 4.

On some specimens, particularly young birds, a fine white powder, like pulverized chalk, is observable on the feathers of the head and hind neck. I noticed this in a very pronounced degree, and extending to the back, in a bird which I shot on the Pirongia Range in the month of November—so much so as to leave a distinct chalk-mark on any dark object brought into contact with it. It is evidently an emanation from the skin, and doubtless serves some useful purpose in the natural economy of the bird.

Of the large and well-defined group of fruit-eating Pigeons found dispersed over the sea-girt lands of the southern hemisphere, the single species inhabiting New Zealand is undoubtedly one of the finest both for size and brilliancy of plumage.

In its native country it is less esteemed for its beauty than for its value as an article of food; and to both Maoris and colonists, in every part of New Zealand, pigeon-shooting, at certain seasons of the year, affords agreeable recreation, while to many it is a source of profitable employment. Owing to the loud beating of its wings in its laboured flight it is readily found, even in the thickest part of the bush, and being naturally a stupid bird it is very easily shot; so that in a favourable locality it is not an unusual thing for a sportsman single-handed to bag fifty or more in the course of a morning. In some districts the slaughter has been so great during a productive season that the Pigeons have never afterwards recovered their numbers; but in most of our woods, notwithstanding this persistent persecution, they reappear in each successive year in undiminished plenty. The

Bush and presented by him to the Colonial Museum:—"Head, neck, and fore part of breast, which in ordinary specimens are shining gold-green, are here thickly strewn with white feathers. On the fore neck the coppery purple band is replaced by a large patch of pure white feathers. The nape, shoulder, and upper surface of wings are also thickly strewn with white feathers; back and uropygium have likewise very white patches, but getting fewer towards the latter portion. The bright green of the breast is succeeded by a band of pale grey, which fades as it approaches the abdomen. Quills and tail-feathers normal colour. In no instance is a parti-coloured feather to be found, the white feathers being pure; even the shafts are destitute of colour. Eyes pink, with carmine-red as is usual; feet paler than customary; the soles flesh-colour rather than yellow; bill normal colour." (Trans. N.-Z. Inst. vol. xviii. p. 120.)

"season" is indicated by the ripening of certain berries on which this species subsists; and the abundance of the birds is regulated to a great extent by that of the food-supply, which is more or less variable. A sporting gentleman pointed out to me a taraire grove at Ramarama, near Auckland, where in 1860 he found the Pigeons so numerous that he shot eighty-five in the course of two mornings; but in the following year, owing to the partial failure of the taraire berry, there was hardly one to be seen there.

It is to be met with on the Little Barrier, and more plentifully on the Hen and Chickens, just before the large-leaved parapara (*Pisonia umbellifera*) has ripened its fruit. This bird seems to be fond of the green berries; and it is accordingly very difficult to obtain ripe seed of this valuable tree.

It is said to have made its first appearance at the Chatham Islands about the year 1855. Be that as it may, it is now comparatively plentiful on all the islands of the group, and has been found breeding on Mangare.

In the spring and early summer it is generally very lean and unfit for the table; but as autumn advances and its favourite berries ripen, it rapidly improves in condition, till it becomes extremely fat. It is esteemed most by epicures when feeding on the mast of the miro, which imparts a peculiar richness to the flesh. In January the berries of the kohutuhutu, poroporo, kaiwiria, puriri, mangiao, and tupakihi constitute its ordinary bill of fare. From February to April their place is supplied by those of the tawa, matai, kahikatea, mapau, titoki, and maire. It is worth remarking that in localities where it happens to be feeding exclusively on the pulpy fruit of the kahikatea, it is not only in very poor condition, but acquires a disagreeable flavour from the turpentine contained in the seeds. Towards the close of this period also, the ti-palm, which comes into full bearing only at intervals of three or four years, occasionally supplies this bird with an abundant feast. These tropical-looking palms often form extensive groves in the open country or in swampy situations; and when the Pigeons resort to them they are speared and snared in great numbers by the Maoris, an expert hand sometimes taking as many as sixty in a single day. In May and June it feeds chiefly on the miro and pate, when it reaches its prime and is much sought after. From July to September it lives almost entirely on taraire in the north, and on hinau, koeka, ramarama, and other smaller berries in the south. During the months of October, November, and December it is compelled to subsist in a great measure upon the green leaves of the kowhai (*Sophora tetraptera*), whauhi, and of several creeping plants. It also feeds on the tender shoots of the puwha, a kind of sow-thistle; and the flesh then partakes of the bitterness of that plant. When the bird is feeding wholly on the dark berries of the whao the colour of its flesh is said to become affected by that of the food.

The Pigeon-season, however, is to some extent contingent on locality; for example, in the spring of 1863 I found these birds in the Upper Manawatu living on kowhai-leaves, and so lean in body as to be scarcely worth powder and shot, while in the low timbered flats under the ranges, where they were feeding on the ripe berries of the karaka (*Corynocarpus lævigata*), they were in excellent condition.

At the Rev. Mr. Chapman's old mission station at Te Ngae (Rotorua), formed in 1835, and now much out of repair and overgrown, there are several hundred acres of sweet-briars, run wild and presenting quite an impenetrable thicket. During the autumn months, when the red berries of the briars are fully ripe, large numbers of the Wood-Pigeon resort to these grounds to feed on this fruit, and at this season become exceedingly fat.

Captain Mair, who kept a winged bird in his possession for about eight months, informs me that it fed readily on boiled potato, rice, wheat, and berries of every kind, and that it ultimately died of sheer fatness. It continued shy and untamable to the last, and on being handled would strike fiercely with its wings. The late Dr. Allison, of Wanganui, however, succeeded in rearing a young one which became perfectly tame and associated with his domestic Pigeons. I may also mention here, as a somewhat curious fact, that at the Chatham Islands, in 1855, I observed one of these birds flying and consorting with a flock of common dove-cot Pigeons which had taken to the woods and become partially wild.

There is probably no New-Zealand bird that could be domesticated to greater advantage than this Pigeon. Some years ago a tame, healthy, and remarkably handsome one was exhibited at the Wellington Pigeon and Poultry Show, and carried off the palm against every competitor in that department. Another, which lived for many months in the Acclimatization Gardens at Christchurch, was shipped home to the Zoological Society, but did not long survive the change of climate.

The New-Zealand Pigeon is strictly arboreal, and appears, as a rule, to prefer the densest foliage. When not engaged in filling its capacious crop with fruit or berries, it generally reposes on a thick limb, with the tail drooping and half-spread, the wings closely folded, and the head drawn in; but on the slightest alarm it stretches up its lustrous neck, and gently sways its head to and fro, uttering a scarcely audible *coo*, slowly repeated. It rises with an awkward flapping, and flies direct, with a rapid opening and closing of its wings, producing the sound so familiar to the gunner's ear. In the bush it generally flies low, but when settling it habitually makes a graceful upward sweep in its course.

When seen from the front its ample white breast is a very conspicuous object in the bush, and the woodcut at the end of this article (from one of my own sketches) will recall its showy appearance to those who are familiar with the bird in its native haunts.

I have remarked a peculiar soaring habit which this bird indulges in during the breeding-season. Mounting high in the air, in a direct upward course, it suddenly opens its wings and tail to their full extent, and glides slowly downwards in an oblique direction, and without any apparent movement of those members. I very frequently observed this peculiar soaring flight during my ascent of the Upper Wairoa river, north of Auckland, where the solitudes of the endless pine-forests afford this species a secure and quiet breeding-place.

On the wing the whiteness of the underparts is very conspicuous, owing to the manner in which the body is swayed from side to side.

This species retires to the high wooded lands of the interior to breed; and its nest is therefore seldom met with. It is a very rude, flat structure, composed of twigs loosely placed together, and containing generally only one, but sometimes two eggs. These are perfectly oval in form, measuring 1·9 inch in length by 1·4 in breadth; the surface is smooth without being glossy, and of the purest white. Mr. J. D. Enys informs me that on the 8th of January, 1862, he found a nest containing one egg perfectly fresh, on the 31st of the same month another containing a young Pigeon fully fledged, and on the 3rd of February two more nests, in both of which there was a solitary half-grown bird.

A nest in the Canterbury Museum (received from Milford Sound) consists merely of a layer of dry twigs, so loosely put together that the eggs are visible from beneath.

There is another nest, from Little River (April 1875), which forms a very pretty object. It is placed on the lateral fork of a branch of totara, supported underneath by an epiphytic growth of native mistletoe (*Loranthus micranthus*), which, although dried, still retains its leaves. The nest (which contains a single egg) is very slight, and admits the light through its foundations, being formed of slender dry twigs of *Leptospermum* laid across each other and forming a shallow depression, with the ends of the twigs projecting all round. Slight as the structure is, however, there is some appearance of finish about it.

In the Rev. Mr. Spencer's fine old garden at Tarawera, where well-grown specimens of English oak, elm, and walnut mingle in rich profusion with almost every kind of native tree and shrub, a pair of these birds some time ago took up their abode and bred for two successive years, at a spot not fifty feet from the reverend pastor's study windows. And they would doubtless have continued to breed in this quiet retreat had not one of the Maori school-boys, anxious to try his fowling-piece and wholly unmindful of the consequences, shot both birds during the breeding-season, leaving a pair of callow young to perish miserably in their nest.

Some colonists are of opinion that this fine Pigeon is less plentiful than it was formerly; but I do not think there is much fear of its becoming extinct so long as the native forests remain.

Its relative abundance may be inferred from the fact that in July and August 1882, Rawiri Kahia and his people snared no less than eight thousand of them in a single strip of miro bush, about two miles in extent by half a mile in width, at Opawa, near Lake Taupo. The birds thus snared are preserved in their own fat and potted as "huahua kereru." Food of this kind is esteemed a great delicacy and elaborately carved *kumetes* are sometimes used for serving it at the tribal feasts.

Notwithstanding its uncertain seasonal movements, there is perhaps no bird so characteristic of the native woods, for, at one time or another, it is met with everywhere. But there are certain tracks of forest which the Pigeon specially affects, the preference being of course due to the predominance of particular fruit-bearing trees. One of these favourite districts is the extensive forest track known as the "Forty-mile Bush," lying between the townships of Masterton and Woodville, and extending thence eastward towards Napier under the name of the "Seventy-mile Bush." A good macadamized road passes through this bush-land, a great portion of which is perfectly level; and perhaps in no part of New Zealand can the transcendent beauty of the native woods be seen to greater advantage. Coming from the Wairarapa side, you first of all pass through some magnificent clumps of rimu, many hundreds of acres in extent, with just a sufficient admixture of kahikatea and rata to set off the peculiar softness of the former, with its "fountain of foliage" and its uniform tint of yellowish green, the young trees gracefully drooping their tasselled branchlets of still paler green. Then, fringing the road on the upper or hill-side, for miles together are glorious beds of *Lomaria procera*, their fronds from three to five feet long, on gracefully pendent stalks, and so closely set that a whole regiment of soldiers might lie in ambush there; then a sudden turn in the road brings you into dense bush again, with its ever-varying shades of green and yellow and brown, blended together in one picturesque and harmonious whole. The tree-fern with its spreading crown is always present—the shapely form of *Cyathea dealbata* with its large umbrella top, the taller *Cyathea medullaris* rearing its head some forty feet or more, and *Dicksonia antarctica* on its massive stem hung round with a brown garment of withered fronds—the lofty dark green rewarewa and brighter kohekohe mingle with the titoki and miro, and the tawa with its light green foliage stands out in bold relief against a dark background of kahikatea trees standing close together. Then you come upon an old Maori clearing abutting on the road, presenting a tangle of new-grown shrubs and saplings, close and compact, and all of the freshest green; and from the very midst of this there rises, like a silent monitor with its bleached arms pointing heavenward, a former monarch of the forest, long since dead and withered and now decaying slowly under the crumbling hand of time, but bearing on its lower forks huge bunches of green *Astelia*, and quite hidden at its base by a luxuriant growth of underwood. In the more recent openings in the bush the trunks of the trees may be seen laden with tons of climbing plants and epiphytic vegetation of various kinds—the kiekie with its hydra-headed branches of waving tufts, the akakura and the waxy *Metrosideros*—and the ground below them covered with ferns and mosses and cryptogams in amazing variety. And so, on and on, through endless changes of timber-growths and woodland scenery, as the coach rattles along the road, disclosing new beauties at every turn—now through a river-bottom filled with close-growing kahikatea, then over a ridge covered with *Fagus*, dark and sombre; now past a wide opening caused years since by the ravages of fire or flood and overgrown with the red-stemmed mako, the native myrtle, and a hundred other less conspicuous shrubs, bound and matted together by masses of tataramoa or covered with a spreading network of pohuehue; then entering again a stretch of what is known as "mixed bush," where all kinds of New-Zealand trees and shrubs and ferns are crowded together in harmonious confusion, presenting a study to which no pen, however gifted, could do adequate justice. As you gaze upon this sylvan picture you are forced to admit that there is nothing in the world more beautiful!

As we approach the river-banks, the low bushes are covered with a thick mantle of convolvulus, closely studded or spangled with the pure white flowers, like innumerable luminous stars on a cloth of vivid green; and the tree night-shade (*Solanum nigrum*) grows in wild luxuriance, its pale blue bells having a pretty effect against the sombre foliage.

But the principal charm of these woods is the rapid change in their aspect as one season succeeds another. In the autumn months, when the berries of various trees have ripened, they are swarming with Pigeons, especially in the more fruitful seasons which occur at intervals of two or three years. In the winter they are deserted, and you may travel for a whole day without seeing or hearing a bird of any kind, except those that commonly frequent the road. But this lifeless season is of short duration, and is followed by the gladness of the early spring-time. The whole bush is then decked out with the beautiful star-like *Clematis*, hanging in garlands round the trees, festooned in clumps among the lower vegetation along the open roadside, and displaying its petals of snowy white in great profusion. The pukapuka, which is abundant everywhere, supports on its poisonous stems a crown of creamy blossoms in clusters so thick as quite to conceal the leafy top; the kowhai, having shed its leaves, is transformed into a glory of golden yellow, each branchlet bending under a cluster of horn-shaped flowers of uniform pale yellow with a green peduncle.

Visit these woods again at the commencement of summer and the whole scene has changed. The hanging festoons of *Clematis* have disappeared and in their place may be seen bunches of green silky tassels, containing the seed-vessels of this plant and possessing a characteristic beauty of their own; and underneath the golden kowhai trees the ground is carpeted with fallen petals. But the crowns of *Cordyline* are now bearing, in rich plenty, their drooping branchlets of fragrant flowers; the tawhero, of which the lower forest is largely composed, is covered with bottle-brush flowers of delicate waxy white; the miro is one mass of whitish inflorescence, intermixed with the pale green foliage of that tree; whilst clinging to the underwood and hanging from almost every branch the kohia creeper exhibits its minute pearly bells in rank profusion. Then every here and there may be seen, placed high up in some sturdy fork, a bunch of *Loranthus* ablaze with its crimson flowers and forming a picturesque object amidst its green surroundings. The rata, or Christmas-flower, as it is called, is just making its appearance; here and there a vigorous young tree in advance of the rest has swathed itself in colour, but for the most part the only indication at present is a crimson blush on some of the branches. Before the ides of December have passed these noble rata trees will be enveloped in a mantle of fiery red. But the whole woodland already seems abloom and the air is laden with a faint but pleasant perfume. As a consequence of this, and the abundance of insect life which it betokens, the bush is again alive with Tuis and other birds.

We rest for awhile in a lovely wooded valley which is illumined by the bright afternoon sun and exhibits some wonderful effects of light and shade. The road lies before us, straight as an arrow, through a wooded vista nearly two miles in length; fringing it, where we stand, is a grove of the beautiful silver-leaved *Pittosporum*, with shapely tops as if specially trained for some ornamental garden; beyond this a clump of its broad-leaved cousin (*Pittosporum undulatum*) closely commingled with the [illegible] and many other stately shrubs, whilst in the shaded hollow below us are some splendid specimens of the native fuchsia, attaining to the size of veritable trees, some having trunks two feet in diameter and branches laden with moss; then behind comes the low forest all abloom as described, and beyond that, far and away, the rolling "forest primeval" of rimu and rata and kahikatea. To add to the enchantment of the spot, there is a whole choir of singing Tuis, who, having regaled themselves among the flowers, are now piping and sobbing in chorus; a wandering flock of *Zosterops*, quite concealed from view, are warbling a low, pathetic lay; a solitary Warauroa from a lofty tree-top emits his plaintive call, with none to answer; and heedless of all the rest a tiny Riroriro, hiding in a bramble-bush, trills its silvery note with untiring energy. Then, as we move forward, a Parrakeet, startled by our approach, rises from the low underwood with laboured and zig-zag flight and settling on a branch near the roadside adds its lively chatter to the other sounds of this sylvan valley. As the sun goes down and the shades of evening advance all these voices are silenced; but the Tui continues still to flit across our path, and the Flycatcher to display its pretty fan as it hawks for invisible flies. Then comes the scream of the Kaka as it wings its distant way high above the tree-tops; after which, with scarcely a moment of twilight between, the woods are plunged in gloom, the Owl comes out from its hiding-place, and the glow-worms shine on the damp roadside.

Such is the New-Zealand bush, replete, as it is, with a flora entirely its own—charmingly green in summer and winter alike—the pride and glory of the land and the natural home of the birds whose life-history I have endeavoured to portray in the foregoing pages.

SUPPLEMENTARY NOTES

TO THE

'BIRDS OF NEW ZEALAND.'

VOL. I.

The following additional notes on some of the Families treated of in the present volume may be of interest to the general reader.

Fam. **CORVIDÆ.**—The only representative of this family in New Zealand belongs to the somewhat aberrant genus *Glaucopis*. At page 4, in my account of *G. wilsoni*, I have stated my reasons for placing this form at the head of the New-Zealand Avifauna; and at p. 30 I have given the result of Dr. Gadow's careful examination of a skeleton which I had submitted to him.

Fam. **TURNAGRIDÆ.**—At pp. 26–30 I have given what may be considered the final record of the North-Island Piopio, a species now on the verge of extinction. Its South-Island representative (*Turnagra crassirostris*) is still to be met with in certain wooded districts, but in rapidly diminishing numbers, and, with other interesting forms that still linger, its doom is sealed.

As recently as December 1887 last one of my New-Zealand correspondents, writing to me from the west coast, says:—"Since I came here I have formed the acquaintance of several old gold-diggers, from whom I have gathered much information on the haunts and habits of many of the species. All of them agree that certain birds are disappearing fast, viz. the Crow, the Saddle-back, the Thrush, the Robin, the Kakapo, the Woodhen, and the Kiwi. Fifteen years ago all these birds existed here in abundance. Every digger keeps a gun and a dog, besides, as a rule, having one or more cats in their huts. All the birds I have mentioned, either from their tameness, their incapacity for flight, or their habit of feeding on the ground, would fall an easy prey to dogs and cats, both of which animals often stray away from the diggers' camp and become wild. Man also contributes to the work of wholesale destruction. Last Sunday I dined on stewed Kiwi at the hut of a lonely gold-digger, who, besides the three cooked for dinner, had four other fat Kiwis hanging on the wall, to serve through the week. My host informed me that he varied his bill of fare with Wekas and Kakapos. These men lead lonely lives in the bush, and only emerge once in a week or fortnight to get stores from some central point whither the trader brings them on a pack-horse. The well-trained dog of the gold-digger is perhaps the most destructive agent as regards the Kakapo and Kiwi. The felling and clearing of forests and the consequent diminution in the supply of honey-producing flowers will account to some extent for the present scarcity of the Bell-bird and the Tui. But, as you have already pointed out, the chief factor in this work of extermination is no doubt the introduced rat, which now exists in immense numbers over the whole extent of the west coast, from the gold-diggers' townships to the remote bush-covered ranges. Added to all these potent causes, I have no doubt that, owing to the changed conditions under which they exist, and the more scanty supply of food, disease in various forms contributes to the general sum of destruction."

Fam. **STURNIDÆ.**—The Saddle-back (*Creadion carunculatus*), which was extremely common in all suitable localities fifteen to twenty years ago, has now disappeared from the North Island, and is becoming scarce in the South, although both this and the allied species (*C. cinereus*) are still plentiful on certain small islands in the Hauraki Gulf. Professor Hutton was the first to discover the nest of this bird on the Little Barrier Island, where he found it lodged in the hollow stem of a tree-fern.

The accompanying figure appeared in the 'Transactions of the New-Zealand Institute' (vol. v. pl. 17), together with the following descriptive notes by Mr. Potts:—

"For its nesting-places a hollow or decayed tree is usually selected; sometimes the top of a tree-fern is preferred. The first nest we knew of was found by an old friend in a hole about four feet from the ground in a huge white pine (*Podocarpus dacrydioides*), close to the bank of the Ahaura river; it contained three eggs hard-set. We found a nest in a dead tree-fern not far from Lake Mapourika, Westland. This was of slight construction, built principally of fern-roots, deftly woven into rather a deep-shaped nest with thin walls; for as the structure just filled the hollow top of the tree-fern thick walls were unnecessary. Another nest (the one figured), found in a small-sized decayed tree in the Okarito bush, was in a hole not more than three feet from the ground. It was roughly constructed, principally of fibres and midribs of decayed leaves of the kiekie (*Freycinetia banksii*), with a few tufts of moss, leaves of rimu, lined with moss and down of tree-ferns; it measured across from outside to outside of wall 12 inches 6 lines, the cavity 3 inches in diameter with a depth of 2 inches."

Fam. **SYLVIIDÆ.**—When I was engaged on my former edition, Mr. T. H. Potts sent me a large series of pen-and-ink sketches of nests which he had collected at various times, all executed by himself and exhibiting the characteristics of each in a very happy manner. These were afterwards published in illustration of that gentleman's "Notes on the breeding-habits of New-Zealand Birds,"

which appeared from time to time in the 'Transactions of the New-Zealand Institute.' But as some of the original sketches are still in my possession, I have much pleasure in reproducing them here, on a somewhat reduced scale. One of the most interesting of these is a representation of the nest of the South-Island Robin (*Miro australis*), which differs from the typical form in its more slender walls and thinner foundation. Mr. Potts, who has collected a large number of these nests, says:—"Its nest is wider and larger altogether than that of *Myiomoira macrocephala*, but not so closely interwoven; moss, sprays, leaves, fine fibres, and grass enter into its construction. Diameter of nest from 5 to 6 inches, of the cavity 3 inches, with a depth of 1 inch 3 lines. A favourite situation appears to be behind such protuberances as are to be found on the huge gnarled trunks of *Griselinia litoralis*, very often not more than 3 feet from the ground."

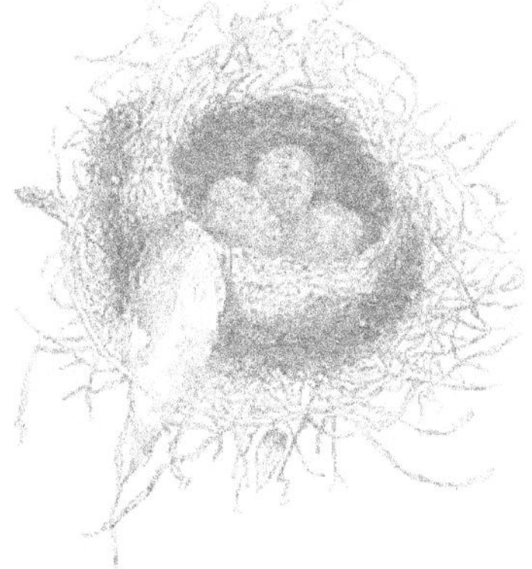

The South Tomtit (*Myiomoira macrocephala*) is somewhat eccentric in its mode of nidification. Among the sketches mentioned above there are representations of four of the nests of this species from one locality, near Ohinitahi, and as they exhibit very different types of architecture I have given woodcuts of all of them.

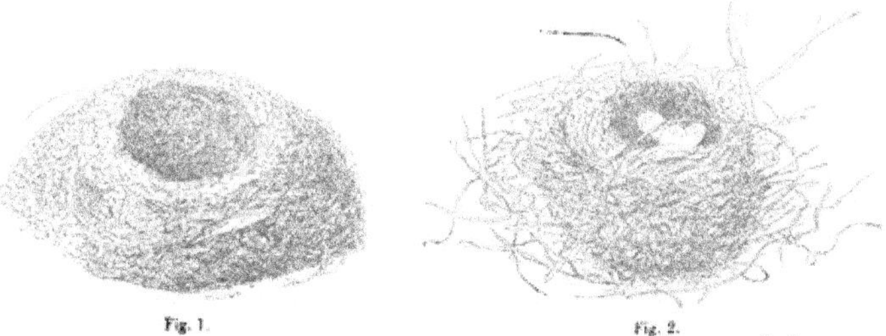

Fig. 1.

Fig. 2.

No. 1 was built of dry sprigs of climbing-plants intermixed with grass-bents and strengthened by means of split shreds of ti-palm leaf, the cavity being lined, as usual, with soft moss. This structure, which appeared to be more loosely put together than usual, was discovered in the head of a ti-palm and contained, in addition to two unfledged young birds, three bad eggs. No. 2 was composed almost entirely of dry moss with a few slender strips of bark fixed to the outer surface, in order to give it stability, and in the lining of the cup could be seen a few green Parrakeet feathers. This nest was placed in a mossy recess on a rocky ledge in thick bush, and when found contained four eggs.

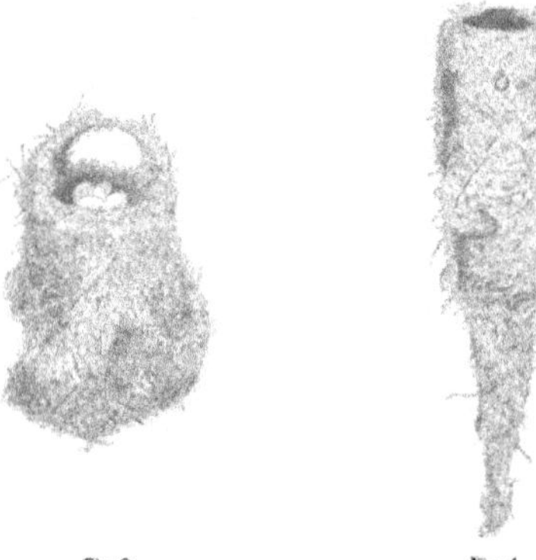

Fig. 3. Fig. 4.

Figures 3 and 4 represent very unusual forms—one of them having an exact resemblance to a moss-basket, with a profusion of tree-fern down in the centre and cavity; the other being of a long tapering form and measuring fully fifteen inches in length from the rim of the cup to the lower extremity of the nest.

I have given, at page 48, some pocket-book sketches showing a considerable amount of variation in the nests of *Gerygone flaviventris*. The following are further illustrations of the kind, the one exhibiting a side view being ornamented with *Acæna*-burrs.

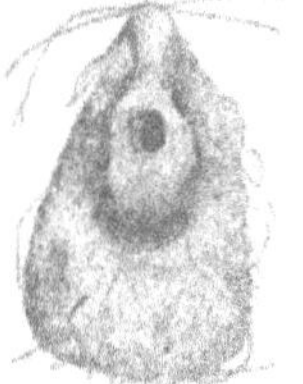

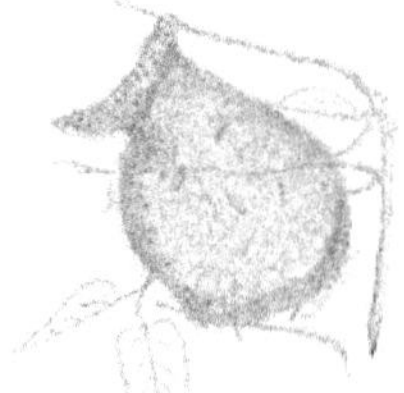

At p. 50 I have stated my reasons for giving *Gerygone sylvestris* (erroneously referred to in the first paragraph of that article as *G. flaviventris*) a place among the birds of New Zealand, although, so far as is known, no specimen of it exists. Mr. Potts, who described the species under that name, seems very positive that the bird which he killed in the dense bush between Okarito and Lake Mapourika was quite distinct from our common species; and Mr. Reischek's report of a small bird on the west coast whose notes he could hear, although he could not see it, may perhaps be confirmatory of its existence.

Fam. **TIMELIIDÆ.**—I have given, at page 58, my reasons for insisting on the association of *Clitonyx albicapilla* and *C. ochrocephala* in one and the same genus. The subjoined woodcuts of the nests show very plainly that the architecture of both species is the same.

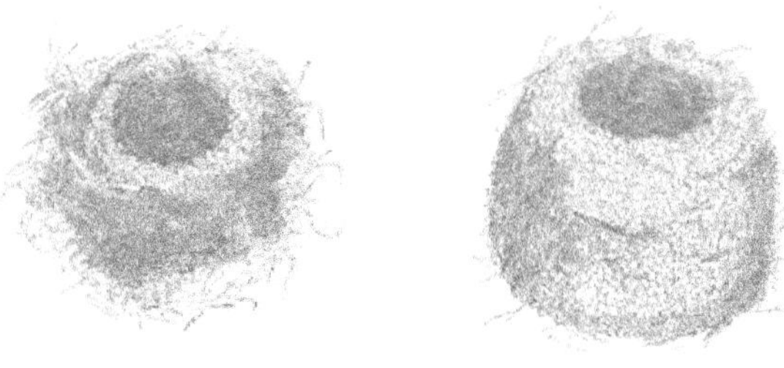

Clitonyx albicapilla. *Clitonyx ochrocephala.*

As already stated at p. 60, the Grass-bird (*Sphenœacus punctatus*) attaches its slender nest to thin reed-stems standing in close proximity to each other, but it is sometimes placed on the ground under shelter of a tussock or tuft of rushes. One of the latter kind (formed entirely of dried grass-leaves) is here depicted.

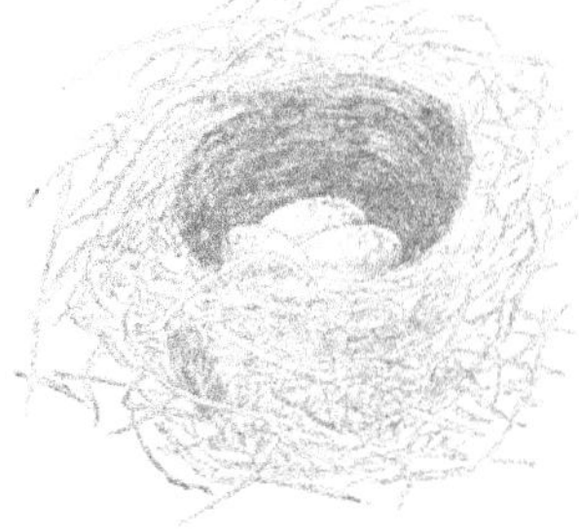

Fam. **MELIPHAGIDÆ.**—Mr. T. Hunt, who has lived on Pitt Island for more than thirty years, in a letter to the press dated the 5th of September last, states that the *Zosterops* (to which he applies the name of Fish-eye) appeared there and on Chatham Island about three weeks after the great Australian fire known in local history as Black Thursday *.

As fully explained at pp. 83, 84, the nests of this species exhibit a considerable amount of individual variation, but the typical character is always the same, and this is well illustrated in the subjoined drawing of one of these pensile cups fixed in a sprig of fern.

As will be seen from the accompanying sketches, the Tui and the Korimako construct their nests on the same principle; but the fondness for gaily coloured feathers (as specially mentioned at p. 91) is confined to the latter bird.

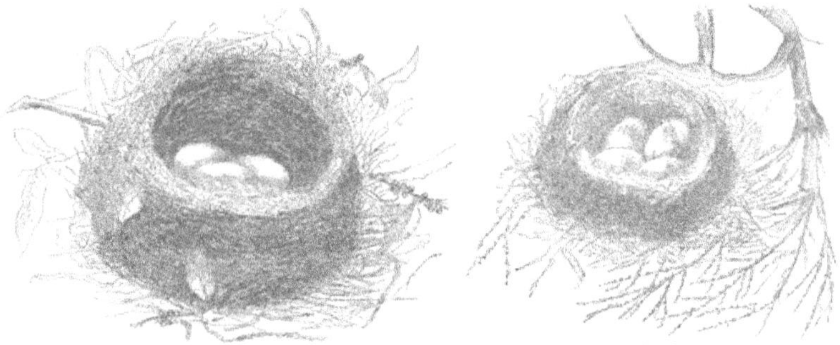

Prosthemadera novæ zealandiæ. *Anthornis melanura.*

The Stitch-bird (*Pogonornis cincta*), which less than fifteen years ago was comparatively plentiful

* He mentions the further circumstance that the House-Sparrow, the Linnet, and the Blackbird have all come over to the Chatham Islands from New Zealand (a distance of 300 miles) and are now so numerous as to threaten to become a nuisance to the agriculturist.

in the southern portions of the North Island, is now quite extinct on the mainland, being met with only on a small wooded island in the Hauraki Gulf. As stated at p. 105, a nest of this species is preserved in the Colonial Museum; and I have much pleasure in reproducing here a sketch of this specimen which appeared in the 'Transactions of the New-Zealand Institute' (vol. iii. pl. 12).

Fam. **XENICIDÆ.**—The recent discovery of the true relations of the New-Zealand genera *Xenicus* and *Acanthidositta* is extremely interesting from a biological point of view; and my own belief is that as we become better acquainted with the anatomy or internal organs of our many endemic forms other equally important alterations will require to be made in our present classification of the genera. In my account of *Xenicus longipes* I have given all the information I have been able to collect respecting it. I have shown, I think, conclusively that *Xenicus stokesii* is a myth, the creation of this new species having been due to an erroneous figure. In company with the late Mr. G. R. Gray, I examined the original drawing at the British Museum, in which I found the bill depicted as straight, and a mere indication given of the white superciliary streak. Mr. Gray told me that his artist was responsible for the alterations in the published figure, and that his own description of the species was inadvertently taken from the latter.

The nest of *Xenicus gilviventris* mentioned at page 112 is now in my collection, and on account of its extreme rarity I have had it photographed and carefully drawn for reproduction here; but, being to a larger scale than the other woodcuts, I have placed it at the end of these 'Notes' on page 250.

At page 115 I have described some peculiar conditions under which the nest of *Acanthidositta chloris* has been found at different times. By way of adding another curious instance, Mr. W. W. Smith has sent me the following note:—"I lately procured an egg of *A. chloris* under peculiar circumstances. One of the men in the garden, when moving some broken pipes formerly belonging to the hot-water apparatus in the vinery, noticed a nest in one of them. Thinking it to be the nest of a mouse, he tore it out, when the tiny egg dropped upon the ground, but escaped injury. I am sending you the specimen, together with the materials composing the nest."

Fam. **PLATYCERCIDÆ.**—As will be seen at page 149, an interesting addition has been made to our Avifauna by the rediscovery on Antipodes Island of *Platycercus unicolor*, a species hitherto without any known habitat. The unique specimen upon which Mr. Vigors founded the species was more than half a century ago living in the Zoological Society's Menagerie at Regent's Park. On the 11th January, 1831, the above-named naturalist exhibited the bird at a Meeting of the Society and made some remarks upon it, stating that, although its native place had not been ascertained, "from the more graduated form of the tail and the plumbeous colour of the bill it was conjectured to have belonged to some of the Australian islands, the Parrakeets of which are distinguished by these characters from the allied groups of the same genus, *Platycercus*, of the Australian continent." The lively and active gait of this bird, as distinguished from the slow and climbing motions of the Parrots, was particularly noticed *. On the death of this rarity it was skilfully mounted and placed in the bird-gallery at the British Museum, where it has remained to the present day. Its bill is conspicuously larger than in the specimens recently brought by Captain Fairchild from Antipodes Island, but this was doubtless due to its having been kept for a long time in confinement. In other respects it corresponds exactly with the specimen forwarded to me by Sir James Hector, and which I have had the pleasure of presenting to the Cambridge Museum. The discovery of the home of *Platycercus unicolor*, after so long a lapse of time, is just one of those events in Ornithology that serve to stimulate and reward the labours of our naturalists abroad.

The subjoined figures of the heads (natural size) of *Platycercus novæ zealandiæ* (fig. 1) and *P. unicolor* (fig. 2) will show, at a glance, how much these species differ from each other in size; whilst the uniform green plumage of the latter readily distinguishes it from all other members of the group. Fig. 2 is taken from the British Museum specimen, in which the bill is rather larger than in mine, owing perhaps to the long captivity of the bird, and the consequent tendency to abnormal growth.

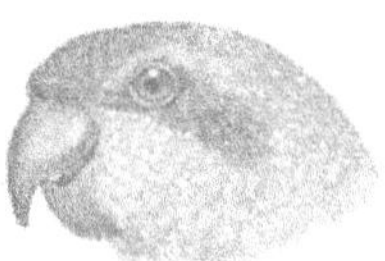

Fig. 1.

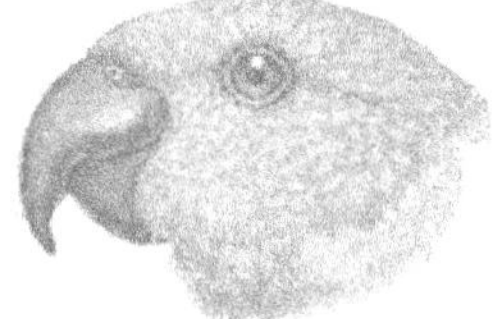

Fig. 2.

I stated at page 149, on the authority of Sir George Grey, that the northern Maoris have a tradition of some very remarkable kind of Parrot as inhabiting Cuvier Island, a high wooded islet near the entrance to the Hauraki Gulf. It may be of interest to mention that this locality has very recently been thoroughly explored by Mr. Adams, a collector employed by the Auckland Museum, and that, although he met with the common New-Zealand Parrakeet and several other familiar species, he found no strange birds there.

Fam. **NESTORIDÆ**.—It will be seen that I have given a full account of the Kea, or Sheep-killing Parrot, with a history of its development into a carnivorous bird. The extraordinary habit which it has so speedily acquired of attacking live sheep, for the purpose of feasting on the kidney-fat, has been the subject of much discussion among naturalists †. By its addiction to this vicious habit

* Proc. Zool. Soc. 1830–32, pp. 23, 24.

† At the Colonial and Indian Exhibition, 1886, a painting in oils of considerable merit was exhibited by Mr. George Sheriff, of Wanganui, showing a pair of Keas at work on a sheep; but the artist has made the mistake of substituting a dead animal for a live one, thus falsifying the record in its most essential feature. As mentioned at p. 170, a pen-and-ink sketch by Mr. Potts in 'Out in the Open' represents the incident correctly in this respect, but the figure of the animal operated upon is devoid of all expression, just as if the sheep submitted to the vivisection as a matter of course or treated the whole thing as a joke. A large drawing in my possession, from the talented pencil of Mr. J. Wolf, gives an admirable idea of the subject. The scene selected is the gorge of the Rangitata, under moonlight, showing the far-off snow-capped peaks of the Southern Alps, flanked by enormous

the bird is doomed, and at the present rate of destruction a few years at most will witness its extermination. For this reason, if for no other, it was incumbent on me as its historian to give the fullest possible account of its natural history.

Mr. Walter Chamberlain, of Harborne Hall, Birmingham, in an interesting paper read before the Largo Field-Naturalists' Society last year, makes the following observations on this remarkable Parrot:—

"Between 1865 and 1870 the shepherds who were pushing their flocks in the south further and further up the slopes of the central range began to complain that the Keas visited their huts and ate the hanging meat, more particularly the kidneys and fat. Here, then, we have the first evidence of the pernicious and, to them, fatal taste for kidneys which has since so rapidly developed. They found the meat hanging with the kidneys *in situ*. They took a special liking to the latter and sought for them high and low, all the more zealously no doubt that the shepherds took counter precautions to preserve the delicacies for themselves. It is most likely that they soon began to find and tear open with their strong bills the sheep that died among the hills, and were thus guided by degrees to the actual seat of the kidneys in the living animals and the readiest way of approaching them. At any rate, about the year 1875, the first sheep—still in the far south—were found wounded just over the loins. There was much puzzling over these wounds, and not unnaturally they were at first ascribed to wild dogs, that is dogs run wild; but at last all doubts were set at rest by a shepherd actually catching a bird on the back of a live sheep hacking at its loins in order to reach the kidneys. Gradually since then the habit has travelled northwards, until only in 1885 the first sheep was attacked in the Rakaia district, not very far south of the extreme northern range, as at present known, of the species. In the meantime the southern birds that had already learnt the trick, commenced to follow the flocks lower down during the winter, and to carry on their devastations more systematically. . . . . . Now, this sudden acquisition of an altogether strange habit by birds in a state of nature is, I think, absolutely unique, and it is certainly a case of great interest to naturalists everywhere as well as to New Zealand run-holders. I cannot call to mind any instance that I have heard or read of which at all runs parallel with it. New habits when they are acquired by species in a natural state have always hitherto, so far as I know, been very slowly developed, and the habit itself, as a rule, is little more than a modification of some previous one performed by instinct, as for instance the painfully acquired experience which teaches wild creatures to avoid a new form of danger, or the easily made experiments which teach them that some crop newly grown by man near their haunts is suitable for food as it stands.

"Consider for a moment the sequence of events and the extraordinary change of habit involved to the parrot. Between 1865 and 1870 the Kea first comes in contact with the shepherd, and commences to steal his meat with a marked preference for the kidneys. This is natural enough, and any other parrot with a tendency to animal food might do the same, but here the matter would ordinarily rest. The shepherds would protect their meat, and the parrots would return to their usual food. Not so with the Keas. Between five and ten years later they have found out not only that kidneys are somewhere inside living sheep, but whereabouts inside and the nearest point on the back from which to reach them. A few years more and they have learnt further, not only that sheep are incapable of defence and unable to hurt their aggressors, but that they are singularly stupid animals, and may be reduced to a still further state of impotence by the simple expedient of worrying, and, moreover, they have worked out a plan of thus worrying the sheep by combining together and attacking the unfortunate animals one after another in succession.

"In the first part of these notes I have stated that I see no reason to rate the intelligence of the *Psittacidæ* generally above the average of other families of birds, but certainly if we were to meet with a few more instances among the former of habits acquired by a process which bears such a striking resemblance to inductive reasoning, or at least to the putting of two and two together, we should not be able to deny them possession of intelligence which, were they a more powerful family, might be dangerous to man himself. . . . . . . I have stated elsewhere on the authority of Dr. Karl Russ, and as a matter of common observation, parrots are not flesh-eaters, and in confinement even the Nestor can be kept in health without it, whilst a moderate amount only is apt to cause disease. Yet the Kea seems able suddenly to abandon to a large extent its natural food and to gorge itself incessantly on raw meat, like a Hawk. Altogether the matter is one well worthy the attention of ornithologists, and it is to be regretted that the too probable extermination of the species may prevent the present or succeeding generations of naturalists

glaciers; and the middle distance is veiled in mist, partly obscuring the stunted *Fagus*-forest which clothes the lower ranges. In the foreground, below the gorge, a sheep attacked by a Kea is writhing its body in agony and kicking up the loose snow from the ground in its frantic efforts to rid itself of this cruel tormentor, which clings tenaciously to its back. Two other Keas on the wing are coming to assist in this work of torture. A small mob of sheep are huddled together under a projecting cliff, trying to obtain a little warmth, while one sheep, more inquisitive than the rest, has advanced a hundred yards or more towards the suffering victim, and is looking on, in silent wonderment, showing that these animals have hardly yet learnt to regard this Parrot as their natural enemy.

witnessing the ultimate development of the habit, which one would expect to result in the production of a purely carnivorous parrot, with modifications of foot and digestive organs in accordance.

"In conclusion, I may remark that the Kea has not yet taken to flesh-eating throughout its range—possibly only from want of opportunity. Further north it still keeps well up in the mountains, and seems content with the diet that satisfied its predecessors; but as the habit commenced in the south and travelled northwards, so fresh cases keep occurring one beyond another, and it seems certain that the necessary information is passed onwards and northwards."

Fam. **STRINGOPIDÆ.**—At page 180 I have mentioned some structural peculiarities in the osseous frame of *Stringops habroptilus.* I have since had the pleasure of presenting a skeleton of this bird to the British Museum, and it is now exhibited in one of the wall-cases in the main hall of the Natural History section. The subjoined woodcuts (after Meyer) will show how widely it differs from the skeleton of *Nestor.*

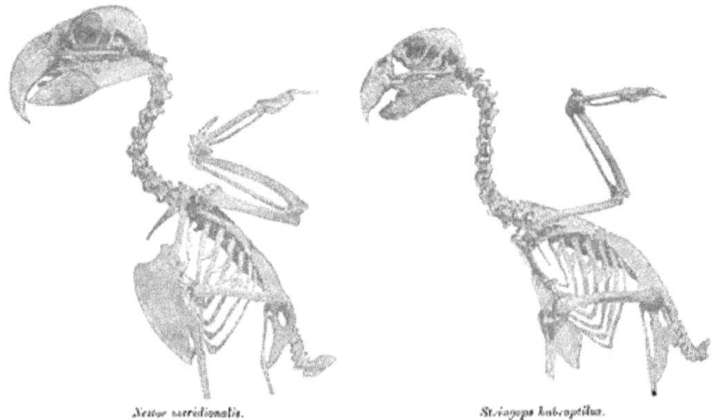

Nestor meridionalis. *Stringops habroptilus.*

Fam. **STRIGIDÆ.**—It has long been supposed that an Owl of much smaller size than the well-known Morepork exists in New Zealand, but I have never myself met with any positive evidence respecting it. Mr. Ellman, as far back as 1861, describing it as "not larger than a Starling," gave it the name of *Strix parvissima*, and Mr. Sharpe, in the British Museum Catalogue (Birds, vol. ii. p. 43), refers the species, without any apparent hesitation, to *Scops novæ zealandiæ*, Bonaparte, of which he gives a full description.

I have stated at p. 205 that the only authority for regarding the unique specimen in the Leyden Museum as a New-Zealand bird is a label in Temminck's handwriting. Deeming this, in itself, insufficient evidence, I sent Mr. Keulemans over to Leyden to make a drawing of the bird in water-colour. He brought back a beautiful picture, of life-size, showing the mottled markings of its plumage in marvellous detail. But I saw, at a glance, that this *Scops* equalling in size small examples of *Spiloglaux novæ zealandiæ*, and with strikingly prominent "horns," could never have been the bird intended by those who have described an Owl "about the size of a Kingfisher." The occipital tufts, characteristic of the genus *Scops*, are so strongly developed in this species that they could not have escaped the notice of the most casual observer, and yet we have no mention of them in any of the hearsay accounts that have been recorded from time to time. In addition to the instances mentioned in the body of this work, the following are taken from 'Out in the Open,' p. 127:—

"Another specimen was procured by a gentleman in one of the forests far above the Rangitata gorge; on being observed on a branch of a tree, it was knocked down and caught during its fall. There was fur on its beak, as though it had not long before devoured a mouse. This bird also was set at liberty.

"It has been taken at the Waimate, where it remained for a day in the roof of a hut. Mr. M. Studholme had it in his hands, but permitted it to escape.

"The late Mr. Phillips, of Rockwood, one moonlight night captured a specimen by taking it quietly off a bough of an apple-tree. Mr. Phillips, like Mr. Studholme with his bird, carried it between his hands and allowed it liberty. He described it as being about the size of our Kingfisher. Note that each observer of this pretty Owl was impressed with its gentleness and its fearless confidence. Both had enjoyed long colonial experience, were accustomed to birds, men of position and well-known beyond their own districts. *Athene parvissima* must not be given up, even to satisfy the most erudite of ornithologists."

Professor Newton, to whom I submitted the drawing, writes to me:—"I certainly admit that your caution has been justified, for it is almost impossible to suppose that the wonderful *Strix parvissima* (!) could have been a bird of the same species." And Mr. J. H. Gurney, whose opinion on such a point is of the utmost value, sends me the following report:—

"I have carefully compared Mr. Keulemans's drawing of the type specimen of *Scops novæ zealandiæ* with the series of Scops Owls preserved in the Norwich Museum, and after doing this, and also referring to the late Professor Schlegel's descriptions of the specimen in his 'Muséum des Pays-Bas,' Oti, p. 27, and 'Revue,' Noctuæ, p. 13, and to Mr. Sharpe's description in a footnote at p. 44, vol. ii. of his 'Catalogue of Birds in the British Museum,' I concur in the belief there expressed by Mr. Sharpe, that *Scops novæ zealandiæ* is a distinct species; but if it be so, two questions will still remain undecided—1st, whether the locality of New Zealand assigned to the Leyden specimen by Temminck's label is correct; and 2nd, if so, whether the species is, or is not, identical with the New-Zealand bird for which Mr. Ellman proposed the name of '*Strix parvissima*.'

"The type specimen of *Scops novæ zealandiæ*, judging from the materials before me, appears to approach most nearly to *Scops menadensis*, Sharpe, a native of the islands of Morty and Ternate, described and figured in Mr. Sharpe's Catalogue of Birds, vol. ii. p. 75, pl. 7. fig. 1: but it would seem to differ from that species in having a somewhat conspicuous nuchal collar, in the under wing-coverts being 'almost entirely ochraceous,' and (to quote Schlegel's words) 'par le manque de taches claires aux plumes scapulaires.'

"I return Mr. Keulemans's beautiful portrait of the Leyden specimen by parcel-post and thank you much for the opportunity of examining it."

In my account of the Laughing-Owl (*Sceloglaux albifacies*) I have mentioned a tendency to variation in the plumage. I have since examined very carefully Mr. G. R. Gray's type (brought to England by Mr. Percy Earl in 1846), and it seems to be a case of partial albinism, for the face is so white as fully to justify the specific name bestowed by him. It has the forehead, cheeks, lower sides of the head, and the whole of the throat conspicuously white; the feathers composing the facial disk and the rictal plumes with black shafts, and those on the lower parts of the face with a central streak of brown widening towards the base.

Fam. FALCONIDÆ.—Mr. J. H. Gurney writes to me (under date March 29, 1888):—"The Australian Harrier found in the Celebes is not *Circus approximans* = *C. gouldi*, but *C. assimilis* = *jardinii*, and neither of these species occurs in the Malay Archipelago" *. It will be seen that in my account of *Circus gouldi* I have limited the eastward range of this species to the Fiji Islands.

* Cf. also Gurney's 'Diurnal Birds of Prey,' p. 22, footnote.

As mentioned in the Introduction to my former edition, Mr. Gurney having sent to the Norwich Museum for a specimen of his *Circus wolfi* of New Caledonia (P. Z. S. 1865, p. 823) for my inspection, I felt no hesitation, after comparing it with adult examples of *Circus gouldi*, in accepting it as a good species, notwithstanding the opinions to the contrary of Professor Schlegel and other continental ornithologists. It appears to me to be readily separable from our bird by its blackish crown and ear-coverts,

I have given an exhaustive account of this fine Hawk because it is one of our most conspicuous birds, being met with in all localities; but, I am sorry to say, it is becoming perceptibly scarcer in many parts of the country, owing to its wholesale destruction by farmers. On one occasion I counted no less than ninety-six heads nailed up in imposing rows against the wall of an outhouse on a small sheep-station. This crusade arises from the popular belief that the Harrier attacks and kills young lambs. That it occasionally does so in the case of weaklings is beyond doubt, but I am of opinion that the mischief done is very much exaggerated. In my history of the species I have endeavoured to vindicate its character as a useful bird.

Fam. **CUCULIDÆ.**—Of this family we have in New Zealand one representative of each of the two well-known genera *Eudynamis* and *Chrysococcyx*. Like Cuckoos in general, both of these species are parasitic in their habits of nidification, and, as a rule, both of them find their dupe in the Grey Warbler (*Gerygone flaviventris*), the builder of a pensile, dome-shaped nest. Mr. Potts has called attention to the frequency with which torn nests of this species are met with, and suggests that this may be due to the endeavours of the Cuckoo to make these nests available for their purpose; yet this view is hardly compatible with the fact that whenever the Cuckoo's egg is found among those of the Warbler, the nest is always in perfect condition. But how the intrusive egg is deposited by its owner is certainly a mystery, particularly in the case of such a bird as the Long-tailed Cuckoo *. Mr. Rainbow writes that, as the result of much observation, he is firmly convinced that the English Cuckoo, after laying its egg, takes it up *in its foot* and deposits it in the nest of its victim. The process, he argues, cannot be satisfactorily accounted for in any other way. In connection with this I may mention the circumstance that a friend of mine in New Zealand shot a Long-tailed Cuckoo which appeared to be carrying some object in its bill. On picking up the bird, he found a broken egg, of a creamy-white colour and, so far as he could judge, of a size corresponding to its own.

Mr. Rainbow's view seems to find confirmation in the following statement, which appeared in 'The Ibis,' 1867, p. 374:—"The long-presumed opinion of the Cuckoo first laying her egg on the ground and then carrying it off for deposition in the nest of some other bird, has of late been singularly confirmed by actual observation. In the German periodical 'Der Zoologische Garten' for 1866 (pp. 374, 375) appears a note by Herr A. Müller, stating that the author watched a *Cuculus canorus* through a telescope, saw her lay an egg on the grass, take it in her bill, and deposit it in the nest of a *Motacilla alba*!"

The feeding of the unwieldy young of the Long-tailed Cuckoo by the diminutive foster-parent and the appropriation of the Warbler's nest by the young of the Shining Cuckoo are such droll phases of bird-life that I have introduced both incidents into the Plates illustrating those species.

One of my best correspondents says:—"It is not a difficult task to find the Warbler's nest when the Long-tailed Cuckoo is about to lay her egg or immediately afterwards. It is laid very early in

---

and likewise by the much darker colour of its wing-coverts. In the otherwise excellent drawing, from the pencil of Mr. Wolf, which appeared in the 'Proceedings' (*l. c.*), these distinguishing features are not sufficiently shown; nor does Mr. Gurney give the necessary prominence to them in his descriptive account, his object having been (as he has since informed me) to point out the distinguishing characters of the species as compared with *C. maillardi* (Verreaux), rather than with *C. gouldi*.

* The late Mr. Henry Mair met with this species, with which he was quite familiar in New Zealand, during a visit to Danger Island. It also occurs at Samoa, but according to the Rev. S. Whitmee, it is less abundant there than in many of the Polynesian Islands.

the morning, and after the female has deposited it her mate may be seen flying rapidly after her, dashing through the trees and sometimes in the open, the female uttering a distressed, peevish cry. On such occasions I have sometimes seen them fly long distances outside the bush." Referring to the foster-parents' nest he says:—"I may inform you of a curious position in which the *Gerygone* occasionally builds its nest. Of course you have seen the grand old white-pines whose trunks are thickly 'bearded with moss.' Selecting a spot where this moss is four or five inches thick, it constructs its nest among it, leaving the entrance just flush with the outer fringes. I was shown one by an old gold-digger, who assured me that during the last 25 years he has on frequent occasions found the nest of the Warbler in such situations."

It is singular that in the same way that the Tui persecutes this Cuckoo in New Zealand, it is the victim in Samoa of another Honey-eater, a much smaller bird *. It is difficult to account for this unless it be due to the Hawk-like markings of its plumage.

Fam. **TETRAONIDÆ.**—The rapid and total disappearance of such a bird as the New-Zealand Quail is very remarkable, when we consider that the members of the restricted group to which it belongs have an almost universal diffusion, and continue to exist, under somewhat similar conditions, in other countries in undiminished plenty †. The causes to which we are accustomed to attribute the extirpation of the Quail (the introduction of sheep and the prevalence of bush-fires) ought to operate with equal effect on such a bird as the Woodhen (*Ocydromus australis*), which, being utterly incapable of flight, is placed at a greater disadvantage even than the Quail; yet this species, instead of being exterminated, continues to thrive and multiply, and is even more numerous than formerly in the settled districts of the South Island. Some have endeavoured to account for the disappearance of the Quail on the theory of migration; but situated as New Zealand is in the great waste of the Pacific Ocean, such a theory seems to me quite untenable. It is true that, as stated at page 228, the bird has recently been found, apparently in considerable numbers, on the Three Kings; but I take it that this is a mere outlying refuge of the species, and that the birds to be found on these small islands are the only survivors of a race now extinct on the mainland, and not to be met with in any other part of the world.

The extreme fecundity of the Quail tribe ought, one would have thought, to have saved this species from such rapid extinction. Mr. J. R. Hill, of Christchurch, kept some California Quails (*Lophortyx californicus*) in his aviary, and was perfectly amazed with their productiveness. One of the hens laid in a single season no less than 80 eggs, forming several new nests during that period. At length she discontinued laying, and collecting 23 of the eggs into one nest commenced to incubate. She brought out all but one, and reared the 22 young ones to maturity.

* The bird is chiefly known to the Samoans as an example of arrant cowardice, owing to the fact that when seen it is almost always chased by a number of Iaos (*Ptilotis carunculata*), from which it tries to escape in the most precipitate manner. I scarcely ever hear the name of the Aleva mentioned by a native without some such remark as this:—"The big bird that is chased by the little Iao!"

† A correspondent who has carefully noted the disappearance of the Quail writes to me:—"It seems to me to be of importance that the life-history of this bird should be correctly recorded; for the story of its rapid extinction will possess much interest for future naturalists. It cannot be said, as in other cases, that this species was exterminated by the introduction of other birds into its natural habitat, because it had almost disappeared before any acclimatized birds had reached the grass-covered downs where formerly it was so abundant. The tussock-fires have been the prime cause of the annihilation of this useful bird by destroying the seeds and insects on which it subsisted. So far as I have been able to discover, the last specimen seen alive was on the Raincliffe Station, in the year 1878."

Fam. **COLUMBIDÆ.**—The New-Zealand representative of this group is perhaps the finest of the whole. It is scarcely inferior in size to *Carpophaga galeata* of the Marquesas Islands, fully as large as *C. goliath* of the Isle of Pines, larger than *C. concinna* of the Moluccas, and far more beautiful in plumage than all three of them. The Nicobar Pigeon (*Calœnas nicobarica*), from the Louisiade Archipelago, is certainly more brilliant, but it is much inferior to our bird in size. The largest Wood-Pigeon in Australia, the Wongawonga (*Leucosarcia picata*), is not to be compared with the New-Zealand bird.

The genus *Carpophaga* is confined to the Philippines, New Guinea, Australia, and New Zealand; the nearest ally of our bird being probably *Carpophaga forsteri* of the Celebes, on the western limit of the Austro-Malayan subregion.

In my account of the species I have mentioned (at p. 234) the immense numbers that are annually killed without any appreciable effect on their abundance, in suitable localities, on the recurrence of each season. In further illustration of this I may add that in a small area of bush between Nukumaru and Weraroa, places of historic interest in connection with the Maori war, four young settlers shot upwards of 400 in the course of two days!

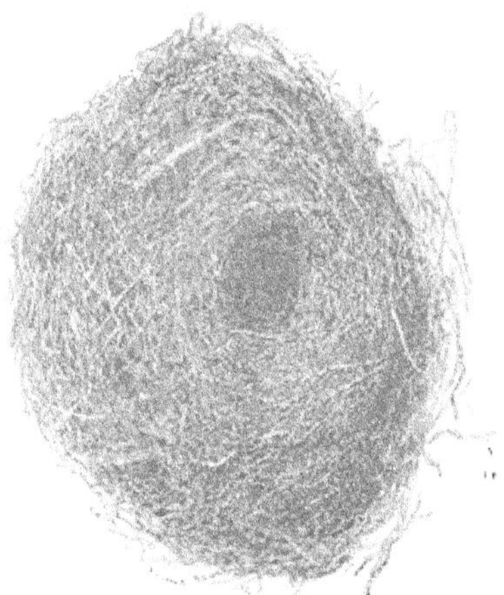

Nest of *Acanthisitta giliviventris* (four-fifths natural size). See page 213.

THE BIRDS OF NEW ZEALAND.

EXTRACTS FROM REVIEWS OF FIRST EDITION.

"Birds, as most people know, or ought to know, form the most important part of the vertebrate fauna of New Zealand; and their importance is maintained not only when they are compared with their compatriots of other classes, but when regarded in reference to members of their own class in the world at large . . . .

"The birds of New Zealand, therefore, merit especial attention, and we are happy to say they receive it at the hands of the authors whose works are above cited. Taking the field in or about the year 1865, Mr. Buller, till then unknown to fame beyond the limits of his native colony, brought out an 'Essay on the Ornithology of New Zealand,' which at once attracted notice in this old world of ours. Some of his views were challenged by Dr. Finsch, then of Leyden, who had paid attention to this extraordinary avifauna; and a controversy ensued. This, to the credit of the controversialists, was carried on in a spirit very different from that in which many another war in natural-history circles has been waged; and the happy result is that on most points the combatants have arrived at the same conclusion, thereby giving assurance to the general public of its being the right one. The essay we have mentioned may be regarded as the preliminary canter which a race-horse takes before he puts forth his full strength; and Mr. Buller's book, or that part of it which is as yet published, shows what he can do now that the colonial authorities have allowed him to come to England for the express purpose of completing his design.

"Captain Hutton is known as an observer who, during several long voyages, had proved that some rational occupation could be found at sea even by a landsman; for, instead of devoting his energies to the ordinary time-killing amusements of shipboard, he watched the flight of the various oceanic birds which presented themselves, and speculated on the mode in which it was performed and the forces it brought into operation—to some purpose, as the Duke of Argyll and Dr. Pettigrew have testified. The pamphlet whose title we give is in some respects a not less significant, if a less ambitious, work than Mr. Buller's; and though to the last must belong the crown of glory, we by no means wish to overlook the useful part which Captain Hutton's publication will play. If here we do not notice it further, it is because its value will be most appreciated in the colony itself, while Mr. Buller's beautiful book appeals to a larger public . . . .

"Of the Kakas (*Nestor*) Mr. Buller admits three species—*Nestor meridionalis*, *N. occidentalis*, and *N. notabilis*—the two first of which, we think, are barely separable. This very remarkable genus of Parrots includes some two or three other species, one of which, the *N. productus* of Phillip Island, is believed to have gone the way of so many animals that only inhabit small islands; and the same fate in all likelihood awaits its congeners. Most animals suffer from not being able to accommodate themselves to change of circumstances; but the very adaptability of the Mountain-Kaka, or Kea, will lead to its early destruction; for, though belonging to the group of Parrots distinguished by their brush-like tongue, and deriving a considerable portion of their subsistence in a manner worthy of the Golden Age, from the nectar of flowers, the wretched Kea (*N. notabilis*), since the introduction of sheep to New Zealand, has incurred the imputation of a fondness for mutton-cutlets *à la Abyssinie*; and the charge, whether true or false, is likely to bring about its doom, since the shepherd is apt to practise what in good old times was called 'border justice,' and the species will probably suffer extinction before its guilt is fully proved or extenuating circumstances admitted. The Common Kaka (*N. meridionalis*), on the other hand, is ably defended by Mr. Buller as one of the most useful birds in the country; yet this also is rapidly diminishing. 'In some districts,' he says, 'where in former years they were excessively abundant, their cry is now seldom or never heard;' and though he adds that 'in the wooded parts of the interior they are as plentiful as ever,' it requires no prophetic eye to see that, with the extension of settlement, the Kaka must succumb . . . .

"Here we must pause. Mr. Buller's book is in every way worthy of its subject; and we trust that we have shown that the subject is worthy of close attention—whether we regard the various forms of New-Zealand birds from the point of view of their intrinsic interest, or from that of so many being now on the verge of extinction. It is easy to be wise after the event, and ornithologists at home do not in these days look back affectionately towards their predecessors who have let so many species pass away without tracing the process of extermination."—*Nature* (July 18, 1872).

EXTRACTS FROM REVIEWS.

"New Zealand is especially fortunate in the possession of many admirable Naturalists, including geologists, botanists, and zoologists. One of the latter (born and bred in the colony), a gentleman who has made many zoological contributions to the 'Transactions of the New-Zealand Institute,' and whose acquirements, more especially as an ornithologist, have been recognized by his having had conferred upon him the Degree of Doctor of Science, as well as the Fellowship of the Linnean, Geological, Royal Geographical, and Zoological Societies, is at present in London passing through the press a magnificent work on the Birds of New Zealand, one that cannot fail to bring prominently into notice the present aspects of scientific culture in that colony."—*Constitutional* (Nov. 18, 1872).

"It is not often that thorough practical knowledge, both in the field and at home, is possessed by the author of a work like the present; but Dr. Buller has studied his subject in both aspects, and the value of his book is clearly enhanced thereby. Moreover he has set about his task in a way that shows us that he thoroughly appreciates the difficulties surrounding it. His personal acquaintance with the birds themselves has been followed up by a critical and impartial investigation of the writings of previous authors; and, lastly, an independent examination of many of the typical specimens in England has placed him in a position to speak with great precision upon intricate points of synonymy. The consequences to many of the indigenous birds of New Zealand, arising out of its colonization by Europeans, seem likely to be so disastrous, that it is high time that authentic histories of them should be put on record before they finally disappear. Dr. Buller's work, therefore, supplies what might have proved a serious omission in ornithological literature. It is not too late to write a full life-history of those New-Zealand birds whose numbers are rapidly diminishing; but a few years hence it is more than probable such a task could not be accomplished. Though the present active causes may be novel, the rapid destruction of the indigenous fauna of New Zealand dates back to far beyond historic times; for though Maori tradition may give an approximately recent time when the Moa still survived, numbers of other similar forms have succumbed whose remains are now found in a semi-fossilized state, and of these we have not another vestige of record. They, like the Dodo and the Solitaire, seem to have fallen victims to some enemy suddenly introduced into their domain, against which they were powerless to make successful resistance. The remains of these extinct birds have furnished the materials for Prof. Owen's series of exhaustive memoirs on *Dinornis* and its allies. Dr. Buller's will form a fit companion work, and thus provide us with a very complete record of the birds of New Zealand both past and present."—*The Ibis.*

"The first work professing to give a complete account of the ornithology of New Zealand must needs be an important one. This ornithic fauna presents so many points of general biological interest, that only those of the islands east of Africa can be compared with it. . . . . . It was high time that a complete account of this fauna should be given by a competent naturalist. Some of the most interesting forms have already become almost, if not quite, extinct; others are fast expiring, or obliged to accommodate themselves to the changed conditions of the country. . . . . . We do not say that the majority of the native species will not survive, though in diminished numbers of individuals; but it is quite probable that some of these survivors will be preserved by accommodating themselves to the new state of things, modifying in a more or less perceptible manner their nidification, food, or some other part of their mode of life; and if such changes should occur, the student of a future generation will find in Dr. Buller's work the means of comparing the birds of his time with those of the past. . . . . . The author has shown unremitting care in adducing all the information that can possibly throw light on his subject; he has spared no pains in illustrating it in the most perfect manner; and the result is that a most valuable work is placed before the student of ornithology, which will offer to every lover of natural history real and permanent enjoyment, and which, by its attractive form, will allure many a young man in that colony from the pursuit of other branches into the camp of ornithology."—*Annals and Magazine of Natural History.*

"A mind may be so imbued with the views of Darwin as to be blind to the evidence of his eyesight, deaf to the logic of facts; but there is no proof that Dr. Buller is either: he is evidently friendly to Darwin's celebrated hypothesis, but sees, hears, and thinks for himself. . . . . . Happily for Science, the author for twelve years has held an official position in New Zealand which has enabled him to visit every part of the country, while his frequent intercourse with the natives has greatly assisted him in acquiring the information required for making such a work complete. . . . . . It contains a vast amount of the soundest natural-history teaching, and seems to combine in an eminent degree the new with the true. . . . . . The illustrations in the first number, the only one yet published, are excellently drawn by M. Keulemans, who always aims at the representation of living birds rather than the conventional attitudes of birdstuffer's specimens. They are well coloured by hand, and thus the work is rendered as ornamental as useful. We cordially recommend the 'History of the Birds of New Zealand' to the readers of the 'Zoologist,' and we sincerely wish it every success."—*Zoologist.*

"The accounts which naturalists from time to time have given to the world of the birds inhabiting New Zealand have been hitherto but fragmentary and incomplete; and although forty years have elapsed since the first of such publications made its appearance, the available sources of information on this subject are still so few in number, that they may be enumerated almost in a breath. The late Mr. George Gray might deservedly be regarded as the pioneer of New-Zealand ornithologists; for, although never an actual explorer of the country himself, his official position gave him unusual facilities for studying its avifauna by means of the numerous collections which from time to time passed through his hands, and not a few of those antipodean species were originally described by him. . . . . .

"When Mr. Gould, in 1865, published his 'Handbook to the Birds of Australia,' he gave, by way of appendix to his second volume, an account of various New-Zealand species which were scarcely known to English readers, save in name; and in point of date this would seem to have been the latest publication on the subject in this country until a few months since, when Part I. of Mr. Buller's splendid work made its appearance. But, although so little, comparatively, has of late been published here, naturalists in New Zealand have been actively engaged for some years past in working out the natural history of their adopted country; and the transactions of two of their scientific societies contain many excellent contributions on ornithology from such able naturalists as Dr. Haast, Dr. Hector, Mr. Potts, and the author of the work now before us. Nor have our friends in Germany been behindhand in their zeal to become acquainted with an avifauna perhaps the most remarkable in the world. . . . . . .

"We recognise in Mr. Buller's publication, however, the first attempt which has been made to give anything like a complete history of the birds of New Zealand; and it would not be easy to overrate the importance which attaches to such an undertaking. . . . . . .

"Those who had an opportunity of seeing the Huia, which lived for some time in the Parrot-house in the Zoological Gardens, could scarcely have noticed it without wishing to learn something of its haunts and habits; and to them Mr. Buller's account of it will prove most entertaining. In the following extract we seem to get a peep of the country which it inhabits, as we search for and find this very curious bird. . . . . . . Such sketches as these go far to enliven a comprehensive work on birds, which, in other respects, is strictly scientific. As regards the illustrations Mr. Buller has been most fortunate; for, under his direction, his artist, Mr. Keulemans, has produced some of the most life-like and beautiful pictures of birds which we have seen. We understand the work is to be completed in five Parts, two of which have already appeared, and a third is in active preparation. It will assuredly become the text-book for all students of New-Zealand ornithology."—*The Field.*

"Dr. Buller has just produced Part IV. of his great work on the ornithology of New Zealand; and we may now fairly say that the high anticipations we had formed as to the author's capabilities have been fully realized. In the book before us we find the two great requirements of science combined—namely, a thorough appreciation of the necessary details which are expected of a scientific work in the present day, and the ability to write in appropriate and entertaining language the life-histories of the birds of which the author has to treat. So rarely are these two qualities found combined in a scientific writer, that the greatest credit is due to the learned author for the admirable manner in which he has performed his task.

"The ornithology of New Zealand is especially interesting, from the fact that the indigenous species are being gradually extinguished; and we read with regret that even within the memory of the author certain birds, which were formerly common, have almost ceased to exist.

"The work contains elaborate scientific diagnoses of the various birds, and a classification of the different names by which they have been known to different writers.

"The coloured plates are really exquisite examples of the lithographer's skill. In every respect the work is a most valuable addition to the scientific student's library, as well as to that of the more general reader, and seems to contain the fullest information on every point connected with this interesting study."—*Land and Water.*

"Although several more or less complete treatises on and lists of the Birds of New Zealand have been published, they were rather of a tentative and preliminary character; and the work before us is the first which gives a full account of this ornithic fauna, which, in zoological interest, is not excelled by that of any other country. . . . . . . There can be no doubt that Dr. Buller, well known in Europe by his preliminary ornithological publications, is eminently qualified to fulfil this task. His long residence in the colony and his official position have given him rare opportunities of making observations and collecting materials; and by a lengthened visit to England he has derived the great advantages of studying typical examples and of availing himself of that typographic and artistic skill in which this country excels. To judge by the first part issued, Dr. Buller has succeeded in producing a work of real excellence. The text is clear, instructive, and not overloaded with unnecessary detail; while the illustrations are beautiful and life-like."—*The Academy.*

"New Zealand may be congratulated on having outstripped the other colonies in the race for scientific honours. Even Canada, with all the resources at her command, has produced nothing at all comparable with the 'Transactions of the New-Zealand Institute.' Now we have before us something of a far more ambitious kind—namely, a complete life-history of the birds of New Zealand, adapted to the present advanced state of ornithological science, and most beautifully illustrated with coloured plates. . . . The descriptive part of the text is very carefully worked out, both in English and Latin; and the history of each species is given in the most complete and exhaustive manner. The plates are extremely beautiful, and are rendered more attractive by the introduction of botanical accessories, representing the indigenous flora of the country. The volume, when complete, will not only be a valuable contribution to scientific literature, but will be an elegant drawing-room companion; for, to adopt the language of a leading scientific journal respecting it, 'the plates are as beautiful in execution as the text is excellent in quality.'"—*Home News.*

"The lamentable way in which the indigenous birds are expiring in that country before the progress of civilization and other natural causes, has rendered it a necessity that a work should be prepared that will rescue from oblivion the feathered denizens of those places which in a short lapse of time 'shall know them no more.' This it has fallen to Dr. Buller's lot to accomplish; and it were small praise indeed to say that his task is executed in an admirable manner. Few ornithological works that have been written come up to the standard of the subject of this notice; and none have yet surpassed it, nor will it be possible to do so. Certainly the author brings to his aid unusual advantages; but even these might fail in the hands of a less conscientious person than Dr. Buller has shown himself to be. . . . . In the Part now before us the history is given of thirteen birds, ten of which are figured; and this brings us down to the end of *Accipitres, Psittaci,* and *Picariæ.* . . . . No one since the time of the late Professor Macgillivray has so successfully combined the two branches of cabinet and field ornithology as Dr. Buller; and his experiences, and those of his numerous coadjutors, are told in a pleasing and instructive manner, which cannot fail to interest and amuse his readers. Indeed it is seldom that we have seen a book which so thoroughly calls for unqualified praise as the present. . . . . . . We have only, in conclusion, to perform the pleasing duty of offering our congratulations to the inhabitants of New Zealand on their possession of so distinguished a naturalist as Dr. Buller, and to the author on the complete success with which his arduous task promises to be crowned."—*European Mail.*

"This admirable work, which places New Zealand in the front rank of countries, from an ornithological point of view, does credit to all concerned in it. Nothing seems to have been spared to make it as good as possible: and this fact is the more gratifying as in a few years many of the native species will probably have become extinct, and the opportunity of observing their habits, which are in most cases very fully described, will be lost for ever. The selection of the species for illustration is judicious, and the Plates are good."—*Zoological Record.*

"Before entering upon my own researches and a dissertation on the species, I will briefly refer to the Ornithological literature that has been published during the last two years; and on this occasion I seize with pleasure the opportunity of drawing attention to an undertaking which I desire to recommend most warmly to all friends of and experts in Ornithology. It is the beautiful work entitled 'A History of the Birds of New Zealand,' upon the publication of which my friend Dr. Buller is at present engaged, having come from New Zealand to London for that special purpose. This work, as is proved by the First Part, which I have now before me, very worthily links itself in with Sharpe and Dresser's 'Birds of Europe,' Sharpe's 'Kingfishers,' and Marshall's 'Capitonidæ.' As with the last-mentioned works, the execution of the Plates has been entrusted to the clever pencil of Keulemans, whose masterly work has long since gained universal acknowledgment, and does not stand in need of any further recommendation. Thus we shall before long be in the enjoyment of an exhaustive description of the Birds of New Zealand, equally perfect in text and illustration, and every person whose means will permit of it ought, without delay, to obtain possession of this beautiful book, the more so as its publication in Parts greatly facilitates the acquisition."—*Dr. Finsch in 'Journal für Ornithologie,'* 1872.

"That New Zealand contains more than an average number of persons interested in the advancement of science is evident, not only from the large number of members belonging to the various scientific Societies in the Colony, but also from the liberal way in which the Legislature votes money for scientific purposes; and to all of those who wish to see an intelligent interest taken in the subject of Natural History by our rising generation, Dr. Buller's beautiful work on the Birds of New Zealand cannot fail to be most welcome. . . . . . A book of this nature can be looked at either from a scientific or from a popular point of view—the nomenclature, descriptions, &c. forming the strictly scientific part, and the life-history of the birds the popular part; each being, in its own way, of equal importance. It is very rare indeed to find the qualifications necessary for the pursuit of both branches of Ornithology combined in one individual; and although we do not consider Dr. Buller's book irreproachable from either aspect, still we know of no other work on Ornithology, the product of a single author, in which both branches are so successfully combined, as in the book before us. . . . . . Dr. Buller's style is exceedingly good, clear, and to the point. Without wasting words, he brings out in a few graphic touches the salient points of whatever he may be describing, and it is easy to see that he is a real lover of nature and delights in a camp-out in the bush. . . . . . The descriptions of the species are excellent. Indeed, we think that these are the best portions of the book; and it is evident that a great deal of labour has been expended over them. In very few books on Natural History do we get such detailed descriptions of the adult, the young, and the varieties of the species, and the methodical manner in which they have been drawn up adds greatly to their value."—*Review by Prof. Hutton in New-Zealand Magazine'* (January 1876).

EXTRACTS FROM REVIEWS.

EXTRACTS FROM REVIEWS OF NEW EDITION.

"There can be no question as to the completeness with which the author treats his familiar subject, nor as to the excellence of the illustrations prepared by the pencil of Mr. Keulemans."—*The Ibis.*

"The Plates are absolutely perfect. . . . . The birds are reproduced in colours on as large a scale as practicable, and with a truth to nature which reflects great credit on the skill of the artist, Mr. Keulemans. So far as the letterpress goes, it ought to satisfy all wants. Sir Walter Buller gives a very complete synonymy of each species from the earliest systematic writers down to the present day. There are also full descriptions of both sexes and of every condition of plumage, with explanatory notes where necessary on the nomenclature and classification, and a complete life-history of each bird from Sir Walter Buller's personal observations, made in the field and forest, and recorded with scrupulous fidelity over a period of 20 years. The technical part (Latin and English descriptive matter) is in smaller type than the popular history, which any one may read with understanding and pleasure."—*The Times.*

"When, in 1873, Dr. (now Sir Walter) Buller, brought out his first edition of the Birds of New Zealand, it took people in this country by surprise; for it seemed an extraordinary thing that a man who had lived the best part of his life at the Antipodes—far removed from the great scientific centres of thought, from the libraries and museums of Europe, and from all those opportunities of fellowship which are considered so necessary to scientific workers—should be able to produce a high-class book, strictly scientific and quite abreast of the time. It needs scarcely to be pointed out that any author, whatever his position, who essayed to produce such a work, without possessing the necessary qualifications—an intimate knowledge of his subject and a well-trained scientific mind—would have been promptly cut to pieces by the reviewers, who are rightly regarded as the guardians of science, in its more technical sense. So far from injurious comment, every science-review of acknowledged standing in London gave the book unqualified praise. Copious extracts from all these reviews were given at the end of the work, and are worth perusal. The author was elected F.R.S., the highest distinction open to a scientific man in this country, and Her Majesty conferred upon him an 'imperial distinction' in recognition of the great value of his work to science.' But apart from the technical knowledge exhibited in this book, regarded from the scientist's point of view, the author was exceptionally fortunate in being able to portray the life-histories of the various species in happy language, and thus to make what otherwise might have been a dry scientific dissertation, pleasant and attractive to the casual reader. *The Daily Telegraph*, in its leading columns, referring to this, described the author as 'the Audubon of New Zealand,' whilst another reviewer said of the life-histories that they were 'quite as seductive as novel-reading.' . . . . . But undoubtedly the best proof that the book was appreciated by the general public was afforded by the rapid manner in which the edition of 500 was subscribed for, and the price to which it afterwards rose . . . . . . It will hardly be a matter of surprise, therefore, that Sir Walter Buller has taken the opportunity, after a lapse of thirteen years, of his visit to England to bring out a new and much enlarged edition. Of this, six parts have now been issued, containing 24 beautiful illustrations in colours. The next part will contain a General Introduction to the whole subject, embellished with numerous woodcuts and lithographs, and this will complete Vol. I. It is believed that the second volume, finishing the work, will be out before the end of the year. It is on a much larger scale than the former edition; and the letterpress is so amplified and added to that the book is practically a new one. Apart from the value of the work as a contribution to the scientific literature of the day, it will form a very beautiful drawing-room exhibit; and all who take an interest in New Zealand, with its curious forms of bird life and its quasi-tropical vegetation, ought certainly to possess themselves of a copy of this book before it is too late, because the number of copies available for Europe and America is strictly limited to 250, three-fifths of which have already been subscribed for."—*Anglo New-Zealander.*

"The specimen Part is illustrated by the most perfect coloured figure of a beautiful bird that I ever saw in any work."—*Prof. Owen's letter to 'Australian Times.'*

"That the author possessed the true instincts of a naturalist in his early days is shown by the excellent accounts of the habits of the birds, which proves that he must have spent a great deal of time in studying the different species in the field. There is, in fact, no more interesting portion of Sir Walter Buller's book than his own personal observation of habits; and in future days, when the avifauna of New Zealand shall have been changed, as it will assuredly be, by the introduction of foreign species to supply the place of the indigenous birds so fast disappearing, our author's work will possess undying attraction, as being the embodiment of the observations of men who saw these interesting New Zealand birds in a state of nature, and much that they saw will read in ages to come like an ornithological romance. . . . . . This melancholy knowledge that so many of the indigenous birds of New Zealand are undergoing a process of extermination lends an increased value to Sir Walter Buller's book; but even this might not be forthcoming in the work of a less erudite author. Long study among the museums of Europe, and acquaintance with the literature of the subject on which he writes, have rendered Sir Walter Buller absolutely the first

EXTRACTS FROM REVIEWS.

authority in the world on the New-Zealand avifauna, both as regards its scientific relations and the life-history of the species which inhabit that wonderful country. In this second edition of his great work the whole subject is treated with a fulness and accuracy certainly not excelled by any modern production: and it will ever remain not only a credit to the author but to the Colony which gave him birth. All that energy of purpose and a thorough knowledge of his subject could do to render his book perfect has been done by the author, who has been lavish to a degree in his efforts to render the illustrations the best on record, and no book of its size has probably had so much money spent upon the plates and woodcuts."—(*R. B. Sharpe*) '*The Colonies and India.*'

I write to congratulate you on the admirable way in which you are carrying out the work. I must say that I envy you the opportunity which you are so well turning to account. There is no other Ornis in the world which so much needs an historian, and no other historian can ever again enjoy your opportunities of the past. Any naturalist who attempts to follow you must write largely on hearsay, and, as I always tell my friends, your book can never be superseded. The letterpress is all that a naturalist could desire, and in the Plates (though these are not unnaturally of unequal merit) Mr. Keulemans has, I think, exceeded by many degrees anything which he has yet done; and I speak from considerable experience of his Plates, since I dabble myself in Ornithological drawing in oils, and refer constantly to his work and that of other delineators of bird-life."—*Letter from a British Ornithologist to the Author.*

PRINTED BY TAYLOR AND FRANCIS, RED LION COURT, FLEET STREET.

www.ingramcontent.com/pod-product-compliance
Lightning Source LLC
Chambersburg PA
CBHW021355210326
41599CB00011B/883
* 9 7 8 3 7 4 4 7 6 2 7 0 0 *